SPECIATION IN BIRDS

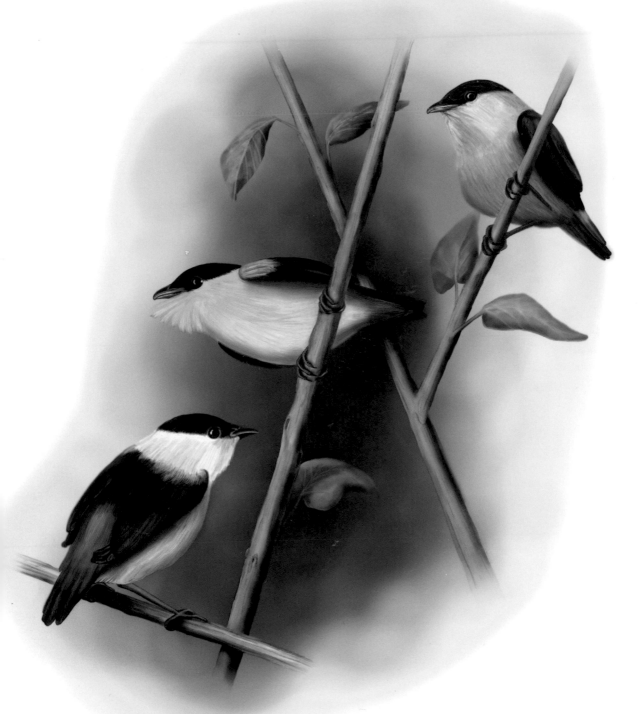

TREVOR PRICE

Speciation in Birds

Speciation
in Birds

Trevor Price

Roberts and Company
Greenwood Village, Colorado

Roberts and Company Publishers
4950 South Yosemite Street, F2 #197
Greenwood Village, Colorado 80111 USA
Internet: www.roberts-publishers.com
Telephone: (303) 221-3325
Facsimile: (303) 221-3326

ORDER INFORMATION
Telephone: (800) 351-1161 or (516) 422-4050
Facsimile: (516) 422-4097
Internet: www.roberts-publishers.com

Publisher: Ben Roberts
Artists: Emiko-Rose Paul and Susan Young
Cover illustration: Emiko-Rose Paul
Copyeditors: Gunder Hefta and Joyce Besch
Designer: Mark Ong at Side by Side Studios
Cover designer: Mark Ong at Side by Side Studios
Compositor: Side by Side Studios
Proofreader: Lee A. Young

ISBN: 0-9747077-8-3

Library of Congress Cataloging-in-Publication Data

Price, Trevor, 1953-
  Speciation in birds / Trevor Price. — 1st ed.
      p. cm.
  Includes bibliographical references and index.
  ISBN 0-9747077-8-3
  1. Birds—Speciation. I. Title.
  QL677.3.P75 2007
  598.13'8—dc22

2007002565

10 9 8 7 6 5 4 3 2 1

Contents

Acknowledgements ix

1 **Introduction** 1
  What is a bird species? 2
  The progress towards species 8
  Appendix 1.1: Using DNA sequences
         to classify populations to species 11

2 **Geography and Ecology** 13
  Ecological divergence with no geographical separation 19
  Ecological divergence with limited geographical separation 21
  Geographical isolation with limited ecological divergence 25
  Geographical isolation with no ecological divergence 29
  Time to speciation 31
  Conclusions 33
  Summary 35
  Appendix 2.1: Molecular dating methods 36
  Appendix 2.2: Phylogeny estimation 38

3 **Geographical Variation** 41
  Factors promoting geographical variation 42
  Factors retarding geographical variation 57
  Conclusions 63
  Summary 64

4 **Parapatric Speciation** 65
  Reduced fitness of immigrants 67
  Reduced immigration 70
  Conclusions 71
  Summary 73

5 **Ecological Speciation** 75
  Models of resources and competition 76
  Reproductive isolation 82
  Different ecological dimensions 85
  Is ecological speciation generally important? 89
  Seasonal migration 92

Conclusions 95
Summary 95

6 **Ecological Controls and Speciation on Continents** 97
Regional variation in species numbers 98
Speciation through time 101
Ecological controls on speciation 108
South America's huge bird diversity 115
Comparison of the ecological controls model
        with ecological speciation 120
Conclusions 121
Summary 123

7 **Behavior and Ecology** 125
Foraging behaviors 127
Correlates of feeding innovation with speciation 129
Predation and nest building 137
Conclusions 139
Summary 139

8 **Geographical Isolation and the Causes of Island Endemism** . . 141
Correlates of endemism with area and isolation 142
Island-biogeography theory of endemism 144
Unusual environments on isolated islands 152
Conclusions 154
Summary 154

9 **Social Selection** 157
Population divergence 168
Constraints on the invasion of new signals 178
Coevolution of signaler and receiver 182
Conclusions 185
Summary 186

10 **Social Selection and the Evolution of Song** 187
Geographical variation 192
Cultural transmission and cultural mutation 198
Population divergence 204
Similarities with human language 208
Genetic divergence 211

Conclusions 212
Summary 213
Appendix 10.1 Song of the Indigo Bunting 215
Appendix 10.2. Song of the Chaffinch 216

11 **Divergence in Response to Increased Sexual Selection** 219
Divergence among domestic breeds 221
Species divergence and breed divergence 228
Divergence under parallel selection pressures:
        Song variation in a ring species 236
Parallel evolution vs. divergence 238
Conclusions 240
Summary 241

12 **Social Selection and Ecology** 243
Background 244
Resource dispersion 248
Adaptive radiation and the evolution of socially selected traits 252
Island patterns 261
General conclusions on social selection 265
Summary 272

13 **Species Recognition** 273
Sexual imprinting 273
Filial imprinting 281
Individual recognition 283
Species recognition 284
Species recognition, sexual selection, and speciation 290
Cultural speciation 293
Conclusions 296
Summary 297

14 **Mate Choice at the End of Speciation** 299
Background 301
Signal similarity as a cause of hybridization 306
Benefits to mating heterospecifically as a cause of hybridization 310
Reinforcement 313
Sexual selection against hybrids 317
Conclusions 320
Summary 321

15 Hybrid Zones 323
Background 325
Patterns 334
Zone movement 340
Differential introgression 342
Does hybridization have a creative role? 345
Conclusions 347
Summary 348
Appendix 15.1 List of hybrid zones 350

16 Genetic Incompatibility 367
Background 367
Accumulation of incompatibilities in birds 371
Chromosomes 383
Conclusions 390
Summary 392

17 Conclusions 395

Glossary 403
References 410
Index of common names 458
Index of scientific names 463
Subject Index 467

Acknowledgments

My thanks go to friends Mark Pavelka and Dolph Schluter for their encouragement and support throughout this endeavor, and especially to Tina Harr during the last two-thirds of it. I thank Jerry Coyne, John Endler, Peter Grant, Darren Irwin, Mark Kirkpatrick, Josh Kohn, Russell Lande, Robert Ricklefs, Kaustuv Roy, and Dolph Schluter for disagreements. Dorrie Jones, Katrina Kelly, Bill Newmark and my parents gave me space in their homes when it was much needed. Nick Davies (Cambridge University), Karl Schuchman (Museum König, Bonn), and Eric Rickart (University of Utah) helped me with various logistical issues when using the libraries at their institutions.

Douglas Futuyma, Peter Grant, Rosemary Grant, Robert Payne, and Dolph Schluter read the entire manuscript. In addition, the following colleagues were kind enough to read one or more chapters: Bruce Beehler (5 chapters), Craig Benkman (2), Staffan Bensch, Tim Birkhead, Jenny Boughman, Gonçalo Cardoso, Nicky Clayton, Jerry Coyne, John Endler, Vicki Friesen, Lisle Gibbs, Emma Greig (2), Sean Gross, Bengt Hanson, Tina Harr (3), Mark Hauber (2), Chris Hill (5), Darren Irwin (2 twice, and another one too), Mark Kirkpatrick, Jill Mateo, Doug Nelson (2), Ken Petren (2), Albert Phillimore (6), Steve Pruett-Jones, Anna Qvarnström (2, one of these 3 times), Pamela Rasmussen (5), Michael Turelli (2), Van Remsen (3), Robert Ricklefs (2), Locke Rowe (2), Daniel Sol, Al Uy (5), Thor Veen, Jason Weir (3), Mary Jane West-Eberhard, Carmen Wheatcroft, David Wheatcroft, and Robert Zink (5). Ryan Calsbeek, Jaime Chaves, Alex Kirschel, John McCormack, Borja Milá and Tom Smith provided a thorough overview of Chapters 1–3, exactly when it was needed. I am particularly grateful to Robert Ricklefs for his generous help with island biogeography theory, to Jason Weir for his deconstruction of Chapter 6, and to Robert Zink for his forceful suggestions of the need to consider alternative viewpoints. More general overviews came from anonymous reviewers and from Allan Baker, David McDonald, Robert Moyle, Gary Ritchison and Haven Wiley. Many others were quick to respond with points of information, including Per Alström, Jessica Eberhard, Chris Filardi, Paul Handford, Don Kroodsma, Kevin Omland and Robert Wilson; I hope I have acknowledged others at relevant places in the text. Adam Stein and Robb Brumfield gave much patient help with the cover picture. Joyce Besch and Gunder Hefta were great editors, catching many errors, and making many clarifications, and Alexa Bontrager was kind enough to read the whole book in proof. I am grateful to Emiko Paul for her tremendous illustrations, and to Ben Roberts for his enthusiastic support. Grants from the Guggenheim Foundation and the Mercator Foundation, the latter sponsored by Diethard Tautz, helped me to get time away from other responsibilities during the final stages of writing.

I would like to take this opportunity to acknowledge the contributions of Andreas Helbig, who died on October 15th, 2005 at the age of 48. Andreas worked in three disparate areas relevant to bird speciation: the genetics of bird migration, molecular systematics, and gene flow across hybrid zones. I enjoyed my one collaboration with him, and he and his work will be missed.

The idea for this book originated with an invitation from the organizers of a Royal Society of London meeting to review sexual selection and speciation at a symposium in 1997. I realized then that there was a diverse literature from other fields that could be brought to bear on the subject of speciation. It was too scattered for me to attempt more than a synthesis for the group of organisms I know most about—the birds—and I set about doing this in a more systematic way.

CHAPTER ONE

Introduction

For as long as it has been appreciated that one species can split into two, birds have been at the center of speciation theories. In the twentieth century, Ernst Mayr, an ornithologist by training, wrote classic volumes on speciation in which he emphasized the importance of geographical isolation: populations in separate locations begin a process of differentiation, and once differentiation is sufficient the populations have become two species (Mayr 1942, 1963). Mayr's main field research was in northern Melanesia (Mayr and Diamond 2001), where closely related species are found on different islands. These species are ecologically and morphologically similar and their main differences are in coloration (e.g., Figure 1.1). In contrast to Mayr, David Lack (1947, 1976) emphasized the role of ecological factors in the origin of species, focusing entirely on birds. Lack had spent some time in the Galápagos Islands and was impressed with the manifest ecological and morphological differences among closely related species of Darwin's finches, which are often found on the same island. Lack's famous predecessor to the islands, Charles Darwin, had been similarly impressed. The complementary roles of geographical isolation and ecological differentiation in speciation have been around for a long time (e.g., Huxley 1942; Mayr 1947; Endler 1982; Smith et al. 2001), and sorting out their contributions is a goal of this book.

Birds are easy to observe in the field and have been widely studied, making them one of the best taxonomic groups with which to review the importance of ecological factors in speciation. But birds have much more to offer. In particular, they are an excellent group with which to investigate the role of behavior, especially those factors involved in choice of mate and the process of sexual selection. Choosing to mate with one's own species is a critical feature of speciation, and birds are the taxon for which the criteria involved in mate choice are best known.

While bird studies have much to say about ecology and behavior in speciation, they have contributed little to genetics. A third speciation worker from the first half of the twentieth century, Theodosius Dobzhansky (1937, 1951), emphasized genetics. Dobzhansky mostly studied fruit flies *(Drosophila)* and never studied birds. The definitive recent book on speciation, by Coyne and Orr (2004), was also by two workers on *Drosophila,* and it was primarily concerned with genetics. Bird genetics is difficult because birds have long generation times and are not easy to breed in large numbers. Although not much is known about the genetics of bird speciation, the available information is sufficient that it can be tied to the ecological and behavioral findings and to results from other groups (as reviewed by Coyne and Orr 2004). Thus, on balance, birds probably provide the best taxonomic group to synthesize the roles of ecology, behavior and genetics in speciation.

In 1990, Sibley and Monroe compiled a list of the world's birds. That list contained 9,672 species. Although some of the species on the list are extinct, in what has been something of a taxonomic revolution, many more have been added as vocalizations have been studied and DNA sequenced (Krabbe and Schulenberg 1997; Alström and Olsson 1999; König et al. 1999; Irwin et al. 2001a; Isler et al. 2001). Owls lead this increase. Because they mostly come out at night and have dull, similar-looking plumages, many owl species were not recognized until recently; the number went up by almost 20% (178 to 212) over ten years and it is likely to go up further (König et al. 1999). Now probably close to 10,000 extant species of birds are recognized, and many times that number have gone extinct over the past 145 million years or so since the first known fossil bird, *Archaeopteryx*. The questions addressed in this book revolve around how all these species originated. Do different ecological or geographical factors make speciation more or less likely in different places or at different times? Are some species more likely than others to produce daughter species? What are the roles of sexual selection and natural selection in speciation? To begin answering questions such as these, I first consider what we mean by a bird species.

WHAT IS A BIRD SPECIES?

Bird watchers identify individuals to species based on plumages, behaviors, vocalizations, and size and shape, and so do the birds themselves. Some species are very similar to each other, making individuals tricky to identify, but there is little doubt in a bird watcher's mind that virtually every individual does belong to a species. The odd exception is attributed to hybridization. In a field guide, the bird watcher is pointed to one or more unique characteristics that diagnose a particular species.

Two or more species that occur in the same locality must remain distinct from each other because they do not interbreed, or if they do interbreed, their offspring are unfit. Much genetic exchange would lead to continuous variation, and individuals would be impossible to classify into discrete groups. Genetic interchange between species can be cut off either through a failure to mate with members of other species or, more rarely, after mating. This is embodied in one common definition of species as "groups of interbreeding populations reproductively isolated from other such groups" (Mayr 1982, p. 273), the so-called biological species concept. Under this concept, the origin of reproductive isolation is synonymous with the origin of species. This means that the study of reproductive isolating mechanisms forms the basis of a research program underlying the study of speciation (Coyne 1992; Coyne and Orr 2004, their Chapter 2). As one focus of this book, I analyze the factors that deter individuals of one species from mating with those of another (premating reproductive isolation), as well as the causes of low success of cross-species matings if they occur (postmating reproductive isolation).

In birds, virtually all speciation involves some spatial separation of populations, which then diverge in morphology, songs, colors, etc. (Chapter 2). If a population (or group of populations) becomes sufficiently distinct from the rest of the species it may be classified as a different subspecies (i.e., geographical race). The subspecies category has been abused, with different intergrading populations sometimes given subspecies rank (Wilson and Brown 1953; Zink 2004), but remains useful if its application is restricted to taxa that are diagnostically distinct (Patten and Unitt 2002; Remsen 2005). After further divergence, subspecies become allopatric (i.e., geographically separated) species, termed allospecies. The collection of allospecies together is termed a superspecies (Amadon 1966; see Glossary).

When taxa remain completely separated in space, the criterion of reproductive isolation cannot generally be used to decide if they should be considered different species. If individuals from different taxa are crossed in the lab, and offspring are missing or infertile, the different taxa are certainly good species, for no genetic exchange would be possible in nature. However, many species that coexist in the same place can be induced to mate in captivity, and when they do, they often produce fertile hybrids (Chapter 16). Thus, we may expect many allopatric taxa to be successfully crossed in captivity, even if they would not interbreed in nature. In short, the criterion used to determine species rank when taxa are found in the same place (i.e., reproductive isolation) cannot be used for taxa that are allopatric. The criteria that should be used to assign allopatric taxa to species remain—to put it mildly—controversial.

After a considered analysis, Helbig et al. (2002) decided to give species status to closely related allopatric taxa if (1) at least one character is fully diagnostic (i.e., it forms a fixed difference between the taxa) and differences between the taxa are of the same order as those seen among pairs of sympatric species, or (2) the taxa are statistically distinguishable by a combination of two or three functionally independent characters (e.g., color, body size, vocalizations, or DNA sequences are functionally independent, but wing length and tail length are not, both being correlated with size). Again, the level of divergence needs to be taken into account through comparisons of differences between sympatric species.

Increasingly, DNA sequences are being used in this assessment. Based on measures of sequence divergence (Chapter 2), many closely related sympatric species appear to have been separated for upwards of 2 million years, and assignment of taxa to allospecies is strengthened when they show similar separation times.

Recent taxonomic treatments without the benefit of DNA sequence data have adopted criteria similar to those suggested by Helbig et al. (2002). Isler et al. (1998) determined that at least three song characters were required to separate sympatric species of antbirds and they classified allopatric taxa on this basis. Populations that show smaller, but diagnosable, differences in song are given subspecies status (Remsen 2005). Mayr and Diamond (2001) assigned populations of northern Melanesian birds to species mainly on the basis of plumage variation. If differences between a pair of taxa are similar to differences seen among sympatric taxa, Mayr and Diamond considered the taxa to be different allospecies; if not they were relegated to subspecies (Figure 1.1). Rasmussen and Anderton (2005) reclassified 83 pairs of subspecies on the Indian subcontinent as allospecies. The newly classified allospecies usually occupy ranges that are well separated from each other (e.g., Western Ghats in southern India vs. Himalayas), and they usually differ both in vocalizations and in plumage, although the differences may be subtle. Other reclassifications of specific taxa have used not only vocalizations and plumage but also recognition of vocalizations as determined by song playbacks (e.g., Irwin et al. 2001a; Alström and Ranft 2003).

Helbig et al. (2002, p. 519) came up with their recommendations because they thought "taxa should only be assigned species rank if they have diverged to the extent that merging of their gene pools at some time in the future is unlikely." Because the future encompasses the possibility that the taxa will expand their ranges and come into sympatry, this philosophy is based in the biological species concept. Although the level of divergence between allospecies may indicate the probability of interbreeding should they come into contact, there is no easy way to predict what will happen. First, levels of divergence in plumage and other traits seem to be a poor indicator of reproductive isolation, for some hybrid swarms are formed between quite different taxa, whereas other very similar species coexist without interbreeding (Chapter 14). Second, levels of reproductive isolation can both increase and decrease as a result of interactions in sympatry (Chapter 14).

There are thus two operational definitions of species used in most taxonomic treatments; in sympatry the biological species concept works well, but for allopatric taxa, practitioners of the biological species concept are confined to asking if differences between populations in traits that are likely to be important in reproductive isolation are diagnostic and sufficiently large. In this book, I follow Sibley and Monroe's (1990) decisions as to whether to classify taxa as

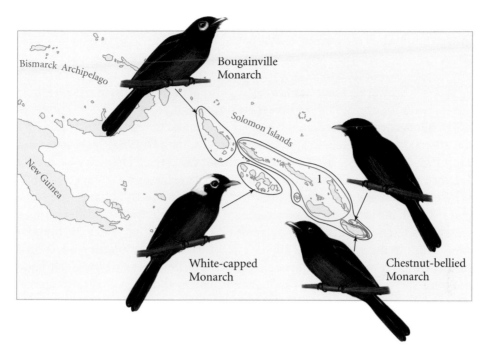

Figure 1.1 Three allospecies in the monarch flycatcher complex (redrawn from Mayr and Diamond 2001, and based on their taxonomy). Distributions of each allospecies are indicated by the red lines. Green lines encircle distributions of different subspecies of the Chestnut-bellied Monarch, which, apart from the all-black form on the small islands of Ugi and Santa Ana, resemble the chestnut-bellied form illustrated [however, the different subspecies do differ morphologically and behaviorally; Filardi and Smith (2005), C. Filardi *personal communication*]. The all-black form is considered to be a subspecies of the Chestnut-bellied Monarch because all-black individuals regularly occur in the range of the other subspecies [Mayr and Diamond 2001; Mayr (1942) previously gave the black form species status]. The subspecies of the Chestnut-bellied Monarch with the distribution indicated by a "1" contains populations whose phylogenetic relationships are illustrated in Figure 1.2.

species, except for species that have been identified since their classification, in which case I follow the primary literature. Sibley and Monroe's criteria are similar to those explicitly stated by Helbig et al. (2002), relying on plumage and, to a lesser extent, morphological (size and shape) differences. It is worth noting that the way borderline species are classified is irrelevant to many questions about speciation [Mayr (1982), quoted in McKitrick and Zink (1988)]. These questions include: (1) What leads to differentiation in traits that are likely to be involved in reproductive isolation, whether or not the differences are sufficient for allopatric taxa to be labeled as different species? (2) What causes reproductive isolation when the taxa are in sympatry?

Phylogenetic species

Some evolutionary biologists object to the fuzziness of the criteria used to assign allopatric taxa based on the biological species concept and instead prefer the so-called "phylogenetic species concept" (Cracraft 1983; McKitrick and Zink 1988; Zink 2006). In its original formulation, a fixed character difference between populations is both necessary and sufficient to assign species rank. Whether or not the taxa would interbreed if they were to occur in the same place is immaterial. In this concept, the focus is on whether the different taxa have had a recent independent history and it is accepted that taxa may merge at some point in the future (Zink 2006). The study of speciation then becomes a study of what restricts gene flow between populations (e.g., Wiens 2004a) and not necessarily what causes reproductive isolation should populations spread into each other's range.

The original definition of a phylogenetic species was "the smallest diagnosable cluster of organisms within which there is a parental pattern of ancestry and descent" (Cracraft 1983). The phrase "parental pattern of ancestry and descent" implies a population that is *monophyletic*. The principle of monophyly is illustrated in Figure 1.2, which shows a tree of phylogenetic relationships, based on mitochondrial DNA sequences, for individuals belonging to two of the northern Melanesian monarch species. Based on the evidence from this limited sample, the White-capped Monarch is monophyletic for this sequence because the two individuals studied share a recent common ancestor and no individuals of the Chestnut-bellied Monarch have this ancestor. But the Chestnut-bellied Monarch is *paraphyletic* because the most recent common ancestor to all individuals belonging to this species also has descendants that are White-capped Monarchs. (In fact, Filardi and Smith (2005) suggest that the Malaitan island population, which is a genetically distinct form, should be reclassified as a species. The result would then be three monophyletic species. Filardi and Smith strengthen the argument for species status of the Malaitan form by noting that it differs in behavior and in body size and shape from the other forms).

In the phylogenetic species concept, diagnosability and monophyly might reasonably be considered to be the same. For example, Cracraft (1992) assigned a population of birds of paradise to a new species because individuals had metallic-blue breast colors, whereas members of the other population formerly considered to be conspecific had violet-purple breast colors. This assumes that a mutation causing the distinctive coloration arose and became established in one population. The population is both diagnosable, because all individuals

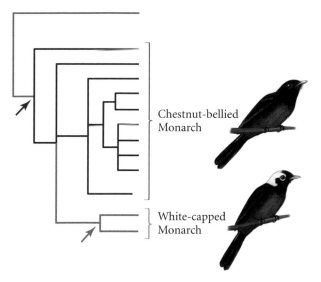

Figure 1.2 Relationships among individuals of two species of monarch flycatchers based on mitochondrial-DNA sequences (sequences from Filardi and Smith 2005; phylogeny reconstructed using maximum likelihood methods described in Chapter 2, Appendix 2.2). The *black line* connects a single Black-winged Monarch from New Guinea, which is also an allospecies in this group. Individuals of the Chestnut-bellied Monarch are drawn from several populations across the range of the subspecies indicated with a "1" in Figure 1.1; they are connected by *red lines* (the other subspecies have not yet been sequenced). Arrows indicate the most recent common ancestor of individuals belonging to the Chestnut-bellied Monarch and White-capped Monarch. According to this reconstruction, the White-capped Monarch (*blue lines*) from the islands of the New Georgia group (see Figure 1.1) is nested within the phylogenetic tree of the Chestnut-bellied Monarch. The depth of the tree is estimated to be about 1 million years, based on the amount of sequence divergence and the methods described in Chapter 2.

have the color, and monophyletic for the gene or genes determining color differences. While the phylogenetic species concept seems to deal with the problem of how to classify allopatric taxa in an unambiguous way, the difficulty comes in deciding which genes to study (Remsen 2005). Provided enough DNA is sequenced, and especially if only a few individuals are sampled, it is clear that many populations will show at least one diagnosable difference (e.g., a single base in a DNA sequence may be fixed for alternative forms in the two samples). No good way to resolve this difficulty has been suggested, and the few taxonomic revisions that have tried to follow this strict application of the phylogenetic species concept (e.g., Cracraft 1992) have not been widely accepted. A more recent emphasis in phylogeny-based species concepts classifies taxa as different species if they are monophyletic for certain DNA sequences. This approach is considered further in the Appendix to this chapter.

Arguments in favor of phylogeny-based concepts lie not only in their applicability to both allopatric and sympatric taxa, but also in their help in the understanding of how characters evolve. This depends primarily on knowledge of evolutionary relationships among taxa and not on whether the taxa would interbreed should they be in the same place (McKitrick and Zink 1988; Harrison 1998). Thus, if we observe several taxa that have a black cap and several that have a blue cap, understanding the evolution of cap color depends on such questions as what color was ancestral and how many times cap color has evolved. Answering such questions requires an accurate estimate of phylogenetic relationships among taxa and much less on whether black- and blue-capped individuals would interbreed if they were found in the same place (Harrison 1998). Although I will address some questions of this nature, they are not the main focus of this book. I am mostly concerned with asking how reproductive isolation arises. Reproductive isolation is perhaps the most crucial component of evolutionary history, for it is the necessary requirement for species to coexist (McKitrick and Zink 1988). Without reproductive isolation, only one species would be found at each location in the world, placing a severe limit on the total number of species.

THE PROGRESS TOWARDS SPECIES

The essential way in which it is thought one species becomes two is outlined in Figure 1.3, reproduced from Mayr (1942). First, the ancestral species is subdivided, either because a rare dispersal event across a barrier leads to the founding of a new population, or because a barrier appears (e.g., the sea rises and causes a peninsula to become an island). Then the divided populations diverge in various traits. After a period of divergence, the barrier may disappear, or be crossed again, and if the populations come into contact the criterion of interbreeding is tested. At this point, individuals from each population may either fail to interbreed (Figure 1.3, 4b), or they hybridize (Figure 1.3, 4a). If they hybridize, the populations collapse back to one species, or establish a stable hybrid zone, or diverge further to the level of full species. With few exceptions the need for geographical separation in the initial stages of divergence appears critical (Chapter 2). This is one of a limited number of general rules of speciation (Coyne and Orr 2004).

Most species that occur in sympatry are separated from each other by strong premating isolation, so understanding why females choose to mate with some individuals but not with others is essential to understanding speciation. Learning is important. Typically, birds pair up with individuals who look and sound like their parents, although this may be modified by other social interactions.

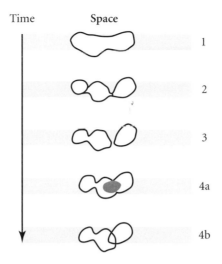

Figure 1.3 Basic model of range changes during speciation in birds (after Mayr 1942, p. 160). Populations become geographically isolated (1–3). After a period of divergence, if the populations come back into contact, they may hybridize (4a) or be completely reproductively isolated (4b).

The result of learning is that assortative mating can rather easily develop as populations diverge in such traits as vocalizations and plumage coloration.

One would think that premating isolation and speciation could develop almost too readily, but individuals do generalize (i.e., they use some features of their parents to identify an individual as a conspecific, even if that individual differs in other traits, and they then use these other traits to identify other conspecifics). This is particularly obvious for species that are especially variable in color, morphology, or song (Chapters 13 and 14). Generalization can lead to the acceptance of similar-looking heterospecifics as mates, especially if it is difficult to find members of one's own species. The result is that some hybridization between species occurs and the species may collapse back into one, unless gene flow between species is cut off at the postmating stage. I will argue that both premating and postmating isolation are likely to be present between species that occur in sympatry.

This book contains three sections. In Chapters 2–7, I investigate the role of ecology and natural selection in speciation. What are the connections between natural selection and reproductive isolation? What features of the biotic and physical environment promote speciation? What ecological attributes of a species may make them most likely to produce daughter species? In these chapters, I begin with the way ecological factors affect divergence between populations (i.e., create geographical variation within species) and end with a consideration of the contribution of ecology and speciation to some of the great patterns in the distributions of species, such as the latitudinal diversity gradient (more species in the tropics than in the temperate regions) and the vast number of bird species in South America.

In Chapters 8–12, I describe the role of sexual selection and other forms of selection arising out of social interactions. Many allospecies differ in colors and songs, and these are likely to have evolved as a result of such selection pressures. Sexual selection is more complicated than natural selection, because it can lead to divergence not only when populations occupy different environments, but also in the absence of any ecological differences between populations. These chapters start by asking how sexual selection alone could drive divergence between geographically separated populations, and they end by investigating the interaction between ecological and social factors. Some striking patterns in bird species diversity, such as the large numbers of endemics and differentiated populations on islands, can be related to differentiation in color patterns and vocalizations, and this may be explained by a strong interaction between ecological differences and geographical isolation. Finally, in Chapters 13–16, I consider further the origin of reproductive isolation. What determines why females of one species choose conspecifics as mates? What causes hybridization? What factors prevent the merging of two species back into one if hybridization is going on?

Note on presentation: Throughout this book I use common names of birds. The corresponding scientific names are in the index.

APPENDIX 1.1
USING DNA SEQUENCES
TO CLASSIFY POPULATIONS TO SPECIES

In a useful review, de Queiroz (1998) noted that under all species concepts, a species is "a lineage that maintains its integrity with respect to other lineages in both time and space." Lineage integrity can be maintained either because the different taxa are in different places, or by various premating and postmating isolating mechanisms when the taxa occur in sympatry. This is an important insight, for it elevates two allopatric taxa that scarcely differ phenotypically to the same rank as two sympatric species, if both have been evolving independently for the same length of time. This lineage-based concept does not directly address the question of how to determine integrity. DNA sequences seem to be the ideal way to do this. DNA sequences are assumed to evolve mainly as a result of substitution of new mutations by genetic drift—changes in gene frequency produced by accidents of birth and death—and thus they can recapture the history of population subdivision effectively. With such a weak evolutionary force as drift, a strong barrier to gene exchange is needed if the taxa are to diverge.

This reasoning underlies updated versions of the phylogenetic species concept. In this case, two taxa are considered different species if they are *reciprocally monophyletic* (i.e., if all individuals in one taxon are monophyletic with respect to all individuals belonging to the other taxon), and reciprocal monophyly is determined using DNA sequences. However, as mentioned in the body of this chapter, identifying monophyletic groups runs into the practical difficulty of which gene sequences to use. Different genes often show different relationships (Jennings and Edwards 2005; Bensch et al. 2006). To see why this should be so, note that an individual descends from two parents and from up to four grandparents. (An individual can have less than four grandparents, because he/she may have only one grandfather and/or grandmother. For example, if she has only one grandfather, she is related to her grandfather through two lines, i.e., her mother and father are half-siblings.) Continuing back in time, an individual has genes from up to eight great-grandparents, and from up to 4096 ancestors from just 12 generations back. Consider an island population that was founded by 100 colonists just 12 generations ago. Where are the 4096 potential contributors to each individual in the population? The question is exacerbated because some founding individuals, or all their offspring, or all their grandchildren, likely leave no progeny. Clearly, each founder that is related to a present-day individual must be related through multiple lines. Chang (1999) shows that on average about 80% of the ancestors are related to present-day individuals and that in an idealized population that maintains a size of 100 for 12 generations, it is a near certainty that every one of these ancestors is connected to every descendant, through an average of five lines. Thus, as a result of random accidents of birth, death, and transmission from parent to offspring, one gene in the current population could have been derived from the same copy in an ancestor (i.e., be monophyletic), but another gene have copies derived from many ancestors (i.e., up to about 80). In fact, the time to monophyly varies greatly from one gene to another (Edwards and Beerli 2000; Hudson and Coyne 2002; Hudson and Turelli 2003; Edwards et al. 2005). Under an idealized model of genetic drift, somewhat less than 5% of all nuclear genes become monophyletic within about N generations (N is the population size), and just over 5% take more than seven times longer than this (Hudson and Coyne 2002).

One way to circumvent the question of which molecule to use is simply to emphasize results from mitochondrial DNA (e.g., Hebert et al. 2004; and, for a practical application, see Pavlova et al. 2003). Because mitochondrial DNA is inherited only through the female line, the number of copies transmitted across generations in the population as a whole is smaller than the number of copies of nuclear DNA. This results in a relatively short time to monophyly, and it leads to relatively high concordance of biological species with reciprocally monophyletic groups as determined by mitochondrial gene trees (Hebert et al. 2004). However, mitochondrial DNA may be quite unrepresentative of the genome as a whole (e.g., Hudson and Coyne 2002; Hudson and Turelli 2003; Bensch et al. 2006), and, in addition, limited samples from two populations can be reciprocally monophyletic, even if populations are still exchanging genes (Irwin 2002). The current emphasis on mitochondrial DNA may more reflect the technological state of the field than any compelling philosophical reason. It is only now that multiple nuclear genes are being sequenced (Jennings and Edwards 2005; Bensch et al. 2006).

In addition to choice of gene, a second difficulty in applying a phylogenetic criterion is that the average time it takes until all individuals in a population are monophyletic for a given gene depends on the size of the population. Many good biological species that have either maintained large populations, or have arisen relatively quickly, are unlikely to be monophyletic at many genes, including mitochondrial DNA (Hickerson et al. 2006). On the other hand, small populations quickly become monophyletic, so many island populations with little evidence of any reproductive isolation between them are likely to be phylogenetic species (Hickerson et al. 2006). Mitochondrial DNA in particular, with its smaller population size, is likely to lead to the recognition of a large number of reciprocally monophyletic populations across groups of islands. Given the criterion of reciprocal monophyly in mitochondrial DNA, several island populations of the Chestnut-bellied Monarch (Figure 1.2) would probably be classified as different phylogenetic species.

The phylogenetic species concept is useful for the investigation of many systematic questions, but one basic argument in its favor, that it provides an objective criterion for delimiting species, appears not to be met. Although further discussion and technological advances might resolve this issue, for the present the philosophy of Helbig et al. (2002) seems the most useful for addressing many questions about the origin of species, especially the origin of sympatric species. In this view, allopatric populations are considered to be species when they are diagnostically different, and when the differences approach a level of differentiation similar to that seen between sympatric species.

CHAPTER TWO

Geography and Ecology

Related species differ from each other ecologically. This is especially apparent when different species are found in the same place. Ecological differences include foraging methods, habitats occupied, and prey type. For example, large species generally consume large prey and small species consume small prey (e.g., Hespenheide 1975; Richman and Price 1992). Even morphologically similar species differ in subtle ecological ways. Although five species of New World warblers forage in the same tree during the breeding season, each species forages in a different part of the tree (MacArthur 1958). The goal of the first part of this book is to ask whether ecological differences between closely related species are connected to speciation, and if so, how? In this chapter, I present an overview of the ways in which ecological factors contribute to speciation. I address the complementary roles played by geographical separation of populations, which prevents or reduces gene flow between them, and different selection pressures in different populations, which arise because the populations occupy different environments.

Ecological factors influence the rate of speciation at every stage of the speciation process. I consider three main elements (Figure 2.1).

Stage 1. *Populations become geographically isolated.* Geographical barriers, such as water gaps for land birds, are a common cause of geographical isolation, but intervening ecologically unsuitable habitats in which it is difficult to survive and reproduce are another. Wiens (2004b) noted that shifting habitats in response to climate change in mountainous regions leads to range fragmentation (Figure 2.2). This is an ecologically imposed barrier to dispersal. Whatever the barrier, if it is not absolute, individuals may occasionally disperse between populations. In this case various ecological factors can cause gene flow to be reduced to zero, if immigrants have low fitness. The development of reproductive isolation in the face of ongoing dispersal between populations is termed parapatric speciation, and is considered further in Chapter 4.

Stage 2. *Geographically separated populations diverge in response to ecologically based selection pressures.* Adaptation to different environments involves divergence not only in naturally selected traits, such as beak size and pre-

ular dates has proven difficult over such long timescales (Graur and Martin 2004), and an alternative calibration, based on the discovery of fossil penguins dating back 61–62 million years, places many of the divergences in the 10 million years before the mass extinction (Slack et al. 2006). This may be more realistic, as the penguin fossils date so close to the mass extinction, and fossil evidence shows that at least the duck/chicken split pre-dates the mass extinction (Clarke et al. 2005). Thus, the ecological explanation as a cause of the speciation events that led to at least some of the orders appears to be rejected. The rarity of modern orders in the fossil record prior to the mass extinction may be because individuals were rare (e.g., as a result of competition with pterosaurs and other taxa of birds that have gone extinct; Harrison et al. 2004; Slack et al. 2006), as well as the general paucity of fossils from the late Cretaceous (Padian and Chiappe 1998; Clarke et al. 2005).

Other examples in which ecological and geographical explanations have been given for some patterns in speciation are listed in Table 2.1. These examples will be revisited elsewhere in this book.

In the rest of this chapter, I present a general overview of the role of ecological differences and geographical separation in promoting population differenti-

Table 2.1: Geographical and ecological explanations
for some patterns in bird speciation

Pattern	Geographical explanation	Ecological explanation	References
Many endemics on islands	Isolation	Unusual biota affect selection	Mayr 1965a; Lack 1976, p. 227
High species numbers in mountainous regions	Multiple barriers	Unusual environments	Simpson 1964; Lack 1976
High species diversity in the tropics	Large area, more barriers to dispersal	High productivity	Rosenzweig 1992; MacArthur 1969
Speciation in the tropics	Refugia during climate change	Selection across ecotones	Haffer 1969, 1997; Smith et al. 1997
Passerine diversity	Neural capacity promotes successful colonization	Small body size	Fitzpatrick 1988; Kochmer and Wagner 1988
Speciation in New World *Dendroica* warblers	Forest fragmentation	Appearance of forest	Lovette and Bermingham 1999; Price et al. 1998

ation. According to Huxley (1942, p. 154) "with geographical differentiation, spatial separation is the primary factor, paving the way for biological divergence and subsequent discontinuity. With ecological differentiation the primary factor in divergence is divergence in functional specialization, which could even lead to speciation in one and the same place." Geographical separation and ecological differentiation are likely to be involved in most speciation events and this dichotomy is rendered as a continuous gradation in Figure 2.3. It is assumed that geographical separation promotes divergence by limiting gene exchange, but the rate of divergence is accelerated when populations occupy different environments, which lead to different selection pressures. I use the framework of Figure 2.3 to illustrate patterns in bird speciation attributable, in varying degrees, to differences between environments and time in geographical isolation. Moving along the *x* axis from left to right, I start with ecological contributions and end by considering the possibility of divergence with no ecological differentiation whatsoever.

Perhaps the big advance of the past 20 years has been our ability to add the dimension of time to that of space via comparisons of DNA sequences, as has already been mentioned with respect to dating the divergence among the orders of birds. Dates can be related to the appearance of geographical barriers. Dates can also be used to test the prediction that speciation is faster when many resources are available.

As described in Appendix 2.1, genetic differences between species of known ages (e.g., from the fossil record, dated appearance of oceanic islands) can be used to place order-of-magnitude estimates on rates of molecular evolution.

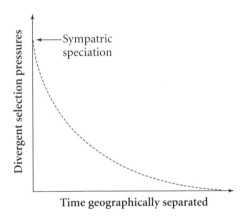

Figure 2.3 A contour of "equal probability of speciation": either strong divergent selection pressures between populations or a long time in geographical isolation, or some combination, leads to a similar probability of reproductive isolation developing between the populations. The area inside the curve (i.e., closer to the origin) has a relatively low probability of speciation and the area outside the curve a relatively high probability of speciation.

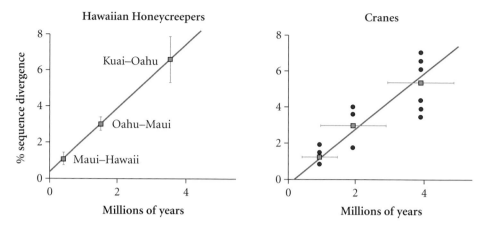

Figure 2.4 Percent divergence in the mitochondrial cytochrome *b* gene plotted against estimates of time since separation. *Left:* Hawaiian Honeycreepers (Fleischer et al. 1998). In this case, estimates of sequence divergence have been further modified by subtracting estimates of within-population variation and making a correction for among-site rate variation and multiple substitutions (see Appendix 2.1). Divergence times are based on the time of island appearance; error bars on the points are an estimate of the standard error. *Right:* Cranes (from Krajewski and King 1996). Divergence times are based on the fossil record. Horizontal bars indicate levels of uncertainty associated with the fossil dates. The three groups represent, respectively, distances between subspecies, species, and the five main clades (i.e., monophyletic groups) within the crane phylogeny. The regression line is fit to the averages (blue points) for each group.

This has proven reasonably successful over the last 10 million years or so. Independent assessments of the rate of molecular evolution are roughly concordant (Weir 2006), as shown in Figure 2.4 for the mitochondrial cytochrome *b* gene of Hawaiian honeycreepers (Fleischer et al. 1998) and cranes (Krajewski and King 1996). For protein-coding mitochondrial DNA, a rate of divergence of about 2% (i.e., two DNA bases in every 100 change) per million years has become the consensus estimate (Klicka and Zink 1997; García-Moreno 2004; Lovette 2004; Weir 2006; Weir and Schluter 2007). This estimate applies to a period spanning from 11 million years ago up to 1 million years ago (García-Moreno 2004; Weir 2006) and is currently based on 47 separate calibrations (Weir 2006; Weir and Schluter 2007). The rate of 2% per million years is used throughout this book in order to obtain a rough estimate of the time it takes to produce species, and of the times of various speciation events in the past 15 million years. Some readers may be uncomfortable with this widespread use of dates without confidence limits, given the many uncertainties involved in calibrations of rates of sequence divergence (Graur and Martin 2004; Appendix 2.1). The dates can be back-translated into sequence divergence by multiplying by two.

ECOLOGICAL DIVERGENCE
WITH NO GEOGRAPHICAL SEPARATION

In theory, speciation can occur rapidly in the absence of any geographical separation if a great diversity of resources, such as food types or habitats, is available (Huxley 1942, p. 154; Rosenzweig 1978; Seger 1985; Dieckmann and Doebeli 1999). In this case, different forms evolve to exploit different resources, and interbreeding among these forms is reduced and finally eliminated because intermediates have low fitness. This is sympatric speciation: the point where the graph meets the y axis in Figure 2.3. In birds, sympatric speciation is extremely rare. Even on isolated oceanic islands each species appears to have been formed as the result of an independent colonization from another land mass and not in situ. To show this, Coyne and Price (2000) compiled species lists for 46 of the most isolated islands in the world. Although the median number of land bird species on these islands was nine, implying many different ecological resources, we found only six cases of two island endemics that belonged to the same genus. Because one species appears to be more closely related to a mainland congener than the other, these cases are thought to reflect separate long-distance invasions from the mainland source (Coyne and Price 2000).

Darwin's finches provide the classic demonstration of the importance to speciation of geographical separation of populations (Mayr 1947). On 47 km^2 Cocos Island, 500 km southwest of Costa Rica, there is only one species of Darwin's finch, the Cocos Finch. This is in contrast to the 14 species of Darwin's finches that are present in the Galápagos Islands. The Cocos Finch has a broad diet, feeding on nectar, fruit, seeds, arthropods, mollusks, and perhaps lizards (Werner and Sherry 1987). It co-occurs with just three other resident species of land birds in a tropical forest. Thus, there seem to be plenty of available ecological niches, but the Cocos Finch has not split into multiple species.

An "island" case that comes closest to what many would consider to be sympatric speciation is that of the flightless Junin Grebe, which is endemic to Lake Junin in the Andes (described further in Figure 3.1, p. 52). Its flying relative, the Silvery Grebe, occurs in sympatry on Lake Junin (Fjeldså 1983), as well as on other lakes through the Andes, and is likely similar to the ancestor of the Junin Grebe. It is possible that a flying form remained on the lake during the formation of the flightless species. A perhaps more likely scenario is that, during the early stages of the evolution of the flightless form, a separate population that retained flight did not breed regularly on the lake. Any flying individuals, including immigrants, may have been at an ecological disadvantage and also interbred with the resident population during initial stages of divergence. However, once the Junin population was sufficiently divergent from populations on other lakes, one of these populations could spread into sympatry, being both reproductively isolated and ecologically differentiated.

Although sympatric speciation is theoretically possible, it is theoretically difficult (Gavrilets 2004, 2005). To see this, consider two kinds of females, A and a, that tend to mate with two kinds of males, B and b, respectively. Speciation occurs if A females mate solely with B males and a females solely with b males. In a single population, one of the easiest ways this process could be initiated is if the female and male traits are the same—e.g., time of breeding (females and males that breed early will tend to mate together, as will females and males that breed late), or habitat preference (individuals that breed in the same habitat will tend to mate together). If the female and male traits are different (e.g., a females prefer to mate with males of one color and A females prefer to mate with those of another color), the two traits (preference and color) will be brought together whenever individuals from one type happen to mate with individuals from the other, and then recombine in the offspring. This means that grand offspring will often carry the preference of one type and the color of the other, resulting in further mixing.

Even if the female and male traits are the same, it is very unlikely that A females will mate spontaneously *only* with B males. For speciation to occur, what is needed in addition is some way in which offspring from matings between the alternative types are unfit. This may happen if offspring from the cross-type pairs show intermediate characteristics, resulting in lower fitness. The lowered fitness of intermediates then favors those individuals that mate with their own type. While an initially strong tendency to mate assortatively, as well as strong selection against intermediates, are both generally required for speciation to be completed (Gavrilets 2004, p. 411; Chapter 14), the process is much easier if the trait that affects mating also affects the fitness of offspring, again because recombination is prevented. For example, individuals with intermediate breeding times or intermediate habitat preference may have low fitness. Gavrilets (2004, pp. 368–387) calls a trait that influences both premating isolation and the fitness of intermediates a "magic trait."

Two bird examples support the idea that sympatric speciation is easiest when one trait has multiple functions. One case involves speciation among African parasitic finches (Payne et al. 2000; Sorenson et al. 2003; discussed further in Chapter 13). These are species that lay their eggs in the nests of other species. The young are raised by the host species, with each parasitic species being largely host specific. A male finch learns the song of his foster species. A female finch learns to recognize the songs of her foster species and uses them to identify suitable mates. The result is strong assortative mating; a single trait (host song) drives both male song and female preference for the song. Perhaps assortment is sufficiently strong to lead to speciation following the colonization of a new host without any accompanying postmating isolation. Whether this is what actually happens remains to be established, and it is certainly possible that there is some spatial separation if different hosts predominate in different locations.

A second possible example of sympatric speciation is based on differences in breeding time, which is an excellent candidate for a magic trait (see Hendry and Day 2005). The Band-rumped Storm-Petrel breeds on islands in the warmer parts of both the Pacific and Atlantic oceans. In many localities, breeding is restricted to a single season or is protracted across seasons. However, in three locations (the Azores, Madeira, and the Galápagos Islands), breeding takes place in two seasons ("cool" and "hot") with little overlap (Monteiro and Furness 1998; Smith et al. 2007; Friesen et al. 2007). On the Azores, hot-season birds commence breeding about four months after the cold-season birds (Monteiro and Furness 1998). This is the time the chicks from the cold-season breeders are leaving their nests, freeing up those same burrows for the hot-season breeders. Hot- and cold-season birds differ morphologically as well as in vocalizations, and they do not overlap in their egg-laying dates—differences that are sufficient to allow ranking the two populations as separate species (Monteiro and Furness 1998; Smith et al. 2007). On the basis of divergence in mitochondrial DNA, Smith et al. (2007) estimate the two species separated about 80,000 years ago. On the Galápagos Islands and in Madeira, hot-season and cool-season breeding populations differ, but the differences are less striking than they are on the Azores, and estimated separation times are at most a few thousand years (Smith et al. 2007; Smith and Friesen 2007). The genetic evidence shows that, within each of the three archipelagoes, hot-season and cool-season birds are more closely related to each other than they are to populations from other areas, and, hence, they are likely to have diverged in sympatry (Friesen et al. 2007). Differences in breeding time cause premating reproductive isolation, and probably also postmating isolation; "hybrids" with intermediate breeding times may have relatively low fitness, for example, because of competition for nest sites (Monteiro and Furness 1998; Friesen et al. 2007).

These unusual cases of sympatric speciation in birds serve to underscore the rarity of sympatric speciation driven by resource diversity.

ECOLOGICAL DIVERGENCE WITH LIMITED GEOGRAPHICAL SEPARATION

The best examples in which ecological differentiation has been linked to speciation are in young adaptive radiations (Schluter 2000, 2001). Simpson (1953, p. 223) defined adaptive radiation as the "more or less simultaneous divergence of numerous lines from much the same adaptive type into different, also diverging, adaptive zones." A more recent definition is "the evolution of ecological and morphological diversity within a rapidly multiplying lineage" (Schluter 2000, p. 10). Both definitions imply that species are rapidly produced in response to the availability of a range of resources. The isolated archipelagoes of Hawaii and the Galápagos provide great examples of recent adaptive radiations. The Hawaiian honeycreeper radiation has given rise to at least 52 species with many different feeding methods and associated beak shapes (Pratt 2005). Their common ancestor dates to perhaps 4–5 million years ago (Tarr and Fleischer 1995; R. Fleischer, *personal communication*).

The adaptive radiation of Darwin's finches (Figure 2.5) is younger and less spectacular than that of the Hawaiian honeycreepers, but it has been better studied. The 14 species of Darwin's finches on the Galápagos Islands are ecologically diverse, with seed-eating ground forms, warbler-like species, and fruit- and leaf-eating tree-dwelling species. They include—in the absence of woodpeckers—the famous Woodpecker Finch, which uses a twig to extract grubs (Grant 1999; illustrated in Figure 6.8, p. 113).

Darwin's finches provide a particularly good example of ecology's role in driving reproductive isolation because the species that have recently diverged from one another are those that are strongly ecologically differentiated (Figure 2.5; Schluter 2001). This applies to the Small Ground-Finch, the Medium Ground-Finch, and the Large Ground-Finch, which coexist on many islands, with each species feeding on a different range of seed sizes. These three species are so recently diverged that they are not reciprocally monophyletic in mitochondrial DNA: two heterospecifics may be more similar in DNA sequence than two conspecifics (Freeland and Boag 1999a; Sato et al. 1999; Zink 2002; Petren et al. 2005). Although the high genetic similarity of these species may partly reflect ongoing hybridization (Freeland and Boag 1999a; Grant et al. 2005), the molecular evidence is consistent with a hypothesis that speciation happened within the past few thousand years. However, error associated with the estimate is large, and it is probably also consistent with speciation on the order of tens of thousands up to perhaps hundreds of thousands of years.

Because of the great similarity in mitochondrial DNA sequences among some species of Darwin's finches, proponents of phylogenetic species concepts do not like the current classification (Zink 2002). Based on mitochondrial DNA, the Small, Medium and Large Ground-Finches are one phylogenetic species. Under the biological species concept the Small, Medium and Large Ground-Finches are three different species. First, when found on the same island these species mate assortatively, with hybridization typically on the order of a few percent of all pairs or less (Grant and Grant 1992; Grant et al. 2005). Second, different populations are assigned to the same biological species based on similar morphology (beak shape and body size). For example, populations

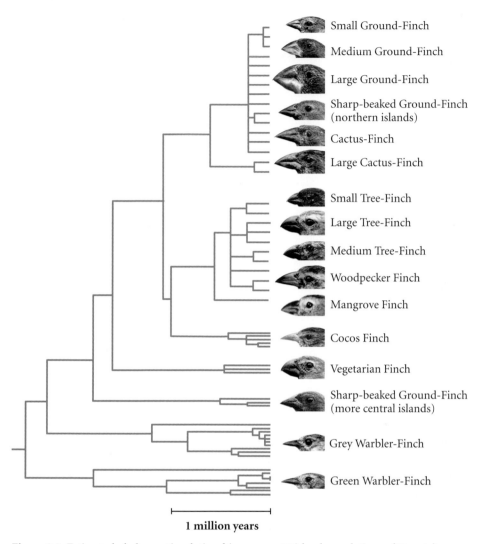

Small Ground-Finch

Medium Ground-Finch

Large Ground-Finch

Sharp-beaked Ground-Finch
(northern islands)

Cactus-Finch

Large Cactus-Finch

Small Tree-Finch

Large Tree-Finch

Medium Tree-Finch

Woodpecker Finch

Mangrove Finch

Cocos Finch

Vegetarian Finch

Sharp-beaked Ground-Finch
(more central islands)

Grey Warbler-Finch

Green Warbler-Finch

1 million years

Figure 2.5 Estimated phylogenetic relationships among 57 island populations of Darwin's finches (using mitochondrial DNA sequences from Sato et al. 1999 and Petren et al. 2005). In this case, the tree topology (branch order) was taken from Petren et al. (2005) and a molecular clock was enforced on this topology using maximum likelihood in PAUP* (Swofford 1998, see Appendix 2.2). The scalebar is based on a rate of sequence divergence of 2% per million years, and was estimated from the distance to the root of the tree in a separate analysis in Tree-Puzzle, based on the HKY-Γ model (Schmidt et al. 2002, Appendix 2.2). Note that different populations of what are currently classified as Sharp-beaked Ground-Finches are widely dispersed on the tree, and that some populations of both species of warbler finches are quite distantly related to each other. The figure was kindly provided by Ken Petren.

of Small Ground-Finches from all islands are classified as conspecific because they are small, with similarly shaped beaks. Morphology is an important species recognition cue and it is likely to indicate the probability of interbreeding in sympatry (Ratcliffe and Grant 1983). The current classification, based on the biological species concept, is useful if we are to understand the evolution of reproductive isolation and, in particular, the link between reproductive isolation and ecological differentiation.

In contrast to the young, ecologically-differentiated, ground-finches, populations of warbler-finches on different islands are ecologically similar but divergent in DNA sequences (Freeland and Boag 1999b; Petren et al. 1999, 2005; Figure 2.5). Indeed, the two different species of warbler-finch, which appear to have been separated for more than 2 million years, are ecologically and morphologically so similar that they were considered conspecific prior to the DNA studies. They respond to each other's songs, implying weak premating isolation (Grant and Grant 2002a). These results suggest that in the absence of large ecological differences, the generation of reproductive isolation has proceeded slowly. In fact, within the warbler-finches some island populations are so strongly differentiated in mitochondrial DNA that they are different phylogenetic species. Thus, in Darwin's finches there are biological species that are not phylogenetic species (e.g., the Small, Medium and Large Ground-Finches), and phylogenetic species that are not biological species (populations of warbler-finches).

Besides the Hawaiian Islands and the Galápagos Islands only one other archipelago seems to have led to a substantial adaptive radiation. In the West Indies a group of thrashers and tremblers has radiated from a common ancestor dating back an estimated 4 million years (Hunt et al. 2001). Five species belonging to three different genera are endemic to the Lesser Antilles, and four co-occur on the islands of Martinique and St. Lucia. The West Indian grassquits may also have radiated within the archipelago, but they currently include continental representatives (Burns et al. 2002). No other archipelago radiations are known that have led to more than two co-occurring species on an island. In northern Melanesia for example, with the possible exception of a pair of white-eyes on Kulambangra, all species on any given island can be traced to separate invasions from outside the archipelago, mostly nearby New Guinea (Mayr and Diamond 2001); even the white-eyes seem likely to follow this pattern (Mayr and Diamond 2001; C. Filardi, *personal communication*). In northern Melanesia ecological and morphological diversification has been limited, and instead, Mayr and Diamond (2001) emphasize the role of geographical isolation among islands as the main cause of speciation (Chapter 8). The proximity of northern Melanesia to other sources, especially New Guinea, explains the absence of adaptive radiation on this archipelago (niches are filled by immigration from New Guinea). Conversely, the isolation of Hawaii and the Galápagos explains the presence of adaptive radiation on these archipelagoes. Similarly, recent adaptive radiations on continents are less spectacular than those on the Hawaiian Islands and the Galápagos Islands, because diverse ecological niches can be filled by immigration of species occupying these niches elsewhere rather than by radiation from a single ancestor (Chapter 6).

GEOGRAPHICAL ISOLATION WITH LIMITED ECOLOGICAL DIVERGENCE

Further along the x axis in Figure 2.3, the time a population persists in isolation, rather than ecological differentiation, becomes the essential ingredient to the production of new species. Superspecies are common (Table 2.2). They contain allospecies that are morphologically and ecologically similar but geographically separated. Many of these allospecies diverged from each other quite long ago (e.g., Klicka and Zink 1997; Weir and Schluter 2007). The two species of warbler-finches on the Galápagos provide one example. These two species are paraphyletic, because one lineage leads to the Green Warbler-Finch, and the other to the Grey Warbler-Finch plus all the other species of Darwin's finches (Figure 2.5), so they would not be classified as a superspecies under some definitions (see Glossary). However the two species are equivalent to allospecies in that they are ecologically similar allopatric replacements.

The Western Meadowlark and the Eastern Meadowlark of North America together form a continental superspecies (Figure 2.6). Despite morphological and plumage similarity, this pair of sister species is estimated to have been separated for about 2.5 million years (S. Lanyon and Omland 1999)—roughly the same time as the whole Darwin's finch radiation. The two species sing very different songs, but W. Lanyon (1979a) was able to cross them successfully in captivity after constructing aviaries large enough that the birds conducted song-flight displays. Six hybrid birds (three males and three females) were all healthy and displayed normally but just one hybrid female gave any fertile eggs in subsequent crosses and only about 10% of all eggs from pairs that included a hybrid bird were fertile. Hybridization between the two species of meadowlarks occurs infrequently in the wild, with the few well-documented cases at the eastern edge of the Western Meadowlark's range (W. Lanyon 1966). The presence of both strong premating isolation (based on song recognition) and postmating isolation (based on hybrid infertility) implies that reproductive isolation is essentially complete, despite the low level of hybridization.

The formation of allospecies has often been related to geographical barriers. Examples include the way by which glacial cycles in north temperate regions and wet–dry cycles in the tropics created habitat refugia. Keast (1961) and

Table 2.2: Examples of superspecies* across different taxa and locations

	Number of genera	Number of species	Number of superspecies	Number of species that are in superspecies	Percentage of all species that are in superspecies	Reference
Taxon						
Corvini (Crows)	25	117	17	38	32%	Sibley and Monroe 1990
Paridae (Titmice)	17	65	11	35	53%	Sibley and Monroe 1990; Gill et al. 2005
Furnariinae (Ovenbirds)	53	231	29	64	28%	Sibley and Monroe 1990
Pycnonotidae (Bulbuls)	21	137	16	48	35%	Sibley and Monroe 1990
Locality						
northern Melanesia[1] (land and freshwater birds)		251	105	165	66%	Mayr and Diamond 2001, p. 127
Patagonia[2] (forest birds)		46	17	18	39%	Vuilleumier 1985
sub-Sahara Africa[3] (passerines)		962	169	486	51%	Hall and Moreau 1970, p. 382
Australia[3] (land and freshwater birds)		538	Not available	183	34%	Keast 1974
North America[4] (land and freshwater birds)		567	127	176	31%	Mayr and Short 1970

* Superspecies include at least two allospecies.

[1] Solomons and Bismarcks. Of the 105 superspecies, only 35 have more than one allospecies in northern Melanesia (average 2.7 allospecies in northern Melanesia in these 35).

[2] Sixteen of the superspecies have only one allospecies in Patagonia, and one superspecies has two.

[3] These entries consider only those superspecies that contain more than one allospecies within the designated geographical region.

[4] Of the 127 superspecies only 44 of them contain more than one species within North America (average 2.1 allospecies in North America in these 44).

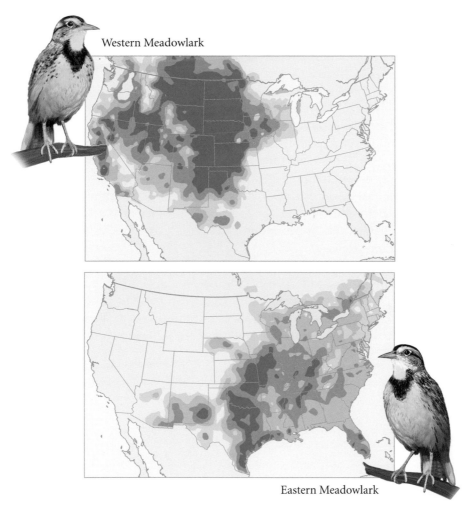

Western Meadowlark

Eastern Meadowlark

Figure 2.6 Distributions of the Western Meadowlark and the Eastern Meadowlark (redrawn from J. Price et al. 1995). Shading reflects density of breeding birds.

Cracraft (1986) showed how distributions of related species fit postulated climatic refugia in Australia. Haffer (1969) famously proposed a series of Pleistocene habitat refugia to account for distributional patterns of closely related species in the Amazon basin. His proposal is controversial, partly because some species divergences pre-date the Pleistocene—they occurred earlier than 2 million years ago (Moritz et al. 2000). Geographical factors that may have caused speciation in the Amazon include barriers resulting from wide rivers, islands created by a rise in sea level, and pre-Pleistocene climatic refugia (reviewed by Cracraft and Prum 1988; Moritz et al. 2000; Haffer 1997, 2002; Newton 2003, his Chapters 10 and 11; Aleixo 2004).

Our ability to add the dimension of time has weakened support for the importance of Pleistocene barriers in the Amazon but has strengthened support for proposed barriers elsewhere. In the boreal zone of North America, Weir and Schluter (2004) identified ten superspecies consisting of two to three allospecies each: one in the taiga, plus one in the Rockies and/or one on the Pacific coast (two are in Figures 2.7). Typically, the taiga species split from the other two an estimated 1.2 million years ago and the Rocky Mountain/Pacific Coast pair split around 0.6 million years ago; the order and relative timing of splits are similar across most groups. These splits are in accord with known biogeographical history and, in particular, with the creation of absolute barriers through the progression of ice sheets. For example, although glaciations commenced about 2.4 million years before the present, the first major advance of ice southward that could have separated Rocky Mountain and Pacific Coast species only occurred about 0.7 million years ago.

Ecological factors should normally interact with geographical isolation to drive divergence among allospecies. The two species of Galápagos warbler-finches differ in that one occupies mostly upland mesic habitats on large islands, and the other mostly drier lowland habitats on other islands (Tonnis et al. 2005). The Western Meadowlark and the Eastern Meadowlark occupy drier

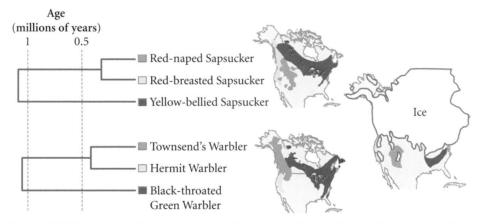

Figure 2.7 Phylogenetic relationships among three species of sapsuckers and three species of warblers, based on mitochondrial DNA cytochrome *b* sequences (Weir and Schluter 2004). These taxa are two of the ten superspecies studied by Weir and Schluter (2004). The right-hand map gives the distribution of three boreal forest refugia at the glacial maximum, and the center maps give the current distributions of species presumed to be associated with each refuge. It is thought that the two westernmost refugia became separated from each other only in association with increased glacial advances commencing 0.7 million years ago. Rohwer et al. (2001) describe the inferred biogeographical history of the Townsend Warbler and Hermit Warbler in response to climate change (see Figure 15.1, p. 330 for illustrations of these species). Estimated divergence times are from Weir and Schluter (2004).

and wetter habitats respectively (W. Lanyon 1956; Rohwer 1976; Chapter 6). Pleistocene speciation among the North American boreal allospecies is likely to have had ecological input. First, there were obvious climatic changes at this time. One effect of such disturbances might have been to cause extinction of predators, prey, or competitors in some places but not others, generating divergent selection pressures (Chapter 3). Second, nine of the ten boreal superspecies groups are migratory—at least one species in the group has disjunct breeding and wintering areas, and different migratory directions often contribute to reproductive isolation (Irwin and Irwin 2004; Chapter 5).

GEOGRAPHICAL ISOLATION WITH NO ECOLOGICAL DIVERGENCE

The extreme point in Figure 2.3 is where the curve meets the *x* axis. This represents divergence that results if populations persist for long enough, without any ecological input whatsoever. That this might happen seems to me to be at least as interesting as the possibility of sympatric speciation, but has been much less discussed. The idea was first championed by Gulick (1890a, b): "I have accumulated a large body of facts indicating that separated fragments of species, though exposed to the same environment, will in time become divergent" (Gulick 1890b, p. 369). Gulick studied land snails of Hawaii and found a correspondence of geographic distance with population differentiation, despite apparent environmental similarities among locations. In response, Darwin noted that different populations must always experience some environmental differences (see Gulick 1890b), a point re-made by Mayr (1947, p. 280): "all geographic races are also ecological races," and others. While this may usually be the case, substantial divergence between populations may be promoted by nonecological factors.

What nonecological factors might promote divergence? Several have been suggested, including genetic drift (e.g., Lande 1981a; Gavrilets 2004) and behavioral innovations that become established in a population and as a result generate novel selection pressures (Gulick 1905). These mechanisms are considered elsewhere in this book, but perhaps the most important mode of nonecological divergence is through the occasional production of a mutation that is favored on the genetic background on which it arises. I emphasize this process. For example, consider a new mutation that results in a new color ornament in males that makes them particularly attractive to females (i.e., is subject to sexual selection). Males carrying the mutation mate with more females and sire more offspring, so the mutation spreads. If different attractive mutations arise in different populations, the populations diverge. Alternatively, consider a new mutation that increases the frequency at which it is itself transmitted from parent to offspring, for example, by killing sperm carrying the other non-mutated form (Hurst et al. 1996). Such a mutation may rapidly increase in a population, and again different mutations may spread in different populations. Interference between such mutations established in different populations could cause post-mating reproductive isolation (Chapter 16), and they may be very important in speciation, but we have little evidence so far. Nonecological processes such as these imply that divergence between isolated populations is inevitable given enough time, but they depend on the accumulation of new, different mutations in different populations. The nonecological model is considered in Chapters 8 through 12 and in Chapter 16.

The nonecological model described in the previous paragraph differs in several ways from the ecological model. One is with respect to the population size that is most conducive to speciation. In the nonecological model, a new favored mutation is more likely to arise in a large population. In the ecological model, a large population may be subject to little overall selection because it is distributed across many different environments, with the effects averaging out (Gavrilets 2004, p. 410). On the other hand, in the ecological model, small populations find themselves in unique environments and are potentially subject to special selection pressures, as pursued further in the next chapter.

Probably the most important difference between the nonecological and ecological models is the influence of gene flow. Under the nonecological model, a new favored mutation may be favored everywhere in the species range, and even a trickle of gene flow between two populations greatly slows the rate at which they diverge. The favored mutation rapidly increases in the population it arises in, and once it is established, it will be introduced into the other with every immigrant, giving it a high chance of spreading through the whole species (Barton 1979). In order for populations to become substantially different in a nonecological model, either a long period of complete separation is needed so many different mutations can accumulate in each population which then interfere with each other's spread, or else the mutation rate has to be so high that different mutations arise more-or-less simultaneously in different parts of the species range (Kondrashov 2003). On the other hand, in the ecological model, populations occupy different environments and genes that are favored in one place may be disfavored in another place. Then, all that is needed for populations to diverge is that selection is strong enough to overcome any retarding effect of gene flow (Chapter 3).

Because divergent selection pressures can be strong, ecological speciation can happen quickly, whereas nonecological speciation should generally take longer. Indeed, the requirement that divergent selection has to be strong enough to overcome gene flow is the reason why sympatric speciation in response to resource diversity (the ultimate form of ecological speciation), if it were to occur, should occur very rapidly. If a single population is to split into two species, divergent selection has to be very strong to overcome a high "migration" rate between the two gene pools (Kondrashov et al. 1998).

TIME TO SPECIATION

Potentially, a way to distinguish the alternative ecological and nonecological models is by the rate at which speciation occurs. Although ecological speciation may be fast, under the nonecological model described here, speciation should generally take a long time. Up until the late 1980s we had little idea about times to speciation, but growth of molecular studies means we can now make educated guesses. For example, MacArthur (1972, p. 241) suggested that it normally takes a few thousands of years for a species to split and diverge sufficiently to be regarded as two species, an estimate that now appears to be quite low. In the case of Darwin's finches, some ecologically-differentiated species are very recently diverged, and it is possible that the rate of speciation is of this order of magnitude. Other examples of young species associated with strong ecological differentiation will be described in Chapter 4.

With respect to divergence among those geographically separated species with less obvious ecological differentiation, Avise and Walker (1998) first used differences between genetically distinct populations within species to place a lower bound on the time to speciation. They defined *phylogroups* as reciprocally monophyletic geographic subdivisions within a species, with monophyly based on mitochondrial DNA sequences [equivalent to some recent definitions of phylogenetic species (Appendix 1.1)]. Phylogroups thus appear to have been separated from each other by a barrier to gene flow for a substantial period of time, but are not usually considered species. Avise and Walker (1998) converted genetic distance between phylogroups into time using the standard 2% divergence for every million years of separation. Using this approach for 68 pairs of New World phylogroups, Weir and Schluter (2007) found an average divergence of about 1.5 million years in more temperate latitudes (centered at 40°N) and about 2 million years at the equator (Figure 2.8). In particular, some populations of tropical species have been separated for a very long time. Marks et al. (2002) identified subspecies of the Wedge-billed Woodcreeper that may have been separated for more than 5 million years (however, it is possible that subspecies such as these will be elevated to species status after further study). Other estimates of within-species divergence times of more than 1 million years include island populations in the West Indies (Lovette et al. 1998; Hunt et al. 2001) and subspecies of tits across Eurasia (Gill et al. 2005). Clearly, divergence to the level of recognized species often takes longer than 1 million years.

Despite these long periods of separation without divergence to the level of full species, it is evident that rates of speciation can be variable, and often substantially less than 1 million years (Figure 2.8). This was shown in the North American boreal zone study (Figure 2.7), which implies that allospecies form on the order of 0.6 million years or less. South of the boreal zone through both North and South America, sister taxa show a great variety of divergence times and some sister species are extremely young (Figure 2.8). Johnson and Cicero (2004) identified nine North American pairs of sister species inferred to have diverged within the past 200,000 years; two were thought to have diverged within the past 35,000 years. Mitochondrial gene exchange as a result of hybridization could lead to an underestimate of time since divergence, so these

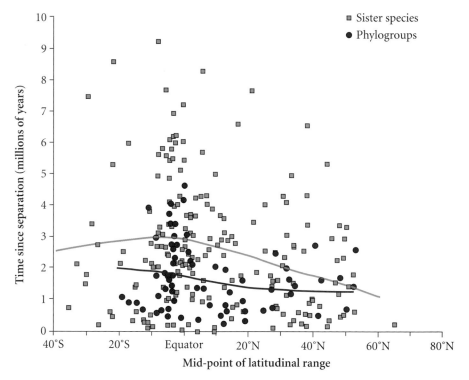

Figure 2.8 Plot of time since separation of phylogroups ($N = 68$) and sister species ($N = 191$) for New World birds (after Weir and Schluter 2007). Time is based on the standard rate of mitochondrial sequence divergence of 2% per million years (Appendix 2.2). Latitude is the midpoint of the maximum and minimum latitudinal extents of the species range (for phylogroups) or the average of the mid-point of the range of each species (for sister species). Lines are nonparametric smoothed regression lines. Note that many sister species are younger than phylogroups within other species. Figure is drawn based on data in Weir and Schluter (2007).

times are lower bounds. In addition, not all of the young pairs are recognized as full species by all authors. Nevertheless, the short times of divergence suggest relatively rapid speciation.

An example of rapid divergence among geographically separated taxa is that of the juncos of North America. This is a complex of extensively hybridizing taxa that are listed as one species, the Dark-eyed Junco, by Sibley and Monroe (1990), but that were listed as five species (and some well-marked subspecies) by Miller (1941). Although fossils of juncos dating to 2 million years ago have been found in Florida (Nolan et al. 2002), Milá et al. (2007a) present molecular evidence suggesting that the current taxa in the Dark-eyed Junco complex have diversified within just the past 10,000 years. The data imply that the Dark-eyed Junco taxa descended from the Mexican Yellow-eyed Junco and rapidly diversified into different forms as they expanded into temperate North America following the last glacial maximum. Milá et al. (2007a) reject hybridization as a cause for the genetic similarity of the different taxa. Two of the taxa occur sympatrically in Nevada apparently without interbreeding, and sequenced individuals in these two taxa are identical in their mitochondrial cytochrome b gene sequence (Johnson and Cicero 2004), although in this particular case it is difficult to exclude the possibility of hybridization as the cause.

CONCLUSIONS

The general importance of geography to speciation is illustrated by the size of the smallest islands that show any signs of within-island speciation. With the exception of a pair of geographically abutting species of hummingbirds on Jamaica (11,000 km²), the smallest islands with obvious in situ speciation are Madagascar (587,000 km²) and New Guinea (786,000 km²) (Diamond 1977). Other observations in support of a critical role for geographical separation include (1) species with larger ranges are divided into more subspecies than those with smaller ranges (Graves 1985; Møller and Cuervo 1998; Belliure et al. 2000); (2) species that occupy mountainous areas, which are presumed to have more barriers to dispersal, are often split into many subspecies (Aldrich 1984; Graves 1985); (3) families of birds with wider ranges and those that span many islands have more species than their sister families (Owens et al. 1999).

Ecological influences on speciation are expected to be strongest early in adaptive radiation, when a large diversity of resources is available. Such periods of "ecological opportunity" may be long in the past with ecological niches now largely filled. This means that if the process of speciation is fundamentally different at different stages of an adaptive radiation the use of present-day distributions to infer the geography of speciation could be misleading (Rosenzweig

1995). For example, Barraclough and Vogler (2000) deduced that geographical separation was the dominant mode in bird speciation, largely because allospecies are common (roughly 40% of all species belong to superspecies; see Table 2.2). But geographical isolation per se may indeed be the main requirement for the production of allospecies if at the late stage in adaptive radiation ecological niches are filled and divergent selection pressures are weak. In the early stages of adaptive radiation, rapid divergence in response to a great diversity of unexploited resources creates conditions most favorable for sympatric speciation (Rosenzweig 1995; Dieckmann and Doebeli 1999). These early stages are generally expected to be difficult to observe, but in birds we have the outstanding example, that of Darwin's finches. Within this group, the presence of speciation on an archipelago (Galápagos), but not on an isolated island (Cocos), suggests that geographical isolation is required even when ecological opportunity is high. Thus, although some models predict that sympatric speciation could occur in response to a wide diversity of unexploited resources, the absence of any examples on isolated islands implies that it is very difficult.

Gavrilets (2004, 2005) has argued that theoretical models of sympatric speciation in response to resource diversity make assumptions that are not generally met in nature. Unless reproductive isolation is strong, populations are prone to collapse back into one species, rather than diverge further. The easiest way to generate strong reproductive isolation is as a result of the incidental accumulation of multiple differences in geographical isolation. For example, in Darwin's finches, both songs and morphology (beak and body size and shape) are important cues used in choosing mates (Ratcliffe and Grant 1983; Grant and Grant 1996c). Generating an association between song and morphology is inevitable if both diverge between a pair of populations, but is more difficult within a single population.

Divergence times between different populations of the same species can be longer than divergence times between some closely related species (Figure 2.8). Populations of what is recognized as one species can be separated by more than 1 million years, but some species are separated by less than 100,000 years, perhaps much less. Clearly the time it takes for populations to diverge to the level at which they are recognized as full species varies by at least an order of magnitude. Some of this rate variation is presumably attributable to stochastic factors, such as the appearance of different mutations. Some, such as the very short times in the junco radiation, remains unexplained. Nevertheless, differences between environments do promote divergence and the generation of reproductive isolation, as is especially clear for the very young species of Darwin's Ground-Finches. In the next chapter I continue the theme of the role of ecology in driving differentiation by investigating ecological causes of geographical variation within species.

SUMMARY

Ecological factors affect speciation rates by influencing (1) the degree to which populations exchange genes, (2) the rate and directions at which populations diverge from each other in ecologically relevant traits, and (3) the probability of successful range expansions. Each forms the topic of a separate chapter (Chapters 4–6). The term "ecological speciation" is used to describe the origin of reproductive isolation between populations in association with their differentiation in response to divergent environmental selection pressures (point (2) above). Ecological speciation is likely to be most rapid when different populations occupy different environments. At one extreme, much environmental variation could lead to rapid speciation in the same location (sympatric speciation). This is rare and the only possible examples of sympatric speciation in birds reflect unusual ways of establishing strong premating reproductive isolation. At the other extreme, given no ecological differences between isolated populations, speciation should eventually result from an accumulation of different mutations, but this may take a very long time. Most speciation involves geographical separation and ecological influences. The time it takes for isolated populations to diverge to the level of recognized species varies from perhaps as little as a few thousand years up to more than a million years. Much of this variation remains unexplained, but the formation of some very young species, such as Darwin's ground finches, can be clearly related to factors promoting divergence to exploit different foraging niches.

APPENDIX 2.1: MOLECULAR DATING METHODS

The basic principle is that, once gene flow ceases between two populations, their DNA sequences diverge as a result of fixation of different mutations. Differences in DNA sequences should therefore be correlated with the time since their divergence from a common ancestor. There are two applications. The first is to obtain relative dates of speciation events, and the second is to obtain absolute dates. These applications depend on a molecular clock: sequence divergence occurs at a constant rate through time. A very simple model of molecular evolution predicts a molecular clock. If the spread of new mutations is due primarily to the chance process of genetic drift rather than as a result of selection, the number becoming fixed per unit time depends only on the mutation rate and is independent of population size. This is because, although the number of new mutations arising in a large population is high, the probability that the descendants of any single mutation spread through a large population by drift is low. The effects cancel out so that the number of new mutations arising in any generation that will eventually spread to fixation is the same whatever the population size.

The molecular clock comes with many caveats (Fleischer et al. 1998; Lovette 2004; García-Moreno 2004; Graur and Martin 2004). Some of the main ones are mentioned here.

1. The mutation rate must be constant across lineages. For example, it may operate more on a *per-generation* than *per-year* scale, so that rates of evolution are slower for species with long generation times (e.g., Sibley and Ahlquist 1990, their Chapter 12). Bleiweiss (1998) showed that in hummingbirds the rate of molecular evolution is negatively correlated with the elevation at which they breed. This may be attributable to effects of nighttime torpor at high altitudes slowing the hummingbirds' metabolic rate.

2. Mutations should be fixed by drift because, if selection is important, the strength of selection may vary. In addition, population size then affects the rate of fixation (e.g., Woolfit and Bromham 2005). Sorenson and Payne (2001) found that parasitic finches have a rate of molecular evolution in mitochondrial DNA that is ~60% faster than that of related non-parasitic finches. A possible explanation is that many mutations are not neutral but are instead slightly deleterious and that they have become fixed in the parasitic species as a result of periodic population size bottlenecks (Sorenson and Payne 2001).

3. There should be little "saturation" as a result of sequential fixation of mutations at the same site. Suppose that a particular site in the DNA sequence is an "A" (i.e., adenine), and that this is then replaced by a "T" (i.e., thymine), which in turn gets replaced by another "A". These two substitutions would go completely unrecorded; instead, one would consider the second "A" to be the same as the ancestral one. Correction for multiple substitutions at the same site requires either the choice of the right molecule for the timescale being studied, or various statistical modifications based on specific models of molecular evolution (Yang 1996; Appendix 2.2).

4. The dating methods themselves assess the time at which a pair of sequences diverged, and this will pre-date the time the two lineages separated (Moore 1995; Edwards and Beerli 2000; Figure 2.9). One method to correct for this is to use an estimate of the genetic distance among individuals sampled at the present day (Figure 2.9), and to subtract the estimate from the genetic difference between species (Moore 1995). This is a very uncertain correction because of the great variation in the time that it can take for all individuals in a population to trace their ancestry back to a single gene copy (Appendix 1.1; Edwards and Beerli 2000), but can be improved by using multiple genes (Edwards and Beerli 2000).

Rates of divergence are affected by other factors. Mutation and drift are stochastic processes, and it may require a long string of DNA-sequence data to get a good estimate of genomewide divergence. Occasional hybridization events may lead to introgression of DNA from one species to another, leading to underestimates in the time since divergence.

A whole suite of additional assumptions is required if species divergence is to be calibrated in terms of absolute time (Fleischer et al. 1998). Molecular divergence between taxa can be compared with absolute time of divergence, as estimated using either the

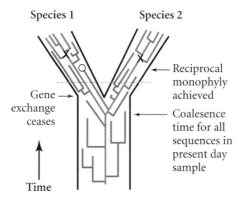

Figure 2.9 Relationships between gene trees and species lineages (modified from Avise and Wollenberg 1997; Klicka and Zink 1999). At any slice in time, several individuals are assumed to be sampled. *Colored lines* represent connections of a single gene sequence in each individual to its ancestor. For example, in the lineage representing species 1, the dashed line indicates a point where four individuals are sampled, as indicated by the intersection of the dashed line with the four colored lines. *Red lines* represent connections among individuals sampled at the present day. Phylogroups form when copies of the sequences sampled within each population are reciprocally monophyletic (i.e., they exclusively share a common ancestor). This is achieved in the diagram when the lineage whose tip is marked with an "o" goes extinct in population 1. The coalescent time for sequences belonging to each species lies earlier than the time of geographical separation, making estimates of divergence based on gene sequences greater than species divergence times (Moore 1995; Edwards and Beerli 2000). A crude way to correct for this is to subtract the time to diverge among the present-day individuals sampled within each lineage (i.e., the average of depth of the two points marked with an X).

fossil record (Helm-Bychowski and Wilson 1986) or biogeographical methods, such as the appearance of a new island (Fleischer et al. 1998) or continental drift (Sibley and Ahlquist 1983, 1990). There are many reasons why the estimated dates should not correspond precisely with the time of population separation (Helm-Bychowski and Wilson 1986; Fleischer et al. 1998; Lovette 2004).

Attempts to calibrate the rate of molecular divergence have been made a few times (García-Moreno 2004; Lovette 2004; Weir 2006). DNA–DNA hybridization distances provide a measure of divergence in single-copy nuclear DNA (Sibley and Ahlquist 1990). These distances were used to construct the famous "tapestry" depicting phylogenetic relationships among 1700 bird species (Sibley and Ahlquist 1990), which has proven remarkably valuable for subsequent bird research. Although many of the details of the branching order of the phylogeny have proven to be inaccurate given recent reconstructions based on DNA sequences, estimated dates of some key transitions have remained roughly consistent (Barker et al. 2004). The commonly used metric for DNA–DNA hybridization distances is "$T_{50}H$", which can be thought of as the temperature at which 50% of the DNA from two species remains annealed when placed in a test tube (T stands for temperature and H for hybrid DNA), and hence is a measure of the similarity of the two DNA sequences (Sibley and Ahlquist 1990). Differences in $T_{50}H$ between own-species annealing and annealing with the DNA of another species measure the amount of divergence between the two species. Sibley and Ahlquist (1990, p. 703) presented a calibration that sets a rate of divergence of 1 °C in $T_{50}H$ as $\Delta T_{50}H = 2.3$ million years for small-bodied species, and perhaps half that for large-bodied non-passerines, which have longer generation times

Other analyses of rates of molecular evolution were based on polymorphisms discovered using restriction enzymes (which cut the DNA at specific sites, depending on the DNA sequence). These methods have now been replaced by DNA sequencing, and the primary approach currently being used is based on long sequences of both mitochondrial and nuclear DNA. The earliest estimate of passerine sequence divergence was that of Fleischer et al. (1998) for three pairs of Hawaiian honeycreepers, which came to 1.6% per million years for the mitochondrial cytochrome b gene (Figure 2.4). Using a large dataset, Weir (2006) obtained a fairly robust estimate of about 2% (i.e., two in 100 bases differ) per million years for divergences over about 1 million years, and that estimate will be used in many places in this book. Despite the large uncertainty associated with this estimate, it allows novel conclusions about rates of speciation to be drawn.

APPENDIX 2.2: PHYLOGENY ESTIMATION

Estimation of phylogenetic relationships among bird species is now usually done based on the similarity of their DNA sequences. It will probably not be long before every bird species has been sequenced at least a little bit. Methods of analysis are also undergoing continual overhaul, with new and better approaches being discovered, and with computers becoming ever faster.

Unless indicated, I have generated all phylogenies in this book using mitochondrial protein-coding DNA sequences, usually about 1000 bases of the mitochondrial cytochrome *b* gene. Nuclear DNA sequences are not generally available. Even when nuclear sequences do become widely available it is not clear that they will add much to the information gained from mitochondrial sequences for the sort of analyses included in this book, which address the tempo of diversification among recently separated species. Mitochondrial DNA has a smaller population size than nuclear DNA, making the time it takes for a new mutation to become fixed in the species relatively short (Moore 1995; Hudson and Coyne 2002; Appendix 1.1). In addition mitochondrial DNA has a much higher mutation rate than nuclear DNA, so that even relatively short divergence times are recorded by a high number of mutational hits. Much longer sequences of nuclear DNA would be needed to get the same amount of information. In further support of the use of mitochondrial DNA, some recent comparative analyses based on mitochondrial DNA have yielded compelling patterns (Weir and Schluter 2004; Weir 2006). Note that the use of mitochondrial sequences to assess patterns of diversification among species is quite separate from the use of mitochondrial sequences to *define* species, a use that remains controversial (see Appendix 1.1)

I am grateful to the many researchers who have posted their sequence data on the public database (GENBANK). From these sequences, I first estimated phylogenetic relationships using a maximum-likelihood method, with the HKY-Γ model of sequence evolution, implemented in *Tree-Puzzle* (Schmidt et al. 2002). The "HKY" component allows for some differences in rates of substitution [e.g., from an A $\rightarrow$ T cf. A $\rightarrow$ G (guanine)]. It is named after Hasegawa, Kishino, and Yano, who first introduced this model in 1985 in a paper studying the divergence of humans from apes. The gamma (Γ) component allows for different rates of substitution across different sites in the DNA sequence (Yang 1996). The actual values for the substitution rates and the Γ parameter to be used are optimized in *Tree-Puzzle*. Typically the Γ parameter comes out to be small (on the order of 0.2), which implies substantial differences in the rate of substitution among sites. One effect of this is that at high levels of DNA divergence, say 7%, the corrected distance comes out to be much larger, often on the order of 12%. This currently seems to be the best estimate available, but it is possible that, when more slowly evolving nuclear genes are sequenced and included, the divergence estimated with this correction will turn out to be too high.

Under the method of maximum likelihood, alternative trees and branch lengths (a branch length is an estimate of the number of substitutions that have occurred along that branch) are fit in an attempt to find the arrangement that has the highest probability (i.e., the maximum likelihood), given the substitution parameters. Because all possible combinations can never be searched, it is difficult to be sure one has got the maximum. Bayesian methods have been developed, partly to deal with this problem. In these methods a large sample of different trees is generated using a specific algorithm. The likelihood of each sampled tree is transformed into a probability, and a consensus tree is built, based on the probability distribution of the sample.

In all trees presented in this book, I wanted branch lengths to reflect the time since separation of a pair of taxa and not just the number of substitutions. This is difficult to achieve. If the rate of molecular evolution has been higher in one lineage than in another lineage, different branch lengths reflect the same time. I used two ways to get around this. The first was simply to enforce a "molecular clock", assuming that the rate of evolution has been constant, as estimated by an average across the whole tree. For large phylogenies the assumption of rate constancy is usually rejected by statistical tests. Unless otherwise stated, in cases with more than 15 species in the analysis, I used a smoothing algorithm (in the program *r8s*, Sanderson 2003) that assumes that rates vary continuously through the tree, and that more closely related lineages share more similar rates.

For the smaller trees I generally used *Tree-Puzzle* with a molecular clock enforced. This prints a tree in which the order of branching is only partially resolved, reflecting genuine uncertainty in branch order (e.g., Figure 1.2), but identifies major groups. The programs I used for the larger trees in a typical analysis are listed here in the order of which I used them, to acknowledge their authors and to illustrate the methods employed, which mostly follow those of Weir (2006). First, I downloaded sequences and aligned them using *Muscle* (Edgar 2004) and *Se-al* (Rambaut 1996). I then used the Bayesian method in *MrBayes* (Huelsenbeck and Ronquist 2001) to obtain a topology (order of branching), based on the GTR-Γ + I substitution model, which is similar to the HKY-Γ model, but the GTR component (general time reversible) allows for different substitution rates between any pair of bases, and the "I" includes an estimate of (invariant) sites that have not changed at all. I imported the topology and sequences into *PAUP** (Swofford 1998) to estimate branch lengths under maximum likelihood, using the optimal substitution parameters as estimated in *MrBayes*. I then entered the branch lengths and topology into *r8s* to produce a clock-like tree.

I calibrated all trees in terms of absolute time by estimating the depth of the tree using the maximum likelihood method in *Tree-Puzzle* with a molecular clock enforced and the HKY-Γ model, and I assumed an evolutionary rate of 1% per million years along a lineage (i.e., 2% divergence between taxa). Occasionally, where indicated, tree depth was taken from the original paper that described the phylogenetic relationships among the species in question. The tree was actually drawn using the program *TreeView* (Page 1996).

CHAPTER THREE

Geographical Variation

Probably more than 99% of all bird speciation events begin with the generation of differences between geographically-separated populations. The establishment of geographical variation is thus the first step on the way to the production of new species, and forms an obvious place to being an investigation of the role of ecological factors in speciation. Vocalizations, colors and color patterns, and morphology vary geographically. In this chapter I concentrate on causes of variation in morphology, which a priori seems the most likely to diverge in response to different natural selection pressures (although morphology is sometimes taken to include coloration, in this book I restrict the term to size and shape). Factors promoting divergence in vocalizations and plumages, which are subject to various forms of sexual selection, are considered in later chapters. However, I include here several studies on the role of gene flow in limiting variation that have been based on analyses of the number of subspecies in a species. Many subspecies are delineated at least partly based on the basis of differences in color patterns, and some on the basis of vocalizations.

Geographical variation is widespread. Mayr and Short (1970) found that two-thirds of all North American bird species contained multiple subspecies. Monotypic species (i.e., those that arc not divided into subspecies) also frequently show some geographical variation (e.g., Johnson 1966). Eighty-four percent (77/92) of studied North American species show identifiable latitudinal variation in wing length (Zink and Remsen 1986), as do 73% (35/48) of some common Andean species (N. Remsen cited in Zink and Remsen 1986). Not only is geographical variation common, it can be large. The subspecies of Song Sparrow breeding in Alaska is dark and weighs ~46 grams whereas the subspecies from the California desert is pale and weighs ~20 grams (Zink and Dittmann 1993). The Song Sparrow is divided into 34 subspecies, most of them clustered in the western United States. Many more examples are given throughout the book.

I first assess the factors that favor geographical variation and then assess the role of gene flow in limiting population divergence. In the case of morphology, four main explanations for geographical variation have been proposed: (1) environmentally induced effects on development that have no genetic basis (i.e., phenotypic plasticity), (2) natural selection, (3) genetic drift, and (4) an interaction between genetic drift and natural selection (Smith et al. 1997;

Schluter 2000). The main factors proposed to prevent divergence are (1) selection favoring similar phenotypes in different places and (2) the retarding effects of gene flow. I conclude that phenotypic plasticity makes some contribution to morphological differentiation, but the role of genetic drift is probably minor. Much geographical variation is likely to be genetically based and established as a result of diversifying natural selection in response to differences between environments. Sometimes this diversification may be limited by gene flow.

FACTORS PROMOTING GEOGRAPHICAL VARIATION

I consider, in turn, the evidence for phenotypic plasticity, for natural selection (cf. drift), and for an interaction between drift and natural selection.

Phenotypic plasticity

Experimental studies testing for a genetic basis to geographical variation in morphology usually indicate a genetic component (Table 3.1) and often reject the possibility that the variation is due solely to direct environmental influences on development. For example, James (1983) and James and Nesmith (1986) cross-fostered Red-winged Blackbird chicks between nests in different regions of North America. Although these studies are often cited as demonstrating the importance of environmental influences on growth patterns, the main result was that rearing site did not affect geographical differences in toe and tarsus length. Even at small spatial scales (3 km in one study), some shape differences seem to be the result of genetic differentiation, although in transplantation experiments such as these, influences of the environment prior to transfer are difficult to rule out (Shapiro et al. 2006). The most striking exception to the rule of genetic differentiation is that of Canada Geese, breeding at a high density on an island, that had skulls that were about 7% shorter than those of the mainland geese. Hand-reared island goslings grew larger, and they equaled the size of mainland birds (Leafloor et al. 1998). In another study two populations of Mallards in a common artificial environment responded to ad-lib feeding by increased tarsus length, but the population with the smaller tarsus length showed a greater increase (Rhymer 1992). This is an interaction between the locality of origin and the conditions in the captive environment, which may have a genetic basis.

Natural Selection

Much evidence implicates the role of selection in the establishment of genetically based geographical variation. First, it is common to find mismatches between classifications based on such traits as color and morphology, and groups defined by shared mitochondrial DNA sequences (Zink 2004). For example, in the Song Sparrow, geographical variation in morphology is not reflected in mitochondrial DNA, which shows little association with geography (Zink and Dittmann 1993). In the Carolina Chickadee, some individuals from a western mitochondrial clade have the phenotypic characteristics of two different subspecies, one of which lies to the north of the other; individuals from the eastern mitochondrial clade likewise occur in both subspecies (Gill et al. 1993). Mismatches such as these likely reflect different processes affecting the evolution of phenotypic traits and molecular markers. Molecular markers may be more subject to genetic drift, which can be easily disrupted by gene flow, whereas phenotypic traits can rapidly evolve under selection. Thus, an island population may scarcely differ in DNA markers from the mainland, as a result of recent colonization or ongoing exchange of a few migrants, but the same population may rapidly evolve larger body size and different coloration given sufficiently strong selection. Alternatively, phenotypic similarity across locations can be maintained by selection when ecological conditions are similar, even if gene exchange between locations has long ceased, and DNA sequences have drifted far apart. As already noted in Chapter 2, rates of morphological and molecular evolution, at least over the timescale of speciation, are decoupled.

Examples of rapid establishment of geographical variation (over tens to hundreds of generations) are usually best attributed to novel selection pressures, although some might reflect nonrandom patterns of dispersal into new areas (Garant et al. 2005). Many of the cases have resulted from introductions or habitat alteration, i.e., the population that has undergone rapid evolution occupies an unusual environment (Table 3.2). The evidence that these changes are a result of natural selection comes from the identification of a plausible causal agent in the environment and the rapidity of the change; some have also been experimentally demonstrated to have a genetic underpinning (Table 3.1). The average phenotype commonly shifts by about 1 or 2 standard deviations of the within-population variation. Changes of this magnitude seem to be largely independent of the timescale over which they have been measured, at least up to a few hundred generations. Evolution of more than 2 standard deviations has been observed mostly on the timescale of thousands of years, inferred on the basis of geographical variation between populations thought to have become established since the Pleistocene (Merilä 1997; Benkman 1999; Clegg et al. 2002a).

Generating a convincing adaptive explanation for any geographical variation is probably the strongest evidence for selection. Explanations for some widespread patterns are typically couched in an adaptive framework. For example, Bergmann (1847, quoted by James 1970) and James (1970) suggested latitudinal variation in body size was directly related to temperature, associated with the ability of large-bodied individuals to more efficiently conserve heat than small-

Table 3.1 Tests of genetic differentiation in morphology or life-history among populations [expanded from Merilä and Sheldon (2001)]

Species	Locations	Comparison between subspecies?	Trait	Method	Outcome* (% genetic)	Early maternal effects checked?[1]	Reference
Canada Goose	Island vs. mainland, Canada, 200 km	No	Skull length	Captive	1	No	Leafloor et al. 1998
Mallard[2]	California vs. Manitoba, 2500 km	No	Tarsus	Reciprocal, captive	3	No	Rhymer 1992
Eurasian Blackbird	Urban vs. rural, 40 km	No	Migratory tendency	Captive	G	No	Partecke and Gwinner 2007
Common Stonechat	Austria vs. Kenya, 6200 km	Yes	Wing length, molt time	Captive breeding	G	Yes	Gwinner and Neusser 1985, Starck et al. 1995
Coal Tit	Swedish mainland vs. Gotland, 100 km	No	Tarsus	Reciprocal, native	98	No	Alatalo and Gustafsson 1988
Great Tit	Oxford wood, 3 km	No	Nestling size, shape	Reciprocal, native	1, 93	No	Shapiro et al. 2006
European Blue Tit[3]	Corsica deciduous woodland vs. coniferous woodland, 25 km	No	Lay date	Captive	0	No	Blondel et al. 1999
European Blue Tit	Corsica vs. mainland, 80 km	Yes	Lay date	Captive	95	No	Lambrechts and Dias 1993
Blackcap	Germany vs. France, 700 km and vs. Canary islands, 3200 km	Yes	Wing length, molt time	Captive breeding	G	Yes	Berthold and Querner 1982

Blackcap	Germany vs. Austria, 500 km	No	Migratory direction	Captive	100	Yes	Helbig 1991
House Sparrow[4]	Costa Rica vs. New York, 3700 km	No	Clutch size	Captive	0	No	Baker 1995
Zebra Finch	Australia vs. Indonesian islands, 600 km	Yes	Wing length	Captive breeding	100	Yes	Clayton 1990a
Dark-eyed Junco	California mountains vs. California coast, 70 km	No	Wing length	Captive	75	No	Rasner et al. 2004
Red-winged Blackbird[5]	Northern vs. southern Florida, 600 km	Yes	Nestling beak length, toe	Reciprocal, native	11, 99	No	James 1983
Red-winged Blackbird[5]	Coastal vs. inland Mexico, 400 km	Yes	Nestling beak length, tarsus	Reciprocal, native	2, 99	No	James and Nesmith 1986

* Reciprocal transplants: estimated as the ratio of variance between column means and the sum of variance between column means and between row means. Columns define nest of rearing, and rows nest of origin in a 2 x 2 table. Others were reared in common environment: estimated as the ratio of the difference in means in the experiment to the difference in means in nature. "G" indicates evidence for genetic differences, but not quantifiable because natural populations not measured.

[1] Maternal effects may be present if nestlings were taken from the nest after hatching. A maternal effect is a direct influence of the mother (e.g., through nutrients she places in the egg). Breeding experiments in the laboratory imply that these were controlled for.

[2] In males the original population means differed by 1.4 mm (0.95 standard deviations). Ad-lib feeding resulted in males 2.2 mm longer (with respect to wild birds) in Manitoba and 2.7 mm. longer in California. Manitoban birds in California were 0.7 mm longer and Californian birds in Manitoba 3.4 mm longer. This suggests an interaction between place of origin and captive environment.

[3] There was an interaction with one population delaying lay date dramatically in captivity and the other not. This may be an unusual response to captivity (Caro et al. 2007).

[4] House Sparrows are recently introduced into North America. They may have been present in Costa Rica for 20 years, and in New York for 140 years. This study involved transfer of adults and did not look at development. It is included because there was convergence in clutch size in captivity, indicating an absence of genetic effects.

[5] Beak length may not have been fully grown at the time of measurement.

Table 3.2: Examples of recent evolution in bird populations.

	Example character	Change (standard deviations)	Time (years)	Size/shape*	Comparison group	Causal factor	Reference
Studies across <300 generations							
Silvery Grebe	Bill length	1.0	80	Only size examined	Museum specimens	Character displacement	Fjeldså 1981, 1983
Scarlet-bibbed Myzomela	Wing length	−0.3 to −0.9	<300	Only size examined	Presumed sources	Character displacement	Diamond et al. 1989
Ebony Myzomela	Wing length	1.0	<300	Only size examined	Presumed source	Character displacement	Diamond et al. 1989
Common Starling (North America)	Beak length	1.7	80	Mostly size	Geographical variation in introduced population	Not known	Blem 1981
Common Myna (New Zealand)	Male tibiotarsus	0.8	100	Only size detected	Geographical variation in introduced population	Not known	Baker and Moeed 1979
Silvereye[1]	Tarsus	1.0, −0.3	150, 100	Size, shape	Source	Not known	Clegg et al. 2002a
Blackcap	Migratory direction	1.5	30	–	Source	Favorable winter grounds	Berthold et al. 1992; Helbig et al. 1994
House Sparrow (New Zealand)	Male tibiotarsus	0.6	100	Shape	Geographical variation in introduced population	Not known	Baker 1980
House Sparrow (North America)	Beak length, femur length	2.1, 1.2	110	Shape	Geographical variation in introduced population	Thermoregulation?	Selander and Johnston 1967; Johnston and Selander 1971

Species	Trait	Value	Generations	Size/shape	Source	Cause	Reference
House Finch	Beak length	1.4	>120	Shape	Source	Not known	Hill 1993
House Finch	Tarsus, beak length	−0.5, 0.5	40	Size, shape	Source	Changed foraging habits	Aldrich 1982
Laysan Finch[2]	Male beak depth	0.9	22	Shape	Known source[2]	Diet switch	Conant 1988
Laysan Finch[3]	Male beak length	1.9	22	Shape	Known source[3]	Diet switch	Conant 1988
I'iwi	Upper mandible length	−0.7	100	Shape	Museum specimens	Diet switch	Smith et al. 1995
Dark-eyed Junco	Wing length	1.4	20	Size	Presumed source	Not known	Rasner et al. 2004
Darwin's Medium Ground-Finch	Beak size (three studies)	0.9, −0.4, 0.7	1	Mostly size	Population study	Type of food available, character displacement	Boag and Grant 1981; Grant and Grant 1995, 2002b, 2006

Studies across >1000 generations

Species	Trait	Value	Generations	Size/shape	Source	Cause	Reference
Silvereye	Tarsus	5.1	4000	Size and shape	Presumed source: time of vegetation on island	Absence of competitors	Clegg et al. 2002a
European Greenfinch[4]	Scapular length, mandible width	2.7, 1.6	5–8000?	Shape	Geographical variation: expansion post-glaciation	Not known	Merilä 1997
Red Crossbill[5]	Beak depth	3.2	6800	Shape	Geographical variation: pollen record of pines[5]	Adaptation to pinecones	Benkman 1993, 1999

NOTE When sexes were separated, only males are shown in the table.
* Shape changes are inferred if the amounts of change in different traits differ greatly.
[1] Colonization of Chatham Island and Norfolk Island, respectively, from New Zealand; dates are known.
[2] Laysan (source) vs. S.E. island in Pearl and Hermes reef.
[3] Laysan vs. N. island in Pearl and Hermes reef.
[4] The variation is thought to have accumulated since the last ice age, and variation in the mitochondrial control region is small (maximum 0.8% difference, Merilä et al. 1997). Nevertheless, two subspecies are involved, and some of the variation may be older.
[5] The source is not confirmed. The other possible source gives a larger difference (but in a different trait).

bodied individuals. Other explanations for the same pattern have been based on a varying biotic environment, notably altered selection pressures brought about by changes in the suite of competitors for resources such as food (e.g., Darwin 1859, p. 68; McNab 1971). Here I will emphasize a role for differences in the biotic environment, including resources, competitors, predators and parasites, as an important factor that drives geographical variation within species. This follows Darwin, who, in a particularly relevant passage, stated (1859, p. 400):

> ... how has it happened in several [Galápagos] islands situated within sight of each other, having the same geological nature, the same height, climate, etc., that many of the immigrants should have been differently modified, though only in a small degree. This long appeared to me a great difficulty; but it arises in chief part from the deep seated error of considering the physical conditions of a country as the most important for its inhabitants; whereas it cannot be disputed that the nature of the other inhabitants with which it has to compete is at least as important, and generally a far more important element of success.

Variation in resources and/or the presence of competitors accounts for many patterns of morphological variation (Table 3.3). Fjeldså (1983) studied geographic variation among all grebe species in the world. He showed that a population of the Silvery Grebe on Lake Junin has a particularly short beak length when compared with populations of the Silvery Grebe on other lakes in the Andes (Figure 3.1). On Lake Junin the Silvery Grebe co-occurs with the large flightless Junin Grebe, and Fjeldså used evidence from diet analyses to demonstrate that the small beak length of the Silvery Grebe on Lake Junin likely reflects an evolved response to avoid competition for food with the Junin Grebe. Fjeldså found a similar pattern of divergence in sympatry for five other species of grebes that have both sympatric and allopatric populations with respect to another species. Diamond (1970) describes cases of niche shifts in New Guinea and surrounding islands. These include a warbler that eats seeds in the absence of the finch with which it usually co-occurs, a flycatcher that feeds in the manner of a swallow in the absence of swallows or swifts, and a swift that feeds like a flycatcher in the absence of the flycatcher. More examples are given in Chapter 6.

When island populations are compared with presumed source populations on the mainland, bill and body size often increase (Table 3.3). Most explanations for this pattern have been related to selection pressures associated with an altered biotic environment. In a comparison of the Tres Marías Islands with mainland Mexico, Grant (1965a) was able to show that increases in bill size and tarsus length of several species on the islands were associated with expanded foraging habits attributed to the absence of competitors. Case (1978) suggested an additional competition-related factor that could drive increased body size on islands. He viewed body size as resulting from a balance between potential advantages accruing through social dominance (i.e., the ability to defend a territory) and metabolic costs (more food is required to maintain a larger size). Because fewer species are present on islands, more resources are available for those that are present, and their population densities should be higher. This places a greater premium on being able to defend a territory, favoring increased body size. In addition, the presence of more resources in a territory of a given size could reduce the costs. The net result is an increase in body size. In birds, direct support for this idea is lacking.

Selection and genetic drift

Peripheral populations and small island populations sometimes show characteristic differences from other populations. For example, Mayr (1942, 1954) noted that islands off New Guinea contain distinctive populations or species of Paradise-Kingfisher. However, geographical variation is much less within the Common Paradise-Kingfisher, which has a large range across the island of New Guinea itself. Observations such as these have led to theories of population differentiation and speciation in small and outlying populations through "founder effects" and "genetic revolutions" whereby genetic drift in small founder populations creates new and unusual gene combinations and these combinations are sorted by selection (Mayr 1954; Templeton 1980). Models of the interaction between genetic drift and selection have received much attention in the speciation literature, even though theory suggests that drift is likely to be much less important in driving diversification than selection in response to environmental differences (Lande 1980; Coyne et al. 1997; Whitlock 1997). First, even weak selection favoring a particular phenotype greatly restricts the possibility for drift away from that phenotype (Lande 1980). Second, small changes in the environment can rapidly produce large changes in the phenotype that would take a very long time to occur by drift (Whitlock 1997). In addition, there is very little empirical evidence supporting an important role for genetic drift in the evolution of phenotypic traits (Coyne et al. 1997).

This leaves us with the question of why island populations should differ from each other. One explanation is that populations on small and/or isolated islands experience particularly unusual biotic environments and hence unusual selection pressures; that natural selection, without contributions from drift, is the prime cause of the variation. Mayr (1954) qualitatively assessed the proposition that islands differ in their biotic environment and dismissed it, but analyses that are more quantitative are needed. In the following sections I examine differences in the biotic environment. I compare environmental differences among small islands to environmental differences among large islands, with the prediction that differences should be greater among smaller islands than among larger islands. I first examine resources and then potential competitors.

Table 3.3: Comparative studies of geographical variation in ecology or morphology.*

	Increase (%)	Decrease (%)	Number of comparisons	Comparison	References
Absence of congener[1]					
Altitudinal range	22	0	49	Offshore island (Karkar) with New Guinea	Diamond 1970
Altitudinal and habitat range[2]	75	0	53	Summit of isolated mountain compared with mountain in main Andes	Terborgh and Weske 1975
Altitudinal range	100	0	1	Single species of brush finch on 3 mountains where it occurs alone cf. others where it occurs with congener	Remsen and Graves 1995
Foraging position in tree	100	0	7	Tits (*Parus* in allopatry and sympatry)	Alatalo et al. 1986a
Difference in bill length	100	0	5	Pairs of species on North American offshore islands compared with same pair on mainland	Grant 1968
Bill length	100	0	6	Grebes (Podicipedidae). In 5 species parallel differences were observed in more than one population	Fjeldså 1983
Trunk and branch feeding	100	0	3	Three unrelated species on Tasmania (all trunk feeders occurring on the mainland are absent)	Keast 1968
Size differences between congeners[3]	68	32	122	Islands vs. mainland	Grant 1968, after Schoener 1965

Increased size on islands

Bill length	78	22	27	Subspecies on North American offshore islands compared with mainland race	Murphy 1938
Bill length	100	0	9	Species on North American offshore islands compared with mainland congener	Murphy 1938
Bill length	69	31	47	Subspecies on Mexican offshore islands compared with mainland race	Grant 1968
Bill length	75	25	20	Drongos (Dicruridae) of S.E. Asia, island vs. source	Mayr and Vaurie 1948
Bill length[4]	28	0	29	Oceanic islands cf. continent (subspecies comparisons)	Clegg and Owens 2002; S. Clegg (*personal communication*)
Body weight[4]	24	0	16	Oceanic islands cf. continent (subspecies comparisons)	Clegg and Owens 2002; S. Clegg (*personal communication*)

* A few species-level differences are also included.
[1] Increase means the change was in the direction predicted by the competition hypothesis.
[2] The patterns may not be as strong as suggested in the original article (Remsen and Graves 1995a).
[3] The difference between a pair of species in a genus on an island is compared with the difference between a pair of species in the same genus on the mainland.
[4] Only differences greater than 10% were counted.

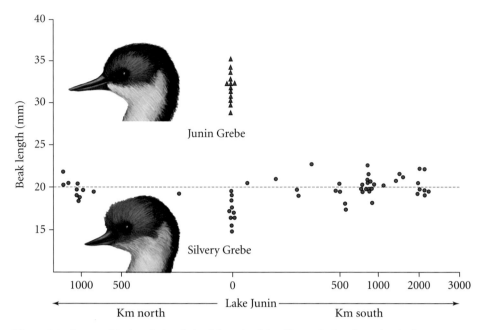

Figure 3.1 Geographical variation in beak length of the Silvery Grebe along the Andes, compared with morphology of the Junin Grebe (from Fjeldså 1983). Measurements are of individual males; the dashed line is an average for all Silvery Grebe populations away from Lake Junin. The endemic subspecies of the Silvery Grebe at Lake Junin has the shortest beak length of all populations of Silvery Grebes, and on this lake, only it co-occurs with its close relative, the larger, flightless, Junin Grebe.

Resources

In the Galápagos, resources for Darwin's ground finches have been measured, and, as suggested by Darwin's quote on p. 48, islands differ from one to the other. In studies across 16 islands, Schluter and Grant (1984a) and Schluter et al. (1985) measured the size and hardness of the common seeds consumed by ground finches on each of the islands. Schluter and Grant (1984a) translated the size and hardness of a seed into a single dimension: the average beak depth of the ground finch population that most efficiently utilizes that seed (the way in which they did this is described in Chapter 5). The actual average beak depth of a population of finches is strongly correlated with the size/hardness of the seeds eaten (Figure 3.2). Thus, geographical variation in beak size is largely accounted for by differences in the kinds of resources available.

All of the Galápagos Islands have small seeds, such as grass seeds, but islands differ in the size/hardness of the largest/hardest seed available (Figure 3.3). I divided the islands into two groups: those that are less than 10 km² in area and

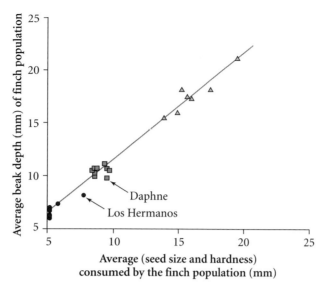

Figure 3.2 Scatter plot of mean beak depth of populations of three species of Darwin's finches [the Small Ground-Finch (●), the Medium Ground-Finch (■), and the Large Ground-Finch (▲)] against seeds of a particular size and hardness consumed by the population. Size and hardness are here cast into a single dimension, as the units of the average beak depth of a finch population best suited to feed on that kind of seed, according to methods described further in Chapter 5, pp. 76–77. On Isla Daphne and the islands of Los Hermanos one species consumes two discretely different seed classes, but in the plot they are matched to the larger/harder class on each island. The regression line is fit to all points except Daphne and Los Hermanos (Pearson's correlation coefficient, $r = 0.99$). Data from Schluter and Grant (1984a) and Schluter et al. (1985).

those that are more than 10 km² in area. The variance among the 8 smaller islands in the largest/hardest seed available is about 22 times greater than the variance among the 6 larger islands. All the large islands, but only some of the small islands, have large/hard seeds (Figure 3.3). Thus, for the ground finches of the Galápagos Islands, small islands are more variable in the kinds of resources available than large ones.

Competitors

In the Galápagos Islands, the association between seed size/hardness and the mean beak depth of the ground finches that consume those seeds is strong (Figure 3.2), but the morphology of these finches also depends on what other species are present. The two outliers in Figure 3.2 are the mean beak depths of the finch populations on the small islands of Los Hermanos and on Daphne.

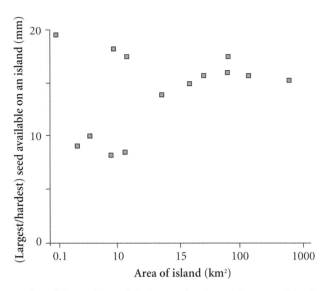

Figure 3.3 Scatter plot of the position of the largest, hardest seeds on any island against island size in the Galápagos Islands. As in Figure 3.2, seed size and hardness are here cast into a single dimension, the units of the average beak depth of a finch population best suited to feed on that kind of seed. Data for the seeds are from Schluter and Grant (1984a) and Schluter et al. (1985) and for island area from Snell et al. (1995).

On these islands a single population consumes the range of seeds that are normally eaten by two different finch species on other islands, and beak size has evolved accordingly (Schluter et al. 1985). (The reason only one species is present in each locality is probably because the islands are so small that they cannot maintain viable populations of both species.) Other competitive effects have been identified. The Small Ground-Finch is particularly small on the northern islands of Pinta and Marchena. This has been attributed to the absence of the bee *Xylocopa darwini* on those islands and the consequent large consumption of nectar by finches (Schluter 1986). The selection pressures that might favor small size in association with nectar feeding have not been identified, but within populations of nectar feeders the smallest individuals are more likely to consume nectar than the larger individuals (Schluter 1986).

To test the prediction that small islands differ strongly in their competitive environment, I used all land bird species as potential competitors with each other and compared the composition of bird communities across islands. The prediction is that small islands differ more in community composition than do large islands. Here, data are available for many archipelagoes in addition to the Galápagos Islands. I used species lists mostly referenced in Boecklen (1997) and the methods of Lande (1996) to partition variation among islands in species numbers (see the worked example in Table 3.4). First, for each archipelago, I divided islands into small, medium, or large, based on order-of-magnitude categories. For example, surveys in the Aland archipelago off Finland covered 24 islands less than 1 km², 13 islands between 1 km² and 10 km², and 7 islands greater than 10 km². Within each group of island sizes, I then calculated the total number of bird species (S_{TOT}), when all islands are considered together, and the average number of bird species on an island (S). The between-island component of species numbers is the total number minus the average number ($S_B = S_{TOT} - S$). The *proportion* of the total species numbers that is partitioned between islands is then S_B / S_{TOT} (Lande 1996).

Results are shown for 11 archipelagoes in Table 3.5. With one exception, smaller islands show a greater between-island component than larger islands in the same archipelago. This is true whether the archipelagoes are tropical or temperate, whether they contain on average very small islands or relatively large islands, or whether they are true islands or mountaintops on continents. Large islands hold more species than small islands, and the loss of any one species from a large island makes relatively little difference to the competitive milieu. On the other hand, islands that hold only a few species vary greatly even if they differ only by one or two species. The one exception is the West Indies. In this archipelago, large islands harbor endemic species. The slightly higher variance among large islands than among small ones may reflect the fact that only on these islands can populations persist long enough to differentiate to the level of

Table 3.4: Example of Lande's (1996) method for partitioning species numbers among and within islands.

Species	Island		
	1	2	3
A	+	−	+
B	−	+	+
C	−	−	+

Three species (A, B, C) are scored for presence (+) or absence (−) on three islands (1, 2, 3).

In this example the total number of species, $S_{TOT} = 3$.

Three species occur on one island and one on each of the other two, so the average number of species on an island is $S = (3 + 1 + 1)/3 = 1.67$.

The between-island component of species diversity, $S_B = S_{TOT} - S = 3 - 1.67 = 1.33$, so the proportion between islands is $S_B / S_{TOT} = 1.33/3 = 0.44$.

Table 3.5: Variation in bird species composition among small, medium and large islands within archipelagoes.*

Location	Type	Proportion between islands[1]			Smaller > larger?	Partitions (km²)	Average number of species (number of islands)		
		Small	Medium	Large			Small	Medium	Large
Finnish archipelago	Temperate islands	76	63	36	YES	1,10	5.0 (24)	17.5 (13)	41.6 (7)
Faeroe Islands	Temperate islands	55	33	12	YES	10,100	11.2 (8)	22.3 (8)	35.2 (5)
Minnesota lakes	Islands in lakes	79	76	45	YES	0.01,0.1	4.7 (22)	11.9 (26)	20.3 (4)
California Islands	Mid-latitude islands	63	55	–	YES	10	9.3 (9)	18.6 (8)	–
Galápagos Islands	Tropical islands	42	16	–	YES	100	11.6 (8)	18.6 (7)	–
Vanuatu[2]	Tropical islands	46	31	33	YES	10,100	21.0 (6)	28.9 (7)	37.3 (15)
Pearl Islands, Panama	Tropical islands	33	24	–	YES	5	24.2 (6)	36.5 (4)	–
West Indies	Tropical islands	67	71	–	NO	500	25.7 (11)	56.0 (8)	–
California mountains	Mountaintops	58	52	35	YES	10,100	13.8 (5)	20.7 (10)	33.6 (5)
Páramo, S. America	Mountaintops	68	64	41	YES	100,1000	10.9 (9)	20.1 (9)	31.8 (5)
Great Basin islands, US	Mountaintops	56	41	–	YES	500	26.6 (12)	41.2 (8)	–

* For original species lists, see citations in Boecklen (1997), except for the Great Basin (Johnson 1975).

[1] The fraction of the species diversity (as a percentage) that is partitioned between islands of different size classes. Calculations follow the methods of Lande (1996). See text and Table 3.4 for further explanation.

[2] Formerly New Hebrides.

full species (Chapter 8), but it may also be because large islands contain some particularly unusual environments creating novel selection pressures (Lack 1976).

In conclusion, the impacts of different resources, different sets of competitor species, and other differences in the biotic environment (predators and parasites) are likely to be major causes of geographical variation across populations, as emphasized by Darwin. Biotic differences may themselves be driven, at least in part, by chance processes of colonization and extinction (Grant 2001), resulting in haphazard patterns of geographical variation across different islands that appear superficially similar. It is the community that "drifts" in species composition. Small islands show greater community drift than larger ones, so a single species should show larger geographical variation in response to a variable biotic make-up among small islands than among large ones. Apart from the beaks of Darwin's finches, quantitative studies that demonstrate the link between biotic variation and geographical variation across islands are missing. Drawing possible connections between differences in the biological make-up across islands and diversifying selection pressures these differences generate is an outstanding research question. A few attempts to do this for sexually selected traits are described in Chapter 12.

FACTORS RETARDING GEOGRAPHICAL VARIATION

Geographical variation is common and often likely to be a result of local adaptation. This implies a strong role for natural selection. The close fit of the beaks of Darwin's ground finches to resources (Figure 3.2) suggests that, in this group, even though individuals regularly disperse, gene flow among island populations is having little influence on diversifying natural selection as the predominant cause of geographical variation in beak size. Even in this group, however, recent ongoing introgression in one island population has had substantial effects on beak shape (Grant et al. 2004, 2005).

More generally, the impact of gene flow on population differentiation depends both on the amount of gene flow and on the strength of divergent selection pressures. Much immigration can lead to a complete swamping of local adaptation, but even with weak selection, migration rates need to be quite high for this to happen. In an example from Li (1976) given by Hedrick (1983), suppose that the selection coefficient, $s = 0.02$ (i.e., the resident genotype has 102 surviving offspring for every 100 offspring produced by the immigrant genotype), and the difference between residents and immigrants is due to variation in a single gene. An immigration rate of 1% (one immigrant for every 99 residents) results in substantial differentiation, but an immigration rate 15 times this results in very little differentiation. Intermediate immigration rates result in an equilibrium between selection favoring differentiation and gene flow opposing it; hence geographic variation (Haldane 1930; Hedrick 1983, pp. 295–305; Hendry et al. 2001). These figures hide many assumptions, and it is possible that if different genes interact, gene flow will be more homogenizing (e.g., Barton 1992).

A retarding effect of gene flow on local adaptation has been demonstrated in the Great Tit on the small island of Vlieland off the coast of Holland (Postma and van Noordwijk 2005). Over 20 years females in the western woodlands, closest to the mainland, had a clutch size just over one egg more than females in the eastern woodlands. This is partly due to the presence of first generation immigrants, but individuals hatched in the western woodlands also lay a significantly larger clutch than those hatched in the east (on average, 0.75 eggs more). The western and eastern woodlands are separated by only about 5 km. The western woodlands receive three times as many immigrants from the mainland (43%) than those in the east (13%). Postma and van Noordwijk conclude that migration rates of 43% are sufficient to disrupt adaptation to island conditions, but those of 13% enable much local adaptation to proceed. In this case, the balance between selection and immigration was detected not only because immigration is high, but also because selection to match the island conditions is strong.

Gene flow appears to limit differentiation over spatial scales that are more relevant to speciation. Belliure et al. (2000) showed that species that have low dispersal distances (from natal site to breeding site, as measured by banding records in the British Isles) are divided into more subspecies (Figure 3.4)[1]. The relationship is strengthened after correcting for differences between species in geographic range size and population size (population size is the estimated number of breeding pairs in Britain; Figure 3.4). Species with larger population sizes tend to have more subspecies, and the effect is attributed to a lower risk of extinction, enabling populations to persist for long enough to differentiate (Chapter 8). Species with particularly high dispersal distances that have rela-

---

[1] In several places in this book, I present analyses such as this one that use a correlation between characteristics of species in an effort to infer causality (the so-called comparative method). The basic procedure in such a test is to assume the species are sampled at random from some larger universe. The correlation between traits in the larger universe is assumed to be zero, and a significance test gives the probability that the correlation between traits in a randomly drawn sample would be larger than the correlation we observe in our sample. A low probability (i.e., low P-value) implies our sample is unlikely to have come from a larger universe in which the correlation is zero. It is commonly argued that species cannot be considered independent of each other (i.e., to have been drawn at random from the larger universe) because some species are more closely related to each other than others are (Felsenstein 1985; Harvey and Pagel 1991). Lack of independence means that if one species is measured, we have some information about related species. This in turn means that probability values are too low, and we are more likely to infer that there is a real correlation in the larger universe. To see this, consider two human families: in one family all 6 children have a genetic disease and are tall; in the other family, all 6

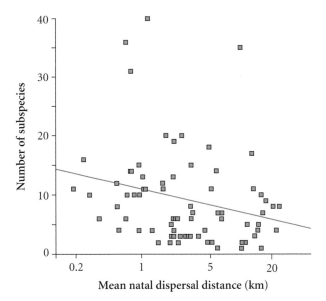

Figure 3.4 Association of number of subspecies in a species with mean dispersal distance from birthplace to breeding site: $r = -0.25$, $P = 0.03$, $N = 75$ (after Belliure et al. 2000). The four outliers (with number of subspecies) are the Winter Wren (40), Barn Owl (35), Great Tit (31), and Eurasian Jay (36), all of which have very large geographical ranges. When range size, body mass and an estimate of population size are included in a multiple regression the association between dispersal and number of subspecies is highly significant $(P = 0.005)$. Significance of other factors is: range size (positive effect on the number of subspecies, $P = 0.004$), body mass (negative effect, $P = 0.07$) and population size (negative effect, $P = 0.02$) (Belliure et al. 2000). Dispersal is the geometric mean of records from banding studies in Britain. Dispersal data are taken from Paradis et al. (1998) and the number of subspecies is from Howard and Moore (1991).

children do not have the disease and are small: the correlation of disease and height would be significant $(P < 0.05)$ if we used a sample size of 12, but not significant (or even testable) if we correctly used a sample size of 2. Felsenstein (1985) devised a model that corrects for the lack of independence given estimates of phylogenetic relationships among the species. However, the whole issue is not straightforward because it depends on the definition of the "larger universe." Some models have been proposed where species can be considered independent of each other. In these models, the characteristics of species present at one slice of evolutionary time give no information about the characteristics of a species that originates later (Harvey and Rambaut 2000; Kelly and Price 2004). They are most applicable to the study of traits involved in an adaptive radiation and have not been widely accepted. Most comparative studies (including the one that referenced this footnote) reported from the original literature use Felsenstein's (1985) method to correct for phylogeny. All those that used Felsenstein's method found similar significance levels to the P-values based on species mean values. Because the significance values are similar, I have not presented them.

tively low numbers of subspecies for their large geographical ranges include the Blackcap (seven subspecies), which is a seasonal migrant, and the large-bodied Grey Heron (four subspecies). It is plausible that in species such as these, dispersal promotes enough gene exchange to limit differentiation.

Gene flow also appears to affect differentiation among bird populations across northern Melanesia (Mayr and Diamond 2001; see Figure 3.5). For each island or island-group in this archipelago, Mayr and Diamond (2001) constructed an endemism index, which measures the degree to which the island harbors differentiated populations (Figure 3.6). First, island endemism is correlated with island area (Figure 3.6). The positive effect of island size is usually attributed to a lower extinction probability for populations that maintain high numbers, as discussed further in Chapter 8. However the correlation of island size with endemism may also reflect the presence of unusual environments on large islands (Lack 1976).

Second, as also shown in Figure 3.6, the endemism index is positively correlated with the distance from a major source: more remote islands contain more

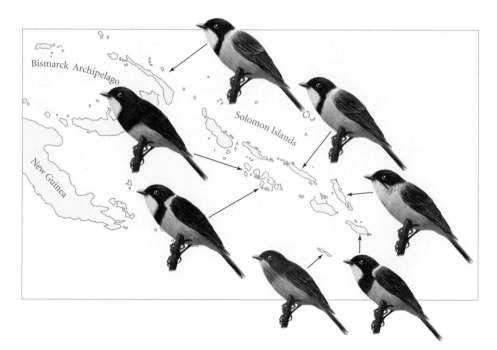

Figure 3.5 The Golden Whistler is the bird world's "greatest speciator" (Mayr and Diamond 2001, p.143), including 66 subspecies falling into five allospecies. Sixteen subspecies, all belonging to one allospecies, are found in northern Melanesia. The males of some of these subspecies are illustrated here; females also vary, and in some cases differences between the females are larger than those between males. Isolated Rennell Island contains a particularly distinctive form *(left at bottom),* and in this form, the male and female are similar. Redrawn from Mayr and Diamond (2001).

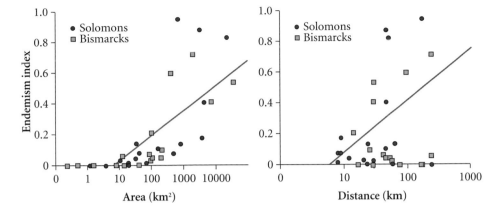

Figure 3.6 Associations of endemism of islands or island groups with island area and distance from a major source in northern Melanesia (from Mayr and Diamond 2001, pp. 195–196). Classifications of islands to groups are as indicated by Mayr and Diamond, land-bridge islands and recent volcanoes are omitted, and only lowland species are considered for islands with a highland zone. The endemism index is defined as $(\sum_{j=1}^{n} i_j)/n$, where n is the number of species on the island and i takes on a value from 0 (not endemic) to 5 (a value of 1 indicates an endemic, weakly differentiated, subspecies; a value of 3, an endemic allospecies; and a value of 5 an endemic genus). Points for the Solomons and Bismarcks (see Figure 3.5 for map) are separated. The Bismarcks cover a greater total area but are closer to New Guinea; when considered as a group they have a lower endemism index (1.28) than the Solomons (1.81), which harbor all five endemic genera. The least squares regression line for area is: Endemism $= -0.12 + 0.16 \log_{10}$area ($P < 0.0001$) ($N = 35$). For distance the line is fit omitting islands with zero endemics which tend to be small and confound the pattern: Endemism $= -0.26 + 0.34 \log_{10}$ distance ($P = 0.014$) ($N = 26$). The multiple regression (including zero endemics) is Endemism $= 0.18 \log_{10}$area ($P < 0.0001$) $+ 0.23 \log_{10}$ distance ($P = 0.0018$) ($N = 35$); the P-values indicate significance of the partial regression coefficients.

differentiated populations. To investigate this pattern further Mayr and Diamond (2001) studied variation within species. A "zoogeographic" species is a class that counts all allospecies in a superspecies as a single species (Mayr and Diamond 2001; see Glossary). Although the decision on whether an allospecies should be considered a true species may be questioned (Chapter 1), the classification of zoogeographic species is not controversial because closely related zoogeographic species are tested by the criterion of coexistence in sympatry. Mayr and Diamond categorized each of the 191 zoogeographic species on northern Melanesia according to its dispersal propensity (Figure 3.7). The most vagile zoogeographic species are those that have historically been observed dispersing (e.g., individuals seen flying across water, or found on an island where they do not normally occur); those of intermediate vagility refer mainly to species that have colonized islands across permanent water gaps (i.e., there has been no land bridge, at least during the history of the species); and those of low vagility are species for which there is no evidence of over-water dispersal during the lifetime of the species.

In northern Melanesia zoogeographic species that are more abundant and that have intermediate dispersal show the highest geographic differentiation (Figure 3.7) of which the Golden Whistler, illustrated in Figure 3.5, is the outstanding example. The correlation with abundance is attributed to longer persistence times for populations of more abundant species, giving them time to differentiate (Chapter 8). With respect to dispersal, Mayr and Diamond suggest that some dispersal is needed to colonize islands, but, beyond that, restricted dispersal aids differentiation. Besides the effect of isolation on limiting gene

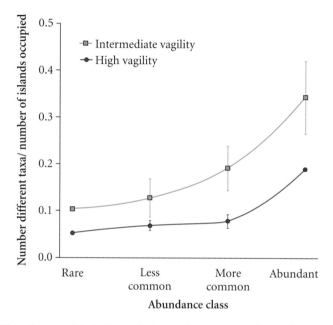

Figure 3.7 Plot of geographical differentiation against vagility and abundance for 50 zoogeographic species occupying at least 30 islands in northern Melanesia, after Mayr and Diamond (2001, p.139). The index of geographical differentiation for a given zoogeographic species is the total number of taxa (subspecies and allospecies) the zoogeographic species contains divided by the total number of islands it occupies. Standard errors are for points with more than one species. *High vagility:* There are confirmed visual records of over-water dispersal. *Intermediate vagility:* Zoogeographic species not in the previous class that occupy more than one oceanic or recently defaunated island, implying over-water dispersal at some time in their history. Abundance estimates are crude and are therefore graded from rare to abundant, with the rare class being less than 1 pair/km^2 and the abundant class more than 100 pairs/km^2. Both lower vagility and higher abundance are significant predictors of geographical differentiation in a multiple regression [*t-tests*, vagility: $t_{47} = 4.5$, $P < 0.0001$; abundance: $t_{47} = 3.2$, $P = 0.026$ ($N = 50$ zoogeographic species), from data in Mayr and Diamond (2001, Appendix 5)]. Sample sizes for the four abundance classes, ranked from rarest to commonest are, for intermediate vagility: 1, 4, 5, 2 zoogeographic species and for high vagility: 1, 24, 12, 1 zoogeographic species.

flow, a second explanation for the high differentiation of isolated populations is that they persist longer, in the absence of a high influx of biological invaders (Chapter 8). However, some species of the least differentiated populations are highly dispersive and it seems likely that extensive gene flow plays an important role in preventing their differentiation. For example, the Cardinal Lory is often seen flying over water, it colonizes new localities easily, and it is not differentiated across its range (Mayr and Diamond 2001, p. 136).

Both the continental and the northern Melanesian analyses imply that gene flow slows the rate of differentiation among populations. A problem of interpretation is that causality may be reversed—differentiation among populations results in reduced dispersal. It is probable that some species experience stronger diversifying selection pressures than others, resulting in higher geographical variation. If this is the case, in these species, individuals that disperse from one place to another will have relatively low fitness because of the different environments they experience, and this should favor a reduction in dispersal propensity (Fisher 1930, pp. 140–142; Endler 1977). Together these factors might cause a negative correlation between dispersal and differentiation across species, not because dispersal retards differentiation but because differentiation favors reduced dispersal.

It seems unlikely that selection favoring reduced dispersal in response to environmental heterogeneity is sufficient to create the observed patterns, because so many other factors besides the possibility of entering an unusual environment influence the costs and benefits of dispersal. In the next chapter, I present evidence that dispersal regularly evolves in response to conditions in the source population. In a specific test of the direction of causality, Smith et al. (1997) found a negative correlation between estimates of gene flow and morphological differentiation among populations of the Little Greenbul in West Africa. Because all comparisons were between ecotone and forest populations, ecological differences between pairs of populations seem unlikely to cause differences in dispersal, so Smith et al. (2005) attributed the correlation to the retarding effects of gene flow on differentiation. More generally, and with respect to the negative correlation between subspecies number and dispersal, it is difficult to see how the large geographical scale over which different subspecies replace one another could have a large impact on the evolution of dispersal distances, as measured within a subspecies.

CONCLUSIONS

The main conclusion is that geographical variation in morphology (size and shape) is usually the result of adaptation to different environments, probably including both abiotic and biotic differences. The presence of consistent differences between populations in morphology and other phenotypic traits implies that individuals could use these differences to identify, and thereby mate with, members of their own population, an essential requirement for the development of reproductive isolation. Differences that are rooted in adaptation also imply that immigrants have lower fitness than residents. This sets the stage for a reduction, and perhaps the eventual elimination, of gene flow between populations, the subject of the next chapter.

Gene flow does seem to retard differentiation, but some of the strongest evidence for this comes from analyses of population differences that have been identified using coloration in addition to size and shape variation, such as the assignment of populations to subspecies. It may be that color pattern divergence is less subject to divergent selection pressures, and more easily disrupted by gene flow. Further discussion of this possibility is deferred to the chapters on social selection (Chapters 9–12).

In the following three chapters I continue the ecological theme of this chapter, asking how variation in the abiotic or biotic environment affects the progress of speciation. In the next chapter, I investigate the ways in which gene flow between populations might be reduced. Then, I ask how divergence between populations due to ecological factors is connected to premating and postmating reproductive isolation (ecological speciation), and in the final chapter of the three, I consider how environmental conditions affect the probability of a species expanding its range, needed for renewed bouts of geographical differentiation.

SUMMARY

Geographical variation in size and shape is common, attributable to varying selection pressures in response to different environments occupied by different populations. Little evidence exists for a role for genetic drift and founder effects in the establishment of geographical variation for ecologically relevant traits. In particular, large differences in morphology among small island populations are likely to be a consequence of selection in response to different biotic environments on different islands. Geographical variation may sometimes reflect a balance between diversifying selection and gene flow, but other times it is likely to be largely independent of gene flow. Some evidence for the retarding effect of gene flow comes from the finding that the number of subspecies in a species is negatively correlated with measures of dispersal in both a continental (the Palearctic) and an archipelago (northern Melanesia) analysis. The demonstration that geographical variation is often adaptive but sometimes opposed by gene flow sets up the conditions for gene flow to be reduced still further, as immigrants should have relatively low fitness in competition with residents.

CHAPTER FOUR

Parapatric Speciation

Geographical barriers that restrict dispersal between populations initiate many speciation events. Such barriers may be areas of completely inhospitable or unsuitable habitat. Divergence on opposite sides of a barrier produces geographically differentiated populations, subspecies, and, later, allospecies. In this chapter, I ask how divergence to the level of species might proceed if such barriers are permeable and individuals disperse between populations. This is parapatric speciation, defined by Coyne and Orr (2004, p. 111) as the origin of reproductive isolation between two populations that initially exchange genes. One way in which gene flow between populations can be cut off is if, once geographical variation has become established, adaptation to different environments results in low reproductive success of immigrants, given they are in competition with well-adapted residents (Nosil et al. 2005); and a second way is if the number of immigrants is itself reduced.

Parapatric speciation takes one of three forms (Table 4.1). First, gene flow can be reduced to negligible levels as a result of *speciation by force of distance* (Mayr 1942). It is plausible that, over long time periods and over large geographical distances, distant populations become strongly differentiated from each other, even if they are connected by a chain of populations that continue to exchange migrants. Although all populations would be lumped as the same species under phylogenetic species concepts (Chapter 1), distant populations could have diverged sufficiently that they would not interbreed—i.e., they are biological species. Smith et al. (1997, 2005) have explicitly promoted this mode of biological speciation, noting that geographical variation within one species is often much larger than differences between sympatric species.

The criterion of reproductive isolation generally cannot be directly tested in the speciation-by-distance model because the different "species" are in different

Table 4.1: Kinds of parapatric speciation

Mode of speciation	Examples
Speciation by distance	Greenish Warbler ring species (Irwin et al. 2005)
Clinal speciation	No clear examples
Island model	Some endemic species on islands and lakes

places. However, in a few unusual examples of speciation by distance, one taxon has spread around a geographical barrier, the two ends of the range have come together, and the populations at the ends do not interbreed (so-called ring species, Irwin et al. 2001c). The best example of this in birds is the Greenish Warbler, which has a more or less continuous range encircling the Tibetan plateau (maps are given later in this book: Figure 5.7, p.93; and Figure 11.5, p. 237). Two reproductively isolated forms to the north of the plateau are connected by gene flow around the west, south, and east of Tibet through a chain of populations, as shown by molecular studies (Irwin et al. 2005). The expansion of geographical range around the western and eastern sides of Tibet out of an inferred southern refuge was accompanied by divergence in songs and song recognition (Chapter 11, pp. 236–238). As ranges expanded, populations came into contact north of Tibet, where they do not interbreed. Geographical variation across the ring implies that much of the reproductive isolation developed prior to contact of the end populations (Irwin et al. 2001b). The origin of reproductive isolation in the Greenish Warbler is considered further in Chapters 5 and 11.

It is unclear how often speciation arises by distance, rather than by strict allopatry, and several formerly classic examples of ring species in birds, such as the Herring Gull complex round the Arctic (Liebers et al. 2004) and the Great Tit through the Palearctic region (Päckert et al. 2005), seem to have had gaps in their distribution during their formation, as ascertained by recent DNA studies. Periods of geographical separation are also likely during the history of the Greenish Warbler (Irwin et al. 2005). Apart from ring species, the result of speciation by distance should be well-differentiated taxa in geographically distant locations connected by other intermediate populations. I am not aware of any examples in which speciation has been demonstrated to have arisen in this way (e.g., by examining results of crosses between taxa at the opposite ends of the geographical range). Thus, despite examples of substantial geographic variation within species, direct evidence that this variation arose in populations continuously connected by gene flow and is coupled with reproductive isolation, is currently limited.

In the rest of this chapter, I ask how gene flow can cease between adjacent populations. Coyne and Orr (2004, pp. 113–117) briefly review models addressing the plausibility of this process. In the theory of *clinal speciation,* a single population becomes divided into two in response to gradual spatial variation in ecological conditions. Ecological differences cause selection pressures that favor individuals who remain near their birthplace. These same selection pressures also favor mating with other residents, because offspring from resident–resident matings are well adapted to local conditions. Theoretically, one population experiencing continuous variation in the environment can split into two or more species (Fisher 1930, pp. 140–142; Endler 1977; Doebeli and Dieckmann 2003). The biological plausibility of these models remains uncertain (Turelli et al. 2001; Gavrilets 2003; Coyne and Orr 2004, pp. 113–114; Polechova and Barton 2005). Potential advantages of vacating one's natal area when it is occupied by superior conspecifics must also counter any selection against dispersal.

Although it is difficult to tell for certain (Endler 1977), clinal speciation does not seem to have occurred commonly in birds, if at all. For example, Ripley and Beehler (1990) found that 50 sister-species pairs of Indian birds could be clearly related to geographical barriers, but that six others met in adjacent habitats at the interface of the Himalayas with the plains of India. Ripley and Beehler (1990) suggested that these six pairs had arisen in situ, but their particular spatial arrangement could easily have arisen as a result of spread from ancient refugia, with the abutting ranges a result of ecological competition between the closely related pairs (Chapter 6). The widespread evidence for the importance of refugia in other speciation events makes this a likely scenario. Most examples in which species currently have abutting ranges on continents (e.g., in hybrid zones; Barton and Hewitt 1985) are probably a result of secondary contact following divergence in geographically different places, rather than divergence in situ.

The third model of parapatric speciation depends on a geographical barrier somewhere in the species range, which restricts migrant exchange between two populations that are occupying different environments. Unlike the complete allopatric model, some dispersal continues across the barrier. Despite dispersal, differentiation proceeds to the level of full species. This is the *island model* of parapatric speciation (Table 4.1). Some endemic species that occupy discrete habitat islands (including true islands, lakes, and mountains) may have originated in this way. It is not always clear that, during the early stages of divergence, populations were exchanging individuals, but the examples are characterized by present-day dispersal between the areas occupied by the different species, and it is plausible that such dispersal was always present. These examples are described in this chapter, and the mechanism by which the species formed is considered. First, immigrants may continue to arrive, but fail to reproduce successfully. Second, the number of immigrants may decline.

REDUCED FITNESS OF IMMIGRANTS

The idea is that residents are better adapted to the prevailing conditions than immigrants. In competition with the residents, immigrants suffer relatively high mortality and low reproductive success, as do any offspring they manage to produce, so fewer of their genes get into the resident population. The reduced gene flow enables greater local adaptation and more divergence, thereby further lowering the fitness of immigrants in competition with residents, in an ongoing feedback loop.

Some evidence indicates that immigrants are often at a disadvantage with respect to residents (Bélichon et al. 1996; Spear et al. 1998; Marr et al. 2002; Hansson et al. 2004; Postma and van Noordwijk 2005). Experimental evidence has not been obtained, and even direct measurements of fitness are not straightforward, because of the difficulty of identifying the immigrants. Of 20 studies (counting each species once) reviewed in the cited papers, 17 compared individuals that dispersed different distances within a population, and only three compared immigrants arriving into an isolated population from outside. Fifteen of the 20 studies (75%) found differences between dispersers and nondispersers in at least one fitness component (e.g., mating success, clutch size). In eight cases, the dispersers had inferred lower fitness, in two, they had inferred higher fitness, and in five, they had higher fitness in some components and lower fitness in others. In four out of six cases in which a measure approximating lifetime reproductive success was obtained, dispersers had lower fitness than nondispersers, and in the other two cases there was no significant difference.

Immigrants may have lower fitness than residents for several reasons. These include a lack of familiarity with the area (e.g., Pärt 1991, 1994) and costs expended in getting there. Bensch et al. (1998) and Hansson et al. (2004) suggested that the lower fitness of immigrant male Great Reed Warblers might be related to differences in song between residents and immigrants. Ecological differences between areas may also be a common cause of reduced fitness and a few studies have addressed this possibility (Marr et al. 2002). Among the Great Tits on the island of Vlieland, Holland, immigrant females lay clutches that are larger than those of the residents. These appear to be larger than optimal, given the rigors of the environment, and the young probably have a lower rate of survival than the residents (Postma and van Noordwijk 2005). In the European Blue Tit on Corsica, Blondel et al. (1999) compared a population in an evergreen–oak woodland with one in a deciduous–oak woodland in separate valleys 25 km apart. The terrain is mountainous, and dispersal between sites may be limited, but individuals do differ substantially in shape: birds in the deciduous woodland have longer tarsi and shorter beaks than those in the evergreen woodland. Laying dates are also more than one month apart, corresponding to the time of the peak food supply in each of the two valleys. These differences likely lead to low fitness of immigrants; however, the evidence that the difference in laying date has a genetic basis is equivocal (Blondel at al. 1999; see Table 3.1). A final example comes from the flightless rails of the southwestern Pacific. These are derived from a widespread ancestral flying form that apparently did not persist on islands that have a flightless species [because most flightless rails are extinct, this has been inferred from fossils (Trewick 1997b)]. Trewick (1997b) suggests that the flightless rails outcompeted the flying form because resources normally used for flight could be used instead for maintenance and reproduction. Thus, during evolution of the flightless forms, any flying immigrants were likely at a disadvantage in competition with a resident form. The flightless rails are described in more detail later in this chapter.

Overall, the evidence on the low fitness of immigrants is somewhat equivocal and always correlative. However, the studies described, except perhaps that of Hansson et al. (2004), are over a much smaller spatial scale than those over which geographical variation and speciation are established. While more research is needed, immigrants might often have lower fitness than the residents, and hence, at large spatial scales, immigrants may be at a considerable disadvantage in competition with local conspecifics. Some exceptions to this must exist, and, in very small island populations, residents that mate among themselves can suffer inbreeding depression, placing immigrants at an advantage (Keller et al. 2001).

In several examples, differentiation to full species can reasonably be inferred to have happened in the face of ongoing immigration. This is likely for Darwin's finches (Petren et al. 2005): immigrants on islands are regularly observed and they occasionally hybridize, and the hybrids are sometimes of low fitness (e.g., Grant and Grant 1993, 1996a). In another example, a population of Red Crossbills occupying an isolated mountain range in western North America is morphologically different from the rest of the species and is adapted to the unusual cones of the lodgepole pines that are found there (Benkman et al. 2001; described further in Figure 5.2, Chapter 5). The population has a distinctive call, suggesting that gene flow into the population is low despite the regular presence of migrants, and Smith and Benkman (2007) found matings with other crossbill call types to be rare. Speciation in Darwin's finches and Red Crossbills is described further in the next chapter.

Other likely examples of differentiation in the face of immigration can be found where a seasonal migrant has given rise to a resident species. Several populations have recently become established as a result of birds breeding in their traditional nonbreeding quarters (Sutherland 1998; Petracci and Delhey 2004). Thus, populations of seasonally migratory Barn Swallows, which normally breed in the Northern Hemisphere and overwinter in the Southern Hemisphere, now breed in both South Africa and Argentina during the northern winter (references in Petracci and Delhey 2004; although in Argentina the birds may still migrate back to North America, D. Winkler, *personal communication*). Phylogenetic analyses have also identified many cases in which resident species in tropical regions have been derived from temperate migrants (e.g., Richman 1996; Chesser 2000). For example, a resident species of chat-tyrant in Ecuador is closely related to a migratory species that breeds in Chile and Argentina and spends the nonbreeding season in Ecuador (Chesser 2000). The scenario is as follows. A population becomes established in the winter quarters. A whole suite of adaptations rapidly evolves in association with both the differences in breeding habitat and the loss of migratory ability. Given strong selection pressures, divergence between the migratory and resident populations can proceed, even if the resident gene pool is initially diluted with recruits from the migratory population. Finally, differences are sufficiently great between the two populations that no recruits from the migratory population can reproduce successfully, and reproductive isolation is complete.

In conclusion, the pervasiveness of geographical variation implies that populations are often locally adapted. Locally adapted residents may be competitively and/or socially superior to immigrants, lowering immigrant fitness and reducing gene flow, sometimes to zero.

REDUCED IMMIGRATION

The second way that gene flow can be reduced is if the number of immigrants into the population is itself reduced. Such a reduction should be favored when immigrants have low fitness, as described above. But this kind of selection pressure is likely to be very weak if, say, a large continental population is sending out emigrants to a small island population, and it is only on this island that the emigrants have low fitness. The strength of selection against dispersal in the large population is very small, and perhaps countered by other factors favoring dispersal across the geographical range of that population. In such cases, the number of immigrants the island population receives is unlikely to be reduced, even if the fitness of these immigrants is low.

Dispersal propensity may more often be reduced as a side effect of adaptation to environments experienced by the source population. Tropical species inhabiting dark forest interiors rarely, if ever, cross water gaps (Diamond 1972; Willis 1974). Diamond (1972) notes the reasons are "psychological, not physiological. Some tropical forest species are even reluctant to cross a road a few yards wide." In northern Melanesia, there are several examples in which different subspecies, or even different populations of the same species, show different tendencies to disperse, implying that such tendencies can quickly evolve. For example, six species classified as vagile on the Solomon Islands have never been recorded off their native island in the Bismarck Archipelago (Mayr and Diamond 2001, pp. 69–70). Montane forests contain the highest proportion of species least prone to disperse between islands, and it is in these forests that many of the Melanesian endemics occur.

Because the metabolic costs of maintaining the flying apparatus are high, if flying ability gives little advantage, it should be reduced. The extreme expression of this is the evolution of flightlessness (Olson 1973; Diamond 1991; McCall et al. 1998; Livezey 2003). Flightless rails, pigeons, ducks, geese, cormorants, parrots, ibises, a heron, owls, a hoopoe, New Zealand wrens, and a bunting on the Canary Islands are (or were) found on mammal-free islands of the world (Mike Dickison, *personal communication;* Rando et al. 1999; Steadman 2006, p. 296). Most flightless species are extinct as a result of human activity over the last 2000 years (e.g., the Dodo of Mauritius, a giant relative of the pigeon). In many ways, the evolution of flightlessness within an archipelago is the ultimate speciation machine because a flying ancestor aids dispersal into the archipelago, but subsequent loss of flight eliminates dispersal between the islands of the archipelago so that each island population then evolves independently.

It is possible that every reasonably sized island in the Pacific has or had its own endemic species of flightless rail, and at least five flightless species appear to have been present on Mangaia (Steadman 1995, 2006). Steadman (2006, p. 316) estimates between 500 and 1600 species of flightless rails have gone extinct in the Pacific alone (there are now only 20 species of flightless rails left in the world; Taylor 1998, p. 56). If all those flightless rails were indeed different species, they account for as much as 12% of the recent historical diversity among birds. Among rails of the genus *Gallirallus* of the western Pacific, the accepted scenario is that flightlessness evolved independently from the same flying ancestor on different islands. The ancestor may have resembled the only extant flying member in the genus, the Buff-banded Rail (Trewick 1997a), which is widely dispersed through the southwestern Pacific (Schodde and deNaurois 1982; Diamond 1991). Molecular and biogeographical studies comparing flightless rails with their nearest flying relatives imply that flightlessness, and hence dispersal between island populations, evolves rapidly (Trewick 1997a; Slikas et al. 2002). The island of Laysan in the northern part of the Hawaiian archipelago was probably largely submerged 125,000 years ago, so the flightless Laysan Rail is inferred to have evolved within that time (Slikas et al. 2002).

In summary, the evolution of reduced dispersal is often a result of adaptation in the source population. It occurs when dispersal provides no large advantage and traits associated with reduced dispersal are selectively favored. This means that the source population sends out fewer emigrants. This does not in itself imply that the focal population receives fewer immigrants. However, reduced dispersal among a cluster of populations means that each population receives fewer immigrants from the others, thereby reducing gene exchange among them.

CONCLUSIONS

Some prepublication reviewers of this book thought that parapatric speciation might be the commonest mode of bird speciation, but it is not at all clear that this is the case. As noted in Chapter 2, even a trickle of gene flow is sufficient to prevent or greatly slow differentiation in genes that would otherwise be generally favored throughout the species range. If these kinds of genes are important to speciation, the difference between no immigration at all and an occasional successful immigrant can be large, and perhaps complete geographical isolation is commonly required, at least in the initial stages of divergence.

Geographical separation is the most efficient way to cut off gene flow completely. Indeed, very many speciation events can be related to absolute geographical barriers that likely prevent gene exchange entirely, such as ice sheets or mountain ranges, at least during the early stages of population divergence. However, once divergence has proceeded to a sufficient level, taxa that come into contact may hybridize with ensuing gene flow, and, in this case, a number of additional selection pressures can cause gene flow to be reduced to zero (Chapter 14). This is "allo-parapatric speciation" (Coyne and Orr 2004, p. 112), and it may be common (Chapter 14). Evidence for true parapatric speciation (i.e., population divergence to the level of full species that has proceeded in the face of dispersal throughout history) is less strong. In this chapter, I have given a few plausible examples of parapatric speciation. They are mostly examples of the island model (Table 4.1), where limited dispersal occurs between semi-isolated populations.

In parapatric speciation, one way gene flow between populations can be eliminated is if the fitness of immigrants and their offspring is reduced as populations diverge. A reduction in immigrant fitness will often be associated with divergence in traits that contribute to reproductive isolation between populations, should they spread into sympatry. For example, songs of immigrants may be different from those of residents, so resident females avoid mating with males singing unusual songs (Bensch et al. 1998). Such song differences should also result in females of each population preferring to mate with males from their own population if populations were to expand into sympatry, that is, lead to some premating isolation. Likewise, if local adaptation results in reduced fitness of immigrants, it should reduce the fitness of hybrids (postmating isolation). For example, Darwin's Medium Ground-Finches eat medium-sized seeds and Darwin's Small Ground-Finches eat small seeds. Where one or the other seed size is found, its corresponding finch species is expected to outcompete immigrants of the other species; the local species is better at exploiting the local food source. However, if both seed types occur in the same place, then both species can coexist. In this case, hybrids between them should have relatively low fitness, because they are unable to exploit either seed size efficiently in competition with the parental species, generating some postmating isolation (Chapter 5).

In summary, reduced gene flow between populations will often be associated with divergence in traits that contribute to both premating and postmating isolation. However, it is possible that, under the criterion of what would happen in sympatry, reproductive isolation between geographically separated populations is far from complete, even when gene flow between them has been reduced to negligible levels. If populations are geographically separated, in addition to any factors affecting reproductive isolation, gene flow between residents and immigrants may be low because of costs associated with dispersal. For example, birds may be tired after having made a long flight, or find the area unfamiliar. Factors such as these would not operate if the populations occurred in sympatry.

A reduction of gene flow to zero between geographically-separated populations is sometimes taken to be the defining step in speciation (Wiens 2004a; Nosil et al. 2005). If this is taken literally, the appearance of a geographical barrier that completely cuts off gene flow between populations immediately elevates these populations to species. I consider it to be more useful to envisage reduction in gene flow as a predisposing factor, allowing populations to diverge in traits that contribute to premating and postmating reproductive isolation, should the populations ever occur together in the same place. The build-up of reproductive isolation in association with divergence in ecologically relevant traits (ecological speciation) is the topic of the next chapter.

SUMMARY

Even if populations are connected by dispersal between them, gene flow into a population can be reduced to zero over time. This happens if immigrants come to be at an increasing disadvantage in competition with residents. Ecological differences between populations thus impose a barrier to gene flow additional to that due to any geographical separation. Perhaps, in this way, seasonally migratory species have given rise to year-round resident species, and some endemic species on lakes, islands, and isolated mountains have formed. Gene flow may be reduced between populations, not only because of the low fitness of immigrants but also because the number of immigrants declines over time. This is especially likely if source populations occupy an environment that favors low dispersal propensity, as in the evolution of flightlessness. Although a reduction of gene flow to zero is an important part of the speciation process, it is not necessarily accompanied by a corresponding reduction in the amount of interbreeding between individuals from different populations that would occur should populations expand into sympatry. Instead, reduced gene flow is best considered a factor enabling populations to diverge in traits causing reproductive isolation, as in the complete allopatric model.

CHAPTER FIVE

Ecological Speciation

In *On the Origin of Species* Darwin (1859, Chapter 4) emphasized natural selection and competition as the major causes of biological diversification. Adaptation to different physical and biotic environments (e.g., variation in climate, predators, prey, parasites, competitors) results in the production of varieties. The problem is that these different varieties should converge when put back into the same environment. Sometimes convergence does not happen because the two forms have come to occupy quite distinct ecological niches (e.g., they live in different habitats or they exploit different prey), both of which are present in the area where they co-occur. Here the different varieties have seized upon "widely diversified places in the polity of nature" (Darwin 1859, p. 112). In this case, they may not directly compete over resources and may coexist without further modification. Other times the two varieties overlap in resource use: "competition will generally be most severe between those forms which are most nearly related to each other in habits, constitution and structure" (Darwin 1859, p. 112). The result of competition for resources may be further divergence to utilize different parts of the resource spectrum, a process now termed ecological character displacement. Other ecological factors, such as predation, may also cause divergence (e.g., Martin 1996), with similar consequences for speciation. However, competition for resources is expected to be an important diversifying force because incipient species should be ecologically similar, inasmuch as they inherit many ecological features from their common ancestor.

Both of the preceding two chapters left off with divergent taxa in allopatry. Reproductive isolation becomes manifest when at least one of these taxa spreads into the other's range. In this chapter, I ask how ecologically-based divergence contributes to the build-up of reproductive isolation between such taxa; note that much of the divergence may happen before the taxa come together. The connection between ecological differentiation and reproductive isolation was not explicitly discussed in Darwin's book, but today it forms the major theme in studies on the role of ecology in speciation (Schluter 2001; Rundle and Nosil 2005).

Rundle and Nosil (2005) define ecological speciation as the "process by which barriers to gene flow evolve between populations as a result of ecologically-based divergent selection." This is slightly broader than my use of the term in this book, because, as noted in the last chapter, factors that lead to reduced gene flow between allopatric populations may not always be relevant to reproductive isolation between populations should they expand into sympatry. Here, I consider how ecologically-based divergent selection contributes to any reproductive isolation that would be manifest if populations were to occur in the same general location.

Ecological speciation includes many topics (such as how ecological factors affect sexual selection) that are considered elsewhere in this book and that build on the results from the basic findings described here. In this chapter, I restrict myself to considering divergence in various ecological traits, such as habitat occupied, prey consumed, and migratory habit, along with associated differentiation in morphology. First, I investigate the role of resource distributions and, separately, ecological competition in driving divergence between incipient species. Second, I consider how the resulting ecological differences lead to reproductive isolation.

MODELS OF RESOURCES AND COMPETITION

Darwin's verbal description of the twin roles of adaptation to resources and ecological competition as major factors driving divergence can be quantified using the concept of the *adaptive surface*. The adaptive surface casts resources into units of fitness of the consumers of the resources. It is defined as the relationship of the average fitness of all the individuals in a population (e.g., the fraction that survive a certain time period) to the mean value of a trait (e.g., beak size) measured on the population. That is, the adaptive surface measures the way in which the average fitness would change if the mean value of the population were to shift (Lande 1976; Fear and Price 1998; see the Glossary). Given several assumptions, populations are expected to evolve towards a peak in the adaptive surface, where they become stuck (Lande 1976).

The shape of the adaptive surface reflects the extent to which different resources vary continuously or discretely. The surface is directly relevant to ecological speciation, because it can be used to model how far populations are expected to diverge from each other. It can also be used to predict the fitness of a hybrid as a function of its phenotype, a component of postmating isolation. I develop the concept of the adaptive surface further by example.

Darwin's ground finches are one of the few groups of organisms for which an adaptive surface has been estimated (Schluter and Grant 1984a). Small Ground-Finches eat small seeds because that is all they are able to crack open. Large Ground-Finches eat large seeds, but they avoid small seeds. This is probably because Large Ground-Finches cannot consume small seeds quickly enough to make them worthwhile eating. Schluter and Grant (1984a) and Schluter et al. (1985) measured (1) the size and hardness of seeds observed to be eaten by a finch population of a given average beak and body size, (2) the abundance of seeds that fall into different size/hardness classes, and (3) the relationship between the abundance of seeds in a class and the abundance of finches consuming them. From these associations, for any given island, they translated the biomass of seeds in different size/hardness classes into the number of finches of a particular average beak depth that the island could support. For example, if there were only grass seeds on an island, a population of individuals with small average beak size would have high abundance. A population with a slightly larger average beak size would contain fewer individuals, because each individual would be heavier and would need to consume more seeds. On this island, then, there would be a single peak in population abundance at small mean beak depth, with the island able support progressively fewer numbers of individuals the greater the average beak size of those individuals.

Estimated adaptive surfaces for the islands of Daphne and Santa Cruz, based on the seed distributions present in the early 1980s, are shown in Figure 5.1. These islands have small seeds (such as those of grasses) and medium-sized seeds (including those of the Giant Cactus, *Opuntia echios*, and the Caltrop, *Tribulus cistoides*) that correspond to two conspicuous peaks. One important feature is that the surfaces are rugged, with tall peaks and low valleys, a characteristic of adaptive surfaces on many islands (Schluter and Grant 1984a). Schluter and Grant (1984a) used these adaptive surfaces to confirm quantitatively Darwin's proposition that different species are seizing on "widely diversified places" in nature. Peaks in the adaptive surface are occupied by only a single finch species on each island, and peak positions quite accurately predict the mean beak depths of the finch populations (as shown in Chapter 3; see Figure 3.2, p. 53).

Another estimate of an adaptive surface is diagrammed in Figure 5.2, this time for Red Crossbills in North America (Benkman 1993, 2003). Crossbills are finches defined by their unusual crossed bills. They feed on the seeds of coniferous trees, and the crossed bill is used to separate the scales on the outside of the cone so that the seed can be extracted with the tongue. The Red Crossbill has an enormous range, from the Philippines north throughout the whole Palearctic region and into Central America. However, within continental North America alone, the Red Crossbill consists of a diversity of discrete morphological types, which are distinguishable by the contact calls that each type makes (Groth 1993a, b). Nine different call types have been described (Groth 1993a; Benkman et al. 2001), and at least four of the call types have a large range across North America (Groth 1993a, p. 53). Each call type has a beak morphology adapted to feed on the cones of a particular conifer (Benkman 1993, 1999). In areas in

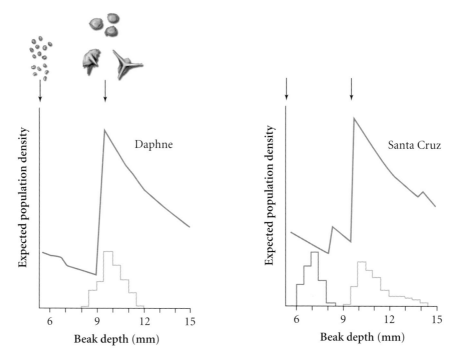

Figure 5.1 Adaptive surfaces, constructed based on seed distributions in the early 1980s, for the neighboring islands of Daphne and Santa Cruz in the Galápagos (Schluter and Grant 1984a; Schluter et al. 1985). The lines give the estimated abundance (in arbitrary units) that a solitary finch population is expected to have as a function of the average beak depth of all the individuals in the population. Two major peaks (*arrowed*) correspond to small seeds and medium-sized seeds, respectively. Medium-sized "seeds" include cactus seeds (*above*) and *Tribulus cistoides* fruits. The histograms give beak-depth distributions for the Medium Ground-Finch (*orange*) and the Small Ground-Finch (*blue*). Although the Small Ground-Finch occurs on Daphne Island, it is rare there, and the Medium Ground-Finch exploits seeds representing both peaks on that island.

which two call types co-occur, they pair assortatively (Groth 1993b; Smith and Benkman 2007). In a sample of 24 pairs in Virginia, assortment was complete (Groth 1993b), but a few cross-type pairings have been observed in another locality (Smith and Benkman 2007). The call types resemble in many ways the different species of Darwin's Ground-Finches, which also hybridize (Grant and Grant 1992, 1994, 1998; Grant et al. 2005). As in Darwin's Ground-Finches, the different call types of crossbills are genetically extremely similar (Questiau et al. 1999; Parchman et al. 2006), suggesting a very recent origin. The estimated adaptive surface for five North American call types of the Red Crossbill also shows peaks and valleys (Figure 5.2).

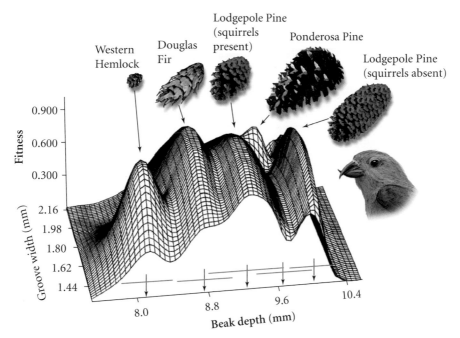

Figure 5.2 A fitness surface for Red Crossbills feeding on cones from different coniferous tree species (from Benkman 2003). This surface differs from the adaptive surface in other figures of this chapter because it is an estimate of an individual's fitness as a function of its phenotype, rather than the average fitness of all individuals in a population as a function of the average phenotype of the population (see Glossary). In one population, Benkman (2003) measured survival of individuals as a function of their morphology. He found that survival was closely correlated with their predicted feeding performance and used this to translate feeding efficiency into individual fitness. He assumed a similar translation around different mean values for the other call types. Each peak in the fitness surface is associated with a distinct crossbill call type; mean beak depths (± 1 standard deviation) for the five types are given along the x axis (from Benkman 1993; Benkman et al. 2001). Groove width refers to a notch in the upper mandible in which the seed is held for husking (Benkman 1993). If the notch is too large or too small, the seed is much less efficiently harvested. The crossbill picture is of a bird from the population at the rightmost peak.

Given adaptive surfaces with tall peaks and low valleys, the question becomes how a species stuck at one peak crosses the valley and gets to another one (i.e., splits into two). How does a Small Ground-Finch consuming small seeds ever get large enough to be able to exploit medium-sized seeds? A small increase in body size is disfavored, because the population is unable to exploit the medium seeds, and less efficient on the small seeds. Several models have been put forward (Fear and Price 1998), but the most common way is likely to be due to geographical and temporal variation in resource distributions. In the case of

Darwin's finches, the standard explanation is that adaptive valleys are crossed via selection on a different island where peaks are in intermediate positions (Grant 1999). In this way, changes in average characteristics can happen in relatively small steps, each one favored by selection. Another potentially important cause of peak shifts results from changes in behavior, and this is discussed in detail in Chapter 7.

A perhaps less frequent cause of peak shifts is the coevolution of the resource with the consumer. This has been suggested for the Red Crossbills of one call type that are restricted to the South Hills of Idaho. These birds have large beaks. The lodgepole pinecones in this isolated area are also large (Figure 5.2). This is an area without squirrels, which elsewhere are important consumers of pine seeds. The cones in the South Hills have apparently evolved new traits, both as a result of relaxation of selection by squirrels and in defense against crossbills (Benkman 1999; Benkman et al. 2001). Lodgepole pine forests in the South Hills probably became fragmented from the main range about 10,000 years ago, and, during that time, it appears that a coevolutionary arms race between the crossbill and the pine has resulted in the rapid evolution of both (Benkman 1999; Benkman et al. 2001). Coevolution has simultaneously created a new peak (a large pine cone) and its occupant (the local crossbill).

Ecological competition

Ecological divergence between incipient species begins in allopatry and may be sufficient that the two taxa do not compete when they spread into sympatry. However, perhaps more usually the two taxa are sufficiently ecologically similar that they do compete for resources to some extent. This can lead to additional divergence. *Ecological character displacement* is defined as evolutionary divergence between species driven by competition between them. Many examples have been given in Chapter 3. As already noted, ecological character displacement, both between distantly related species (e.g., birds and bees in the Galápagos) and between close relatives (e.g., grebes, Darwin's finches) is likely to be an important cause of geographical variation in morphology, in what habitats are occupied, and in what feeding methods are used (Table 3.3).

The role of competition in driving divergence between related species can be visualized most easily using a simple adaptive surface with a single peak and symmetrical tails, as would arise from the resource distributions shown in Figure 5.3. In this case, a single species will evolve to have its mean value centered at or near the peak. [The mean will be displaced from the peak if some phenotypes (e.g., larger individuals) are superior competitors (Taper and Case 1985).] Consider two similar species (or incipient species with substantial reproductive isolation) that have diverged somewhat and have come back into contact.

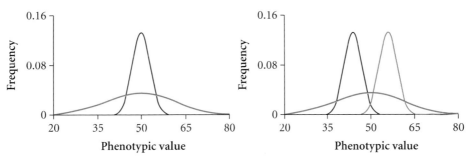

Figure 5.3 Ecological character displacement (after Doebeli 1996). In the left-hand graph the frequency distribution of a trait (e.g., beak depth) is shown in red for a single species. The distribution is assumed to follow a bell-shaped (normal) distribution whose width is measured by its standard deviation (see Glossary; 1 standard deviation = 3 units in this figure). The black line indicates the distribution of the resources (in units of the phenotype most efficient on those resources), also assumed normal and with a standard deviation of 11 units. Any phenotype most efficiently exploits the resources that match its value, so an individual competes most strongly with other individuals whose phenotypes are most similar to its own. In this example, the efficiency of resource exploitation by a particular phenotype declines as a normal distribution with a standard deviation of 5.5 units. A single species evolves to lie under the peak of the resource distribution, but phenotypes at the extremes have higher fitness than the ones in the middle because at the extremes there are many more resources available on a per capita basis. Because of this, if two similar species near the peak differ only slightly in their mean values, they will displace each other to higher and lower mean values. The mean positions eventually reach equilibrium where the intensity of competition between the species is balanced by declining resources (right-hand graph).

Despite the fact that the resource distribution favors convergence, ecological competition drives divergence. Individuals efficiently exploiting the extremes of the resource distribution have more resources available to them (and hence higher fitness) than those that compete with members of the other species. At equilibrium, both species are displaced from the peak in the resource distribution (Figure 5.3, *right*). At this equilibrium, within each species, an individual with the phenotype that matches the species mean value has the highest fitness. An individual that lies to the left or right of the species mean value has lower fitness. In particular, an individual that lies under the peak in the resource distribution has relatively low fitness because it suffers from high competition with members of both its own and the other species (Taper and Case 1985). This implies that, if a hybrid between the two species is intermediate, it too may have low fitness.

Character displacement can occur only if resource distributions are broad with respect to those that can be efficiently utilized by a single species, because only then are unexploited resources available in the nonoverlapping parts of the distributions (Figure 5.3). If the distribution of resources is wide relative to the ability of a species to exploit those resources, two species diverge to consume different parts of the resource spectrum. If the distribution of resources is narrow, character displacement will be limited, and convergence, rather than divergence, may be favored.

The theory of ecological character displacement has been developed largely for simple bell-shaped resource distributions. However, the adaptive surface can be used more generally to assess character displacement. This is shown in Figure 5.1 for the Galápagos finches. On Daphne Island, a single species—the Medium Ground-Finch—consumes resources exploited by two species on larger islands, such as Santa Cruz Island. The result is that the Medium Ground-Finch on Daphne has a relatively small average size and is displaced from a position predicted simply on the basis of peaks in the resource distribution (see also Figure 3.2, p. 53). The implication is that the relatively large size of the Medium Ground-Finch on Santa Cruz is a consequence of ecological character displacement, resulting from the presence of the Small Ground-Finch.

REPRODUCTIVE ISOLATION

Divergence to exploit different resources may lead to reproductive isolation. I consider the contribution of various ecological dimensions, but especially prey size and habitat, to premating and postmating isolation. A relatively unusual ecological dimension that has been related to the generation of both premating and postmating isolation is seasonal migration. The migration examples introduce no new principles, and to keep them together, I have placed them in a separate section at the end of this chapter.

Premating isolation

Premating isolation often results from an individual's selection of a mate based on morphology, plumage, vocalizations, or most usually on some combination (Chapter 13). These traits are subject to sexual selection as well as to natural selection, and the direction and intensity of sexual selection may be affected by ecological differences between populations (Chapter 12). Naturally selected features also result in individuals of one species mating more frequently with individuals from their own species. In one obvious example, different species may occupy different habitats or altitudinal ranges; most individuals of one species simply do not encounter the other when searching for mates.

Much of the ecologically-caused divergence between populations relevant to reproductive isolation is likely to occur before they come into contact. However, if incipient species do come into contact and hybridization is sufficiently low that they maintain their integrity, ecological character displacement between them could lead to additional divergence, strengthening premating isolation. Among Darwin's ground finches, beak shape and body size (as well as songs) are known to be used in species recognition (Ratcliffe and Grant 1983; Grant and Grant 1997a, 1997b). Because beak shape and body size are used as species-recognition cues, Grant and Grant (1997b) suggested that ecological character displacement in these morphological traits has contributed to a strengthening of premating isolation. Character displacement between good species is widespread (Table 3.3), not only in morphology but also in habitat occupation and in altitudinal distribution. Such ecological displacements may contribute to the reduction of hybridization, as suggested for Darwin's finches, but it is also possible that the displacement occurred only after speciation was complete.

Postmating isolation

Postmating isolation arises if individuals from different populations either do not produce offspring when they mate or, if they do, their offspring have low fitness. Low hybrid fitness arises in two ways (Rice and Hostert 1993; Schluter 2001; Coyne and Orr 2004, ch. 7; Rundle and Nosil 2005). In the first, genes in the incipient species interact poorly in the hybrids, resulting in offspring that are inherently unfit whatever environment they are found in. For example, among other deficiencies, hybrid lovebirds (*Agapornis* spp.) have a tendency to get gout (Buckley 1969). Genetic incompatibilities such as these may sometimes result from adaptation to different environments, but may also result from nonecological mechanisms of population divergence. I defer consideration of these causes of postmating isolation to Chapter 16, where I show that complete intrinsic loss of hybrid fitness can take quite a long time to develop (perhaps millions of years). The second cause of low hybrid fitness arises when hybrids differ in phenotype from the two parental species and postmating isolation is contingent on the environment in which the hybrid finds itself. In captivity, for example, hybrids should be viable and fertile. This is the cause of low fitness explicitly considered in models of ecological speciation (e.g., Dieckmann and Doebeli 1999; Gavrilets 2004).

In the character-displacement model of Figure 5.3, when a hybrid between two species has an intermediate phenotype, it has a lower fitness than the average fitness of either of the parental species. With rugged adaptive surfaces, such as those in Figures 5.1 and 5.2, a morphologically intermediate hybrid falls in a valley between two peaks, and thus has lower fitness than the parental species. Among Darwin's finches on Daphne Island, one can imagine that hybrids produced by matings between the Small Ground-Finch and the Medium Ground-Finch, which are intermediate in size (Figure 5.4, *left*), are poorly adapted to feed either on small or on medium-sized seeds, especially when in competition with the parental species. Indeed prior to the El Niño event of 1982–1983, hybrid survival was low (about 50% that of the Medium Ground-Finch), and none of the hybrids bred (Grant and Grant 1993, 1996a). Environmental conditions thus imposed essentially complete postmating reproductive isolation between the Medium Ground-Finch and immigrant Small Ground-Finches.

A few other studies have demonstrated ecological differences between hybrids and the parental species, although the fitness consequences have rarely been measured. Good et al. (2000) found that hybrid gulls on the Pacific coast of North America (described in Chapter 15) showed the habitat preferences of one species in one area, and the diet of the other in another area, both of which led to relatively high fitness for the hybrids. Hybrid wagtails (Sammalisto 1968) and crows (Saino 1992) utilize a wider range of habitats than the parental species, although whether this affects fitness is unknown. In the wagtails, Sammalisto (1968) found a strong correlation between the proportion of hybrids in

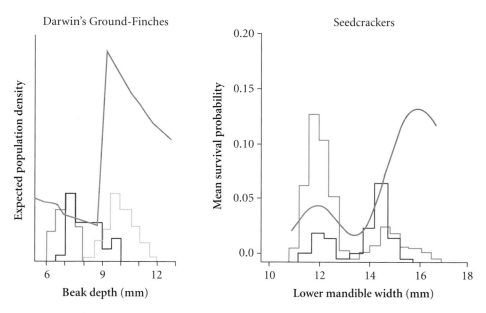

Figure 5.4 *Left:* Adaptive surface for Darwin's ground finches on Daphne Island (as in Figure 5.1). Morphological distributions are of the uncommon immigrant Small Ground-Finch (taken from the Santa Cruz Island distribution in Figure 5.1), the Medium Ground-Finch, plus hybrids between them (in *red*, from Grant and Grant 1994). *Right:* Adaptive surface for the Black-bellied Seedcracker in Cameroon (Schluter 2000, p. 105), calculated from measurements of survival of juveniles of different sizes reported in Smith (1993). Lower mandible width is the strongest correlate of survival. The distribution in *blue* is of adults measured in the field (N = 1252) and the distribution in *red* is of offspring from crosses in captivity between the large and small morph (N = 47 offspring from four families; Smith 1993).

the breeding population and the temperature in the season of the previous year when they were reared; the reasons for this are not clear.

Ecological similarity between the hybrid and one or other parent should lead to high fitness

Given environmentally contingent fitness, the key feature is that the phenotype of the hybrid differs from that of either parent. A logical consequence of this is that, if the hybrid were to resemble one or the other parent, it should suffer no reduced fitness. This is strikingly illustrated by the distinct morphs of the Black-bellied Seedcracker in West Africa (Smith 1987, 1993). The small and large morphs, whose beak sizes differ by about 20% (Figure 5.4, *right*), coexist in south-central Cameroon. The large morph specializes on the hard seeds of one species of sedge, whereas the small morph feeds on soft seeds from another sedge species as well as on a variety of other small seeds (Smith 1987). The morphs mate randomly (Smith 1987) and sing identical songs (Slabbekoorn and Smith 2000).

Smith (1993) used a captive-breeding experiment among the small and large morphs to show that the inheritance of morph size is attributable to a single gene with the large allele dominant to the small one. This means that crosses between the two morphs result in offspring looking like one or the other morph, rather than being intermediate. The example is compelling because an adaptive surface for the bill-size morphs has been estimated through direct measurements of the survival of juveniles in their first year (Figure 5.4, *right*). The fitness surface has a valley, and individuals of intermediate bill size are at a disadvantage. Thus, if a mating between the large and small morph did result in the production of intermediates, there should be some postmating isolation. The contrast between the randomly mating seedcracker morphs and the strongly assortatively mating Darwin's finch species and crossbill call types suggests that postmating isolation may be critical to the maintenance of premating isolation, a theme that is developed further in Chapter 14.

DIFFERENT ECOLOGICAL DIMENSIONS

So far, this chapter has been largely about finches. Darwin's ground finches and Red Crossbill call types are the most prominent examples of young species whose divergence can be clearly related to ecology. The two major adaptive radiations on archipelagoes (Hawaii and the Galápagos), as well as a smaller one on Tristan da Cunha (Abbott 1978; Ryan et al. 2007), were initiated by finchlike forms, despite the fact that other groups, notably the *Myadestes* thrushes of

Hawaii, appear to have arrived contemporaneously (Lovette et al. 2002). These observations suggest that finches are predisposed to ecological speciation. In this section, I ask why finches might radiate more easily than other groups, and whether other groups have also experienced ecological speciation.

Perhaps not coincidentally, the few adaptive surfaces that have been constructed in birds are all for finches. These finches consume discrete resources (seeds); the surfaces are rugged, with peaks and valleys; and different species lie under different peaks (i.e., they consume seeds of different plants). The rugged adaptive surface reflects not only the discreteness of the resources but also the way a finch feeds on those resources. A finch that is efficient at eating large seeds is inefficient at eating small seeds, and one that is efficient at eating small seeds is unable to handle large seeds. An intermediate-sized finch falls between the cracks, so to speak. Thus, one feature of finches that might make them prone to ecological speciation is that individuals are physically constrained to consume a relatively narrow portion of the entire resource spectrum. This means that, even in the presence of one species, resources are generally available to another species, so character displacement and divergence are facilitated.

For groups other than finches ecological speciation may be less easy. Although little evidence can be directly related to the question, the adaptive surface may be less rugged for groups other than seed-eaters. For example, the failure of the two species of warbler finches in the Galápagos to diverge into large and small coexisting species (Chapter 2) could reflect the fact that these finches consume mostly arthropods, and arthropod sizes follow a unimodal, continuous distribution (although this is not known).

Ecological speciation along a prey size and associated body size axis may also be easier than speciation along other dimensions such as habitat. The adaptive surface for prey size may be more rugged, with deeper valleys and taller peaks, than for habitat, so the fitness of hybrids along the prey size dimension is particularly depressed, leading to high postmating isolation. This seems possible. In the gulls, wagtails, and crows noted in the previous section, hybrids showed wider habitat preferences than the parental species, and in the gulls, hybrids do appear to have superior fitness than the parental species in areas where different habitats merge (Good et al. 2000). In the Tristan da Cunha finches, ecological differentiation on the body size axis has led to a large and small pair of coexisting species, but, within the smaller species, ecological differentiation along the altitudinal gradient has led to some phenotypic differentiation, but no strong assortative mating (Ryan et al. 1994, 2007).

Divergence between incipient species via ecological character displacement may more readily develop along the body-size/prey-size dimension than habitat. Schoener (1965) suggested that, even though a species of a certain size is limited in the range of prey sizes it can consume, it may be able to occupy all available habitats. This means that underexploited resources are less available to a second species along the habitat dimension than along the prey-size dimension, restricting character displacement and often favoring convergence. However, it is difficult to compare prey size and habitat in this way until more adaptive surfaces have been constructed, because they are measured on different scales. A habitat at the top of a mountain may present many problems to a species that typically lives in one at the bottom of the mountain. Character displacement does seem to occur regularly between species occupying different altitudinal zones on mountains (Table 3.3, p. 50).

The shape of the adaptive surface for species other than finches and for other ecological dimensions is not known. However, we can ask if certain ecological dimensions are associated with speciation events. If, for example, a pair of sister species differ along the habitat axis and not a body-size/prey-size axis, the body-size/prey-size axis could not have contributed to the speciation event that separates this pair of species. In the next section, I ask how recently diverged species differ ecologically, contrasting the habitat and body-size/prey-size axes.

Habitat and body-size axes

To assess the importance of alternative ecological dimensions to speciation, Richman and Price (1992) studied eight species of Old World leaf-warblers belonging to the genus *Phylloscopus*. These species breed sympatrically along an altitudinal gradient in the western Himalayas and differ in body size (associated with prey size) and types of habitat occupied (Price 1991). Based on a molecular phylogeny, we identified three distinct clades. The first clade consists of the three smallest species, the second of the two medium-sized species, and the third of the three largest species (Figure 5.5). Within the three clades, species differ in altitude, associated with different habitats. For example, within the large-bodied clade, one species, the Western Crowned-Warbler, occupies coniferous woodlands at lower elevation, whereas its close relative, the Greenish Warbler, occupies birch and willow at higher elevations. Diamond (1986) also found that close relatives differ in altitude, based on a taxonomic analysis of the birds of New Guinea. The fact that the most closely related species differ primarily in altitudinal distribution suggests that the body-size/prey-size axis was not involved in the divergence of these species pairs. Diamond (1986) and Richman and Price (1992) suggested that habitat might have generally been the first step in species divergence, and body size differences always arose later, long after speciation is complete. If this is the case, only habitat divergence could have contributed to ecological speciation in these species.

The difficulty with this interpretation is that, although current pairs of sister species are separated by habitat, this does not necessarily mean that earlier spe-

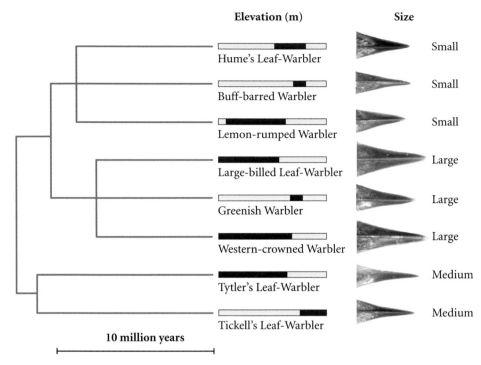

Figure 5.5 Order of niche evolution among the *Phylloscopus* warblers in the western Himalayas of India (Richman and Price 1992). The phylogeny was calculated using maximum likelihood with a molecular clock enforced, based on mitochondrial cytochrome *b* DNA (Appendix 2.2). The timescale is based on the standard calibration of 2% divergence per million years. Ecological and morphological data are from Price (1991) and scanned beaks are from Price et al. (2000); maximum length illustrated is 8 mm. Elevational distributions (*darkened*) are over a gradient from 2100 m to 4000 m. The Buff-barred Warbler and Hume's Warbler overlap in elevation but have different habitat preferences: rhododendron and birch, respectively (Price 1991). Mean body size of each species ranges from 5.1–6.4 g for the small clade, 6.7–7.2 g for the medium clade, and 8.3–10.0 g for the large clade. The Greenish Warbler is illustrated in Figure 11.5.

ciation events during the radiation of the group were also associated with habitat. It is plausible—and perhaps likely—that body-size divergence was the primary axis of ecological divergence during some of these earlier speciation events. Thus, in the Old World leaf-warblers, a reasonable model is that size differences originated within a single habitat, and that species of different body sizes then separately gave rise to additional species (more or less simultaneously; see Figure 5.5) when additional habitats appeared (e.g., in response to climate change or mountain building). [Sympatric speciation along an altitudinal gradient is not implied here, and at least the initial stages of divergence appear to have been between populations separated by barriers that lay between the

Himalayas and other areas (Johansson et al. 2007)]. Under this scenario, body-size differences (arising first) and habitat differences (arising second) are both implicated in separating sister species, but at different speciation events.

A more general observation that body-size differences separate sister species comes from adaptive diversification in many groups, including Darwin's Ground-Finches, white-eyes, flycatchers, and thrushes. On several islands, pairs of closely related species are thought to have been formed as a result of double invasions from a source elsewhere (Mayr 1942; Lack 1944; Grant 1966; Coyne and Price 2000). After the first invasion, the species differentiates sufficiently that a second invasion from similar ancestral stock remains distinct. Lack (1944) analyzed 13 such cases (including Tristan da Cunha), finding that six differ markedly in size (bill-length differences >15%, Grant 1968) but occupy the same habitat and that seven differ in habitat, at least partly distinguished by altitude. However, all seven of these also differ in bill length, by the 15% criterion (*personal observations*). In conclusion, closely related species can differ in body size and/or habitat, so, in different cases, either or both axes could contribute to ecological speciation. In the next section I ask whether such differences do regularly contribute to ecological speciation.

IS ECOLOGICAL SPECIATION GENERALLY IMPORTANT?

Many closely related pairs of species differ in body size and/or habitat and divergence along these axes, as well as feeding method, migration, and other axes, should contribute to both premating and postmating reproductive isolation. But these ecological differences may have arisen independently of the speciation process, and it is difficult to demonstrate a connection between ecological divergence and the initial establishment of reproductive isolation (i.e., ecological speciation) for birds other than Darwin's finches, Tristan finches, and crossbills. This is partly because even closely related sympatric species are often old. The coexisting species of Old World leaf-warblers are separated by many millions of years (Figure 5.5), and divergence times of millions of years apply to most species in ecological communities (Chapter 6). The age of these species means that the alternative nonecological models of speciation introduced in Chapter 2 (p. 29) are difficult to exclude. In an extreme view, speciation is completed in allopatry by nonecological mechanisms, and any ecological differentiation is incidental. For example, it is possible that the ecological differences among coexisting species of Old World leaf-warblers in Kashmir (Figure 5.5) are a result of divergence between completely reproductively isolated species, rather than these ecological differences having directly contributed to speciation. Perhaps songs are the main basis of premating isolation and genetic incompatibilities are a main factor causing postmating isolation, as seems to be the case for the meadowlarks described in Chapter 2 (Figure 2.6, p. 27). Both song divergence (Chapter 10) and genetic incompatibilities (Chapter 16) may be regularly produced by nonecological mechanisms (although, of course, they may also be affected by ecological factors as described in Chapters 12 and 16).

One could also argue that those young sympatric species—Darwin's Ground-Finches and Red Crossbill call types—that provide strong support for the idea of ecological speciation are ephemeral and irrelevant to the formation of the older species that make up the majority of bird communities (proponents of the phylogenetic species concept would probably make this argument). Such young species can clearly originate rapidly under the right ecological conditions—in Darwin's finches, selection on body size was so intense in one episode that the mean size of the Medium Ground-Finch was shifted 5% of the way towards the Large Ground-Finch (Price et al. 1984)—but the problem with maintaining them is that environments are likely to fluctuate. Changing conditions drive species back towards each other, reducing premating isolation and increasing hybrid fitness, and this could result in the two species merging. Indeed, this sort of thing was observed to happen in less than a decade during the long-term study of Darwin's finches on the small island of Daphne (summarized in Figure 5.6). After the drought year of 1977, hybrids between the Small Ground-Finch and the Medium Ground-Finch survived poorly. Conditions changed following the abundant rains during the El Niño event of 1982–1983 (Gibbs and Grant 1987). Hybrids actually survived better than the Medium Ground-Finch, in association with an abundance of small seeds; they also appeared to survive better than the Small Ground-Finch, although sample sizes were small (Grant and Grant 1998). Furthermore, the shift towards larger body size as a result of the previous bout of selection on the Medium Ground-Finch was largely erased as a result of selection for small size (Gibbs and Grant 1987; Grant and Grant 2002b). This example illustrates a serious difficulty for permanent ecological speciation along a single dimension; fluctuating environments can quickly remove any ecologically enforced reproductive isolation, resulting in the collapse of two incipient species back into one (Muller 1942, p. 83).

The formation of permanent species solely from ecological causes may depend on divergence along multiple ecological dimensions (e.g., prey size, feeding method, habitat). For example, a generally rare seed may become common in the environment for a number of generations, providing a bridge to another resource, which, although common, was previously not exploited (Grant 1998). Populations may move from one adaptive peak to another, diverging far from each other along many ecological dimensions. Over time, a population moves along a trajectory (1) from which it is unlikely to return

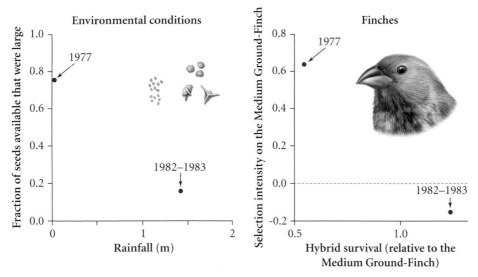

Figure 5.6 Environmental conditions (rainfall and seed composition), hybrid survival, and selection after a severe drought in 1977 and after the extraordinary rainfall of 1982–1983 on Daphne Island in the Galápagos. *Left:* Environmental conditions. *Right:* The change in conditions resulted in high survival of hybrid Small × Medium Ground-Finches. In addition, the direction of selection differed. The dashed line indicates no selection in the population. After the drought of 1977, individuals with large beaks survived better than those with small beaks, whereas, after the El Niño event, individuals with small beaks survived better than those with large beaks. The selection intensity is the difference in mean beak depth (in units of standard deviations) between survivors and the original population over a one-year (after 1977) or two-year (after 1982–1983) period. Based on Grant and Grant (1993, 1995, 1998).

because the exact reverse sequence of environments is very unlikely and (2) that results in sufficient divergence that high levels of reproductive isolation from other populations are generated. This means that ecologically driven divergence could cause speciation over very long timescales, even among taxa that are exploiting quite similar resources, especially if those taxa remain geographically separated so that interbreeding is impossible. However, throughout this time, the taxa are diverging via nonecological mechanisms as well, and, over long timescales, ecological and nonecological factors should both be involved.

Later on in this book, I consider two mechanisms of nonecological divergence that also operate over long timescales. They are the spontaneous appearance of novel behaviors, such as a new way to exploit a food resource (Chapter 7), and the appearance of new genetic mutations, such as a color mutation that makes individuals more attractive (Chapter 9), or a mutation that makes sperm more efficient at fertilizing the egg (Chapter 16). Often—but not always—the chances of a new behavior or mutation becoming established must depend on features of the environment, such as the foods that are potentially available in the case of a new feeding behavior, or the presence of predators who are attracted to a new color. Thus, ecological differences between populations should contribute in many ways to population differentiation and, by implication, to speciation.

Despite the difficulty of unequivocally demonstrating a role for ecology in speciation, it seems likely that ecological factors lead to some reproductive isolation during most speciation events. In the next chapter, I return to this question, and suggest that rapid species radiations deep in the past, as identified using phylogenetic reconstruction, are likely to have involved ecological speciation.

SEASONAL MIGRATION

I conclude this chapter by considering the contribution of seasonal migration to reproductive isolation. Seasonal migrants breed in one location and overwinter in another, and the evolution of seasonal migration has been generally attributed to ecological factors such as the seasonal appearance of food resources (Cox 1985; Alerstam et al. 2003).

If two species migrate in different directions and their hybrids migrate in an intermediate direction, the hybrids may encounter unfavorable conditions and consequently have reduced fitness, imposing some postmating isolation (Helbig 1991; Irwin and Irwin 2004). The difficulty of studying hybrids in the wild has meant that this has yet to be shown in nature, but low hybrid fitness has been inferred both from experiments and from comparative studies. Helbig (1991) conducted a captive breeding experiment in the Blackcap, which breeds commonly in the western Palearctic and typically winters in southern Europe and Africa (Cramp 1992). This species has both southwesterly migrating and southeasterly migrating populations. Individuals from crosses between the two populations orient in an intermediate direction. The inference is that offspring from natural hybridization events might end up stranded in the Sahara.

A number of Siberian species are divided into subspecies, one of which migrates to the Indian subcontinent around the western side of the Tibetan plateau and the other of which migrates to southeast Asia around the eastern side of the Tibetan plateau (Irwin and Irwin 2004). Hybrids between the subspecies are expected to migrate across the plateau, with correspondingly reduced fitness. One example is the Greenish Warbler ring species (Figure 5.7), already mentioned in connection with parapatric speciation (Chapter 4, p. 66). Irwin and Irwin (2004) developed a graphical model for the evolution of migratory direction in this species (Figure 5.7, *right*). Migratory orientation is inferred to have gradually changed in response to divergent selection pressures, as ranges

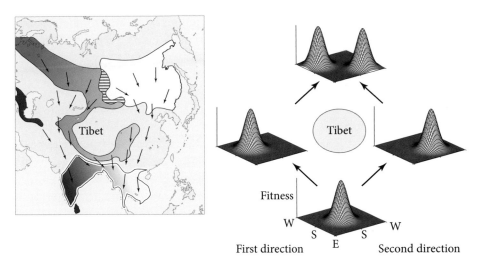

Figure 5.7 *Left:* Breeding and winter (India and southeast Asia) distributions of the Greenish Warbler. Arrows indicate fall migration routes. North of the Tibetan plateau, two reproductively isolated forms overlap and migrate in different directions towards their respective winter quarters. The inference is that hybrid offspring, if they were to migrate in an intermediate direction, would have reduced fitness because they would fly into the Tibetan plateau. *Right:* Model of evolution of migratory direction during spread from a southerly refuge, resulting in discrete differences between species as a result of geographically varying selection pressures (from Irwin and Irwin 2004). Colors indicate different named taxa (see Figure 11.5, p. 237).

expanded around both sides of the Tibetan plateau from a Pleistocene refuge in the south. The result is that migratory direction is dramatically different where the ranges overlap in the north. The model illustrates the general point that widely different adaptive peaks can be reached as a result of gradual change in response to geographically varying selection pressures, without the need to cross the valley of low fitness that the hybrid is expected to fall into.

Within the past 50 years, more than 50 species are known to have developed new migratory distances or directions (Sutherland 1998; Fiedler 2003). A spectacular example is the altered migratory direction of Blackcaps (Berthold et al. 1992). In the 1960s, Blackcaps were first noted wintering in southern England. Banding records indicated that these come from Austria and Germany and belong to a population that normally orients to the southwest. Instead of taking the usual migration route in the fall, some individuals migrate northwest to England. About 10% of the individuals are thought to take this novel migratory route from some central European populations. In a remarkable experiment, Berthold et al. (1992) captured 40 wintering birds in southern England, bred them in captivity, and showed that the young birds oriented in the same direction as their parents. Later, Helbig et al. (1994) crossed British birds with central

European birds and showed that the hybrids oriented in an intermediate direction. Offspring whose mother was from Britain oriented in a similar direction to offspring whose father was from Britain, indicating that no maternal effect was present (i.e., no direct influence of the mother) and convincingly demonstrating that the change was genetic. Furthermore, the inheritance pattern was consistent with that of migration orientation being determined by a single gene (Helbig et al. 1994).

Some premating and postmating reproductive isolation may be arising as a result of the changed migratory route. Experiments with captive birds led to the suggestion that British birds arrive on the breeding grounds in central Europe about ten days earlier than those taking the ancestral route (Terrill and Berthold 1989). Recent field studies have confirmed the difference in arrival date and have shown that this leads to assortative mating, with pairs from the same wintering grounds about 2.5 times more likely to form than would be expected with random mating (Bearhop et al. 2005). In addition, the inferred southwesterly migratory route of hybrids may lead them into unsuitable wintering grounds, thereby causing some postmating isolation. If this is the case, unlike the Greenish Warbler model (Figure 5.7), a valley of low fitness had to be crossed during the origin of the new migratory route; the breeding experiments imply that a mutation in the original central European population would produce an intermediate individual, resembling the orientation of a hybrid and presumably having the fitness of hybrid. Possibly, those first individuals that took a generally intermediate route but still succeeded in making it to England were compensated by high overwinter survival, or their early return to the breeding grounds gave them access to good territories.

If migratory direction affects both premating and postmating isolation, it is by definition a "magic trait" in the terminology of Gavrilets (2004), and hence a particularly suitable ecological trait to drive ecological speciation (see p. 20). However, both premating and postmating isolation need to be strong if complete reproductive isolation is to be generated (Gavrilets 2004, p. 411). It thus seems unlikely that the altered migratory direction alone would eventually lead to complete reproductive isolation in this system.

In addition to migratory direction, migratory propensity may differ between closely related taxa. Crossbreeding experiments among resident and migratory populations of the Blackcap have shown migratory propensity to be genetic and hybrids to be intermediate (Berthold and Querner 1981). Harris et al. (1978) found that hybrids of crosses between Herring Gulls and Lesser Black-backed Gulls in Britain had poor survival. Although the causes are unknown, the Herring Gull is generally resident, and the Lesser Black-backed Gull is migratory, and this seems to be partly genetically determined because cross-fostered Lesser Black-backed Gulls migrated normally. Cross-fostered Herring Gulls showed some tendency to migrate, although not as far as the Lesser Black-backed Gulls (Harris 1970). Rohwer and Manning (1990) and Rohwer and Johnson (1992) argued that unusual migration and associated molt patterns could lower the fitness of hybrids between Baltimore Orioles and Bullock's Orioles.

In summary, migratory differences between populations appear to be a powerful way of generating at least some reproductive isolation. Other features of seasonal migrants, such as the likelihood that they will colonize new locations, should also promote speciation (Phillimore et al. 2006), whereas still others, such as a greater mixing of different populations, might limit differentiation.

CONCLUSIONS

Divergence under natural selection generates differences that contribute to both premating and postmating isolation. Many traits associated with ecological selection pressures, including body size, habitat occupied, and time of breeding, result in some premating isolation. However, in many cases, postmating isolation is likely to be important in the maintenance of species in sympatry (Chapter 14). It is here that any role for ecological speciation needs critical investigation. Competition between the parental forms may mean hybrids have low fitness, and such low fitness will be accentuated whenever the adaptive surface has peaks and valleys, but it is unclear how often postmating reproductive isolation produced in this way is both strong and permanent.

Understanding the contribution of the many ecological differences between species to reproductive isolation requires much more study, and several outstanding questions need to be addressed. What is the shape of the adaptive surface for groups other than finches, and for dimensions other than body size? What are the ecological causes of hybrid unfitness? How does divergence along multiple ecological dimensions occur, and how do temporal fluctuations in the environment contribute to this divergence? Why have most sympatric species diverged from their nearest common ancestor so long ago? This last question is partly addressed in the next chapter, where I consider a third and final way in which environmental factors limit speciation (see Figure 2.1). This is by affecting the chances of successful range expansions, a necessary requirement for a renewed cycle of population fragmentation and differentiation.

SUMMARY

Ecological speciation describes the process whereby population differentiation in various ecological traits, including prey size, habitat occupied, and migratory tendency, generates reproductive isolation. Ecological differences can lead to premating isolation (e.g., time and location of breeding, choice of mate based on body size) and postmating isolation (e.g., hybrids have lower fitness because they have lower feeding efficiency). The classic example in birds is that of Darwin's Ground-Finches in the Galápagos Islands. Extensive field studies have demonstrated that all of the elements of ecological speciation are in place and have shown that it can occur rapidly, but field studies have also highlighted the contingency of reproductive isolation on prevailing conditions. Most co-occurring species of birds are much older than Darwin's Ground-Finches, making it difficult to assess the role of ecological speciation in their production. Manifest ecological differences between related species suggest that most speciation events are likely to have an ecological contribution. However, particularly with respect to postmating reproductive isolation, it is unclear if ecological causes are sufficient or even important in many speciation events.

CHAPTER SIX

Ecological Controls and Speciation on Continents

More species are found in some places in the world than in others (Figure 6.1). Over much of the Sahara, a square the size of about 12,000 square kilometers typically holds about 20 breeding bird species (and a square of similar area in the Somali desert holds just three; Jetz and Rahbek 2002). At the other extreme, many squares in the tropical rainforests contain more than 400 species (Rahbek and Graves 2001; Jetz and Rahbek 2002), and a remarkable location in Ecuador holds the world record of 845 species (Rahbek and Graves 2001). The obvious difference between deserts and rainforests is rainfall, with associated plant productivity. The Ecuador square also straddles the Andes, and it contains a diversity of habitats—from lush tropical forest to tundra above the tree line—as do other regions of the world with an exceptionally high density of species (Orme et al. 2005; Hawkins et al. 2007; see Figure 6.1).

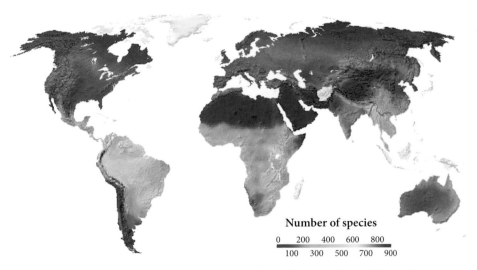

Number of species

0   200   400   600   800
  100   300   500   700   900

Figure 6.1 Patterns of species richness among breeding birds across the world's continents, estimated to a resolution of 27.5 km by 27.5 km, obtained by overlaying range maps (from Hawkins et al. 2007; no data are available for Afghanistan). Figure kindly provided by B. Hawkins.

If environmental factors such as productivity and habitat diversity affect species numbers, they should also affect the rate of speciation. Consider the following thought experiment: Remove all bird species from the Sahara and from tropical South America, but somehow hold all resources fixed. Then drop a single bird species into each area. One might expect speciation to proceed faster, but certainly to continue for longer, in South America than in the Sahara, with the net result again being more species in South America. In this chapter I ask: What is the evidence for environmental limits on the number of species? What are the alternatives? How do such limits affect the rate of speciation? I use the results from this and previous chapters to analyze the underlying causes of the huge bird diversity in South America.

REGIONAL VARIATION IN SPECIES NUMBERS

As expected from the desert/rainforest comparison, regional variation in species numbers is strongly correlated with plant productivity (Hawkins et al. 2003a, b; 2007). Based on estimates of the number of bird species breeding in 10,169 squares, each measuring 110 km by 110 km, across all continents except Antarctica, Hawkins et al. (2007) found the correlation of species number with a measure of plant productivity to be $r = 0.84$. In this study, productivity was measured by estimated evapotranspiration, which is available from a database maintained by the United Nations. Productivity is driven largely by water availability in tropical and subtropical regions and up to as far north as 40° N, but, at higher latitudes, energy input (i.e., temperature) becomes more limiting than water (Hawkins et al. 2003b).

One explanation for the association between productivity and species numbers is via a direct elevation of population size: "A poor environment supplies too meager a resource base for its would-be rarest species, and they become extinct" (Rosenzweig and Abramsky 1993, p. 56; reviewed by Evans et al. 2005). On the other hand, more resources mean that would-be rare resources are sufficiently common to maintain a species that utilizes those resources (MacArthur 1969; Evans et al. 2005). For example, comparing tropical rain forests in Central America with temperate forests across an area seven times larger in North America, Schoener (1971) found twice as many breeding insectivorous species in the tropics. This difference is entirely due to the presence of many more large-billed forms (below a bill length of 14 mm, there are actually slightly more temperate species). Schoener argued that the presence of many large-billed species in the tropics was due to the availability of sufficient numbers of large insects. Other studies comparing tropical localities with temperate localities (e.g., Terborgh 1980; Erard 1989; see Sherry 1990) have been able to relate at least 50% of the increase in species numbers in the tropics to the presence of unusual niches. For example, a number of species in the neotropics follow ant swarms, feeding on animals that are escaping the swarm. Nothing equivalent to this is found in the temperate regions.

Although productivity is a strong correlate of species numbers, it is not the only one. Topography, which reflects habitat diversity, is an important additional predictor of bird-species diversity in tropical regions (Rahbek and Graves 2001; Orme et al. 2005; Hawkins et al. 2006, 2007). Many species occupy relatively narrow altitudinal belts, replacing each other as one moves up mountains (e.g., Terborgh 1977; Graves 1985; Herzog et al. 2005). As is the case with productivity, this suggests that an increase in the number of niches generates increased numbers of species.

Correlations of productivity and habitat diversity with numbers of species in a limited area imply that species number is limited by resource quantity and diversity. The pattern has strong empirical support, but it is correlational. There may be confounding factors, and there certainly are other influences on species numbers besides those driven by productivity and habitat. Some of these are not directly relevant to speciation. For example, Hawkins et al. (2003a) found that central Mexico had more species than predicted from its productivity, and they attributed this to a recent faunal mixing of Neotropical and Nearctic species.

The key alternative explanations for differences in species number in different locations are based on age and area (Wallace 1891; Pianka 1966; Rosenzweig 1992, 1995, pp. 287–288; Ricklefs 2004). The main argument is that over time novel adaptations lead to the exploitation of environments in more efficient ways. For example, the development of the crossed bill of the Red Crossbill (Figure 5.2) resulted in efficient exploitation of pinecones. Such innovations result in a new species doing something better than the old one, and this could drive other species extinct. Thus it is not immediately clear that they will always lead to an increase in species number. Sometimes, however, a new innovation may enable a species to insinuate itself between others in ecological space, resulting in an increase in species number. Area has effects that are similar to those of age: More species originate in geographically separated locations, and each can potentially evolve new adaptations.

Age predates productivity as an explanation for high species diversity in the tropics. Wallace (1891, pp. 310–311) stated: "The equatorial regions . . . are a more ancient world than that represented by the temperate zones, a world in which the laws which have governed the progressive development of life have operated with comparatively little check for countless ages. . . ." Wallace (1891, p. 309) also argued that "the causes [of high species diversity in the tropics] are not to be found in the comparatively simple influence of heat and light." Although the tropical regions are currently large, tropical climates were formerly even more extensive. Sixty million years ago, much of the world experienced a tropical or subtropical climate (Zachos et al. 2001). A general cooling followed. The Antarctic ice sheet first started to form about 37 million years ago, the north Polar ice sheet only about 8 million years ago (Zachos et al. 2001) and even a few million years ago, Germany had a subtropical flora (Utescher et al. 2000). This implies that much time and space has been available for organisms to evolve in tropical environments, more so than in temperate environments. Indeed, at least in the neotropics, families and higher taxa are on average older than they are in temperate regions (i.e., the root of the tree that connects species within a tropical family tends to lie further from the present than the root of the tree that connects species that make up a temperate family; Gaston and Blackburn 1996; Ricklefs 2005a; Hawkins et al. 2006, 2007). So a contribution to the very high species numbers in tropical rainforests may arise out of the longer time and large areas that have been available for organisms to evolve in this environment.

The ecological and age/area explanations for species numbers can be evaluated by comparing different regions of the world. A prediction of the ecological model is that the number of species in similar habitats should be similar in different regions of the world. Conversely, different numbers of species in similar habitats imply that age/area explanations should be explored (Ricklefs 1987; Schluter and Ricklefs 1993). Qualitatively, there is evidence both for similarities and for differences in the numbers of species occupying matched habitats on different continents, depending on the comparison (Ricklefs 1987; Schluter and Ricklefs 1993; Figure 6.2). The data in Figure 6.2 were chosen partly because they illustrate some striking similarities in species numbers in similar habitats, consistent with a model in which ecological conditions limit species numbers. However, it can be seen in this figure that, across tropical rainforests, the number of species in a local area increases with the number in the region, as predicted from age/area models. This suggests that ecological conditions are not the only determining factor. But here it is worth noting that the Neotropical rainforest receives perhaps 50% more rain than the Afrotropical one (Hawkins et al. 2003a), so the habitats are not identical in each region (a general problem with this sort of comparison), and indeed the region with the highest productivity has the highest number of species.

Time must play a role, but it is difficult to establish how and where it has played its part in affecting numbers of species in local communities. The most striking evidence for an influence of time on local species diversity is very recent. The number of species in cities is much lower than in the surrounding natural habitat, but the number of individuals can be much higher (Marzluff 2001). Low species numbers in cities may reflect the short time the habitat has been around, with few species currently well adapted to exploit it. This con-

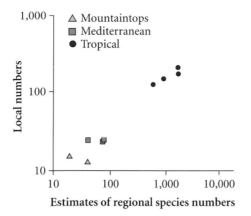

Figure 6.2 Plots of local numbers of species (i.e., the number of species in a small area) vs. regional numbers of species (i.e., the number of species across a large region in which this area is embedded). Neither regional nor local estimates are comparable between the three habitat types. *Local estimates:* Mountain sites are two tropical sites above the tree line (Dorst and Vuilleumier 1986, p. 132), both of whose areas span approximately 300 km² (Mt. Kenya, Africa, and Teta de Niquitao, northern Andes, along the *x* axis). Mediterranean chaparral numbers are based on plots of 2–3 ha from Chile, California, and South Africa, (respectively, along the *x* axis) from Cody and Mooney (1978). The four tropical rainforests are approximately 2–5 km² plots taken from Beehler (1981), Erard (1989), and Schluter and Ricklefs (1993) at sites in New Guinea, Nigeria, Colombia and Peru (respectively, along the *x* axis). *Regional estimates:* For the high-altitude regions, they are based on numbers of species in the Páramo of the northern Andes and in high-altitude regions of East Africa (Dorst and Vuilleumier 1986, p. 135). For the Mediterranean regions, they are based on estimates of the number of species occurring in the total area defined as "Mediterranean" by Cody (1975). For tropical rainforests, they are based on numbers of species in each country, whose land areas vary by a factor of 1.5. In all cases, regional estimates are confounded by area. For the tropical rainforest comparisons at least, differences in area make only a small contribution to the total regional disparity in species numbers (based on published species–area relationships in Rahbek 1997).

trasts with more natural habitats. For example, Pautasso and Gaston (2005) showed that across mostly north temperate censuses of breeding birds on various study plots, productivity was correlated with both the number of individuals and the number of species, as would be expected if resource abundance affects species numbers.

SPECIATION THROUGH TIME

If ecological conditions limit the number of species that can be maintained in a particular habitat or location, the addition of more species becomes increas-

Table 6.1: Patterns of diversification in phylogenies

Taxon	No. species in taxon	No. species on phylogeny	γ statistic	P-value[1]	γ statistic (2%)[2]	P-value (2%)[1,2]	[3]Depth (my)	[3,4]Diversification interval (my)	Reference
Anthus (pipits)	45	41	**−5.58**	**<0.01**	**−3.88**	**<0.01**	11.5	3.7	Voelker 1999
Leaf-warblers[5]	61	55	**−5.32**	**<0.01**	**−3.38**	**<0.01**	10.4	3.0	Olsson et al. 2004, 2005
Tangara (tanagers)	49	43	**−4.62**	**<0.01**	**−2.73**	**<0.01**	8.2	2.6	Burns and Naoki 2004
Dendroica (New World warblers)	27	24	**−4.42**	**<0.01**	**−3.06**	**<0.01**	5.1	2.0	Lovette and Bermingham 1999
Caciques and oropendulas[6]	21	17	**−3.02**	**<0.01**	−1.91	0.03	6.8	2.9	J. Price and Lanyon 2002, 2004
Sylvia (warblers)	24	24	**−2.93**	**<0.01**	−1.40	0.08	17.5	7.0	Böhning-Gaese et al. 2003
Icterus (New World orioles)	28	28	**−2.82**	**<0.01**	−0.62	0.27	6.0	2.3	Omland et al. 1999
Acrocephalus (reed warblers)[7]	36	20	**−2.72**	**<0.01**	−2.24	0.01	12.1	4.2	Helbig and Seibold 1999
Auks[8]	23	23	**−2.51**	**<0.01**	−0.98	0.16	10.3	4.2	Friesen et al. 1996
Andropadus (African greenbulls)[10]	16	16	−2.21	0.01	−1.33	0.09	19.6	9.4	Roy 1997
Storks[9]	19	16	−2.07	0.02	−1.08	0.14	10.8	4.8	Slikas 1997
Amazona (parrots)	31	28	−2.04	0.02	**−3.45**	**<0.01**	6.4	2.3	Russello and Amato 2004
Cranes (family Gruidae)	15	15	−1.38	0.08	0.41	0.66	10.4	5.2	Krajewski and King 1996
Empidonax (flycatchers)	17	17	−0.91	0.18	−0.69	0.25	8.9	4.2	Johnson and Cicero 2002

Carduelis (finches)	31	24	−0.01	0.50	−0.29	0.39	15.3	1.9	Arnaiz-Villena et al. 1998
Myiarchus (flycatchers)	21	19	0.02	0.51	2.39	0.99	11.8	5.0	Joseph et al. 2004
Anas (ducks)[10]	42	41	1.95	0.97	−2.10	10.02	8.5	2.8	Johnson and Sorenson 1999

In some cases, several genera were combined (see footnotes 5, 6, 8, and 9 below). The table has been sorted according to magnitude of the γ statistic.

All trees were calculated using the Bayesian/Likelihood/Rate-smoothing methods on mitochondrial DNA sequences outlined in Appendix 2.2. See also Weir (2006), who uses similar methods.

Tree depth (i.e., the distance from the tips to the root of the tree) was estimated in *Tree-Puzzle* (Schmidt et al. 2002) with a molecular clock enforced, and the HKY-Γ model.

* Number of species in genus is usually taken from Sibley and Monroe (1990), plus any updates.
[1] A low *P*-value indicates strong evidence for a slowdown in diversification towards the present. Values < 0.01 are in bold.
[2] Calculated for a phylogeny that has been truncated at 2% sequence evolution along a branch from the present.
[3] Millions of years, assuming a rate of sequence evolution of 2% divergence/million years.
[4] The average time between branching intervals on the tree, assuming a pure birth model. Calculated as $t/\ln(N/2)$, where t is the depth of the tree and N is the total number of species in the group.
[5] *Phylloscopus* and *Seicercus* (Palearctic and Asian species only).
[6] *Cacicus, Amblycercus, Gymnostinops, Psarocolius,* and *Ocyalus.*
[7] The missing species are from islands across the Pacific.
[8] *Alle, Uria, Alca, Cepphus, Brachyramphus, Synthliboramphus, Ptychoramphus, Cyclorrhynchus, Aethia, Cerorhinca,* and *Fratercula.*
[9] *Mycteria, Anastomus, Ciconia, Ephippiorhynchus, Jabiru,* and *Leptoptilos.* (Note: No species of *Anastomus* are on the tree.)
[10] These genera are known not to be monophyletic (Johnson and Sorenson 1999, U. Johansson, *personal communication*).

ingly difficult as the limit is approached. This leads to what I call an "ecological controls" model (this idea is termed "saturation" in the ecological literature, where the emphasis is on determinants of local species numbers, rather than speciation; Schluter and Ricklefs 1993). In the ecological controls model, speciation is high when many unexploited or underexploited resources are available and becomes increasingly difficult as more species are added. The model leads to the strong prediction that, as a location gets filled up with species, the rate of production of new species should slow down. What should happen under an age/area model is less clear, but it could be argued that in this case, new species arise at a more-or-less constant rate, as new innovations appear. Species numbers may continue to increase (e.g., Ricklefs 2003), or else an equilibrium is reached where the origination of new species occurs at the same rate at which older ones go extinct (e.g., Ricklefs 2004).

Patterns of species diversification through time can be assessed using molecular phylogenies. Mark Pavelka and I did this using the genus, or a closely related group of genera, as the basic unit, because this typically includes ecologically similar species, which are likely to compete with each other. The genus is a classification that has been determined independently of this study, enabling us to be objective in the choice of the unit we analyse. Some of the groups we studied are not monophyletic (see Table 6.1); that is, within the phylogeny, some species are not included because they are currently classified as belonging to different genera. These species are likely to be ecologically different and less likely to compete with the other species. Here I give results for all of the groups we analyzed that contain at least 15 species. Six phylogenies for some of those groups are shown in Figure 6.3.

We used the method of Pybus and Harvey (2000) to ask if the pattern of diversification in these phylogenies shows a slowdown towards the present. The method is described in Figure 6.4, with reference to a phylogeny for the storks. The basic idea is to tally the total number of lineages in the tree as a function of time, starting at the root (where there are two lineages) up to the present day (where the number of lineages equals the number of species). Under a null model of constant probability of branching through time and no extinction (the pure-birth model), the expected increase in the number of lineages against time follows a straight line on a logarithmic plot (Figure 6.4). In the pure-birth model, variation around expectation follows a known distribution and can be summarized by the so-called γ statistic (Pybus and Harvey 2000). In the pure-birth model, the γ statistic is normally distributed with an expected value $= 0$ and a standard deviation $= 1$. A large negative value of γ rejects the null hypothesis of constant birth, and indicates instead that the speciation rate has slowed down towards the present (Figure 6.4). If extinction is present, and the probability of speciation and the probability of extinction both remain constant

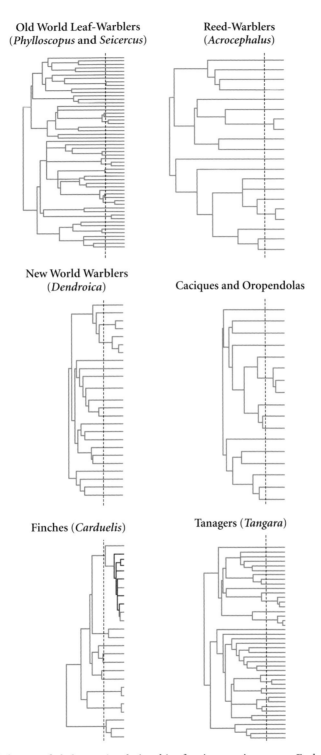

Old World Leaf-Warblers
(*Phylloscopus* and *Seicercus*)

Reed-Warblers
(*Acrocephalus*)

New World Warblers
(*Dendroica*)

Caciques and Oropendolas

Finches (*Carduelis*)

Tanagers (*Tangara*)

Figure 6.3 Estimates of phylogenetic relationships for six passerine genera. Each phylogeny is based on 1000–2000 bases of mitochondrial DNA sequence and has been computed from the original sequence data (references are in Table 6.1) following Bayesian/Likelihood/Rate-smoothing methods described in Appendix 2.2. The dashed line indicates 2 million years before the present according to the standard 2% divergence per million years. Red lines connect species of *Carduelis* finches that live in the Andes (the three other members in this clade are other South American species).

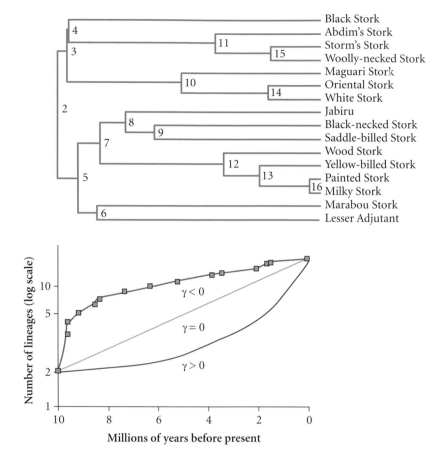

Figure 6.4 The method used to ask if speciation rates are slowing down or speeding up through time, illustrated using a phylogeny for the storks, calculated as in Figure 6.3. The upper curve on the graph plots the total number of lineages present (on a logarithmic scale) against time, with each square corresponding to a node in the phylogeny. Each number on the phylogeny corresponds to the addition of an extra lineage. If the probability of speciation is constant through time, and there is no extinction (the "pure-birth" model), the expected number of lineages through time follows an exponential increase, as illustrated by a straight (*blue*) line on the semilogarithmic plot. The so-called γ statistic is based on the cumulative distribution of the internode distances (Pybus and Harvey 2000). It is a fairly complicated expression, but has the property that, under the pure-birth model, the distribution of the statistic follows a normal distribution with mean = 0 and standard deviation = 1. In the absence of extinction, if $\gamma < 0$, speciation rates are declining towards the present; if $\gamma = 0$, speciation rates fit the pure birth model; and, if $\gamma > 0$, speciation rates are increasing towards the present. In the case of the storks, $\gamma = -2.07$, which is equivalent to a one-tailed probability of just under 0.05 that the pattern of speciation fits a pure-birth model, and it implies some slowdown in speciation rate towards the present. If extinction is also going on, the effect is nearly always to cause a bias towards $\gamma > 0$, i.e., the appearance of an increasing speciation rate towards the present (Pybus and Harvey 2000; Weir 2006), as indicated by the *red* curve. This is because any recent extinction eliminates only one or a few potential descendant species, whereas older extinctions eliminate many more. Thus, extinctions typically leave an imprint that mimics increasing speciation towards the present.

through time, the expected value of γ is positive. This means that the test is biased against the detection of a slowdown in speciation, making it conservative (Pybus and Harvey 2000; see legend to Figure 6.4 for further explanation). Even a pattern of mildly increasing (random) extinction towards the present leads to a bias against the detection of a slowdown (Weir 2006). A few species from some of the genera have not been sequenced, but results are also robust to this (*personal observations;* see also Weir 2006).

Based on the γ statistic, 9/17 (52%) of all genus-level phylogenies show relatively few branching events towards the present (setting statistical significance at $P < 0.01$; Table 6.1). For example, the New World warblers show strong evidence of a slowdown (Figure 6.3, middle left panel; Lovette and Bermingham 1999). In this group, 13 speciation events are recorded in the first 2 million years, but only three in the last 2 million years (Figure 6.3). This is despite the presence of many more lineages in the last 2 million years than in the first 2 million years, each of which is available to produce daughter lineages.

One possible explanation for the slowdown is that it is more apparent than real. Because divergence to the level of a diagnosable species can take a long time, the most recent branching events that will lead to species sometime in the future are not recorded (Avise and Walker 1998). Accordingly, we recomputed the statistics after truncating the phylogenies at 2% down from the tips (dashed line in the phylogenies illustrated in Figure 6.3), which is an estimated 2 million years from the present. This encompasses most differences that have been recorded between taxa within species (Figure 2.8, p. 32). Even after this truncation, 5/17 (29%) of the phylogenies show a significant slowdown towards the present (Table 6.1).

Weir (2006) studied patterns of speciation among 17 different groups, mostly genera, occupying the Amazon basin, containing between five and 19 species. He analyzed 12 additional groups not in Table 6.1, but included only those species that occur within the Amazon basin. 12/17 taxa show a negative γ, with five significant at $P < 0.05$, and a summary statistic across all taxa was highly significant ($P < 0.0001$). The magnitude of the slowdown, as estimated by γ, is negatively correlated with the maximum number of sympatric species in the taxon (Figure 6.5). Finally, Zink et al. (2004) found evidence for a slowdown in some additional North American taxa.

These results seem compelling, but, as this book is going to press, Albert Phillimore has discovered a difficulty with the interpretation of the patterns in Table 6.1 and Figure 6.5 as indicative of a general slowdown in speciation rates. This is because, under the simplest hypothesis, that all lineages have an equal probability of speciating at any time, the events that take place early on in a clade's history are critical to determining a clade's size. Groups consisting of many species will, generally speaking, have experienced high speciation rates early in

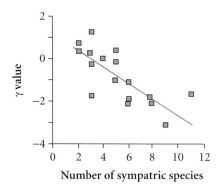

Figure 6.5 In groups from the tropical lowlands of South America, the slowdown in diversification rate towards the present (as measured by the magnitude and sign of the γ statistic; a large negative value reflects a strong slowdown) is negatively correlated with the maximum number of sympatric species in the taxon (least-squares regression, $r = -0.75$, $N = 17$, $P < 0.001$). The maximum number of sympatric species was determined by overlaying range maps. The relation remains significant ($P < 0.05$), even when both the total number of lowland species in the taxon and taxon age are included in a multiple-regression analysis (from Weir 2006).

their evolution. This is a consequence of exponential growth: If, by chance, a few speciation events happen early during diversification, multiple lines are produced. Each of those lines has a chance of diversifying further, thereby resulting in large clades. Now, under a hypothesis of equal probability of speciation throughout history, clades that undergo unusually rapid speciation early in their history should relax back towards the average rate later in diversification. Thus large clades should show a slowdown in speciation rates, whenever the probability of speciation is constant in space and time (conversely small clades should show a speed-up). The conclusion that a slowdown in large clades represents any sort of nonrandom process needs to be treated with circumspection.

Given the current uncertainty about whether patterns of speciation through time do represent an overall nonrandom process, I instead emphasize predictions of the ecological-controls model with respect to patterns of speciation in space. As suggested below, such controls are likely to operate through increasing limitations on the ease of range expansions and populations persisting in new areas.

ECOLOGICAL CONTROLS ON SPECIATION

Species range expansions driven by productivity may be a major determinant of local species abundance. Storch et al. (2006) present a computer simulation of this process, in which they drop a species onto a single square with probability based on that square's productivity, and assume it can expand into any suitable squares (i.e., not into the oceans), with probability of expansion into an adjacent square dependent on that square's productivity. They show that this fits a global model of species distributions very well, and argue that the association between productivity and ability to expand range is causal (e.g., would-be rare resources are commoner in more productive areas, so populations can be more easily established in new locations). As noted by Mayr (1947) and in Chapter 2, range expansions are critical to ongoing speciation.

To consider how ecological controls on range expansions limit speciation, I extend Mayr's basic diagram of speciation in space (illustrated in Figure 1.3, p. 9) to include an ecological dimension. Diamond (1973) presents such a diagram, based on his studies of New Guinea birds (Figure 6.6). Diamond (1973) used altitudinal distribution as his axis of ecological differentiation, but the principles should equally apply to other axes, such as body size. Seven stages of speciation are illustrated, and Diamond cited several examples of each stage among the New Guinea avifauna. Stages 1–3 follow the Mayr model; a continuous population becomes fragmented and divergence proceeds. At stage 3, divergence results in some ecological differentiation (in Figure 6.6, this is represented by slightly different altitudinal ranges). Through stages 4–7, the differentiated forms come into contact again and extend into each other's geographical ranges, coupled with character displacement in occupied altitude. In the end, both species may occur throughout the other's geographical range, subdividing the elevational gradient. The whole cycle could be repeated again to produce three species in one location, and so on.

Figure 6.6 indicates two critical steps in the production of new species. One is range fragmentation (stages 2–3), enabling population divergence to proceed. The other is the ability of the two species to expand into the other's range, so that ranges can be fragmented again. It is at this second step that ecological controls should slow the rate of speciation. As species fill up niche space, it becomes more difficult to expand ranges, reducing opportunities for further rounds of geographical isolation. With reference to Figure 6.6, every time a new species becomes ecologically specialized to part of the altitudinal range, it becomes increasingly difficult for yet another species to insert itself into an even narrower elevation belt. The process becomes stuck at stages 4 through 6.

Theoretical models firm up these arguments. Case and Taper (2000) investigated the role of resource competition in affecting range expansions of ecologically similar species (Figure 6.7). They did this by extending the basic character displacement model described in Figure 5.3 (p. 81) to include the dimension of space. They assumed spatial variation in the optimal value for some ecological trait, and, at each point in space, the resource distribution follows a bell-shaped

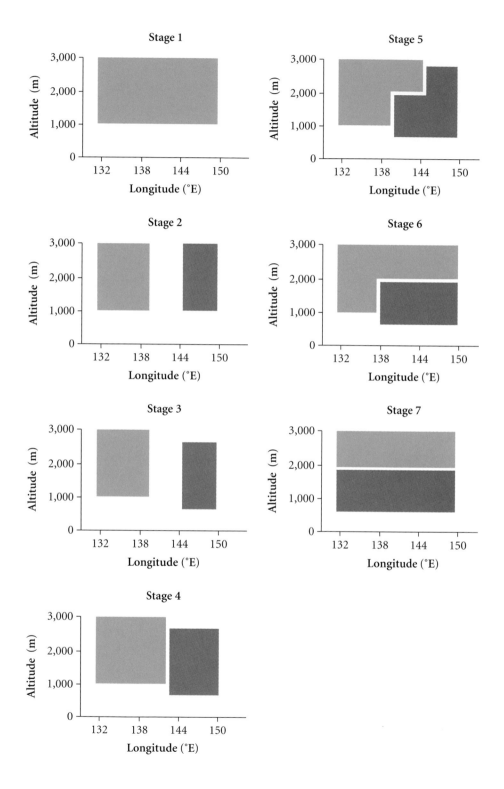

(opposite page)
Figure 6.6 Patterns in the altitudinal distributions of species in New Guinea (from Diamond 1973). Each of the seven panels is represented by two to more than 10 examples in the New Guinea avifauna. The *blue* and *red* represent altitudinal and geographical distributions of pairs of closely related species. The diagrams are highly stylized and, in reality, altitudinal distributions contract towards the geographical range limits of a species. The seven stages illustrated in these panels are taken to indicate "snapshots" of the process of speciation resulting from range fragmentation, leading up to the production of fully sympatric, altitudinally displaced species pairs. Note that extension into each other's range (stage 5) results in inferred character displacement, with each species occupying a different mean elevation as well as occupying a narrower elevational range in sympatry, when compared with allopatry.

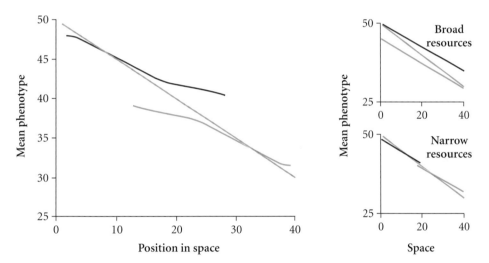

Figure 6.7 Character displacement during range expansion (after Case and Taper 2000; T. Case, *personal communication*). This is the spatial extension of the character displacement model illustrated in Figure 5.3 (p. 81). The mean distribution of resources varies across space, so that the optimal phenotype for a single species differs in different locations (*grey line*). The *colored lines* represent the equilibrium geographical distributions and mean phenotypes of two competing species that have spread into each other's ranges from opposite ends of the gradient. In the area of species overlap, they show character displacement (i.e., the species deviate from each other, as well as the optimum; see Figure 5.3). If, at each locality, resources are widely dispersed around the mean value (i.e., individuals that deviate from the optimum do not suffer a strong fitness penalty), character displacement will be large, and expansion into each other's range may be complete (*top right*). On the other hand, if the resource distribution is narrow, the two species will hardly extend into each other's ranges and will competitively exclude each other to different positions along the resource gradient (*lower right*).

curve around that optimal value (i.e., the width of the resource curve in Figure 5.3 is the same everywhere). A single species occupying the whole area is selected to match the optimal value; it will show geographical variation that tracks the variation in the environment. If, instead, two similar species at different ends of a spatial gradient come into contact and spread into each other's range, they undergo character displacement and diverge to exploit different resources, as illustrated in the New Guinea birds.

Extension of species into each other's range need not be complete. Character displacement causes the mean phenotype to become increasingly displaced from the optimum as one species penetrates further into the range of the other. Eventually, the mean phenotype may be displaced so far from the optimum that the resources become insufficient for the species to sustain itself. Critically, a wide distribution of available resources results in complete overlap of the ranges of the two species, whereas a narrow distribution results in virtually no overlap, and each species competitively excludes the other (see the smaller panels in Figure 6.7).

The model of Case and Taper (2000) shows that successful expansion of one species into the other's range should occur most readily in a resource-rich environment that is free of other competitors, because that is when underutilized resources are readily available. As the ability of a population to persist in a new location is an essential component to ongoing speciation, the model indicates that a resource-rich, competitor-free environment should promote speciation. As species fill up niche space, range expansions become increasingly difficult. The predictions are (1) successful range expansions are easiest in competitor-free locations and (2) closely related species in "full" environments occupy mutually exclusive ranges as a result of competitive exclusion. I consider each in turn.

Range expansions

The most spectacular examples of range expansions facilitated by absent competitors are on oceanic islands where species have been produced in the absence of ecologically similar but very distantly related taxa. Darwin (1859, p. 391) noted "islands are sometimes deficient in certain classes and their places are apparently occupied by other inhabitants; in the Galápagos reptiles and in New Zealand giant wingless birds, take the place of mammals."

The Akaipōla'lāu of Hawaii, the extinct Huia of New Zealand, and the Woodpecker Finch of the Galápagos Islands feed on grubs deep in the tree, a niche occupied by woodpeckers in many other places (Figure 6.8). (On Madagascar, the woodpecker niche is filled by the aye-aye, an unusual primate with a long finger for probing, and, in New Guinea, a "woodpecker" marsupial with a long finger fills a similar niche; Soligo 2005.) Indigenous land mammals are absent

Figure 6.8 Three species that feed (or fed) on grubs deep in trees, associated with the absence of woodpeckers. From left, Woodpecker Finch (Galápagos Islands), Akaipōla'lāu (Hawaiian Islands) and female Huia (New Zealand). The extinct Huia was sexually dimorphic in beak length and the male's beak was much shorter and straight.

from New Zealand, where five extant species of kiwi (Burbidge et al. 2003) eat mainly earthworms and show many features of small insectivorous mammals, including low body temperature and a well-developed sense of smell (Calder 1978). The extinct moa-nalos ("lost-fowl") were large flightless ducks of Hawaii (Olson and James 1991; Sorenson et al. 1999). They had beaks resembling the mouthparts of giant tortoises and apparently filled a grazing niche in the absence of both mammals and large terrestrial reptiles on these islands (Olson and James 1991). Such examples make a clear case for the absence of competitors in enabling a population to persist in an unusual niche, eventually leading to substantial divergence from its ancestor. Woodpecker Finches, kiwis and moa-nalos have been produced in the absence of, respectively, woodpeckers, mammals, and tortoises, and it is difficult to believe they would have been there otherwise.

Range exclusions between close relatives

On continents, allospecies are common, with abutting ranges and minimal geographical overlap (Table 2.2, p. 26; Aliabadian et al. 2005). They are ecologically similar, and may represent the end stage of a niche-filling process. Peterson et al. (1999) showed ecological similarity for geographically separated pairs of related species in Mexico. They were able to predict the distribution of one species based on the ecological attributes (climate, elevation, and vegetation) of where its presumed sister lives.

The Western Meadowlark and the Eastern Meadowlark (Chapter 2; Figure 2.6, p. 27) illustrate the basic pattern of limited geographical overlap associated with ecological similarity. The two species differ somewhat ecologically, because the Western Meadowlark is generally associated with drier prairies of the west, and the Eastern Meadowlark is generally associated with the more forested areas to the east, and adaptations to their respective environments should give the competitive edge to each of the species in its preferred location. They also have greater range overlap than many North American allospecies (Amadon 1968; J. Price et al. 1995). Despite this, the two species appear to be pretty much ecological replacements of each other. It is likely, for example, that a single species of meadowlark could readily extend over the whole of the range that the two now occupy, and it is clear that the Eastern Meadowlark and the Western Meadowlark can easily exploit the minor differences in habitat characteristic of each species because, in parts of its range, the Eastern Meadowlark in fact does so; the dry prairies of Oklahoma, Kansas, and Missouri are occupied by a well-differentiated subspecies of this species (W. Lanyon 1956; Rohwer 1976). Where the two species co-occur in Oklahoma, the Western Meadowlark tends to occupy drier, higher elevations than the Eastern Meadowlark (W. Lanyon 1956; Rohwer 1976), suggesting competitive displacement between the two species, and, where individuals of both species breed in the same place, they are interspecifically territorial (Rohwer 1973).

The best explanation for the failure of allospecies to penetrate into each other's range seems to be competitive exclusion, coupled with adaptation to environmental factors that give the competitive edge to each species in its own range. Some other explanations for limited overlap of related species include hybridization (Goldberg and Lande 2006; Chapter 15) and various forms of social competition (Chapter 15), but these seem less widely applicable.

SOUTH AMERICA'S HUGE BIRD DIVERSITY

Given the arguments presented above, speciation is facilitated by (1) the presence of underexploited resources leading to successful range expansions and (2) range fragmentation. I conclude this chapter by asking how these factors could have contributed to the buildup of the avifauna in South America. The continent contains almost one-third of the world's bird species. This is not simply a consequence of its being large, although it is. In fact, South America is slightly smaller in area than sub-Saharan Africa (17.8 million km² vs. 20.4 million km²) but it has almost twice the number of land and freshwater bird species, estimated as 3,150 for South America and 1,623 for sub-Saharan Africa by Fjeldså (1994). Mountain ranges (Dorst and Vuilleumier 1986), tropical rain forest (Fleming et al. 1987), rainfall in the forest (Hawkins et al. 2003a), and latitude are all more extensive in South America than in Africa, and this must provide some of the explanation for the difference in species numbers between the two continents. Species diversity in the northern Andes is particularly impressive (Rahbek and Graves 2001). In addition to the presence of different species in different habitats at different elevations, there can be quite rapid geographical turnover between mountaintops, or, in the case of species living at lower elevations, on different sides of mountain ranges (Kattan et al. 2004).

Throughout most of the past 65 million years (up to about 3 to 3.5 million years ago; Keigwin 1978), South America has been an island, and most of its species have originated in situ. Among the three major groups of birds—nonpasserines, sub-oscine passerines, and oscine passerines (song birds)—the sub-oscines underwent an endemic radiation within South America to produce 966 of the species currently there (Ricklefs 2002). They include flycatchers in open habitats and the majority of the forest passerines. The 702 oscine passerines in South America (Ricklefs 2002) result from 50 or more separate invasions from the north. This assessment was kindly made by Jason Weir and is based largely on phylogenies using mitochondrial-DNA sequences within genera and families, as well as the recent phylogeny of relationships among these taxa in Barker et al. (2004). Although some affinities are uncertain, the majority of oscine arrivals date to within the past 15 million years, and most of these to the past 10 million years. Some of the first arrivals, such as the tanagers, which contain more than 200 species in South America and may date back 15 million years, have radiated extensively since they invaded (Burns 1997; Ricklefs 2002). Some of the later ones, such as the Barn Swallow that has recently begun breeding in Argentina (Chapter 4), have not diverged sufficiently from their North American ancestor to be considered a different species, let alone diversified within South America.

Fjeldså (1994) compared the history of species diversification for South America and Africa, using Sibley and Ahlquist's (1990) phylogeny (described in Appendix 2.1). Following Fjeldså, I employ a calibration that sets Sibley and Ahlquist's $\Delta T_{50}H$ distance metric to 2.5 million years. This produces dates that roughly correspond to those obtained from some sequence-based estimates, at least with respect to some critical periods of South American bird diversification (Burns 1997; Barker et al. 2004), but the many cautions in translating DNA

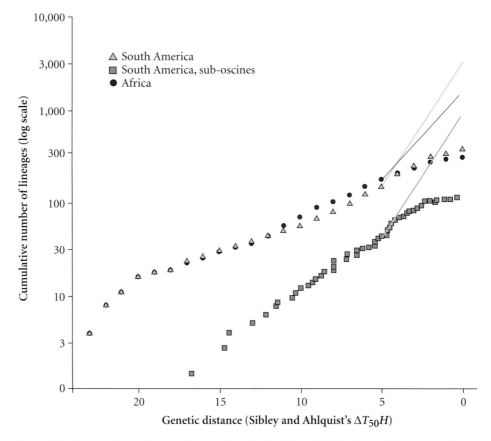

Figure 6.9 Lineage-through-time plots (as described in Figure 6.4) leading to the present-day species of South America and Africa (as estimated by the depth of nodes on the Sibley–Ahlquist phylogeny, redrawn from Fjeldså 1994), and for the sub-oscines in South America separately (from Ricklefs 2005a). Note the logarithmic scale, so that a straight line indicates exponential growth. Virtually all present-day species not on the phylogeny are thought to have diverged since $\Delta T_{50}H = 5.0$ (12.5 million years; Fjeldså 1994), so the number of present-day species is connected by straight lines to $\Delta T_{50}H = 5.0$. In the full data sets for each continent, some of the inferred lineage-splitting events, especially lower down in the tree, have probably not occurred on the respective continent but have resulted from more recent invasions. For example, oscine passerines in South America represent probably about 50 separate invasions from North America (J. Weir, *personal communication*). However, the South American sub-oscine radiation is largely endemic to South America (Ricklefs 2002).

distances into time should be born in mind (Chapter 2, Appendix 2.1). Fjeldså produced lineage-through-time plots for each continent (Figure 6.9). Based on taxonomic considerations, Fjeldså inferred that if species not studied by Sibley and Ahlquist were added to the phylogeny, virtually all the missing nodes would lie within the last 12.5 million years. With the addition of these species, the difference in species numbers between South America and Africa is clearly attributable to higher net diversification rates (speciation minus extinction) in South America over the last 12.5 million years (Figure 6.9). Diversification rates over this time have also been higher in South America than in North America (Ricklefs 2005a; Cardillo et al. 2005). In South America, when the avifauna is considered as a whole, there is little evidence for a slowdown in diversification rate towards the present (Figure 6.9).

What accounts for the rapid, and apparently ongoing, diversification in South America? Fjeldså (1994) distinguished two classes of species: first, those that belong to radiations that contain at least ten species that have diverged over the past 6 million years and, second, single species or superspecies whose node of insertion into the phylogeny lies more than 6 million years ago (Figure 6.10). He was able to classify about 20% of the species to one or other of these classes (the actual numbers in each class are higher, but the age of other species could not be determined because the phylogeny is incomplete). Fjeldså plotted the distribution of young and old species on a 200 by 200 km grid. Results giving the ratio of young to old species, which show the patterns most clearly, are pre-

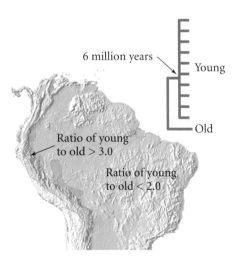

Figure 6.10 Map of northern South America showing the ratio of young to old species (redrawn from Fjeldså 1994). A young species is defined as a species belonging to a clade of at least 10 species sharing a common ancestor at ~6 million years or younger, and an old species is defined as a single species or superspecies that inserts into the Sibley–Ahlquist phylogeny at a later date.

sented in Figure 6.10. Relatively young species are strikingly associated with the Andes.

Case studies of Neotropical genera using mitochondrial DNA sequence data also show that many recent speciation events have occurred in the mountains (Bates and Zink 1994; García-Moreno et al. 1999; Chesser 2000; Weir 2006). One example is the recent radiation of *Carduelis* finches (Figure 6.3). Another is that of the metaltail hummingbirds (genus *Metallura*), in which one species spans the midelevation belt throughout the central Andes, whereas, at higher elevations, a series of species are geographical replacements of one another, having diverged in the Pleistocene to form a superspecies (Figure 6.11; García-Moreno et al. 1999). This example suggests that a new species is able to spread through the novel conditions of higher elevations once such conditions have been created (e.g., by mountain building or climate change). The new species then speciates on different isolated mountain tops, perhaps in response to barriers created by extensive glaciations (Weir 2006). In addition to birds, Hughes and Eastwood (2006) report that some plant taxa have exceptionally high rates

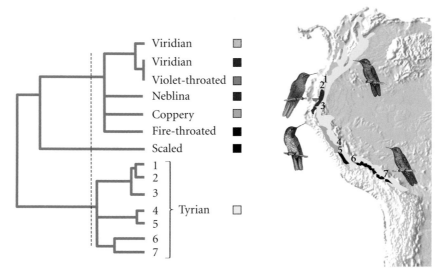

Figure 6.11 A phylogeny of *Metallura* hummingbirds (Metaltails) based on 345 base pairs of the mitochondrial cytochrome *b* gene (García-Moreno et al. 1999; the addition of 510 extra base pairs gives a similar topology and similar branch lengths), using the maximum-likelihood methods described in Appendix 2.2. Two subspecies of the Viridian Metaltail are shown separately; two others are not included. Illustrated (clockwise, from top left) are males of the Violet-throated Metaltail, two different subspecies of the Tyrian Metaltail and the Neblina Metaltail (redrawn from original illustrations by David Alker). Distributions are from del Hoyo et al. (1999). The dashed line indicates 2 million years under the conventional rate of sequence divergence of 2% per million years.

of recent diversification in the Andes (mostly above 3000 m), which they attribute to the recent availability of high-altitude habitat and the appearance of multiple barriers associated with climate change and the steep, dissected topography.

Pleistocene speciation events are common in the mountains and rarer in the lowlands, but the northern Andes may also have been a pre-Pleistocene center of diversification (Burns and Naoki 2004; Weir 2006). Weir (2006) used phylogenetic reconstructions to identify many examples in which lineages have invaded the Andes from lowland sources over the last 10 million years, which may be attributed to new habitats created in association with the rise of the Andes. Although paleoaltitudes are difficult to determine (Rowley et al. 2001; Hay et al. 2002), Gregory-Wodzicki (2000) reviewed paleobotanical and geological studies that imply that the central Andes was half its present height 11 million years ago and that the eastern cordillera of the northern Andes may have tripled its height from 5 to 2 million years ago. The Andes are tectonically active and still growing, especially in the north. By contrast, the major tectonic activity in East Africa was initiated perhaps 25 million years ago, and has proceeded intermittently to the present (Chorowicz 2005). Recent speciation in Africa is also correlated with topographical complexity, at least for forest birds (Fjeldså and Lovett 1997), and this has been attributed to the creation of refugia during climate change (Fjeldså and Lovett 1997; Roy 1997).

The general failure of the South American avifauna in total to show any sign of a slowdown in diversification rates can perhaps be attributed to the relatively recent creation of species as a result of uplift and climate change in the Andes. In the Amazon basin, the genera are often old, and they show an overall slowdown in speciation rate towards the present (Weir 2006; Figure 6.5). An additional explanation for the recent rapid diversification in South America is that the oscine passerines have only recently arrived, and most of their diversification has happened in the past 10 million years. These came from North America, first entering the continent in the north, perhaps in the northern Andes, and this could provide some of the explanation for the high species diversity there (Kattan et al. 2004). However, the rapid diversification of the South American fauna apparently applies to the endemic sub-oscines as well (Figure 6.9).

Whether productivity, barriers, and habitat diversity along elevational gradients can together explain the exceptionally large number of present-day species in South America is unclear. An interesting possibility is that many young species in small geographical isolates (such as some of the hummingbirds in Figure 6.11) are ephemeral; recent climate change and/or the rise of the Andes has created an "overshoot" in the number of species, which may relax back through extinction to more closely match the number of species on other continents. An overshoot has been suggested for specific adaptive radiations in birds, fishes, and spiders (Price et al. 2000; Salzburger and Meyer 2004; Gillespie 2004). It also arises in some models of adaptive radiation (Gavrilets and Vose 2005), in which the progression from rapid to slow speciation is accompanied by extinction that proceeds at a more steady, intermediate rate and takes time to overtake the declining speciation rate (Gavrilets and Vose 2005).

COMPARISON OF THE ECOLOGICAL CONTROLS MODEL WITH ECOLOGICAL SPECIATION

As developed in this chapter and emphasized by Mayr (1947, 1963), speciation rates must be affected by the presence of suitable conditions for range expansions and population persistence in new locations, so that further rounds of population differentiation can take place. This leads to the prediction that speciation will be rapid early in an adaptive radiation, because in environments with few competitors (and enemies such as predators and parasites) the establishment of populations in new locations will be relatively easy.

Yet, there is a second reason why we expect high speciation rates in depauperate environments. This is that strong selection pressures accelerate the rate at which populations diverge to the level of full species, the process of ecological speciation described in the previous chapter. Several factors may contribute to more rapid rates of population divergence early in adaptive radiation rather than later. First, as argued in Chapter 3, in a depauperate environment, there should be greater spatial variation in selection pressures, because the presence or absence of a particular species makes a greater difference to the overall biotic environment than it would in a species-rich environment. Second, in a depauperate environment, incipient species should spread more easily into sympatry and undergo character displacement, driving additional divergence that can contribute to reproductive isolation. Third, as will be considered more fully in Chapter 11, costs of bearing sexually selected traits may be reduced in an environment with few biotic pressures, enabling more rapid divergence in different directions. For reasons such as these, ecological speciation should be most common during the early stages of an adaptive radiation (Rice and Hostert 1993).

In summary, two processes that affect speciation contribute to relatively high speciation rates when new environments become available (Rosenzweig 1975, pp. 132–134; Schluter 1998). The first (ecological controls) is a theory about the most favorable conditions for new populations to become established in new locations. The second (ecological speciation) is a theory about the rate at which reproductive isolation between populations develops once they have become established in different places. In the ecological controls model, there is no prediction that early in an adaptive radiation the rate at which reproductive isolation is achieved is accelerated with respect to late in an adaptive radiation. In this model, a high speciation rate early in adaptive radiation stems from differences in the rate at which populations are established in new locations. On the other hand, in the ecological speciation model, strong divergent selection pressures can result in very rapid acquisition of reproductive isolation (Chapter 2).

In the previous chapter I noted that it was difficult to tell if ecological speciation generally contributes to those speciation events that have led to many present-day species because these species diverged from their closest extant relative a long time ago. However, inspection of the phylogenies illustrated in Figure 6.3 suggests that multiple speciation events early in the diversification of some groups are clustered together in time, much more closely than the minimum 0.6 million or more years it may take to produce modern pairs of allospecies. (See, for example, the New World warblers in Figure 6.3; in these groups, within the limits of phylogenetic resolution, there are several more-or-less simultaneous speciation events deep in the past.) Many of these species currently occur sympatrically. I suggest two explanations for this pattern. First, a single species could have fragmented more-or-less simultaneously into multiple isolated populations (e.g., Lovette and Bermingham 1999). These populations could have followed the course of a prolonged period of differentiation in allopatry, the eventual development of reproductive isolation, and then spread into sympatry. Second, short branch lengths deep in phylogenies may instead reflect rapid ecological speciation coupled with the attainment of sympatry early, rather like the young Darwin's Ground-Finch radiation. It seems likely that this latter mechanism provides part of the explanation for short branch lengths deep in phylogenies, and that ecological speciation has contributed to rapid diversification early in many adaptive radiations.

The evidence points both to range expansions and to divergent selection pressures driving speciation, and to both being stimulated by ecological opportunity, that is, by the availability of underutilized resources.

CONCLUSIONS

As species build up in a region and niches become filled, we expect a general transition from range expansions and the production of ecologically differentiated, co-occurring species, to the production of ecologically similar sister species unable to expand into each other's ranges (allospecies). Eventually, range size becomes small. No barriers arise that can fragment the range, so speciation slows to a virtual halt, waiting for new ecological opportunity (or for new innovations to arise). The predicted slowdown in speciation rate through time is found in several genus-level phylogenies (Table 6.1), but statistical

Table 6.2: Effects of ecological interactions on speciation rate

	Rate of population differentiation	Range expansion subsequent to speciation
Competitive exclusion	Increase[1]	Decrease
Character displacement	Increase[2]	Increase[3]

[1]Ecologically differentiated populations competitively exclude or reduce the fitness of immigrants, reducing gene flow.
[2]Natural selection associated with competition drives incipient species apart.
[3]Character displacement facilitates successful range expansions.

issues preclude interpreting the patterns as strong evidence for a slowdown (see p. 107).

Ecological and geographical limitations on range expansions may account for much of the disparity among similarly-aged taxa in their diversification rate (Ricklefs 2003). Ricklefs (2003) analyzed passerine taxa at the family and tribe level (the tribe is a taxonomic category between genus and family). He argued that those few families with an exceptionally high number of species have resulted from occasional range expansions into new regions following the collapse of geographic barriers (e.g., the escape of the Corvids from Australasia into Asia as the continental plates neared each other). Barker et al. (2004) suggested that among the Passerines, the different times at which species dispersed out of Australia could explain different rates of subsequent diversification, with the species that dispersed earlier able to diversify more.

Families and tribes with exceptionally few species are often ecologically unusual (Ricklefs 2003, 2005b) or restricted to geographically circumscribed regions (Ricklefs 2003), limiting their opportunity to become successfully established in new areas. Because a model of random speciation and extinction, plus geographical and ecological controls on range expansions, together seem to explain much of the differences among similarly-aged taxa in the number of species they contain, Ricklefs (2003) suggested that few characteristics of species directly affect the probability of their producing daughter species. In the next chapter, I will investigate this proposition in more detail. I will show that, in fact, some traits of species do make them more likely to disperse and to colonize new areas successfully, and that these traits are correlated with diversification rates.

As described in this and the two preceding chapters, different kinds of ecological competition affect speciation rates (Table 6.2). First, consider incipient species. If the incipient species are parapatric, resident individuals may competitively exclude immigrants, resulting in a reduction in gene flow and an acceleration of the rate at which reproductive isolation is attained (Chapter 4). When incipient species spread into sympatry, character displacement between them causes divergence in ecological characteristics, again contributing to reproductive isolation (Chapter 5). Thus, both competitive exclusion and character displacement have positive effects on the development of reproductive isolation. Now consider full species. Character displacement is associated with successful range expansions, and this also promotes speciation, because it enables further rounds of range fragmentation and differentiation. However, competitive exclusion between good species has the opposite effect, lowering the opportunity for successful range expansions, and slowing speciation rates. Under the theory of ecological controls, this is the ultimate ecological cause of speciation slowdown.

SUMMARY

The number of bird species found in different parts of the world correlates with plant productivity. One explanation for this is that, in a more productive environment, unusual resources are sufficiently abundant for a specialist on those resources to persist (i.e., more "niches" are available). In addition, at lower latitudes, species numbers are especially high in mountainous regions, which is at least partly attributable to habitat turnover along an altitudinal gradient. Such ecological controls on the number of species in a locality should also be controls on the rate at which species are produced. The actual mechanism whereby speciation is affected by niche availability is likely to be through limits on range expansions. Range expansions are seen as an essential step in speciation because they enable further rounds of population differentiation. The most favorable situation for range expansions is in a resource-rich, competitor-free environment, where inefficiently utilized or unutilized resources are available. As an environment fills up with species, it becomes more difficult for new populations to become established and to persist for long enough to differentiate to the level of full species. I apply these principles to the extraordinarily rich South American avifauna, which is attributable to much diversification over the last 12.5 million years. Much recent speciation has happened in the northern Andes, where species exploit habitats likely to have been created along with the relatively recent rise of these mountains and/or climate change, and ranges were fragmented during the Pleistocene glaciations.

CHAPTER SEVEN

Behavior and Ecology

Hallo the Blue Tits have been at the top of my milk again.

—Tony Hancock in *The Radio Ham*, BBC publications (2003)

Ecological factors influence the ease of range expansions and the rate at which populations diverge from each other. As described in the previous two chapters, because of these factors, speciation rates can vary across both time and space. In this chapter, I consider range expansions and population differentiation from the species point of view. I ask how differences in behavior cause some species to be more likely to colonize new locations or to change genetically, thereby giving rise to daughter species more frequently than other species. First, both dispersal propensity and the ability to survive and persist in a novel environment will affect the chances of successful range expansions and, hence, the probability of speciation. Second, behaviors influence directions of evolution.

Two elements are involved in becoming successfully established in a new location. The first is getting there, and the second is staying there. With respect to getting there, dispersal propensity is an important limitation on range expansions and speciation. A crude measure of dispersal predicts species numbers across families of birds (Phillimore et al. 2006). Families with many species, such as parrots and finches, include species that tend to disperse. In families with only one species, such as the Ostrich, that species has relatively low dispersal capabilities (Phillimore et al. 2006).

The importance of dispersal is most dramatically illustrated by the colonization of oceanic islands. Cooperative breeders (i.e., species in which several individuals help to raise a single brood) have low dispersal, both because they are usually resident species (i.e., do not undertake seasonal migrations) and because at least one sex typically remains on its natal territory, first as a helper, then as a breeder (Cockburn 2003). Probably because of low dispersal tendencies, relatives of cooperative breeders are generally absent from Pacific islands (Cockburn 2003). By contrast, the Silvereye has colonized many islands off Australia in the past 200 years (Clegg et al. 2002b), and relatives of this species are common across the western Pacific Ocean (e.g., Figure 8.5, p. 149). Although

species with high dispersal should reach new locations more easily and this promotes speciation, dispersal is a double-edged sword because, once a new location is colonized, low immigration rates assist speciation by reducing gene flow
(Chapter 3) and by making it less likely that the population goes extinct, for
example, through the introduction of disease (Chapter 8). Dispersal propensity
clearly has multiple influences on speciation probability and I do not consider it
further in this chapter.

Simply reaching a new area is important, but it is only half the battle. The
population has to remain there, and the first few generations are likely to be
critical. Dealing with novel resources, predators, parasites, and competitors
should be more successful if individuals are able to alter their behavior in adaptive ways (Baldwin 1896; MacArthur and Wilson 1967, p. 107). This means that,
if difficulties of range expansion are an important limit on speciation rate,
behaviorally flexible species, such as those able to exploit a new food resource,
should be the ones most likely to give rise to daughter species.

A second influence of altered behaviors is on directions of genetic differentiation. Once a new behavior becomes established in a population, it leads to new
selection pressures, resulting in genetic evolution. The so-called "vampire" finch
on Wenman Island in the Galápagos (actually a population of the Sharp-beaked
Ground-Finch) has the unusual feeding behavior of drinking blood after pierc-

Figure 7.1 Sharp-beaked Ground-Finches feeding on the blood of the Masked Booby on
Wenman Island in the Galápagos. Copyright D. Parer and E. Parer-Cook/AUSCAPE.

ing the bases of wing feathers of boobies (Bowman and Billeb 1968; Figure 7.1). It also has a particularly sharp beak that may have evolved, at least in part, in response to novel selection pressures associated with the blood-drinking habit (Schluter and Grant 1984b). In general, because a great diversity of behaviors can arise, and because each can generate novel selection pressures, behavioral change could be an important driver of genetic divergence between populations.

The idea that behavioral flexibility is involved both in successful range expansions and in leading to genetic differentiation is embedded in one of the more famous speciation quotes: "A shift into a new niche or adaptive zone is almost without exception initiated by a change in behavior. The other adaptations to the new niche, particularly the structural ones, are acquired secondarily" (Mayr 1963, p. 604). The actual mechanisms by which persistence in new localities and genetic change are brought about, and the role of behavior in general, are perhaps some of the least appreciated and least explored areas of speciation (West-Eberhard 2003). In the rest of this chapter, I review the role of foraging behaviors in influencing successful range expansions, in driving genetic differentiation, and in contributing to speciation. Less work has been done on antipredator and antiparasite behaviors. However, nest predation is a major cause of fitness loss in birds (Ricklefs 1969; Martin 1995), and, in a final section, I consider how flexible nesting behavior may have driven some patterns in speciation.

FORAGING BEHAVIORS

Bird watchers often report unusual feeding behaviors in the "short note" section at the back of an ornithological journal, the star journals in this regard being *The Wilson Bulletin* (in the United States) and *British Birds*. Many behaviors might be considered mundane: gull drops mussels on road to break them. Others are quite remarkable: House Sparrows use an automatic sensor to open a bus-station door, Black Kite uses bread as bait to catch fish, gull adapts the shell-dropping technique to kill rabbits (Lefebvre et al. 1997, 1998, 2001, 2002, 2004). Novel feeding behaviors develop spontaneously and often. By 2004, Lefebvre and co-workers had compiled a total of 2,213 *published* reports of unusual or novel behaviors (Lefebvre et al. 2004). The vast majority of these behaviors are likely to be cultural innovations, rather than the result of a genetic mutation.

If a single individual comes up with a new behavior, it will often die out with the individual. For example, on Isla Daphne in the Galápagos Islands, Boag and Grant (1984; P. Boag, P. Grant, and B. Grant, *personal communication*) observed two Cactus Finches that pursued lizards, pecking at their tails. If the lizard jettisoned its tail, the tail was eaten as if it were a caterpillar. The tail-eating habit died with the birds. To have any significance for evolution, new behaviors must become widespread, and this can happen in two ways. Either many individuals adopt the behavior independently, or some individuals copy the behavior from others. First, behaviors may become common in the population because all individuals similarly change their behavior when they encounter a novel feature in the environment. The Coal Tit is small and forages on the outer parts of trees, whereas the Willow Tit and Crested Tit are larger and forage on the inner part of the tree. Experiments in both field and lab show that removal of the larger species results in the Coal Tit moving to the inner parts of the tree, and this is a consistent result across individuals (Alatalo et al. 1985; Alatalo and Moreno 1987).

In other cases, social learning plays an important role. Typically naïve individuals acquire new behaviors more rapidly if others are already performing them (Sasvari 1985b; Lefebvre 2000). A classic example both of independent acquisition and of social learning is that of Blue Tits drinking milk after pecking through the aluminum-foil covering of bottles placed outside people's houses in Britain (Fisher and Hinde 1949). Bottle opening by Blue Tits was first recorded in the village of Swaythling near Southampton in 1921. Twenty years later, it was present at many locations across England (Fisher and Hinde 1949; Lefebvre 1995). Bottles were a new feature of the environment, and bottle opening apparently arose independently in several different places. However, the very rapid spread of the behavior implies social facilitation (Lefebvre 1995), which may be no more complicated than naïve Blue Tits visiting sites where other Blue Tits are already drinking milk (Hinde and Fisher 1951; Lefebvre 1995).

Lefebvre (1986) showed, in Common Pigeons, that the behavior of piercing a paper top covering a dish is transmitted to observing pigeons, both in the laboratory and in the field. Subsequently, Lefebvre and Giraldeau (1994), again working with pigeons, found a positive effect of number of tutors and a negative effect of number of observers on acquisition of a novel feeding behavior: pecking open a stopper on an inverted tube containing seeds. The negative effect of observers was attributed mainly to individuals being distracted by naïve birds (i.e., other observers) and thus viewing the behavior less closely. The implication is that, if a new behavior does spread, it will do so extremely rapidly because of the dual effect of increasing numbers of tutors and decreasing numbers of observers. The few examples of recorded rates of spread of new culturally transmitted feeding behaviors, including that of milk-bottle opening, do show rapid increase (Lefebvre and Giraldeau 1994). Both the simultaneous adoption of novel behaviors by many members of the population and the rapid increase in the behavior associated with observational learning mean that behaviors can quickly become established in populations.

CORRELATES OF FEEDING INNOVATION
WITH SPECIATION

Different taxa of birds have different frequencies of reported unusual feeding methods (Lefebvre et al. 1997, 1998, 2001, 2004). The frequency of reported foraging behaviors has been correlated with forebrain size separately on three different continents (Figure 7.2). In addition, among seven species of passerine birds, the frequency of reported feeding innovations is correlated with their ability to learn a task in the lab ($r = 0.92$; Timmermans et al. 2000). These results suggest feeding innovations are meaningful measures of so-called trial-and-error learning. Learning by observing other individuals is also correlated with trial-and-error learning (Sasvari 1985b; Lefebvre 2000). Together, this implies that we can assign a suite of traits distinguishing "daft" from "smart" birds. Daft birds spontaneously come up with few feeding innovations and are unlikely to copy from other individuals. A good example of a daft bird is the Emu, which has a noticeably small brain for its body. The staff at the San Diego

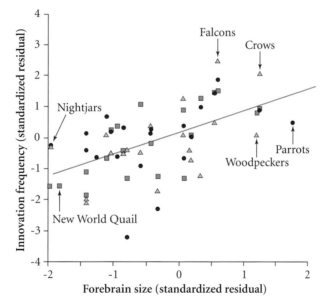

Figure 7.2 Association of feeding innovations [residuals from a regression of log (number of reports) against log (number of species in a region)] against forebrain size (residuals from a regression of log forebrain weight on log body weight) in various taxa of birds. The taxa are defined at or around the level of the order: ■ North America, ● Australia, ▲ Europe. For each continent considered separately, the two variables are significantly associated (rank correlations, all $P < 0.05$). The line is the least-squares regression fit to all the data combined. European data are from Nicolakakis and Lefebvre (2000) and North American and Australian data are from Lefebvre et al. (2001).

Wild Animal Park's Bird Show used to take great delight in explaining that they had not been able to train their tame Emu to go regularly back through the door from which it entered the show, despite many years of daily effort to teach it to do so. Smart birds spontaneously come up with many feeding innovations and are likely to learn rapidly. A good example of a smart bird was the parrot featured in the same show. Apart from being able to talk, the parrot was easily trained to take dollar bills from the outstretched hand of a spectator.

Feeding flexibility or its correlate, brain size, appears to influence speciation rates. First, more species-rich taxa of birds contain species with relatively large brains and many feeding innovations (Nicolakakis et al. 2003; Figure 7.3, *left*). When brain size and feeding innovations are combined in a statistical model, only feeding innovations remain as a significant predictor of species richness (Nicolakakis et al. 2003). Second, species with larger brains are divided into more subspecies, even after controlling for other confounding variables known

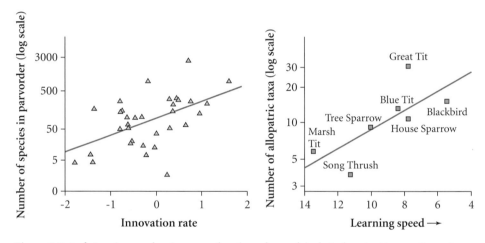

Figure 7.3 *Left:* Species number in parvorders (on a log scale) plotted against innovation rate (from Nicolakakis et al. 2003). A parvorder is a taxonomic level between family and order (Sibley and Monroe 1990). Innovation rate is the residual of number of feeding innovations reported against research effort (i.e., an estimate of the number of research papers for a taxon), which therefore controls for any biases due to numerical abundance of the taxon. There is a significant association ($r = 0.56$, $N = 33$, $P = 0.001$). The association remains if the different ages of the parvorders are controlled for using the depth to the parvorder's root in the Sibley-Ahlquist phylogeny in a multiple regression (*unpublished observations*). *Right:* Number of allopatric taxa (subspecies or allospecies) in a zoogeographic species plotted against the ability to learn (from Sol, Stirling, and Lefebvre *unpublished;* $r = 0.73$, $N = 7$, $P = 0.05$). The ability to learn was measured by Sasvari (1985a). He trained captive birds to peck at a dish containing food, then removed the food and gave a food reward from a remote location. The measure is the number of trials needed before the bird learned to peck at the empty dish to receive the reward. Small values (to right of the graph) indicate fast learning.

to affect numbers of subspecies, such as range size and dispersal propensity (Sol et al. 2005b). Third, species with more feeding innovations and those that learn more quickly in the laboratory contain more subspecies and allospecies (Sol 2003; Sol, Stirling, and Lefebvre, *unpublished observations*; Figure 7.3, *right*). Fourth, Owens et al. (1999) found an association between the number of species in a taxon and a simple measure of habitat generalism among its component species. However, this last result was not confirmed by Phillimore et al. (2006), who instead found diet (rather than habitat) generalism to be a better predictor of species diversity. Although measures of both habitat and diet generalism may be related to foraging innovation, there is no necessary connection: it is possible that birds which show little innovation occupy a diverse range of habitats, or even forage in diverse ways (Greenberg 1990; Sherry 1990). Claudia Mettke-Hofmann (*personal communication*) could find no association between diet generalism and exploratory behavior in her studies of parrots [described in Mettke-Hofmann et al. (2002)].

Feeding flexibility may affect speciation through causing successful range expansions or by affecting the rate of genetic differentiation from the source. I consider each in turn.

Range expansions

As one way to assess the importance of feeding flexibility in range expansions, Sol and Lefebvre (2000), Cassey et al. (2004), and Sol et al. (2002a, 2005a) have studied human-aided bird introductions into exotic locations. They found that, after controlling for the number of introduction attempts, two other variables remain important. First, habitat generalists are more often successfully introduced than habitat specialists (Cassey et al. 2004). Second, species with many reported foraging innovations (as well as its correlate, large brain size) are more often successfully introduced (Sol et al. 2002a, 2005a; Figure 7.4). For example, the Eurasian Blackbird has 23 reported foraging innovations (Sol et al. 2002a) and has been shown to learn quickly in the lab (Sasvari 1985a). It has successfully established itself in half the locations where it has been introduced (Sol et al. 2002a; Figure 7.4). It is one of the few species to invade pristine habitats in its introduced range and to invade human habitats both in its natural range and in its introduced range. Sol et al. (2005a) argue that it is feeding innovations, or at least skills directly related to feeding innovation, that drive invasion success. This is because, when brain size and feeding innovations are entered together into a regression model, feeding innovation remains a significant predictor of invasion success.

These results from human-aided introductions suggest that behavioral flexibility in foraging promotes successful range expansions. Natural colonization

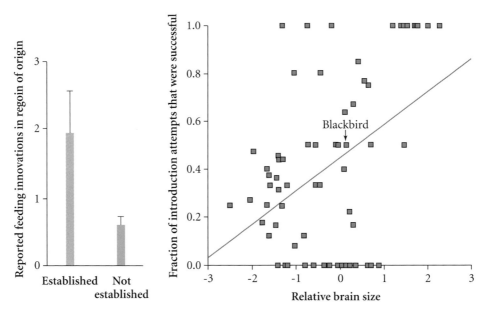

Figure 7.4 *Left:* Number of feeding innovations (mean ± standard error) recorded for species in their ancestral location in comparisons of species that have been successfully and unsuccessfully introduced into New Zealand (from Sol and Lefebvre 2000). *Right:* Number of successful human introductions of birds to localities across the world as a fraction of the number of attempts, plotted against relative brain size (residuals from a regression of log brain weight on log body weight; from Sol et al. 2002a, b). The association is significant ($r = 0.43$, $N = 69$, $P < 0.01$).

events have been less well studied (Greenberg and Mettke-Hofmann 2001). In a qualitative report, Mayr (1965b) listed six features that seemed to be associated with natural expansions of range. One of these was the ability to expand habitat occupancy at the point of colonization. [Other features Mayr considered to be associated with successful dispersal included traveling in flocks, ability to fly over inhospitable areas, and similarity of food or habitat (see Tonnis et al. 2005) in the old and new locations.] Sasvari (1985a) notes that, among the seven species he studied in the lab (Figure 7.3), the fastest learners, such as the Eurasian Blackbird and the Great Tit, were well established in human habitats, whereas the slowest learners, such as the Song Thrush and Marsh Tit, were confined to less disturbed or entirely pristine habitat; however, Lefebvre et al. (2004) state that the evidence connecting their feeding-innovation data to exploitation of urban habitats is equivocal. In studies on 61 parrot species in captivity, Mettke-Hofmann et al. (2002) found that the 12 island species were more exploratory than the 49 species from continents, but it is not clear if this reflects evolution subsequent to island colonization rather than the ability of exploratory species to become established more easily. Greater exploration may evolve on islands because of a greater diversity of available foods in the absence of other competitor species, more intense intraspecific competition, and a release from predation pressures (Mettke-Hofmann et al. 2002).

Genetic divergence

Speciation involves genetic as well as behavioral divergence. If behavioral flexibility drives genetic change, population divergence in ecology and associated morphological traits could be accelerated, leading to an additional or alternative cause of accelerated speciation in behaviorally flexible species (Wyles et al. 1983; Sol 2003). Sol et al. (2005b) explicitly considered this in their study of the correlation of brain size with subspecies diversity. They found that brain size has an effect on subspecies differentiation that is largely independent of species range size. Because range size is likely a measure of the ability to colonize new locations, Sol et al. (2005b) suggest that brain size influences speciation in ways additional to range expansions, that is, by affecting the rate at which populations evolve genetically.

It is simplest to consider two routes to genetic divergence following behavioral change. In the first, the behavior itself becomes genetically determined. This is the process known as genetic assimilation (Waddington 1961; Price et al. 2003). In the second, the behavioral change results in selection pressures on other traits.

The literature on genetic assimilation is quite confusing, which is surprising because the first discussion of the process was one of the clearest. In what became labeled the Crusoe experiment by Haldane (1954), Spalding (1873) suggested a hypothetical experiment: Suppose Robinson Crusoe takes a couple of parrots and teaches them to say "How do you do sir?" He continues his teaching of their offspring and breeds only from those birds that say "How do you do sir" the most frequently and with the best accent. After a sufficient number of generations, young birds repeat these words so soon that an experiment is needed to decide if it is by instinct or imitation; eventually, after further breeding, the instinct is established. The critical feature of this experiment is selection: by breeding from individuals that say "How do you do sir" the most proficiently, Crusoe is selecting those with a genetic tendency to say these words. Eventually, individuals arise that say "How do you do sir" innately because different genes increasing the propensity to learn readily are brought together in the same individual. The process of assimilation can be illustrated using the adaptive surface model (Figure 7.5). Establishment of a population in a new environment results in increased use of a behavior. For example, finches may eat a novel kind of seed. Subsequent selection favoring those individuals that employ the behavior the most efficiently results in genetic evolution.

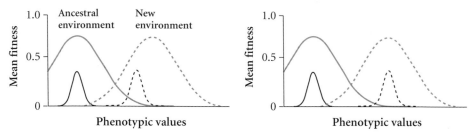

Figure 7.5 The adaptive-surface model of the contribution of behavior to successful colonization of a new environment (after Fear and Price 1998; Price et al. 2003). The *grey* curves are the adaptive surfaces (i.e., mean fitness curves) in the ancestral range and in the new environment (*dashed*). The *red* curves describe the efficiency of individuals in each environment. *Left:* All individuals change their behavior upon entry into the new environment (dashed *red* curve). Note that, in the absence of behavioral change, the mean fitness in the new environment would be zero (i.e., the population would go extinct), so, in this case, changes in behavior are essential to population persistence. *Right:* Selection favoring individuals most efficient in the behavior results in genetic change, and the population moves to lie under the new peak. The populations in the ancestral environment and the new environment are now genetically different.

Behavioral flexibility need not accelerate population divergence, and it can slow the rate of evolution (Huey et al. 2003; Price et al. 2003). This is because individuals may exploit a new environment fairly efficiently as a result of changes in behavior. In this case, selection on the population is less than it would be without any behavioral alterations, and genetic evolution will proceed more slowly, or perhaps not at all (Huey et al. 2003). The slowing of evolution can also be illustrated using the adaptive-surface model (Figure 7.5). If efficiency on the new resource is such that the population moves straight under the new peak (i.e., to the position illustrated in the right-hand graph of Figure 7.5), no selection operates to favor increased efficiency and there will be no evolution; instead, flexibility should persist (Price et al. 2003). This is probably rare. Normally, we expect the exploitation of a new resource to be associated with some selection and subsequent genetic change.

As a likely example of genetic assimilation, consider the Woodpecker Finch in the Galápagos Islands (Figure 6.8, p. 113). This finch extracts grubs from trees, sometimes using a tool (e.g., a cactus spine) to do so. Tebbich et al. (2001) found that 13 young birds held in captivity all developed tool use regardless of whether they were able to see an adult using a tool ($N = 7$) or not ($N = 6$). Thus, the behavior develops without the need to observe other individuals. This is illustrated by the story of Whish, a baby Woodpecker Finch hatched in early March, 1997, who was abandoned by his parents. Gayle and Godfrey Merlin raised Whish and described his development (Merlen and Davis-Merlen 2000). He became explorative: "his interest in all things was extreme" and "any potential crack was worth trying to open," including trying to work a video camera and to open people's mouths. At about the age of eight weeks, he began to pick up sticks, and, with two weeks of practice, he became highly proficient with them. Thus, Whish began his tool-using habits by playing with small sticks and stems, and possibly the evolutionary origin of the behavior lies here. (Tebbich suggests that these skills may themselves have their root in nest-building.) But it is clear that genetic changes have taken place, for it has become a well-developed habit only in the Woodpecker Finch. Two other species of Darwin's finches have been recorded using a tool in this way, each just once (Lefebvre et al. 2002).

Estimates from a phylogeny of Darwin's finches (Figure 2.5, p. 23) imply that the Woodpecker Finch split from the other finches long after the establishment of the original Galápagos finch population (more than 1 million years). Thus, the tool-using behavior probably did not arise soon after the arrival of finches in the Galápagos Islands, even though the resource is likely to have been present. The sequence of events may have been as follows: One bird developed the habit spontaneously as a result of a fortuitous sequence of trials. The habit was culturally transmitted to other individuals and spread through the population. Once the behavior was established, selection for increased efficiency at grub extracting resulted in genetic evolution in the same manner as the Crusoe experiment, and it now develops spontaneously. The main alternative is that a single individual developed the behavior as a result of a genetic mutation and that the mutation spread as a result of the increased fitness of its bearers. Distinguishing these alternatives requires comparisons of rates of genetic mutation, behavioral innovation, and modes of transmission of behaviors from one individual to another (West-Eberhard 2003), but it seems probable that behavioral innovations arise at a much higher frequency than genetic mutations.

The second route to genetic divergence between populations comes when behavioral changes precipitate selection pressures on other traits. A good example of this sort of thing comes from the work of Hill (1993, 1994) on geographical variation in carotenoid coloration. Hill asked how changes in carotenoid intake in the diet might have driven genetic change. It may be a stretch to label carotenoid intake a change in behavior, when the foods eaten are similar, but the principles are the same; an environmentally induced change precipitates genetic change in another character. Different kinds of carotenoids are responsible for most of the red, orange and yellow colors in birds (Brush 1990). Carotenoids cannot be synthesized by birds but must be ingested, after which they can be modified in the body to produce pigments of different colors or deposited directly in the plumage. Canaries exhibited in shows are fed cayenne pepper and other additives to improve their feather coloration, and this has led to much debate about whether it should be allowed, not all of it amicable (Vriends 1992; Birkhead 2003).

Hill studied the red breast patch of the male House Finch. Subspecies from Michigan and Mexico have similar patch colors, but the Mexican subspecies, which appears to be the derived one, has a smaller patch. Hill (1993, 1994) found that carotenoid-manipulated diets prior to molt affected patch color of both subspecies. A low-carotenoid diet results in a loss of the red color (Figure 7.6). However, the size of the patches is not affected by diet and a hybrid between the Michigan and Mexico subspecies had an intermediate patch size, suggesting that patch size is genetically determined.

Using tests in the lab, Hill (1994) showed that females from both populations prefer males with small red patches over large drab ones. When the Mexican population was founded, Hill (1994) argues that a carotenoid-deficient diet would have initially led to large drab patches. Subsequently, sexual selection favored the sequestering of carotenoids over a smaller area of feathers, thereby increasing its brightness. In this way, the patch evolved to a smaller size as a consequence of a direct environmental influence on the development of patch color. The argument can be modeled using the adaptive-surface terminology (Figure 7.6).

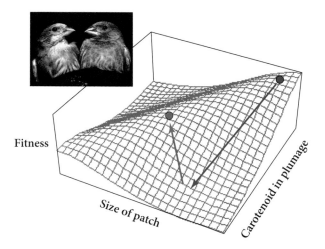

Figure 7.6 Model of the way a change in diet might lead to genetic evolution of a color patch in House Finches [from Price (2006), following the reasoning of Hill (1993, 1994)]. In preference tests, females prefer males if they have bright patches, and small bright patches are favored over large dull ones. Thus, male mean fitness (with respect to sexual selection only and ignoring contributions from natural selection), when described as a function of carotenoid in the plumage and patch size, forms a ridge rising to the right. A decrease in carotenoid availability in the environment (*blue arrow*) results in sexual selection toward a position with a smaller patch size (*red arrow*). The inset shows a bird from Michigan (right), which has a relatively large red patch, and another from Michigan raised on a carotenoid-free diet (provided courtesy of G. Hill and D. Bay).

These observations on Woodpecker Finches and House Finches could probably be expanded many times over. In some cases, genetic change may follow behavioral change in a fairly predictable way. On the island of Gotland, in the absence of the large species of tits present on the Swedish mainland, the Coal Tit forages in the inner parts of the tree (Gustafsson 1988), and it has evolved to become larger (Alatalo and Gustafsson 1988; Table 3.1, p. 44). Many examples of consistent morphological change on islands in response to absent competitors fit the pattern of behavioral change leading to the evolution of morphology in the direction of the absent competitor (Table 3.3, p. 50). In other cases, a considerable element of chance likely affects which behaviors become established and spread through the population, so behavioral changes could play an important role in driving genetic evolution along diverse trajectories. The solution to extracting grubs from inside trees has been quite different in different isolated locations of the world (Figure 6.8, p. 113), and this may be partly a result of different behavioral innovations. In the adaptive-surface framework, there are multiple peaks, and behavioral differentiation determines which peak in the adaptive surface the population evolves towards.

PREDATION AND NEST BUILDING

Individuals entering new environments are likely to encounter new predators and new parasites. Dealing with such enemies may be an important part of persisting in the new environment but has been less investigated than feeding plasticity. In the famous case of the "snake that ate Guam," the introduced, nocturnal Brown Tree Snake (*Boiga irregularis*) has devoured most of the native bird species on Guam (Fritts and Rodda 1998; Fritts and Leasman-Tanner 2001; Wiles et al. 2003). Since the mid-1960s, eight out of 11 native forest species have gone extinct. The Mariana Swiftlet is a survivor, but the only persisting colony roosts and breeds in a location inaccessible to the snake (Fritts and Leasman-Tanner 2001). The Brown Tree Snake is common in its native range (northern and eastern Australia, New Guinea), where the native birds presumably have appropriate antipredator behaviors.

Nest predation is a major cause of fitness loss in bird species (Ricklefs 1969; Martin 1995; Fritts and Rodda 1998). As in feeding behaviors, all individuals in a population could show a similar, adaptive, response in a novel environment, thereby increasing chances of population persistence. For example, Forstmeier and Weiss (2004) showed that Dusky Warblers place their nests higher when ground predators are more abundant. Alternatively, a single individual could come up with a new nesting behavior, which becomes established in the population as a result of being copied. Many examples of unusual nesting behaviors have been reported in the short notes sections of ornithological journals. Nicolakakis and Lefebvre (2000) recorded 176 such cases in European birds (e.g., Eurasian Siskin nesting in a hanging flower pot). It seems plausible that establishment of unusual behaviors in populations by copying could happen fairly readily, if individuals build nests similar to those they were raised in, or nest in similar locations—unlike foraging behaviors, new cultural mutations may be more readily transmitted from parent to offspring rather than to unrelated observers. However, I am unaware of any examples.

Nicolakakis and Lefebvre (2000) could find no correlation of unusual nesting behaviors with a species brain size. However, Madden (2001) showed that bowerbird species that build elaborate bowers have particularly large brains, and Payne (2005a, p. 32) found that cuckoo species that build their own nests have larger brains than the parasitic cuckoos, which do not build their own nests but rather lay eggs in the nests of other species. It seems likely that inventiveness of individuals within species has played a part in driving the evolution of nest building.

Earlier in this chapter (p. 132) I used the contrast between the Song Thrush and the Eurasian Blackbird in learning speed and associated foraging innovation as an explanation for the greater colonization success of the Eurasian Blackbird. Instead, Tomialojic (1992) suggested a role for nest-building differences. He found that Song Thrushes always have an inner layer of mud, wood pulp, or dung in their nests. Tomialojic argued that the need to plaster an inner layer requires wet conditions and the presence of suitable materials, limiting the Song Thrush's ability to colonize urban and dry areas (it has a very patchy distribution in the Mediterranean region). On the other hand, the Eurasian Blackbird in Europe can build nests with no mud layer, and it chooses more solid locations on which to build. The Eurasian Blackbird breeds regularly in urban habitats, and it is one of the most successful colonizers of new areas (Sol et al. 2002a). It has a relatively large brain compared with the Song Thrush, but it remains to be determined whether colonization reflects foraging flexibility, nest building flexibility, or a preadapted nest.

About 60% of all bird species belong to the order Passeriformes (Sibley and Monroe 1990). In Europe, passerines belonging to modern groups appear in the fossil record only from ~25 million years ago (Mayr 2005), and this is about the time passerines in general underwent an explosive radiation (Feduccia 1995). For the 40 million years before that, common species belonged to such orders as the Upupiformes (hoopoes and relatives), the Coliiformes (mousebirds, now represented by only six species in Africa), and relatives of the parrots and swifts, all of which contained species with more diversified foraging habits than are seen among the extant species in these orders. These species apparently occupied many of the foraging niches now utilized by present-day passerines (Mayr

2005). One appealing hypothesis to explain why the passerines were successful is their diversity of nest-building habits (Collias 1997; Olson 2001). Nest-building strategies within the passerine birds include such remarkable behaviors as the stitched nest of the Common Tailorbird (hence its name) and the complex woven structures of weaverbirds (hence their name). Olson (2001) concludes that the entire passerine order is "hard-wired" for nest-building inventiveness. New nest-building techniques may enable species to expand into new environments; according to Eberhard (1998), the ability of a few species of lovebirds (a nonpasserine) to build complex domed nests opens up many more potential nest sites to them. In contrast to passerines, many arboreal nonpasserines (including most species of lovebirds) are stereotyped hole-nesters and/or species that do not dome their nests, and Olson (2001) suggests that this lack of diversity may have doomed them in competition with passerines.

CONCLUSIONS

Once a population has reached a new environment, larger brains and flexible foraging behaviors should enable it to persist, and this may be part of the reason why brain size seems to be associated with speciation rate. Not all of the changes in a new environment need be adaptive; a male House Finch's reduced intake of carotenoids makes him less attractive to females. Whether adaptive or non-adaptive, behavioral changes in response to novel environments affect subsequent directions of evolution, and this may be an additional reason why flexible foraging behaviors are correlated with speciation.

Spontaneous behavioral innovations and their subsequent fixation in the population can promote differentiation even in the absence of environmental differences. For example, behavioral differences, rather than different environments, explain some of the diversity of ways in which the woodpecker niche is exploited. Behavioral innovations and their subsequent increase in a population lead to the potential for great evolutionary diversification. They are an example of a nonecological model of population differentiation. A different nonecological model is more popular and forms the basis of the next four chapters. In that model, sexual selection plays a prominent role in causing population divergence.

SUMMARY

New environments present many challenges, and individuals better able to alter their behavior to deal with those challenges are most likely to survive and reproduce. Among species introduced by humans across the world, those that show more flexible feeding behavior in their native range are the most likely to become successfully established in new locations. Feeding innovation is correlated with brain size, and both feeding innovation and brain size are correlated with the number of species in a taxon. Flexible feeding behaviors may aid in range expansions, thereby promoting speciation. In addition, because new behaviors arise frequently and spread rapidly, populations may easily diverge in behavior, setting the stage for subsequent genetic divergence. Examples in which altered behaviors have likely preceded genetic change include the evolution of morphology on islands in the absence of competitors and the origin of the stick-probing habit of the Woodpecker Finch of the Galápagos Islands, which develops without tutoring. Antipredator and antiparasite behaviors have been less well studied, but an attractive hypothesis for the success of the passerine birds is their diversity in nest placement, which may be related to dealing with predators and parasites. The twin roles of behavioral flexibility in enabling population survival and in driving genetic evolution in novel directions probably play a more important part in speciation than is currently appreciated.

CHAPTER EIGHT

Geographical Isolation and the Causes of Island Endemism

Many species are spatially separated from their closest relatives. They include allospecies belonging to superspecies on continents and nearly all species on isolated oceanic islands. Ecological differences between such species appear to be small. Often the species are defined based on color pattern and/or song differences. They may have been separated for quite a long time and, where tested, hybrids between them may have inherently low fitness (Chapter 16). In this chapter, I begin an exploration of the contribution of nonecological modes of population divergence to the production of these species, by considering the role of geographical isolation in the differentiation of populations. I do this by studying patterns across islands. Islands are ideal for this purpose because degree of isolation is provided by a measure of distance from other islands or from continents. In the following chapters, I address the processes that might cause the patterns and lead to divergence for reasons unrelated to ecology.

As shown in Chapter 2, differentiation of populations to the level of species can take a long time (hundreds of thousands of years). Mayr (1965a) showed that larger islands tend to contain a higher proportion of endemics than smaller ones, and he argued that this is because it is only on large islands that populations persist long enough to differentiate to the level of recognized species. Diamond (1980, 1984) showed that, across the Pacific, more isolated islands have a higher proportion of endemic species than less isolated islands. Because more isolated islands have few species in general, he suggested that reduced competition lowered the risk of extinction on these islands.

In this chapter, I emphasize the roles of area and isolation in increasing persistence times of populations as a cause of island endemism, but I briefly consider the possible role of ecological factors. Unlike most chapters in this book, the results depend critically on deciding whether isolated populations should be given species rank or not. Because nearly all taxa on an isolated island have their closest relative elsewhere, the criterion of reproductive isolation cannot be used. The number of endemic species on an island is essentially the number of phenotypically well-differentiated taxa.

I first show that the proportion of endemics on an island is positively correlated with island area and isolation. Second, I develop a simple graphical model
to illustrate why large, distant islands should have particularly high rates of
population persistence. Finally, I place a timescale on extinction rates based on
molecular data. Results indicate that population persistence on islands can be
long, comparable to times that separate allospecies on continents.

CORRELATES OF ENDEMISM WITH
AREA AND ISOLATION

Endemic species are generally common on oceanic islands. For example, before
humans arrived on Rodriguez Island in the Indian Ocean (~570 km from the
nearest larger island and 1,570 km from Madagascar), it was inhabited by at
least 13 species of land birds, 12 of which were classified as endemics (Gill
1967). In analyses of the relationship between isolation and endemism, Adler
(1994) included known historical extinctions for the Indian Ocean, but fossil
data have not so far been included for the Pacific Ocean (Diamond 1980, 1984;
Adler 1992). The studies have been criticized because of the large number of
unaccounted extinctions (Steadman 1995, 2006). Steadman (2006, p. 409) estimates conservatively that up to about 1700 nonpasserine and 80 passerine
species have been lost from the Pacific islands (and this does not include Hawaii
and New Zealand). Despite the high number of extinctions, it is difficult to see
how extinction could create patterns in the data, and Adler (1992) argues that
enough of the avifauna has remained intact to permit analysis.

A reanalysis of Adler's data shows that the number of endemic species is positively correlated with both area and isolation across both the Indian Ocean and
the Pacific Ocean (Figure 8.1). The pattern for isolation is striking because more
isolated islands have fewer species in general, so, as noted by Adler, the proportion of endemics increases strongly. [A caveat is that the result depends on *isolation* being defined in terms of distance from a continent in the Indian Ocean
and in terms of distance from the nearest large land mass in the Pacific Ocean
(Adler 1992, 1994; see legend to Figure 8.1).] Overall, the result implies that,
from the point of view of species production, the more remote the island the
better. Gene flow is likely to be negligible over such oceanic distances, implying
that the retarding effects of gene flow on differentiation are an unlikely cause of
the correlation between isolation and endemism. Instead, following Diamond
(1980, 1984), I will emphasize the possibility that more isolated populations
persist for longer than less isolated populations, often long enough that they
differentiate to the level of recognized species.

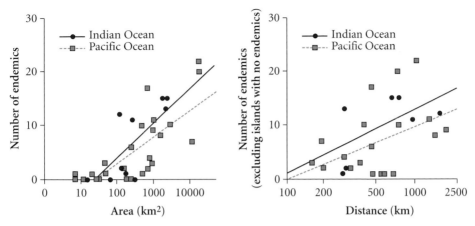

Figure 8.1 Number of endemic species on islands and small archipelagoes in the tropical Pacific and Indian oceans. In the Pacific, only islands west of 120°W were included, and the archipelagoes of Hawaii and northern Melanesia are excluded because there has been much speciation within those archipelagoes (Mayr 1969; Diamond 1977; Mayr and Diamond 2001). Data are from Adler (1992, 1994). *Left:* Plotted as a function of island area (km², $\log_{10}$ transformed). Least-square regression slopes are fit separately for each ocean (Pacific Ocean: $r = 0.74$, $N = 27$, $P < 0.0001$; Indian Ocean: $r = 0.81$, $N = 14$, $P = 0.0004$). *Right:* Plotted as a function of distance (km on $\log_{10}$ scale) from nearest larger land mass (Pacific Ocean) ($r = 0.4$, $N = 19$, $P = 0.07$) or nearest continent (Indian Ocean) ($r = 0.6$, $N = 7$, $P = 0.16$). Least-square regression lines are fit separately for each ocean and only for islands or small archipelagoes with at least one endemic; very small, distant islands have no endemics and confound the pattern. It should be noted that the patterns are less strong if the nearest continent is used for the Pacific Ocean and the nearest larger land mass is used for the Indian Ocean (in the Indian Ocean, this is largely due to the proximity of Mauritius to Reunion). The multiple-regression equations (including islands with no endemics) are, for the Pacific Ocean, number of endemics $= -23 + 4.68 \log_{10}$ area ($P < 0.001$) $+ 6.4 \log_{10}$ distance ($P = 0.034$), $N = 27$, and, for the Indian Ocean, number of endemics $= -30 + 6.7 \log_{10}$ area ($P < 0.001$) $+ 7.7 \log_{10}$ distance ($P = 0.035$), $N = 14$. The *P*-values indicate significance of the partial regression coefficients.

Correlates of endemism with area and isolation are also present in northern Melanesia, as already described in Chapter 3 (Figure 3.6, p. 61). In this case, part of the reason for the correlation is likely to be that gene flow prevents differentiation, with higher rates of dispersal among less isolated islands. Indeed, some particularly dispersive species are scarcely differentiated across the archipelago (Chapter 3). However, even within this archipelago, it is doubtful that the correlation of island isolation with endemism can be completely explained by the retarding effects of gene flow. At the taxonomic level just below that of allospecies, few distinctive subspecies have ever been recorded off their own islands (Mayr and Diamond 2001). Dispersal of these subspecies is so low, and successful immigration to different islands is likely to be so infrequent, that it seems unlikely that gene flow could play a significant part in preventing further differentiation up to the level of allospecies. In addition, the association of isolation with endemism is present at the generic level. All five genera classified as endemic to northern Melanesia occur in the more remote Solomon islands, and none are found in the less remote Bismarcks (Mayr and Diamond 2001). It is hard to see how differentiation above the species level can be affected by gene flow.

ISLAND-BIOGEOGRAPHY THEORY OF ENDEMISM

The correlation of isolation with endemism over large spatial scales is difficult to explain through gene-flow effects that limit differentiation, because successful dispersal is likely to be extremely rare. An alternative explanation is that the rate of population extinction is lower on isolated islands, so populations have time to differentiate to the level at which they are given species status (Diamond 1980, 1984). I develop this idea graphically, starting with MacArthur and Wilson's

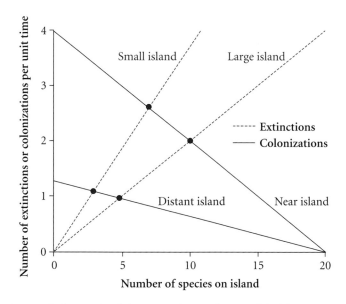

Figure 8.2 A numerical example of the MacArthur–Wilson model of determinants of the number of species on an island. The number of species going extinct per unit of time (*red lines*) increases as a function of the number on the island, simply because more species are available to go extinct. The number of species colonizing per unit time (*black lines*) decreases as a function of the number on the island, simply because many of those in the source are already present. At equilibrium, the number of species going extinct in a given time interval equals the number colonizing, as illustrated by the *red* circles.

(1967) theory of island biogeography (Figure 8.2), which was originally proposed to explain why relatively few species are found on small and isolated islands.

In MacArthur and Wilson's theory the total number of species on an island is attributed to a balance between colonization and extinction. The number of new species that colonize an island in some time interval decreases with the number of species already on the island, simply because many of the species available in the colonist pool are already present. The number of species that go extinct in the same time interval increases with number on the island, simply because more species are available to go extinct. The colonization and extinction processes are independent of each other. At equilibrium, the number of successful colonists in a given time interval balances the number going extinct in that same interval.

More species are found on islands that are closer to the source because birds arrive more frequently and successful colonization rates are higher (*black lines* in Figure 8.2). More species are found on large islands because extinction rates are lower (*red lines* in Figure 8.2), attributed to the fact that population sizes are larger on larger islands. The two colonization lines and two extinction lines illustrated in Figure 8.2 lead to four alternative equilibria. Large islands near to the source have the most species at equilibrium, and small, distant islands have the fewest (Figure 8.2).

Endemism

This simple island-biogeography theory can be adapted to ask why the most *endemic* species are found on larger and *more* distant islands. Assuming that endemism reflects the length of time a population persists once it has become established, the relevant factor is extinction (*red lines* in Figure 8.2). An extinction line is given by the equation: $Y = bX$, where Y is the number of species going extinct in a given time interval, X is the number of species present on the island and b is the slope. Rewriting the equation gives

$$b = \frac{\text{Number of extinctions per unit time interval}}{\text{Number of species present at beginning of time interval}}$$

$$= \text{The proportion of species going extinct over the time interval}$$

Strictly, in this model the time interval has to be sufficiently short so that a species cannot both invade and go extinct in the same interval. However, the point can be illustrated by assuming a time interval of, say, 1 million years. The slope of the line (b) is the extinction rate, so $1-b$ measures the proportion of species present at the beginning of a time interval that persist through the interval. In Figure 8.2 the extinction rate for the large island is $b = 0.2$. Thus, 20% of species present at the beginning of the time interval are expected to go extinct, and 80% will still be present. If we also assume that 1 million years is the average time it takes for differentiation to proceed to the level of an endemic species, an expected 80% of all species on the island will be endemics. Because small islands have higher extinction rates than large islands, not only do small islands have relatively few species, a smaller proportion of these species should be endemic.

In this simple model, extinction depends only on island size, so the theory predicts that the proportion of endemics should be the same on a near as on a distant island of the same size. If population persistence is to explain the positive correlation between isolation and the proportion of endemics, the extinction rate must be lower on distant islands than on near islands. As illustrated in Figure 8.3, two ways in which this might happen have been suggested. In the first, extinction rates increase as a function of the number of species on an island, because, with every additional species, each of the species present is constrained to maintain a lower population size on average (Diamond 1980, 1984).

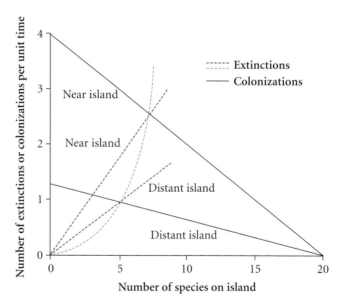

Figure 8.3 A modified version of the simple island-biogeography model of Figure 8.2, which results in relatively higher extinction rates on near islands than on distant islands. Only one island size is considered in this figure. Either (1) the number of species going extinct per unit time increases sharply with the number on the island (*blue line*) because, as more species are added, each species is expected to maintain a lower population size, increasing its extinction risk or (2) the extinction rate is higher on near islands than it is on distant islands because extinction is driven by colonization (*red lines*). In this example, at equilibrium, the fraction of species going extinct per unit time is 1 in 5 (20%) on the distant island but 2.5 in 7 (~40%) on the near island.

In the second, extinction rates are not independent of colonization rates, but instead colonization causes extinctions; the higher colonization rate on near islands causes high turnover and a reduced proportion of endemics (Ricklefs and Cox 1972). I consider each in turn.

If total resource abundances are similar across islands, on islands with few species (i.e., distant islands because they receive fewer colonists), each species will have more resources available, resulting in higher population sizes (Diamond 1980, 1984). Higher population sizes should lower the extinction risk. I illustrate this with the *blue line* in Figure 8.3, which shows an increasing extinction rate as the number of species on an island increases. Under the parameters in this figure, at equilibrium, the extinction rate is about twice as high on the near island than the distant island, and this arises because the near island carries 50% more species than the distant island.

In the second explanation, colonizations drive extinctions (Ricklefs and Cox 1972). The recent decimation of the Hawaiian honeycreepers illustrates how this might happen. The native Hawaiian avifauna is highly susceptible to avian malaria. The spread of the malaria required the presence of an introduced mosquito vector, and then probably arrived along with introductions of exotic bird species (van Riper et al. 1986). Thus, bird introductions have added species to the fauna, but they have also driven native species extinct. In this case, an increase in the colonization rate did not increase the number of species in the manner predicted from the simple theory, but rather increased the rate of species turnover.

Following Wilson's (1961) studies on Melanesian ants, Ricklefs and Cox (1972) and Ricklefs and Bermingham (2002) gathered evidence from the West Indies that natural invasions can drive extinction, leading to a so-called taxon cycle. Species more recently established in the West Indies tend to occur on many islands, and they are often found only in the lowlands of those islands. Species with fragmented distributions are older, they tend to be the endemics, and they occur more often in the interiors of the islands. The older species may have been displaced towards the interiors of the islands, or driven extinct on some islands, by the spread of the younger species. The process can cycle continually, with new invaders proceeding through the taxon cycle to eventual extinction. Patterns weakly consistent with the taxon cycle have also been observed in northern Melanesia (Greenslade 1968; Mayr and Diamond 2001, Chapter 35) and in the Galápagos Islands (Grant et al. 2000).

Ricklefs and Cox (1972, p. 212) suggested that the correlation of endemism with island isolation reflects the taxon cycle over oceanic scales. In the extreme, if a colonization event were to always cause an extinction, an increased colonization rate would lead not to an increase in the number of species maintained on the island but entirely to increased turnover. The main consequence is that,

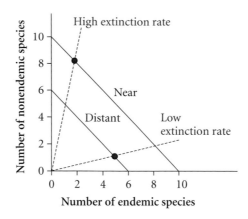

Figure 8.4 This figure illustrates how, if rates of extinction (*red lines*) are much higher on near islands than on distant islands, fewer endemics can occur on these islands even though they contain more species in total. On an island near to the source, the number of species is assumed to be 10, and, on a distant island, 6. A high rate of extinction on the island near to the source results in only 2 of the resident species being endemics (i.e., 20% of the species are endemic). A low rate of extinction on the island far from the source results in 5 of the resident species being endemics (i.e., 83% of the species are endemic).

because near islands receive more colonists, they have higher extinction rates. The effect is illustrated in Figure 8.3, in which the red lines compare extinction rates on near islands with those on distant islands. While the taxon cycle ties extinction to the successful colonization of an island by other bird species, near islands should also suffer increased rates of other kinds of biological invaders, such as predators, parasites, and pathogens, which also increase the chance of extinction. In this case, increased extinction of bird species on near islands might result in increased rates of successful colonization by other species, rather than colonization driving extinction.

Whatever the ultimate cause, low rates of species extinction on distant islands can lead not only to a higher proportion of endemics but also to a higher number of endemics on these islands (Figure 8.4), as is observed.

Timescales

If low extinction rates explain the positive associations of island endemism with area and isolation, colonization and extinction must happen on the same timescale as speciation, that is, much more slowly than originally envisaged by MacArthur and Wilson (Steadman 2006, his chapter 18). Indeed, endemic species in the West Indies (Ricklefs and Bermingham 1999, 2002), in the central Pacific Ocean (Slikas et al. 2000, Figure 8.5; Filardi and Moyle 2005), and in the

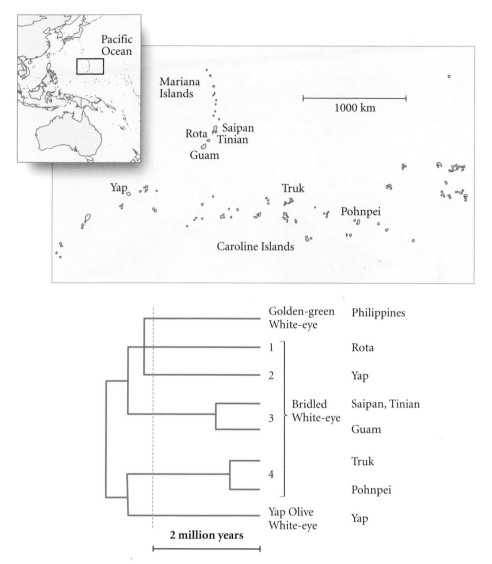

Figure 8.5 Relationships among populations of Bridled White-eyes from Micronesia (after Slikas et al. 2000) as well as two close relatives (the Golden-green White-eye from the Philippines and the distinctive Yap Olive White-eye). Phylogeny was constructed from mitochondrial sequence data using maximum likelihood (see Appendix 2.2, p. 38), with the scalebar assuming the conventional 2% divergence per million years. Previously, the different populations of Bridled White-eye were all classified as separate subspecies, except for the Pohnpei and Truk populations, which were combined as one subspecies. Based on morphological and genetic distinctiveness, Slikas et al. (2000) consider that, in fact, the Bridled White-eye consists of four different species, as indicated by the four deep branches in the phylogeny (1–4 in the figure). Note that small island populations, such as the one on Rota (84 km^2), have persisted for a very long time.

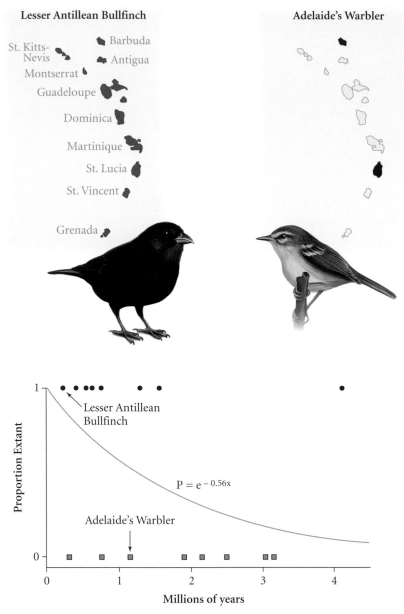

Figure 8.6 Method used by Ricklefs and Bermingham (1999) to calculate an island extinction rate across ten islands in the Lesser Antilles. All species were assumed to once occupy all the islands illustrated, as the Lesser Antillean Bullfinch now does. The Adelaide's Warbler is currently present on only two of the islands and is presumed to have gone extinct on the others. For the island of St. Vincent, the time a species has been present in the Lesser Antilles (based on maximum sequence divergence between any pair of populations in the Antilles, or, in some, cases distance to a mainland relative) is plotted against its presence (●) or absence (■). The Lesser Antillean Bullfinch is present on St. Vincent, and based on mitochondrial DNA sequence divergence it is assigned a time since establishment in the Lesser Antilles of about 150,000 years (R. Ricklefs, *personal communication*). The Adelaide's Warbler is absent from St. Vincent and it is assigned a time since establishment in the Lesser Antilles of 1.2 million years (R. Ricklefs, *personal communication;* Lovette et al. 1998). The exponential curve fit to the data estimates the probability of a species being present on the island as a function of time since its establishment.

Indian Ocean (Warren et al. 2005) can be separated from their relatives by large genetic distances, suggesting that at least some island populations persist for more than 1 million years. Island populations of individual species can also be highly genetically differentiated from each other (Lovette et al. 1998; Slikas et al. 2000; Ricklefs and Bermingham 1999, 2001; Filardi and Moyle 2005).

Ricklefs and Bermingham (1999) used information on genetic distance to estimate rates of extinction of birds of the Lesser Antilles. They restricted their analysis to a sample of species that contain Antillean populations well differentiated from any North American taxa (i.e., relatively recent arrivals in the islands were excluded). First, Ricklefs and Bermingham estimated the time the species had been present in the Lesser Antilles, based on mitochondrial DNA sequences, and the standard translation of 2% sequence divergence per million years. They estimated these dates using either the largest genetic distance among island populations of the species, or the genetic distance from the island species to close relatives on the mainland. As noted by Cherry et al. (2002), the use of these methods means that gene flow among populations underestimates of the length of time a species has been present, which will lead to an overestimate of the extinction rate.

For each island, Ricklefs and Bermingham then plotted the presence or absence of a species as a function of the time it has been in the archipelago. The method is illustrated for the island of St. Vincent in Figure 8.6. Species that are estimated to have spread through the Lesser Antilles relatively long ago are more often absent from St. Vincent today than species that appear to have arrived in the archipelago more recently. Assuming that all of the species were originally present on St. Vincent (this may be a big assumption), the extinction rate can be estimated (Figure 8.6). I show results for all islands in Figure 8.7.

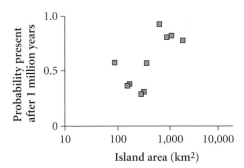

Figure 8.7 Probability that a species is present on an island 1 million years after it became established, plotted against island area, for islands of the Lesser Antilles (after Ricklefs and Bermingham 1999). The probability was calculated using the methods outlined in Figure 8.6, and assumes the conventional rate of mitochondrial sequence divergence of 2% per million years.

The probability of a population persisting for a million years is low up to the size of St. Vincent (350 km²), but approximately 0.75 on one of the four larger islands. For these islands, the timescale of extinction does seem to be roughly comparable to a timescale of speciation estimated for allospecies on continents—on the order of 1 million years (Ricklefs and Bermingham 1999).

Low extinction rates on islands seem to be at variance with more direct observations of extinction over tens (Diamond 1969, 1971) to thousands of years (Diamond 1972). But these occurrences apparently reflect: (1) the effects of human disturbance; (2) small islands where extinction is high and endemic species are absent; (3) the turnover of noncore species on islands—a species may immigrate to an island and persist for a few tens of years, but never become properly established (Diamond 1972); or (4) land-bridge islands created since the Pleistocene that may be too small for some species: in northern Melanesia, Mayr and Diamond (2001) conclude that even large islands of ~1,000 km² have lost ~20% of their species in the last 10,000 years.

Turnover on the order of a million years implies that successful colonization by populations that are destined to evolve into an endemic species must also occur on the same timescale. Depauperate islands near continental sources, such as a new volcano, are rapidly colonized (Diamond 1972). But, as communities build up they should become resistant to invasion (MacArthur 1972, Chapter 5; Lack 1976; Case 1990). First, species that arrive first and grow to large population size have a strong competitive advantage over small groups of individuals of other species that arrive later (MacArthur 1972; Case 1990). Second, the competitive advantage of residents will be accentuated by adaptation to island conditions, as emphasized in Chapter 4.

UNUSUAL ENVIRONMENTS ON ISOLATED ISLANDS

In 1949, Lack promoted the importance of geographical isolation as a cause of differentiation. Because subspecific differentiation is more pronounced across islands than over continental land masses, but the latter provide more varied physical conditions and habitats, Lack (1949) suggested that isolation of populations, and not differences in their environments, is the major factor in subspecific differentiation. Lack's arguments promoting time in isolation as the cause of differentiation follow the line of reasoning developed in this chapter but ignore the possibility that islands vary sufficiently in their biotic environment to drive differentiation (Darwin 1859; Chapter 3).

In later publications, Lack (e.g., 1976) changed his views, and he became one of the strongest proponents of the pervasive role for selection in driving differentiation. He attributed the correlation of endemism with island size partly to the presence of distinctive environments on large islands (e.g., peculiar forests in highland zones). He suggested that differences in the biotic environment were the main cause of differentiation among island populations in the West Indies, although he did not return to consider the general problem of large numbers of subspecies across islands.

Thus, an alternative explanation for the correlation of endemism with isolation is that isolated islands have unusual environments, leading to strong and unusual selection pressures. In Chapter 3, I showed that small islands within archipelagoes do appear to be biotically quite different one from the other. Similarly, it is fairly clear that isolated islands have unusual environments, simply because they have few species in total. Distinctive faunas characterize distant islands and archipelagoes, often including an absence of non-flying mammals. Some of the bird species that have evolved to occupy the niches of absent mammals were described in Chapter 6.

One way the time-in-isolation explanation for endemism can be compared with an unusual-environment explanation is through estimates of ages of endemic species. The time-in-isolation explanation depends on long persistence times for a population to differentiate to the level of a species, and therefore predicts that island endemics should be old, albeit with some variation. In the unusual-environment explanation endemics could be young or old. Thus, very young endemics seem unlikely to have originated solely as a result of geographical isolation without ecological input (Chapter 2). The old age of many endemics has been confirmed in the West Indies (Ricklefs and Bermingham 2002), even though some, such as the Lesser Antillean Bullfinch (Figure 8.6), appear to be young. The few studies to date in northern Melanesia are equivocal on this point, but some allospecies seem to be young (Filardi and Smith 2005).

A second prediction of the island-biogeography model concerns the age of endemics on the most isolated islands. Low colonization rates on isolated islands mean they have fewer species in general, so the effect of isolation on decreased extinction rate has to be very strong to explain the increase in the *number* of endemic species with isolation (Figure 8.4). Thus, the island biogeography model predicts that endemic species on more distant islands should be particularly old, but this has yet to be tested.

The idea that endemism is driven by unusual environments on isolated islands could be examined more directly by comparing the ecology of related species across islands of differing degrees of isolation. However, both geographical and ecological considerations lead to the prediction that isolated islands should contain more endemics, the former because populations on isolated islands have longer persistence times, and the latter because such populations experience unusual selection pressures. It seems likely that the two factors operate together, making conditions on isolated islands particularly favorable for differentiation to proceed up to the level of recognized species.

CONCLUSIONS

The studies reviewed in this chapter have been confined to islands, but they are likely relevant to continents too. Although extinction on continents should be less common than extinction on islands (because population sizes are generally larger), results outlined in this chapter show how a sufficient time in isolation could lead to speciation. On continents, the critical limiting factor may be the length of time that populations remain separated from each other. On islands, geographical isolation is easier, but population persistence is more problematic.

The most frequently discussed mechanism of differentiation due to isolation per se is the spread of new favored mutations and the confinement of these mutations to a portion of the species range (introduced in Chapter 2). As described in the next chapter, sexual selection is a powerful force that could generate divergence between populations, if different favored mutations arise in different populations. Many allospecies and endemic island species differ from their close relatives most conspicuously in songs and in plumage colors, traits typically thought to be subject to sexual and other forms of social selection. Thus, factors affecting the evolution of socially selected traits need to be understood in an effort to ask how divergence may be driven as a result of geographical isolation alone. That is the goal of the following chapters, but I should note in advance that it is difficult to draw firm conclusions. Although we can show ecology is important in some cases, and likely plays roles in both reducing gene flow and driving population divergence, in most examples the relative importance of time in isolation and divergent selection pressures remain uncertain. I revisit this in more detail at the end of Chapter 12, after considering the evolution of socially selected traits.

SUMMARY

Across northern Melanesia, the Indian Ocean, and the Pacific Ocean, the number of endemic species on an island is positively associated both with isolation and with area. The effect of distance on endemism is strong because, across the oceans, the number of endemic species actually increases with distance, even as the total number of species decreases. One explanation for the positive correlations of area and isolation with endemism is that the probability of extinction is lower on large and more distant islands, resulting in longer persistence times of populations and a greater chance that they will differentiate to the level of full species. On large islands, longer persistence times are expected because population sizes are larger. On more distant islands, persistence times may be longer because (1) the presence of relatively few species on such islands results in each having a relatively large population size, and (2) distant islands receive relatively few biological invaders that could cause extinction of the resident populations. These findings lead to an island-biogeography theory of speciation; small or less isolated islands have higher rates of species extinction than large or more isolated islands, resulting in fewer endemics. A second explanation for the positive correlation of island isolation with endemism is almost certainly that isolated islands have unusual biota resulting in unusual selection pressures. These alternative island-biogeography and ecological explanations for endemism remain to be critically evaluated.

CHAPTER NINE

Social Selection

Like humans, most bird species are highly social. Communication between conspecifics generally begins when a chick hatches and solicits food and/or shelter from its parents, and continues throughout life. Individuals display or call to each other when they meet. They also engage in escalated interactions—for example, in disputes over food or in courtship between the sexes. Many displays, vocalizations, and color patterns are employed during such social interactions and these traits are assumed to have evolved because of benefits to the signaler. For example, an individual with a conspicuous plumage patch deters rivals more efficiently than one without such a patch, or a male with a more elaborate courtship display is more attractive to females. I define social selection as selection on traits such as songs and plumage patterns that arises out of competition for resources (including food, shelter, nest sites, and mates; West-Eberhard 1983). The term *sexual selection* is reserved for that subset of social selection in which the resources are mates.

In this and the following three chapters, I ask how social selection drives the evolution of plumage patterns, vocalizations, and displays, thereby causing populations to diverge in those traits. While these four chapters may at times read more like a contribution to the way in which socially selected traits evolve, divergence in socially selected traits is an essential component of most—perhaps all—speciation events. First, the classification of many allospecies as species is based on differences in vocalizations and plumage patterns (Chapter 1). Second, vocalizations and plumage patterns are the major traits used by conspecifics to recognize each other when they occur in sympatry (Chapter 13). Their use implies that divergence in socially selected traits is critical to the establishment of pre-mating reproductive isolation. One reason that socially selected traits are so important in species recognition is that social interactions are primarily among conspecifics, which obviously requires that individuals recognize their own species.

In the past few decades, the mechanisms of sexual selection have been much studied. Darwin (1871, p. 234) introduced the idea of females choosing among males: "females prefer or are unconsciously excited by the more beautiful males." In this case, traits that increase male attractiveness increase in frequency across generations, because the more attractive males gain more mates, or more fertile mates, than the less attractive ones. Darwin (1871) also noted that males may sometimes choose females (p. 117), and that males compete directly among themselves, for example, for resources that form the focus of female choice (p. 152). However, Huxley (1938) argued that other social interactions are as important as sexual selection in affecting the evolution of traits used in displays. First, in species that form pair bonds, displays between the male and female are common after pair formation (Wachtmeister 2001). Second, social signals are widely used in threat situations in both the breeding season and the nonbreeding season. For example, in the European Robin, both females and males have a bright red breast that is displayed prominently in territorial disputes through much of the year (Lack 1943). Pair formation and mating are relatively quiet affairs with limited display, and, in this species, "sex is greatly overshadowed by fighting" (Lack 1943, p. 63). Sexual selection is demonstrably important in many groups (Andersson 1994), but the emphasis placed upon this mode of selection has come at the neglect of other forms of social selection.

In his discussion of sexual selection by female choice, Darwin (1871, p. 230) suggested that novelty per se sometimes stimulates females. This is an important inference because it is one way new male traits could appear and replace old ones in a never-ending cycle, resulting in continuous divergence between populations and species. The bellbirds studied by Snow (1976; Figure 9.1) provide a concrete example. Snow (1976) argued that the different male traits (e.g., three bare wattles, a beard of stringy wattles, a bare throat of colored skin) were the result of sexual selection in arbitrary directions.

Under this idea, and more generally with respect to any form of social selection, a new signal that arises, say, by mutation, may become established in a population because it stimulates receivers more than the old signal. Populations will diverge if, by chance, different stimulatory signals arise in different populations. The mechanism occurs in the absence of any ecological differences between populations. It therefore may be applicable to many allospecies, such as the bellbirds (Figure 9.1), which appear to be ecologically and morphologically quite similar to each other. It is a null model, because divergence by this process should go on whether or not additional ecological influences promote divergence of socially selected traits. I introduced this model in Chapter 2 (see Figure 2.3, p. 17), where I noted that sexual selection is a powerful nonecological driver of speciation.

In this chapter and those that follow, I consider the evolution of socially selected traits, starting with the assumption that ecological influences are absent, and then adding increasingly complex ecological contributions. First, using mostly plumage patterns and displays as examples, I evaluate the suggestion that a signal can become established in a population simply because it has desirable effects on the receivers of the signal. As discussed in the second part of the chapter, a major obstacle to the process is that individuals that respond to new signals may experience higher costs than those that continue to respond to the old one. It is plausible that these costs can be overcome if the stimulatory value of the signal is strong.

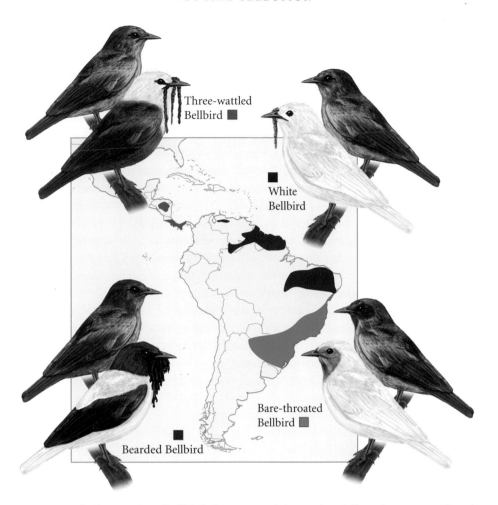

Figure 9.1 The four species of bellbirds (genus *Procnias*). Females of all species are greenish and similar (redrawn from Snow 1982; del Hoyo et al. 2004). The Bearded Bellbird has a disjunct range. Range maps are from del Hoyo et al. (2004).

In the next chapter, I address ways songs might diverge even if populations occupy identical environments. Then I add in ecology. In Chapter 11, I ask how populations may diverge in socially selected traits as a result of experiencing similar, but changing, environments. Finally, in Chapter 12, I return to the question of ecological speciation: How might populations occupying different environments experience divergent selection pressures on socially selected traits? I will conclude that different environments drive divergence in socially selected traits in many ways, and it is useful to bear in mind an example when reading the present chapter. One such example might be that dense forests favor vocal cues over visual ones.

I first describe some specific case studies of social selection, because I will refer to them repeatedly in this and subsequent chapters. I consider three forms of social selection: sexual selection by female choice, mate stimulation, and threat (including competition among males for territories in the breeding season).

Sexual selection by female choice

Various male signals correlate with success in attracting mates (Andersson 1994; Gontard-Danek and Møller 1999), and many experimental manipulations of signaling traits have demonstrated an effect on female choice (Andersson 1994; Hill 1994; Pryke and Andersson 2002; Senar et al. 2005). Therefore, these signals are generally thought to have spread as a result of sexual selection.

The most fantastic male displays are associated with lek mating systems. About 150 bird species are known to form leks (Bleiweiss 1997). In this system, a female chooses from among a group of advertising males and retreats to raise the brood herself. In one lekking species, the Greater Sage Grouse, a typical lek in southern California has a few tens of adult males that defend small territories (Wiley 1973; Gibson et al. 1991). Males have elaborate plumage and structural adornments (Figure 9.2) and a "strut" display that includes mechanical noises produced by brushing the wings against the esophageal pouch, erection of the tail and neck plumage, inflation of the air sacs, and a vocalization consisting of two coos, a pop, a whistle, and another pop (Wiley 1973; Dantzker et al. 1999). A female comes to the lek to observe males and then mates with one male, probably just once. In one study, the number of females a male obtained correlated with the frequency with which he displayed and the length of the interval between the two pops in his vocalization (Gibson et al. 1991).

Polygynous species, such as lekking species, tend to have more developed secondary sexual characters (i.e., sexually selected traits developed in males) than monogamous ones (Darwin 1871, p. 266; Dunn et al. 2001; see Chapter 12), but sexual selection operates in socially monogamous species as well, which form the majority. Some males attract females of higher quality than other males do, some males go unmated, and females may copulate with males who are not their social partners (Kirkpatrick et al. 1990; Møller 1992). Among species that form socially monogamous pair bonds, with one female and one male occupying a territory, the frequency of young resulting from extrapair copulations averages about 11%, but it can be as high as 55% (Griffith et al. 2002). Females may engage in extrapair copulations with males that are more attractive than their social mate [e.g., have more complex songs (Hasselquist et al. 1996, Kempanaers et al. 1997), have notes in the song that are difficult to sing (Forstmeier et al. 2002), have larger plumage patches (Sheldon et al. 1997), or have brighter plumage (e.g., Greene et al. 2000); reviewed by Griffith et al. 2002].

Sexual selection in the socially monogamous Barn Swallow has been studied in Europe and in North America (Møller 1988, 1992; Saino et al. 1997; Safran et

Figure 9.2 Male Greater Sage Grouse displaying. Photograph by Neil Losin.

al. 2005). In European populations, males and females look similar and both help to raise the brood, but the male's tail is about 20% longer than the female's. Longer-tailed males gain reproductive advantages over those with shorter tails by pairing with females in good condition and by siring additional young via extrapair copulations (Møller 1992; Saino et al. 1997). Experiments show that increased tail length affects both flying performance (Evans 1998) and food-gathering ability (Møller 1988). Costs of increased tail length could provide an important check on the exaggeration of this feature, because it affects the ability of the male to contribute to the rearing of his brood. In this way, a male's tail affects the fitness of both himself and his partner. This differs from the lek-mating system described above, in which the male ornaments have no direct influence on a female's fitness.

Displays after pair formation

Most bird species form pair bonds and in these species the male and female continue to display to one another after mate choice has been completed (Huxley 1914, 1923; Wachtmeister 2001). This is a relatively neglected aspect of social selection, but it is an essential component of courtship, promoting reproductive investment into offspring. In Table 9.1, I summarize studies that

Table 9.1: Experiments on mate stimulation

1. Stimulation of females by male traits

Species	Male trait or manipulation	Female response	Reference
Red Junglefowl	Male comb size	Larger clutch size with larger comb	Parker 2003
Indian Peafowl	Train length of males	More eggs laid for males of long train length	Petrie and Williams 1993
Mallard	Paired with "attractive" vs. "unattractive" males[1]	Egg volume larger when paired to attractive males	Cunningham and Russell 2000
Blue-footed Booby	Experimentally reduced foot color	Reduced courtship, fewer copulations	Torres and Volando 2003
Budgerigar	Call type and quantity	Increased follicle size	Brockway 1965
Common Pigeon	Testosterone added to males	Three-day delay of egg laying cf. controls	Murton et al. 1969
Ring Dove	Testosterone added to castrated males	Increased intensity of wing-flap display	Erickson and Lehrmann 1964
Northern Mockingbird	Male song	Male starts to renest[2]	Logan et al. 1990
Barn Swallow	Experimental manipulation of male tail length	More broods, young, and provisioning with males of longer tails	de Lope and Møller 1993
Barn Swallow	Male tail manipulation	More carotenoids in eggs of tail-shortened males	Saino et al. 2002
Zebra Finch	Different color bands on males	Increased share of parental care when paired with attractive male	Burley 1988

Species	Stimulus	Response	Reference
Zebra Finch	Different color bands on males	Higher egg mass when paired with attractive male	Gilbert et al. 2006
Zebra Finch	Song repertoire size in tape playbacks	More pecks at button producing the high-repertoire song	Collins 1999
Zebra Finch	Presentation of model and song	Estrogen increases, egg laying	Tchernichovski et al. 1998
Bengalese Finch[3]	Song complexity	More strings carried, higher estradiol	Okanoya 2004
European Serin	Song playback in the wild	More nest-building	Mota and Depraz 2004
Island Canary	Male song	Number of eggs laid much greater with song	Bentley et al. 2000
Island Canary	Song repertoire size	Nest building, egg laying higher for larger repertoire of songs[4]	Kroodsma 1976
Island Canary	Specific phrase in male song	More copulation solicitation displays in estradiol-implanted females	Vallet et al. 1998
Island Canary	Rapid repetition of high-bandwidth syllable	More copulation-solicitation displays in photoperiod-primed females	Drăgănoiu et al. 2002
Island Canary	"Unattractive" vs. "attractive" song	Increased testosterone in the egg	Gil et al. 2004; Tanvez et al. 2004
Dark-eyed Junco	Testosterone in male	Increased nest attendance, reduced egg size	Ketterson et al. 2001
Eight passerine species	Song repertoire size in tape playbacks	More copulation-solicitation displays in estradiol-implanted females	Searcy 1992

(continued)

Table 9.1: Experiments on mate stimulation (*continued*)

2. Stimulation of males by female traits[5]

Ring Dove	Females present or not present	Testosterone increase	Feder et al. 1977
Ring Dove	Female calls; intact vs. ovariec-tomized; courting vs. incubating	Testosterone increase	O'Connell et al. 1981a, 1981b
Zebra Finch	Females present or another male present	Courtship and testosterone higher when male present	Pröve 1978
Zebra Finch	Live female vs. stuffed female vs. dummy female	Increased song quality with more realistic stimulus	Bischof et al. 1981
Zebra Finch	Different color bands on females	More paternal care when mated to attractive (black-banded) female	Burley 1988
White-crowned Sparrow	Female display induced by estradiol cf. controls	Testosterone increase, testicular size increase	Moore 1983
White-crowned Sparrow	Females present or not present	Song nuclei in brain increased in size when females present	Tramontin et al. 1999
Pied Flycatcher	Free-living females implanted with estradiol	More mate guarding and territoriality	Silverin 1991

[1] Attractiveness determined by number of displays females gave towards males in a group situation.

[2] Playback of taped song in territory causes the male to start building a new nest; this may be mediated by changes in female behavior.

[3] The Bengalese Finch is the domesticated variety of the White-rumped Munia.

[4] This may be a result of the absence of some specific "sexy" phrases from the (artificially) shortened song (Kroodsma 1976), but Leboucher et al. (1998) showed that different phrases stimulate nest building to the same extent. O'Connell et al. (1981a) refer to several earlier papers where the presence of females stimulated male testicular development and circulating hormones. See also Murton and Westwood (1977, their Chapter 5) for a review.

[5] O'Connell et al. (1981a) refer to several earlier papers where the presence of females stimulated male testicular development and circulating hormones. See also Murton and Westwood (1977, their Chapter 5) for a review.

have experimentally demonstrated increased reproductive investment by an individual in response to social stimulation. For example, in the Ring Dove (the domesticated form of the African Collared-Dove), females held in isolation from males do not usually lay eggs (Lehrman 1959; Lehrman et al. 1961; Murton and Westwood 1977). They can be stimulated to lay eggs if they are able to observe males through glass plates. Castrated males, which do not display, are less effective stimuli than displaying males (Figure 9.3).

Passage through the initial stages of the breeding cycle is accompanied by changing frequencies of different displays by both the male and the female (Wachtmeister 2001). In the Common Pigeon, a male first displays to an unfamiliar female with *head-bowing* [the male's head is lowered as he calls; this is an aggressive display used towards all intruders, and it exhibits the metallic green colors of the crown and back of the neck (Fabricius and Jansson 1963; Murton and Westwood 1977)]. This display is gradually replaced by *nest-demonstration* (the male sits down, nods his head, vibrates his wing, and utters a different call).

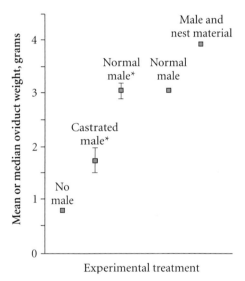

Figure 9.3 Mean or median oviduct weight of female Ring Doves (based on $N = 20$ females) after 6 or 7 days in various treatments. Points with standard errors are from Erickson and Lehrman (1964, after 7 days) and are medians. Points without standard errors are from Lehrman et al. (1961, after 6 days) and are means [Lehrman et al. (1961) show differences over a complete 7-day period, but here I show the 6-day points because at 7 days, differences between the nest-material group and the no nest-material group have disappeared (both are similar to the no nest-material group on day 6)]. The points with an asterisk indicate the male was held in a separate cage and viewed through a glass partition. From left: (A) controls: females held alone; (B) females can view castrated males through a glass partition, castrated males do not display; (C) females can view normal males through a glass partition, these males display; (D) females and normal males held together in a cage; (E) females, males, and nest material in a cage. Standard errors are approximate, being estimated from the published range.

Figure 9.4 Some displays of the Great Crested Grebe. Either sex takes the initiative, and the displays are interchangeable between the sexes. The *cat display* and *head shaking* are part of the so-called discovery ceremony, but *head shaking* is often performed independently, and it is the only display seen later in the breeding cycle (i.e., after egg laying). The famous *weed ceremony* happens late in the prelaying period. The Great Crested Grebe has other displays that are not illustrated. Illustrations after line drawings in Cramp and Simmons (1977).

The change is accompanied by a change in the female from a submissive posture to a dominant posture.

Mutual displays between the sexes reach an extreme in some species in which the male provides considerable parental care (Huxley 1914, 1923), such as the Great Crested Grebe (Figure 9.4). The male and female of a pair display to each other through initial pair formation up to egg laying, typically a period of several weeks. The displays themselves are very similar in the two sexes, and both sexes acquire special plumages in the breeding season (Figure 9.4). The plumage is sufficiently ornate that the population was persecuted in the nineteenth century as a result of the commercial demand for it (only 42 pairs remained in England in 1860, although the population now numbers in the thousands; Cramp and Simmons 1977). Huxley (1923) likens the mutual display of the Great Crested Grebe to that of humans falling in love.

Threat displays

Threat displays are widely used during contests for resources, such as food in the winter or territories in the breeding season. Threat displays include move-

Figure 9.5 Great Tit threat displays (illustrated after line drawings in Cramp and Perrins 1993).

ments of body parts and feathers, and they may be accompanied by a variety of call notes (Andrew 1957; Tinbergen 1951; Balph 1977). Different display elements employed during contests between individuals may be conveying information by convention. According to Hurd and Enquist (2001): "watching birds fight over seeds gives the strong impression of a simple language at work." Some contests may be of relatively short duration, but others, particularly those involved in skirmishes at territorial boundaries, can be greatly prolonged, with many exchanges of display elements between signalers (Hinde 1952; Hurd and Enquist 2001).

Different threat displays are used in different contexts. For the Great Tit, Wilson (1992) listed 12 discrete elements in displays of aggression towards other individuals. Some are illustrated in Figure 9.5. Others include *turning body* (where the head is swayed from side to side) and *vertical flight* (a peculiar display flight with the body held in a nearly vertical position) (Kluijver 1951, p. 16). Many of the displays take place between rivals at very close range. Wilson (1992) maintained captive birds in aviaries for 24 days. He found that *wings out* and *tail fanned* were used together, often over contests for food, and that, about 20% of the time, they were followed by attacks. The frequency of these displays approached zero after ten days (Figure 9.6). This is probably because birds learned to recognize each other individually and respected a dominance hierarchy. By contrast, *head up*, associated with *horizontal body*, and *turning body*, occurred predominantly between males in nonfeeder situations and continued to be used throughout the study (Figure 9.6). Hinde (1952) showed that, in the

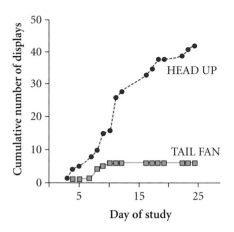

Figure 9.6 Wilson (1992) maintained two pairs of Great Tits in each of three aviaries. He presented food in a small tray and recorded interactions at the tray for a total of two hours a day using video. He found that some displays (e.g., *head up*) continued to be used throughout the captive period, whereas others (e.g., *tail fan*) were only used between unfamiliar birds, and their frequency rapidly declined.

field, these three displays, as well as *vertical flight,* are often used in skirmishes between males at the boundaries of established territories.

In summary, different displays are used in different threat situations and at different stages of courtship during pair bonding. Displays also change through time during the process of mate choice. In Satin Bowerbirds, the kinds of displays favored by females depends on the female's age as well (Coleman et al. 2004). The importance of all this variation to population divergence is that different kinds of displays may come to predominate in different populations. For example, populations and species of some sea birds that breed in high-density colonies have a higher repertoire of threat behaviors than those breeding in low-density colonies (Birkhead 1978; Waas 1990). Ecological differences that might drive such differences are pursued further in Chapter 12.

POPULATION DIVERGENCE

All social situations involve communication between a signaler and a receiver. Novel signals should be favored when they promote a favorable response in the receiver (Darwin 1871; Huxley 1923). In this section, I consider theory, experiments, and comparative studies that show how stimulation of the receiver of the signal could play a major role in the diversification of displays, and then I consider how costs to the receiver may limit the process.

Models of signal perception

In recent years, the ideas of Darwin and Huxley on the stimulatory effects of novelty have been investigated using computer simulations of signal processing. In particular, Enquist and Arak (1993) and Arak and Enquist (1995) developed models of the ways in which perceptual biases influence signal evolution using artificial (and very simple) neural networks (reviewed by Enquist and Arak 1998). The basic simulation is described in Figure 9.7. Three cell layers consist of an input layer (i.e., the retina), a hidden layer, and an output layer that often consists of a single cell. An image is projected onto the input layer in various orientations. Cells in the input layer that fall in the shadow of the image transmit a numerical value to the next layer. The network is trained by mutating the transmission properties from cell to cell, followed by selecting those mutant networks that respond

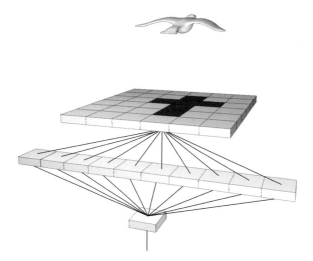

Figure 9.7 Model of neural networks (Enquist and Arak 1993). An image is projected on to an input layer of 6 × 6 receptor cells, which either fire (value of 1) or do not fire (value of 0). Each of the receptor cells is connected to each of ten cells in a hidden layer (for clarity, the connections from only one cell are shown). Input into each of the hidden cells is a weighted sum of the values coming from the 36 cells in the receptor layer, where the weights are values assigned (initially randomly) to each connection, times the value output from the receptor cell. Transmission out of each hidden cell is a sigmoid function of what comes in, and a similar procedure follows through the output cell. The network is trained to respond to an image such as the one shown, by projecting the image on to the network in numerous rotations, and scoring the output from this image when compared with the output from other images. The connection weights are randomly mutated, and networks that give high favorable output values are retained for further training. The network is trained to distinguish between a "correct" image and one or more "incorrect" images, with training completed when output values are always >0.5 for the correct image and <0.5 for incorrect images.

more strongly to the chosen image and less strongly to other images (which may be specific alternatives or random images). The "mutations" may be considered to be genetic, across-generational changes, or changes that result in learning during the lifetime of an individual. The selection plus mutation process is repeated until successful acceptance of the chosen image and rejection of the nonpreferred images reaches some predetermined level. Some relevant results from these simulations are described in the following paragraphs.

NOVELTY AND EXAGGERATION   Because it is easy to think of images that have never been encountered, it seems inevitable that some of these will produce greater responses than anything the sensory system has previously experienced (Enquist and Arak 1993). One particular example of this is when a network is trained to discriminate between two signals and to respond positively to one of them. Other signals that vary further beyond the positive response sometimes elicit a greater response (e.g., larger, louder, brighter signals lead to increased network responses; see Figure 9.8).

POPULATION DIVERGENCE   Networks evolve in response to the same discrimination task in many different ways, because training depends on random mutation (Arak and Enquist 1993). This is shown in Figure 9.8; one of the trained neural networks (*upper panel*) gives a strong response to both tested images, but the other network (*lower panel*) ignores the second test image. For example, as a result of random mutation during training, the network at the end of the second simulation may particularly emphasize the presence of a white center as a way to correctly identify the favored signal. The white center is absent from the second tested image, so this network does not respond. Results show the great potential for signal divergence between populations. First, different kinds of signals could arise in different populations, and both be favored by the same perceptual system (e.g., the two test signals in the *upper panel* of Figure 9.8). Second, different mutations could arise in the perceptual system, causing a signal to be favored in one population but not in another.

THREAT DISPLAYS   A main requirement of signals used in threat is that they convey an unambiguous message (Hurd and Enquist 2001). This is suggested by the observation that threat displays differ discretely and do not lie along a continuous scale (Hurd and Enquist 2001). Because of the need to send a clear message, Huxley (1938, p. 420) reasoned that a threat signal should have few distinctive components of form or color, arranged in a striking way. Hurd et al. (1995) showed this using neural nets. They modeled a network with two output cells, rather than the one in the basic design outlined in Figure 9.7. They selected the network to respond correctly to two different signals (one output

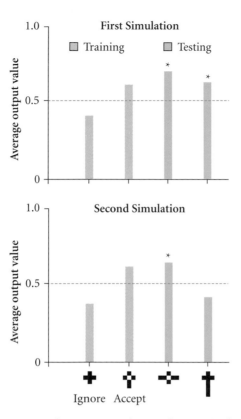

Figure 9.8 In two separate simulations, a neural network was trained to discriminate between two images, indicated as "ignore" and "accept." The response to each image is shown for the resulting networks *(orange bars)*. Each network was then tested against other images *(green bars)*. In three cases, indicated with an asterisk (*), the response to the test is stronger than it is to the image that the network was trained to, and in one of these, the response is 15% greater. In the other case *(lower right)*, one network ignores one image (i.e., shows a response less than 0.5). From Arak and Enquist (1993).

cell for each signal) and required the network to avoid pure black and pure white, but shades of gray were allowed. Starting from initial randomly produced images, the two signals and the neural net were all allowed to evolve, according to favorable effects the signals produce on the network. Figure 9.9 shows the results, separately for two different training runs. In each run, signals evolved that were largely the inverse of each other, and each of the signals was boldly patterned, sometimes with little gray. Hurd et al. (1995) note that bold patterns that are the inverse of each other are farther apart in sensory space and hence may be more easily distinguished than uniform patterns that are different color shades, for example.

Simulation 1

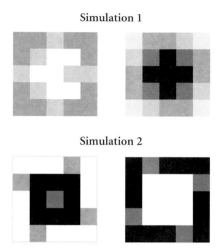

Simulation 2

Figure 9.9 *Upper row:* An example of signals that evolve when a network is trained to respond correctly to two signals (i.e., one of two output cells gives a large response to one signal and the other output cell gives a large response to another signal) and to ignore both a pure black and a pure white background (from Hurd et al. 1995). *Lower row:* Because of random mutation, different results arise with different training runs.

The neural-net models have been criticized for their simplicity (Dawkins and Guilford 1995). However, even the simple models seem to have large explanatory power, and more complex models of perceptual systems are only likely to add more diversity (Enquist and Arak 1998). They show that new signals will occasionally be favored and that different social situations will favor different kinds of signals, both of which could lead to population divergence.

Experimental evidence

Novel and, in particular, extreme traits can elicit extreme reactions in receivers. Most birds use two main sensory channels to communicate: sound and vision. Psychological studies show that a louder sound or a brighter color will often increase the response of the receiver, but increased sound or increased color will magnify the response more than would be expected if the effects of the increases were additive (Rowe 1999). The use of both vocalization and visual displays together makes a signal more easily detected, more easily discriminated from other signals, more easily identified from incomplete information, and more easily learned and remembered (Rowe 1999); this may sometimes simply be a matter of paying more attention to the signals (Rowe and Skelhorn 2004). The implication is that, from the point of view of the signaler, the addition of vocalizations and/or color patterns to its displays should have favorable impacts on receivers.

Within a single modality, experimental studies have demonstrated that some traits make others more easily perceived (for example, stripes on a tail may make tail length more easily judged; Hasson 1991). In addition, bigger, brighter, more conspicuous displays do often produce greater receiver responses. When Guhl and Ortman (1953) stuck red feathers on the tail of one individual in a flock of pure white chickens, for several days all the other chickens ran away from it. While interesting, the phenomenon has been more closely examined in the context of mate choice. Females are often stimulated by traits that lie beyond the natural range (Ryan and Keddy-Hector 1992; Arak and Enquist 1993). A striking example is that of Burley's pioneering experiments on Zebra Finches.

Zebra Finches are small Australian finches that are socially monogamous, with long-term pair bonds (Zann 1996). They have orange-red beaks. The males have redder beaks than the females and also have chestnut flank feathers and golden-brown cheek patches. Burley (1986a) demonstrated in no-choice tests that females preferred to perch within view of males carrying red color bands rather than males without red bands, and males wearing blue bands were avoided (Figure 9.10, *left*). Subsequently, Burley (1986b, 1988) demonstrated that males wearing red bands gained higher reproductive success. In a large

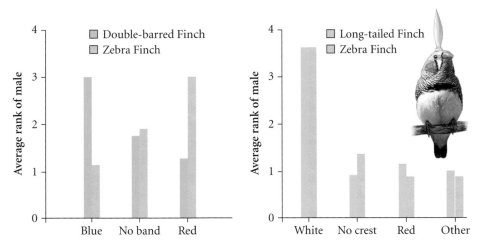

Figure 9.10 Female preferences for unusual traits in Estrildid finches. Female preferences are measured as the fraction of time the female spent perched in view of the male during a 2-hour observation period (only one male type is presented at a time). The data are the average preference ranking of males of different phenotypes. *Left:* comparison of female Double-barred Finches and Zebra Finches viewing color bands on conspecific males (*N* = 8 Double-barred Finch females viewing Double-barred Finch males, *N* = 15 Zebra Finch females viewing Zebra Finch males; Burley 1986b). *Right:* comparison of Long-tailed Finches and Zebra Finches viewing artificial crests on conspecific males (*N* = 13 Long-tailed Finch females and *N* = 15 Zebra Finch females; Burley and Symanski 1998). The illustration is of the white feather attached to a male Zebra Finch (redrawn from a photograph in Burley and Symanksi 1998).

aviary, nests associated with red-banded males produced about twice the number of surviving offspring than nests associated with males wearing green or orange bands. In addition, half of the red-banded males became simultaneously bigamous, but none of the males with the other colors did. Furthermore, red-banded males gained in extrapair fertilizations (Burley et al. 1996). Most remarkably of all, the males carrying red bands survived for longer than the other males during the experiment (Burley 1986b). One reason for this may be that the red-banded males contributed about 25% less effort in parental care than the males banded with other colors. The increased number of offspring coming from the nests associated with red-banded males seems to have resulted from substantially elevated investment by the females (Burley 1988). The results suggest that, if a red band "mutation" were to appear, it would be favored by sexual selection.

Burley extended these studies to a relative of the Zebra Finch, the Double-barred Finch, which has a very different plumage pattern. In this species, the beak is greyish-blue and no red is present in the plumage. Burley (1986a) showed that Double-barred Finch females had preferences for males wearing blue color bands, and they avoided the red-banded males (Figure 9.10, *left*). These experiments clearly demonstrate a preference for more extreme male traits than those present in the population.

Finally, Burley and Symanski (1998) attached conspicuous artificial crests to male Zebra Finches and to male Long-tailed Finches (Figure 9.10, *right)*. The nearest relative to these two species that has a crest may have diverged from them more than 20 million years ago (Burley and Symanski 1998). (Aviculturalists have obtained a crested mutation in the Zebra Finch, but its crest is of a form quite different from naturally occurring crests, and is associated with a skull defect; Chapter 11). Burley and Symanski (1998) found strong female preferences for males with artificially attached white crests, and that the preferences were very similar in each of the two species (Figure 9.10). They argued that the preferences have their origin in "non-adaptive neurophysiological processes." An alternative explanation for the preference of white crests over other colors is that white feathers are used as a nest-lining material by both species, apparently for reasons of camouflage (Burley and Symanski 1998). Thus, females may be particularly tuned to white.

Preferences for extreme traits may often originate as a result of training to discriminate between different images, as in the neural-net model described in Figure 9.8. Ten Cate et al. (2006) demonstrated this experimentally (Figure 9.11). They made use of the fact that in captivity male Zebra Finches prefer to sing to females that resemble their mother and avoid females that resemble their father. They studied an all-white strain of Zebra Finch, in which males and females are indistinguishable based on plumage. For a number of broods, they painted the beak of the father to differ from that of the mother, and studied the

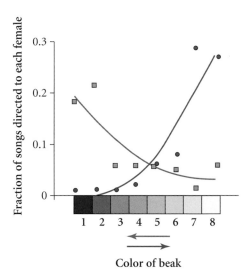

Figure 9.11 An albino male Zebra Finch was placed in a cage surrounded by eight albino females in individual cages with differently painted beaks. Males had been raised by albino mothers with beaks of color 3 and fathers with beaks of color 6 *(black lines* and *squares)* or mothers with beaks of color 6 and fathers with beaks of color 3 *(blue lines* and *circles)*. In both cases, males showed a preference for females that were more extreme than their own mothers, in the direction away from their fathers. The colors along the *x* axis differ from those of the beaks, but they give an approximate impression. A total of 34 males were used in the experiments. The line is a smoothed nonparametric regression; it is possible that in both directions the very extreme beak colors are less preferred than those that are less extreme (ten Cate et al. 2006). Redrawn from ten Cate et al. (2006).

reactions of the young, when they were themselves adult. As expected, when adult, males preferred to sing to females with the beak color of their mother and avoid females with the beak color of their father. However, males showed preferences for females that were even more extreme than those of their mothers, in the direction away from their fathers (Figure 9.11).

Comparative studies of the associations of plumage patterns with displays

As already described for the Greater Sage Grouse, Ring Dove, and Great Crested Grebe, body parts used in displays are often accentuated by a color patch or an ornament, such as a crest or an elongated tail (Darwin 1871, pp. 94–95; Armstrong 1947, his Chapter 21; Tinbergen 1951, 1959; Hinde 1952; Orians and Christman 1968; Goodwin 1983, pp. 32–35), supporting the thesis that the color patches are present because they increase the effectiveness of the display. Tinbergen (1951, pp. 25–27) gives many examples in birds, stating that "one is able to predict the display movement in a newly encountered species after merely viewing its morphological equipment." In the Great Tit (Figure 9.5), *wings out* and *tail fanned* display the white wing and tail patches. During *turning body*, "birds display both their white cheeks, which contrast with the black cap, and also the black stripe on the yellow breast" (Kluijver 1951, pp. 15–16). In discussing *head up* and *vertical flight*, Hinde (1952, p. 79) states the "colour pattern has been evolved to make the display more conspicuous" (the breast stripe is particularly obvious).

Fitzpatrick (1998a) quantitatively evaluated the association between the presence of tail displays and tail patterns. Among western Palearctic birds, she found that 88% of species with tail markings ($N = 285$) have been recorded moving their tail in display (mostly agonistic, but some courtship), whereas only 63% ($N = 235$) of species without tail markings did so—a highly significant difference. Some additional associations of color patches with movement of different body parts in display are shown in Table 9.2, and a more detailed study of a single genus, the Old World leaf-warblers, is described in Table 9.3 (Marchetti 1993). In this study, Marchetti measured threat displays by videotaping males that approached a speaker broadcasting conspecific song. The use of a display and the presence of a color patch are highly correlated, albeit with some exceptions. For example, one species that has a rump patch and white in the tail does not seem to display these traits during aggressive interactions, although the tail patches are conspicuous in flight in the gloomy forests where it lives, and they are likely used in other aspects of intraspecific communication (Marchetti 1993).

Both Fitzpatrick's (1998a) results and the data in Tables 9.2 and 9.3 are significant largely because, when a patch is present, it is displayed. By contrast, body parts are often moved, even when no obvious patch is associated with the movement. Although not formally tested, these results are consistent with the idea that displays are largely ancestral: color patches and feather ornaments have become added to displays if "further stimulation is advantageous" (Huxley 1923, p.277).

If color patches are added to increase the effectiveness of a display, the kinds of color patches that are favored should be highly contingent on the displays that are already employed. In one of the best examples of contingency, Lorenz related the evolution of color patterns in ducks to inferred ancestral displays (reviewed in Lorenz 1971). If a male Mallard is courting a female, he will often reach back with the bill behind a raised wing and scratch the base of the feathers, thereby making a sound. The origins of this ritualized display, known as *sham preening*, appear to come from *displacement preening*, an appeasement display in the face of threat used in many species. In the Wood Duck, Mandarin Duck, Eurasian Wigeon, and Garganey, a very similar display occurs, but here the male points his beak to a distinctive area of plumage. For example, the Garganey touches a bright blue-grey

Table 9.2: Examples of threat displays of some passerine species

	Horizontal Body	Head-Up	Wing-Out	Tail-Fan	Crown-raising[3]	Turning-Body
Common Starling	P	N	P	N	N	N
Great Tit	P*	P*	P*	P*	P	P
Chaffinch	P*	P	P*	P*	P	P
American Goldfinch	P*	P	P*	N	N	P
Common Redpoll	P*	P*	P*	N	N	N
Cassin's Finch	P	N*	P	N	N	N
House Finch	P	P*	P	P	N	N
Red Crossbill	P	N	N	N	N	N
Yellowhammer	P*	P*	N	N*	N	N
Dark-eyed Junco	P*	P*	N	P*	N	P
Rose-breasted Grosbeak[1,2]	P*	P*	N*,2	P[1]	P	N
Yellow-headed Blackbird[4]	P*	P*	P*	P	N	P
Red-winged Blackbird[4]	P	P	P*	P	N	N

P = display present; N = display not observed

Most species have more displays than those listed here (e.g., the Common Starling has twelve listed by Ellis 1966).

Not all species were studied in both the breeding and nonbreeding season. Displays are mainly from Wilson (1992) and references listed therein.

*Indicates that a pattern is associated with the display (patterns were assigned using field guides). *Horizontal Body*: a contrast between patterns when viewed head on. *Head-Up*: a contrast across the breast. *Wing-out*: conspicuous wing-bars. *Tail-Fan*: patches in tail. If species are treated as independent, patch is significantly associated with display, based on the first four columns; χ^2 test, $P < 0.05$
[1] Species has white rump patch. Tail-spreading is rare (Dunham 1966).
[2] This species flicks its wings rather than holding them out. The flicks reveal underwing coverts that are sexually dimorphic: bright red in males, and yellow in females (Dunham 1966).
[3] This is the classic "fear" response, indicating retreat in many species that do not have special crests (see, e.g., Dunham 1966).
[4] From Orians and Christman (1968). Although the Yellow-headed Blackbird and Red-winged Blackbird displays are qualitatively similar, they differ quantitatively in intensity and frequency of use. The authors (p. 72) note that the differences are "related to the location of the bright patches to be displayed."

Table 9.3: Agonistic displays and color patterns of eight different species
of Old World warblers in Kashmir, India (Marchetti 1993)

	Wing	Crown	Rump	Tail
Tickell's Leaf-warbler	A	A	A	A
Tytler's Leaf-warbler	A*	A	A	A
Greenish Warbler	P*	A	A*	A
Large-billed Leaf-warbler	P*	A	A*	A
Hume's Leaf-warbler	P*	A	A	A
Western Crowned-warbler	P*	P*	A*	A
Lemon-rumped Warbler	P*	P*	P*	A
Buff-barred Warbler	P*	P*	P	P

In this group, all pale patches (on a pigmented background) are due to tracts of feathers
lacking melanin in their tips (Price and Pavelka 1996).

A = species does not have the patch; P = species has the patch. *Indicates the patch is
prominently moved during displays (Marchetti 1993). Species are ordered from least
patterned to most patterned. Treating patches and species as independent, presence of
the patch and use of the trait are significantly associated (χ^2 test, $P < 0.001$).

area on the outside of wing, whereas the Mandarin Duck touches the inside
bright-orange part of its single enlarged inner secondary feather (Figure 9.12). A
second display by the Mallard is *nod-swimming*, which is used by groups of males
when being viewed by females. The male circles around the courted female, head
flattened on the water, showing the conspicuous green sheen on the head, except
for a few erected feathers on the nape. Because they are erected, these feathers
appear to be black entirely framed in green. In other species, a similar display is
associated with permanently erected feathers (Wood Duck) or a permanent color
pattern on the head (Common Teal).

CONSTRAINTS ON THE INVASION OF NEW SIGNALS

If a new signal provides an advantage to the signaler, it should spread. Evidence
summarized above implies that this could be an important mode of diversification. But, so far, I have ignored any negative impacts that the signal may have on
the receiver. In fact, costs to receivers of paying attention to a new signal may be

Figure 9.12 Displacement preening in courting ducks (illustrations after drawings in Tinbergen 1951, p. 126). The Mallard (*top*) places its bill behind the wing, the Garganey (*center*) points to an outside area on the wing where the feathers are distinctively greyish-blue, and the Mandarin Duck (*below*) touches a bright-orange elaborated feather.

high; if so, the receivers are under selection to avoid the new signal and to continue to focus on the old one (Bradbury and Vehrencamp 2000). In this section, I consider how costs to the receiver constrain the invasion of new signals. Costs arise in at least two ways. First, the new signal leads to an increased response in the receiver and this response is itself costly. Second, a new signal may be less informative than an old one, so the receiver receives fewer benefits by switching its attention to the new signal.

Experimental increases in a male trait often increase female investment (Table 9.1), presumably above optimal levels. For example, male Zebra Finches wearing red color bands have a sexually selected advantage, but this seems to come at a cost to the female who pairs with such a male because she works harder to raise the brood (Burley 1986b, 1988). This is a conflict between the sexes; traits favored in males have detrimental effects in females [or, conversely, traits favored in females decrease fitness of males (Arnqvist and Rowe 2005)]. In the case of male display, an increase in the display decreases the fitness of the female receivers. A possible consequence is that females evolve to become increasingly resistant to the male's display. Then males add more exaggeration to overcome this resistance (Dawkins and Krebs 1978; Arak and Enquist 1995; Holland and Rice 1998). The process goes to equilibrium, where, for the males, the benefits of carrying exaggerated traits (attracting and stimulating females) are outweighed by various costs (e.g., being attacked by other males or by predators, or metabolic costs associated with an energetic display). At this point, the level of reproductive investment by the female will also evolve to equilibrium, where the male display stimulates the female to invest at a level that is optimal for her fitness (Holland and Rice 1998).

If this were the whole story, one might think that new signals could quite easily spread, for it would only require a trait that stimulates the receiver more than the previous one to set the process in motion again. However, this is not the whole story. Given that signals are expensive for the male to make and display, signalers in good health (or quality) display at higher rates and with more exaggerated ornaments than those in poor health (Figure 9.13). This means that females can gain benefits rather than costs from choosing to mate with males with the greatest exaggeration (Fisher 1915; Zahavi 1975; Andersson 1994, pp. 25–27; Johnstone 1995). The importance of this is that switching to a new trait may often mean switching to a trait that provides less information about its bearer.

Sexually selected traits do often seem to be indicators of quality (Johnstone 1995). These are termed *honest indicators* because they provide information to the receiver about the signaler. For example, increased expression of color in the plumage of several species is negatively correlated with parasite levels during feather growth (Harper 1999; McGraw and Hill 2000). Experimentally manipulated levels of bloodsucking mites affect tail growth in Barn Swallows (Møller 1990). The color of the wattles of male chickens is correlated with the quantity of ingested nematodes (Zuk et al. 1990). These are examples of parasites that affect secondary sexual traits, and perhaps the best-known theory of sexual selection posits that many male traits have evolved because they reveal which males are carrying genes for resistance to parasite infestation, so females mating with them pass these genes on to their offspring (Hamilton and Zuk 1982).

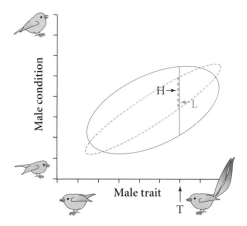

Figure 9.13 The honesty of a trait determines the value of a signal to a receiver. In this example, the development of a male's tail indicates his condition. The ellipses enclose 95% of the points when the correlation = 0.5 (low honesty, L, *blue line*) and when the correlation = 0.9 (high honesty, H, *grey line*). During mate choice, a female choosing a male with tail length T, on average, mates with a male in higher condition if the trait is more honest. Similarly, during a contest, an individual reacting to another with tail length T, on average, responds to a better fighter if the trait is more honest.

However, plenty of other potential benefits could accrue to females that mate with high-quality males. They include higher contributions to parental care, reduced chances of parasite transmission to the female during mating, better access to resources, and other genetic benefits besides parasite resistance (Johnstone 1995; Møller and Thornhill 1998). Møller and Thornhill (1998) found that females paired with attractive males receive less parental care in about half the species they reviewed. Sometimes, this may have been a result of male "manipulation," as in Burley's experiments, whereby the more attractive male stimulates the female to work harder for no obvious benefit. In other cases, reduced care is likely compensated through access to superior resources (Studd and Robertson 1985, 1988), as well as increased offspring fitness because of the genes they receive from their fathers (Møller and Thornhill 1998).

While most theory and experiments have been applied to sexual selection by female choice, socially selected traits used in other contexts can also be honest indicators, although not necessarily of quality. In contests over territories and food, signals and displays indicate fighting ability and/or motivation to fight (Rohwer 1982; Maynard-Smith and Harper 2003, pp. 92–95). For example, in the Great Tit (Figure 9.6), dominant males have larger breast stripes in both experimentally established flocks (Järvi et al. 1987) and at feeders (Maynard-Smith and Harper 1988). In this case, males with large stripes may also be of high quality, making them more attractive to females: the width of the breast stripe is reduced at the next molt when parasite loads are experimentally increased (Fitze and Richner 2002), and, in an Oxford study, males with larger breast stripes were paired to females that laid relatively large clutches (Norris 1990).

In the Great Tit, Maynard-Smith and Harper (1988) found that fights at a feeder were most frequent between birds of similar stripe size. Birds of small stripe size probably defer to those with larger stripes. In this case, the cost of a "cheat"—an individual of inferior fighting ability displaying a large signal—is thought to be that the cheater incurs a high rate of aggression from individuals that have similarly sized signals, but are higher in quality. When the cost is socially imposed in this way, plumage signals are termed *badges of status* (Rohwer 1982; Maynard-Smith and Harper 2003, pp. 92–95). Whether a badge of status can maintained entirely through costs associated with aggression remains controversial, and it is possible that other factors associated growth and development also influence badge size (e.g., Ferns and Hinsley 2004; McGraw 2005).

COEVOLUTION OF SIGNALER AND RECEIVER

The preceding discussion implies that receivers should be selected to focus on signals that are informative about the signaler, and that signalers do often make informative signals, because only those individuals in high quality bear the costs of developing an exaggerated version of the signal. However, a novel, uninformative, and cost-free trait may be favored if it stimulates the receiver sufficiently. This leads to an evolutionary race, with the receivers evolving counteradaptations to resist the new trait in favor of informative, costly traits. Krakauer and Johnstone (1995) made a brilliant and under-cited model of this process using simple neural nets. They considered a group of signalers that randomly vary in quality. Receivers are favored if they can correctly determine the quality of the signaler based on his/her signal, whereas signalers are favored if they can deceive the receivers into perceiving them as being of high quality, whatever their true quality actually is (Figure 9.14).

In Krakauer and Johnstone's model peaks in signaler fitness across time reflect the fact that a new low-cost signaling trait has spread through the population, because it strongly stimulates receivers. Following this, receivers evolve to focus on a trait that reveals the underlying quality of signalers, so receiver fitness increases and signaler fitness decreases. Coevolution of signaler and receiver continually generates new preferences as a side effect of the changes that are going on in the receiver network (Figure 9.15). Cycling can go on for-

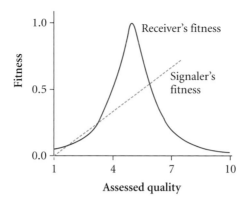

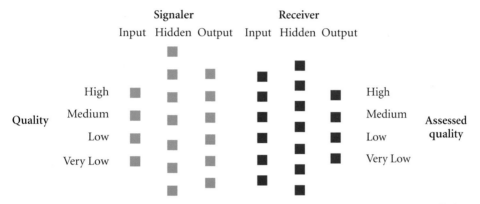

Figure 9.14 A neural net model of exploitation and honesty in communication (from Krakauer and Johnstone 1995). *Above:* Fitness functions that result in a conflict between the signaler and receiver. Here, a signaler has true quality = 5 units. The signaler wants its quality to be perceived to be as high as possible. The receiver maximizes its fitness if it perceives the signaler's quality correctly. *Below:* Populations of two simple neural networks coevolve with each other. Each signaler in the population is assigned a random quality (say, "high"), and the corresponding cell in the input layer fires. Each hidden cell receives and transmits a connection value (initially randomly assigned) from the input cell that fired. Each output cell receives a sum of all the hidden cells' output, weighted by connection values. The output cell's value is then transmitted to the corresponding cell in the receiver network's input layer. The receiver network translates the input into a perceived quality, defined as the cell in the output layer that receives the largest value. Different networks are differentially replicated into the next generation, depending on their fitness (each signaler is tested against each receiver and fitness is summed across tests). The signaler's fitness is its perceived quality minus the cost of making the signal. To model honesty, cost is defined as the (size of the signal)/(quality of the signaler); in other words, for a signal of the same size, signalers of lower quality suffer a higher cost. The size of the signal is set to be the sum of values transmitted by the signaler's six output cells. Quality was assigned values 3 (very low), 5 (low), 7 (medium), or 9 (high).

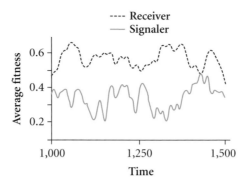

Figure 9.15 Results of simulations of the Krakauer–Johnstone model described in Figure 9.14 (from Krakauer and Johnstone 1995). The figure plots the average fitness of receivers and signalers over 500 generations of the simulation. Occasional mutations result in a signaler being perceived to be of high quality and that signaler's network spreads. Mutations arise in the receiver that correctly perceives signaler quality, and that network spreads. The result is a continual cycling of signals and perceptions of signals.

ever, limited only by mutation to new signals and the presence of undefended biases in receivers. This could produce great divergence as different traits and different preferences arise and spread. However, a quasi-equilibrium may be attained for a long time, where signalers are making costly signals, and receivers have high fitness. This equilibrium may be difficult to disrupt by new signals if the costs to receivers of switching to a noninformative signal are very high (Arak and Enquist 1995; Bradbury and Vehrencamp 2000).

Identifying cases in which new traits have replaced old traits as part of some sort of cycle is not easy, but bowerbirds illustrate the idea. The nineteen species of bowerbirds fall into eight genera. Sixteen of the species build bowers, the most fantastic display of "art" outside of human culture (Diamond 1988). Among the species that build bowers, mating is promiscuous, and males do not help to raise the young. One of the species in the genus *Amblyornis,* the Vogelkop Bowerbird of northern New Guinea, is dull olive brown. Males of some populations build remarkable hutlike structures up to 3 meters across and 1.5 meters tall, adorned with piles of colorful fruits and flowers; males of other populations build maypole-like structures decorated with dried palm leaves (Diamond 1988; Uy and Borgia 2000). [As in all species of bowerbirds that have been studied, the quality of the bower and number of decorations is correlated with the number of matings (Uy and Borgia 2000).] Two other species of *Amblyornis* have more brightly colored plumage but less elegant bowers (Gilliard 1969, p. 299).

In a second genus of bowerbirds (*Chlamydera*), two species have bright crests, but two others do not (Gilliard 1969, p. 384). One of the species without a crest picks up ornaments in its beak (such as berries) and displays them with a head twist in a display similar to that used by the crested species. Thus, within *Amblyornis* and *Chlamydera*, ornaments on the plumage and decorations on the bower are negatively associated. Gilliard suggested a "transference hypothesis," whereby less costly decorations replaced the ornaments as the object of female choice, resulting in selection on the male plumage to become more cryptic. Kusmierski et al. (1997) tested this hypothesis using a molecular phylogeny. They found statistical support for the hypothesis in *Chlamydera*, but not in *Amblyornis*. The youngest pair of species of bowerbirds (in the genus *Chlamydera*, the Fawn-breasted Bowerbird with no crest and the Great Bowerbird with a pronounced crest in the male) are separated by about 4% sequence divergence in mitochondrial DNA (from Kusmierski et al. 1997) or ~2 million years, under the conventional timescale employed in this book. Two populations of the Vogelkop Bowerbird have diverged in bower design and in ornaments used with minimal genetic differentiation and a timescale of less than 300,000 years, perhaps much less (Uy and Borgia 2000). These figures suggest the sort of times over which trait replacements occur. However, it must be noted that these replacements are happening between populations that occupy different environments (Uy and Borgia 2000). For example, Madden and Tanner (2003) found, in captive trials of two species of bowerbirds, that females prefer the same color of grapes for food as those that males use in their bower decorations. They suggest that female sexual preferences have been shaped by foraging preferences.

While bower decorations may originate and replace plumage colors because of their low costs, as they become a focus of female choice, they also become a focus of male competition. Males steal decorations and destroy rivals' bowers (Borgia 1985; Borgia and Gore 1986; Madden 2002). Competition among males imposes new costs, and, in this way, decorations seem to have evolved into honest indicators (Madden 2002).

CONCLUSIONS

Novel signals can spread simply because they appeal to the receiver. Signals may be quite arbitrary in form, with "conspicuousness" or "attractiveness" the prime target of selection. In principle, this could lead to the formation of new species in the absence of any ecological differences between populations, if different signals arise and become restricted to different parts of an ancestral species range. Given the great diversity of possible signals, the different directions in which populations can diverge are very large. Thus, unlike naturally selected traits that are more restricted in their design possibilities, socially selected traits are almost unlimited in their potential diversity. Certainly, sexually selected traits show great variety, as illustrated by the bellbirds (Figure 9.1). Divergence in arbitrary directions, attributable simply to random mutation, is a plausible explanation for this diversity. Ultimately, tests of the idea will come from assessing alternative ecological mechanisms that could drive divergence (Chapter 12).

SUMMARY

Many populations and closely related species occupy superficially similar environments and show little obvious ecological differentiation. Colors, ornaments (crests, tails), and vocalizations distinguish these taxa. Such traits are used in communication between conspecifics, and they are generally thought to have evolved in response to various forms of sexual and social selection. Theory, comparative studies of displays and color patterns, and experimental manipulations all show that a novel signal can increase in frequency if it stimulates the receiver sufficiently. The rate at which new signals accumulate, and hence the rate at which populations diverge, should depend on the rate at which they appear, the extent to which they are favored, and the costs to the receiver of paying attention to them. These costs may be high. Because displays and songs are often energetically expensive to perform, and because ornate plumages are difficult to develop and to maintain, individuals performing more extreme versions of a display or carrying more exaggerated versions of a trait are of higher quality. This means that signals indicate the benefits to receivers of responding to a particular signaler. New signals will often not be such good indicators, and receivers that pay attention to the new signal will have relatively low fitness. This can lead to strong selection against receivers that react to new, less informative traits, but the equilibrium may be broken by the appearance, through mutation, of a particularly attractive trait.

CHAPTER TEN

Social Selection
and the Evolution of Song

... that the sweetness of his notes may yeeld a delectable resonancie.

—Thomas Coryat (1611) on the European Robin (from Lack 1943)

Bird song functions in mate attraction, mate stimulation, and territory defense (Nottebohm 1969; Catchpole and Slater 1995; Collins 2004). Songs are quite varied across species; some are simple, whereas others are complex. In many species, only the male sings, whereas, in many others, both sexes do. Ornithologists use song as a major means of species recognition, as do the birds themselves (Catchpole and Slater 1995). Here, I focus on mechanisms by which populations diverge in song and in other vocalizations. In continuation with the previous chapter, I emphasize the importance of geographical separation and defer ecological influences to subsequent chapters.

Just over half the bird species of the world either learn their songs from conspecifics or they improvise. The other species apparently sing innate and relatively simple songs or else they have no recognizable song, as is true of most large-bodied nonpasserines. Most species have a variety of call notes (Baptista and Kroodsma 2001), but storks and condors do not seem to vocalize at all (P. Rasmussen, *personal communication*). Learning and improvisation are largely confined to the songbirds [the oscine passerines, consisting of 4561 species in the Sibley and Monroe (1990) compilation], the hummingbirds (319 species), and the parrots (358 species), with isolated reports from five other groups (Wagner 1944; Baptista 1996; Baptista and Kroodsma 2001; Kroodsma 2004). Most of the research on birdsong has been done on the songbirds, a group that contains all of the world's finest singers.

Catchpole and Slater (1995) and Baptista and Kroodsma (2001) review methods of analysis of birdsong. Songs consist of a number of "repeat units," which are always sung as a single chunk. Such repeat units are referred to as syllables when they are relatively short. Much of the sound structure of each syllable is captured on a frequency vs. time plot, termed a sonogram (also known as a sonagram or a spectrogram; see Figures 10.1, 10.2). Based on the trace of a sonogram, each syllable can usually be assigned to a discrete class, called here a *syllable type* (this is equivalent to categorizing words: the phrase "sing, sing, song" consists of three "syllables" and two "syllable types"). Some syllable types are quite variable (Baker and Jenkins 1987). In the House Finch, Bitterbaum and Baptista (1979) traced continuous variation among individuals from what they labeled one syllable type to another. Nevertheless, in most species, it is possible to assign syllables to discrete classes, as has been confirmed for four species using computer analysis (Nowicki and Nelson 1990; Baker and Boylan 1995; Irwin 2000).

Songs vary in the form, number, repetition, and order of syllable types. In many species, songs can be classified into different discrete song types. Despite occasional variants, the basic form of each song type is rather stereotyped. In

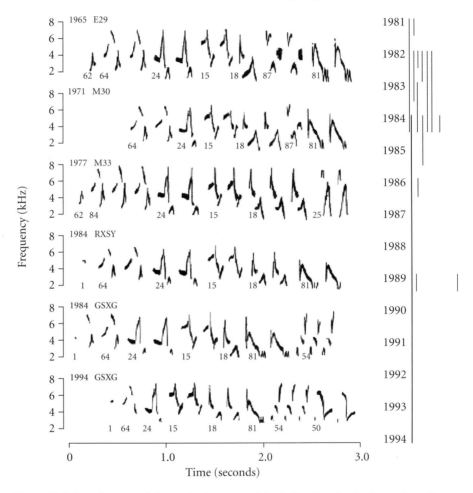

Figure 10.1 Inheritance and change in the songs of the Indigo Bunting (redrawn from Payne 1996). *Left.* Similar songs recorded in different years during a population study in Michigan. A unique letter code identifies a different male, and the numbers on the sonogram a catalogued syllable type. The recording of male GSXG (*red*) is shown for two different years. Whereas the number of repetitions of song elements is generally consistent for most individuals, GSXG varied his repetitions both within a song bout and between seasons. *Right.* The population was studied in depth from 1981–1994, when all males singing the song were recorded (each vertical line represents one male). GSXG is the *red line*. See Appendix 10.1 for additional details.

about 30% of all songbirds studied each male sings a single song type (Beecher and Brenowitz 2005). Some males share a song type, and other males in the population sing other types (e.g., Grant and Grant 1996c; Nelson 1998). An example of a species in which most males sing a single song type is the Indigo Bunting (Figure 10.1; Appendix 10.1). In most other species, individual males sing a few to hundreds or more song types (Catchpole and Slater 1995, pp. 164–172). A male Chaffinch in England typically sings two to three song types (Figure 10.2; Appendix 10.2), the Eurasian Blackbird sings 22 to 48 types, while

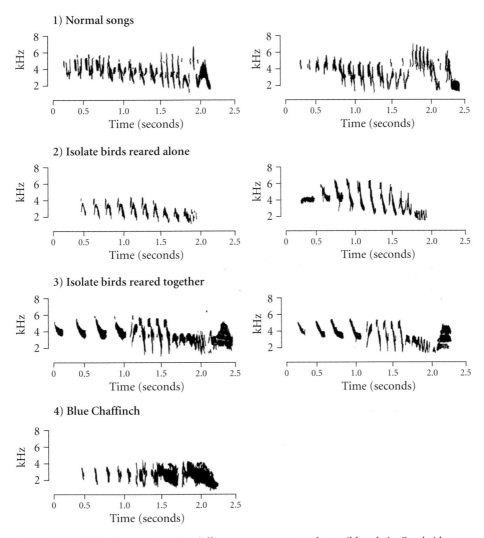

Figure 10.2 Chaffinch songs. (1) Two different song types sung by a wild male in Cambridge, England. (2) Songs sung by males hand-reared after being taken from the nest at 5 days of age and each male subsequently held in visual and auditory isolation from all others. (3) Songs sung by two hand-reared males raised together with three other birds, but isolated from all other birds. (4) Song of the Blue Chaffinch from the island of Teneriffe in the Canaries. All figures are redrawn from Thorpe (1958a).

Table 10.1: Responses of birds to their own versus a foreign dialect

Species	Song repertoire size (mean)	Scale (km)	Strength of preference for local dialect*		
			Songs or calls	Approach	
Male responses					
Yellow-naped Parrot[1,2]	–	50	1.0	∞	Wright and Dorin 2001
Hill Myna[3]	–	180	1.6	2.8	Bertram 1970, p. 147
Short-toed Tree-Creeper[3,4]	1–3	1200	1.6	1.1	Thielcke and Wüstenberg 1985
Short-toed Tree-Creeper[3,4]	1–3	1400	16.0	10.0	Thielcke and Wüstenberg 1985
Winter Wren[1]	4–7 (5)	70	–	5.8	Kreutzer 1974
European Blue Tit[3,4,12]	1–3	100	–	15	Schottler 1995
Goldcrest[3]	18–33 (23.6)	500–1000	–	1.3	Becker 1977
Firecrest[3]	~3	1300	–	1.1	Becker 1977
Orange-tufted Sunbird	1	1	2.2	1.3	Leader et al. 2002
Pine Grosbeak[1,7]	–	1000	–	∞	Adkisson 1981
Corn Bunting[8]	1	1	0.1	3.0	McGregor 1983
Song Sparrow	6–11 (8)	10–20	1.6	1.7	Harris and Lemon 1974
Song Sparrow	6–11 (8)	400	1.6	–	Searcy et al. 1997
Swamp Sparrow[5]	1–7	1500	–	0.6	Balaban 1988
White-crowned Sparrow	1	6	1.6	–	Petrinovich and Patterson 1981
White-crowned Sparrow	1	6–20	1.3	1.5	Tomback et al. 1983
White-crowned Sparrow[4]	1	700	4.9	–	Petrinovich and Patterson 1981

White-crowned Sparrow	1	100	1.7	5.0	Nelson and Soha 2004
Sharp-beaked Ground-finch[13]	1	100	–	3.0	Grant and Grant 2002c
Sharp-beaked Ground-finch[13]	1	100	–	6.7	Grant and Grant 2002c
Northern Cardinal[3]	8–11	11–34	3.5	3.0	Lemon 1967

Female responses

			Copulation Solicitations†	
Song Sparrow[14]	6–11 (8)	400	4.8	Searcy et al. 1997
Song Sparrow[4,10]	6–11 (8)	50	3.4	Patten et al. 2004
Swamp Sparrow[6]	1–7 (3)	1500	1.7	Balaban 1988
White-crowned Sparrow	1	5	2.9	Baker 1983
White-crowned Sparrow[15]	1	1900	12.5	Baker 1983
White-crowned Sparrow	1	–	13.1	MacDougall-Shackleton et al. 2001
Yellowhammer[12]	1–3 (2)	700	1.9	Baker et al. 1987
Brown-headed Cowbird[4,9]	2–7	1600	3.0	King et al. 1980
Brown-headed Cowbird[11]	1	40	2.1	O'Loghlen and Rothstein 1995, their Fig. 4

* Typically, about ten birds were tested. Preference is based on ratio of number of songs or calls given after playback. Approach is the ratio of the mean distance approached in response to foreign song to mean distance in response to own song, unless indicated in left-hand column (superscripts 1,3,8).

† Ratio of the number of copulation solicitations given by a female to own vs. foreign dialect (but see note 11).

1 Approach column gives the ratio of number of individuals responding more strongly to own dialect vs. number responding more strongly to other dialect.

2 The vocalization is the "Pair duets": sex-specific notes and contact calls repeated alternately by members of a pair.

3 Approach column gives ratio of fraction of birds tested that approached speaker in response to local dialect to fraction of birds tested that approached speaker in response to foreign dialect.

4 Comparisons across subspecies.

5 This was a test of syntax (notes from one dialect were rearranged to match the syntax of another). Males gave a weaker response to own syntax.

6 Ratio of rankings of solicitations for own song vs. other song. See also note (5).

7 Vocalization is the "location call": a loud call communicating over distance, and known to be learned in this species.

8 Approach is relative time within 5 m of speaker. McGregor argues that the low singing response to the bird's own dialect is a consequence of the intense approach responses. A test on a dialect 40 km away gave uniformly lower responses.

9 Ratio of percentage of times females gave responses to solitary male "perched songs" of their own subspecies cf. the other subspecies.

10 The two response measures are with a consubspecific mount and with a heterosubspecific mount, respectively.

11 Response to the song termed the flight whistle. This is the ratio of the average duration of response: the length of time the (estradiol-implanted) female holds the copulatory position.

12 See Figure 10.9, p. 212.

13 The first line is Genovesa response to Darwin, and the second is Genovesa response to Wolf (approach based on horizontal distances).

14 Searcy et al. (2002) show a decline in female response over a 500 km transect.

15 Females were held in captivity from a young age; they were played their natal song dialect during their first spring season and tested during their second spring.

the Brown Thrasher sings more than 1000 types (other examples are given in Table 10.1). Although song complexity depends more on the brain than on the voice box (Baptista and Trail 1992), one interesting way in which some songbirds produce such variety is by singing different syllable types from the right and left sides of the syrinx (Suthers 1990).

Geographical variation is most obvious in those species that learn their songs, and it appears to be low in species that do not learn songs [New World flycatchers (Lanyon 1978), shorebirds (Miller 1996), antbirds (Isler et al. 2001, 2005), and doves (de Kort et al. 2002)]. However, even in these groups, different subspecies are sometimes distinguished by discrete song or call differences. Song learning introduces the possibility of much greater error ("cultural mutation") as well as rapid spread of new mutations through copying, which together seem to have led to high levels of geographic variation in many species. In this chapter, I emphasize how geographical variation becomes established as a result of the cultural learning process, and how this might contribute to speciation. Songs may evolve genetically as well as culturally. Causes of genetic differentiation are considered in a final section.

I first describe patterns in the geographical variation of songs. In the following sections, I consider how song variants arise, how they spread, and how they might become restricted to a portion of the species range.

GEOGRAPHICAL VARIATION

Among those species that learn their songs, geographical variation in song is widespread, occurring at all spatial scales. Some species show so much geographical variation that song types can turn over more or less completely in a distance of less than 10 km (e.g., Figure 10.3). Variation in song type may be due to rearrangements of a limited number of syllable types that are themselves geographically widespread (e.g., the Indigo Bunting, Baker and Boylan 1995; Appendix 10.1). In other species, both syllable types and song types both show much variation across space (Figure 10.3). In the Swamp Sparrow, syllable types and song types both show variation, but the basic building blocks of each syllable (termed "notes") are few in number and geographically rather invariant (Marler and Pickert 1984; Balaban 1988). This may apply to other species too (Nelson et al. 2004).

At the level of tens to hundreds of kilometers, geographic variation may be continuous or it may be organized into dialects. Mundinger (1982) defines a dialect as a song tradition shared by members of a local population of birds, with a boundary delineating it from other variant song traditions. Dialects are primarily found in single song-type species and are reported from many species

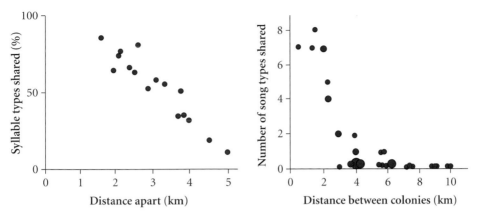

Figure 10.3 *Left:* Syllable-type sharing in the House Finch in southern California (Bitterbaum and Baptista 1979). Each point represents the fraction of a single bird's syllable types that is found in a syllable-type catalogue compiled from a sample of 45 birds at 0 km. Each bird sings, on average, four song types with each song type containing, on average, 13 different syllable types. Note the rapid decline in syllable-type sharing with distance (correlation, $r = -0.95$). *Right:* Song-type sharing among Yellow-rumped Cacique colonies in Panama (Trainer 1989), based on pairwise comparisons between six simultaneously active colonies in each of three years ($N = 45$ comparisons). Circle size approximates the number of comparisons, with the smallest circle indicating one observation. All males in a colony share 5–8 song types. These change rapidly over time, and within a colony 78% of all song types are distinctly different from one year to the next. However, dispersal between colonies, plus copying, results in song sharing among colonies within each year, over a few kilometers.

of songbirds, as well as from hummingbirds and parrots (Wright and Dorin 2001; Table 10.1). The best-studied dialect species is the White-crowned Sparrow of western and northern North America (Marler and Tamura 1962; Baptista 1975). In this species, most males sing just one song type. The first part of the song is a pure-tone whistle that is shared across the whole species (Soha and Marler 2000; see Figure 10.4), and dialects are defined based on variation in the latter part of the song. Seven dialects occur over an area of 180 km² around San Francisco; each dialect is sung by, on average, a few hundred males (Baptista 1975; Trainer 1983). Males of one dialect sing very similar songs to each other, with fairly sharp boundaries between dialects. Not all song characteristics turn over at the same point; some males sing a song that contains syllable types from both dialects, and some are bilingual, singing two songs. Trainer (1983) showed that the same dialects survived 10 years in more or less the same positions, although the boundaries had shifted somewhat.

In a migratory subspecies of White-crowned Sparrow along the coast of the state of Washington, each dialect occupies a larger area (hundreds of square kilometers), with intergradation over tens of square kilometers (Baptista 1977;

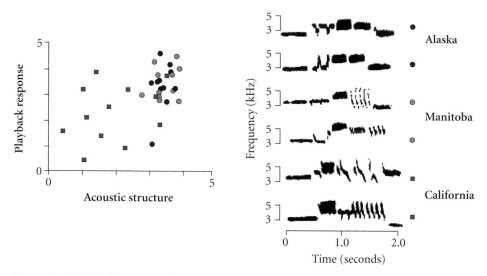

Figure 10.4 Male White-crowned Sparrows (*Zonotrichia leucophrys*) in Churchill, Manitoba (subspecies *Z. l. gambelli*) were played songs from (1) their own population (●), (2) the same subspecies in Alaska 2,600 km distant (●), and (3) a different subspecies, *Z. l. oriantha,* mostly from California (■), but the two rightmost values along the acoustic axis for this subspecies are from Oregon. Two songs from each subspecies are indicated. The acoustic structure of the songs (a composite measure of 34 quantitative measurements taken from sonograms) correlates with the intensity of response to playback (a composite measure based on distance of approach to the speaker and vocalization responses, $r = 0.45$). (From Nelson 1998)

Chilton and Lein 1996; Nelson et al. 2004). Baptista (1977) attributed the larger area and the greater degree of introgression in Washington State when compared with San Francisco to greater dispersal in Washington, associated with seasonal migration. In the highly migratory subspecies of the White-crowned Sparrow (namely, *Zonotricha leucophrys gambelli*) of Alaska and northern Canada, adjacent males sing one of three or four highly distinctive song types, with no tendency for neighbors to share songs let alone to form dialects (Nelson 1998). Geographical variation is present, but over very large distances; populations of *Z. l. gambelli* separated by 2600 km sing similar but significantly different songs (Nelson 1998; see sonograms in Figure 10.4).

Finally, parts of songs, or even whole songs, can be relatively uniform over large geographical ranges. As noted, in the White-crowned Sparrow, all males sing the introductory whistle. The Coal Tit sings much the same territorial advertisement song from North Africa throughout Asia to Japan (Martens 1975, 1996, Martens et al. 2006; Figure 10.5). Across the Himalayan chain, including west China, the vocalizations of several different subspecies are extremely similar, even as morphology and genetics differ greatly (Martens et al.

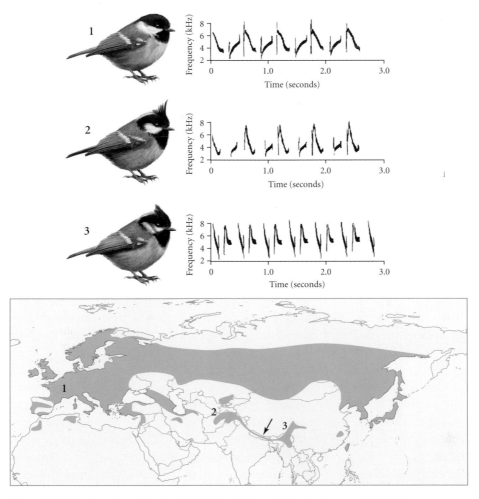

Figure 10.5 Distribution of the Coal Tit (after Martens 1975, Martens et al. 2006). The arrow marks a hybrid zone between two recognized subspecies in Nepal. Song similarity is reflected in song recognition across the range. Thielcke (1969) showed that German Coal Tit males respond aggressively to playback of songs from Afghanistan, and Martens (1975) demonstrated weak responses of Nepal males (on either side of the hybrid zone) to German Coal Tit songs.

2006). The related Black-capped Chickadee of North America sings a characteristic and familiar *fee-bee-ee* whistle throughout most of its range from Nova Scotia to British Columbia (3,500 km) [It also has a "gargle" song that is commonly used in male–male interactions and is geographically variable (Ficken 1981).] Remarkably, at the ends of the range, on offshore islands in Massachusetts as well as in Oregon and Washington, populations of Black-capped Chickadees sing different songs. On the Massachusetts islands, males may sing a repertoire of two or more distinctly different whistle songs, each of which differs from the mainland form (Kroodsma et al. 1999). One island even has dialects (Martha's Vineyard, ca. 150 km²).

Many superspecies contain allospecies or subspecies characterized by highly distinctive song variants that extend over large geographical areas with rapid turnover between them (e.g., Figure 10.6). Where related taxa meet, they form narrow hybrid zones or else have limited geographical overlap without interbreeding. Because the taxa are defined based on song similarity and geographical distribution, Martens (1996) termed them regiolects. They are a step up from dialects on the road to speciation, in that the songs are often (but not always) more distinctive, they define a group that is partially reproductively isolated from another group, and they extend over a relatively large geographical range, comparable to that of many species.

Song recognition

Just as it is difficult for humans to understand different dialects, and impossible to understand different languages, geographical variation in song impedes recognition of birds from different locations. This has been demonstrated many times using song playbacks. In the field, males respond aggressively to playbacks of songs from their own population. To elicit a response from a female, procedures have been more complicated, requiring the playback of songs to a captive bird that is usually implanted with estradiol, and then recording the number of copulation solicitations the female makes. The copulation solicitations are spectacular; the female lowers and vibrates her wings, arches her back, and separates the feathers around the cloaca in an immediate, and often brief, response to playback of a song.

Playback experiments show that individuals typically respond more strongly to songs or calls belonging to their own dialect rather than to those from other dialects (Table 10.1). In the White-crowned Sparrow, females of the subspecies *Z. l. oriantha* more readily gave responses to songs and calls from their own dialect rather than to those from a dialect just 6 km away (Baker 1983). They responded weakly to a dialect from 25 km away and to the songs of another subspecies. On the other hand, within the subspecies *Z. l. gambelli*, males responded similarly to songs from their local area and to songs from 2600 km away, reflecting the limited geographical variation in the song over a large spatial scale in this subspecies (Nelson 1998; Figure 10.4), but they responded only weakly to songs of *Z. l. oriantha* (Figure 10.4). This pattern is general; naïve individuals from one regiolect usually respond to the songs of widely separated individuals within their own regiolect but not to songs from other regiolects (Martens 1996; Irwin

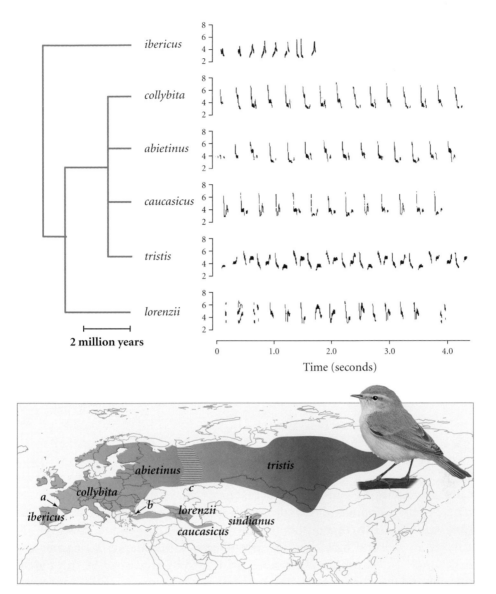

Figure 10.6 Regiolects in the Chiffchaff superspecies (from Helbig et al. 1996; Martens 1996). *Clockwise from top left:* (1) Estimated phylogeny among six continental taxa (Canary islands taxa are omitted). Helbig et al. (1996) consider *lorenzii* (which is a subspecies of *sindianus*) and *ibericus* (this taxon was named *brehmii* in their treatment) to be separate species; *collybita, abietinus,* and *caucasicus* to be subspecies of *collybita*; and the status of *tristis* as a species or subspecies of the *collybita* group to be equivocal. All taxa look very similar, *tristis* is illustrated. The phylogeny was constructed using maximum likelihood with an enforced molecular clock (Appendix 2.2, p. 38), and the time-scale bar assumes the standard rate of molecular evolution of 2% per million years. (2) Representative sonograms. (3) Geographical distributions. Three narrow zones of overlap *(a, b, c)* are identified: *ibericus* and *collybita* are known to hybridize in a narrow zone in the Pyrenees *(a)*, and hybridization may also occur in the other zones.

et al. 2001a). Sometimes responses across regiolects are asymmetrical. In three species, males from subspecies in outlying regions respond strongly to songs from males closer to the center of the range, but those in the center respond weakly or not at all to the outlying songs (Becker 1982, p. 228).

Song recognition may usually result from learning. Features of familiar songs are stored in memory, and individuals singing those features are readily recognized and responded to. In addition to responding strongly to familiar songs, individuals typically do respond somewhat to songs from other dialects (Table 10.1). This is presumably because some features of the foreign songs resemble those of the local song (e.g., the introductory whistle of the White-crowned Sparrow song is common across dialects). Upon repeated exposure to foreign songs, individuals are likely to generalize from the invariant cues to the other song features, and, as a consequence, they respond strongly to the foreign song (Irwin and Price 1999; Chapter 13). MacDougall-Shackleton et al. (2001) found that captive female White-crowned Sparrows showed similar responses to songs from another dialect as to their own if they had been exposed to the foreign song the previous season. The demonstration that learning modifies song recognition does not alter the main conclusion—that song divergence is also divergence in a cue used to identify members of one's own population, and it can therefore develop into a species-recognition cue.

CULTURAL TRANSMISSION AND CULTURAL MUTATION

Cultural mutations—and, in some cases, genetic mutations—are the initial requirement for populations to diverge in song. In this section, I consider how songs are normally copied from one individual to another and how they are affected by cultural mutation. In Appendices to this chapter, I describe the development and evolution of songs in two passerine species, the Indigo Bunting and the Chaffinch. These are representative of many results from songbirds.

Song transmission

In a few species, notably Darwin's ground-finches, sons usually learn their father's song (Grant and Grant 1996c), but this is atypical. In most territorial songbirds unrelated neighbors sing similar songs [(Mundinger (1982) estimated this to be true of more than 95% of all songbirds studied up to the time of his review]. Similarity between neighbors is largely achieved as a result of juveniles copying adults. The copying process begins very early in life, and, in many species, songs of adults do not usually change from one year to the next (Nelson 1997). In the field in Washington State, a young male Song Sparrow learns each part of his repertoire of from five to 13 song types from one to several contiguous males in the first few months after leaving the nest, probably skulking around in the bushes listening. The young male usually settles near them the following spring and may take over the territory of a bird that has died (Beecher et al. 1994; Nordby et al. 1999). He tends to drop some songs and to retain those in his repertoire that are sung by surviving adults (Nordby et al. 1999).

In all species studied, song development begins with a subsong that young birds may start to sing within a few days of leaving the nest (Marler and Peters 1982). The subsong is typically a rambling series of often ill-defined notes. After a period of weeks to months (depending on the species), the first evidence of imitations appear in what is called plastic song. The ordering and structure of syllables is highly variable in plastic song, but, by a process of trial-and-error matching of produced song to stored memory, the adult song crystallizes out of plastic song, accompanied by the deletion of many syllable types. For example, Swamp Sparrows in the laboratory reduced the number of syllable types sung from more than 11 to three during the month before the crystallization of the final song (Marler and Peters 1981, 1982). In some species, different song types from different tutors are recognizable in the plastic song, and several may be deleted during crystallization (Marler and Peters 1981; Nelson and Marler 1994; Nelson 1997).

Many of the syllable types sung in the plastic song do not have counterparts in possible tutors' songs. During the process of crystallization, they are modified and combined to produce syllable types that match those sung by an adult tutor (Margoliash et al. 1994; Tchernichovski et al. 2001). This means that one way of introducing entirely new syllable types into a population is to retain some unusual syllable types from the plastic song into the crystallized song (termed "invention" by Marler and Peters 1982). This is commonly seen in the lab, for example, when young birds are raised in the absence of adult tutors. Young reared in complete isolation develop odd and simple songs, but young raised in peer groups develop recognizable songs that differ strikingly from those of other birds (Figures 10.2, 10.7) although they are often similar among members of the same group. Invention on a smaller scale is also seen in the laboratory when birds are tutored either by tapes (Red-winged Blackbirds, Marler et al. 1972; Swamp Sparrows, Marler and Peters 1981) or by live models (Song Sparrows, Nordby et al. 2000).

In addition to invention of entirely new syllable types, small errors in the copying process can produce new variants (Marler and Peters 1982; Margoliash et al. 1994). Cultural mutation of songs can also arise through the copying of

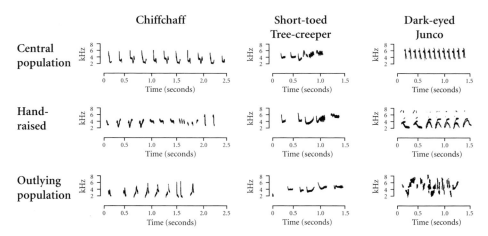

Figure 10.7 Examples of songs of wild birds, hand-raised birds, and wild birds from outlying populations (see also Figure 10.2). *Left:* Chiffchaff: wild bird from southern Germany; song of one of two untutored German males raised together; wild bird from Spain (Thielcke 1983). Thielcke et al. (1978) found that 87% of Spanish birds responded to the song of German birds, but that no German birds responded to song of Spanish birds. Isolate song has not been tested. *Center:* Short-toed Tree-creeper: wild bird from Germany; song of one of two untutored males raised together; wild bird from Cyprus. In playback experiments, ~40% of wild German birds responded to the untutored song and about 15% responded to the song of a bird from Cyprus (Thielcke and Wüstenberg 1985). *Right:* Dark-eyed Junco: wild bird from California (Marler et al. 1962); hand-raised bird from age of seven days (brood of four raised together and within earshot of various wild bird species, but probably not other juncos; Marler et al. 1962); wild bird from Guadalupe Island, 300 km off Baja California (Mirsky 1976). Both hand-raised and Guadalupe island birds have longer syllables and more syllable types in their songs than wild continental birds (see Table 12.4, p. 266). Male Dark-eyed Juncos also sing a "short-range" song (Titus 1998), and the possibility that the Guadalupe Junco is singing a variant of this song needs to be investigated (Nolan et al. 2002). Playback experiments have not been conducted.

different parts of songs from different tutors, or different parts of different songs from the same tutor (e.g., Beecher et al. 1994; Payne and Payne 1997).

Cultural mutation rate

For song differences among populations to be relevant to speciation, different populations need to be characterized by features that are shared by all individuals. Population divergence thus requires, first, the appearance of distinctive new cultural mutations and, second, the spread of these mutations by cultural transmission. However, mutation and cultural transmission counter one another; mutation is, by definition, a failure to copy a tutor. In order to produce dialects or species-specific songs, some mutation is necessary, but not too much.

I define the cultural mutation rate of songs (or syllables) as the fraction of all song or syllable types in a population that were absent in the previous genera tion. A very high mutation rate means that every individual sings a different song, resulting in no transmission. This will slow or even prevent population divergence. In fact, some species do develop songs by invention with little or no copying from conspecifics. The classic example is the Gray Catbird, which sings a long, rambling song of a few hundred different "units." (A unit is equivalent to a syllable type as used here, but some are quite long.) Individuals held in isola tion in the laboratory develop songs similar to wild males (Kroodsma et al. 1997). Playback of these songs to wild males leads to aggressive responses (Kroodsma et al. 1997). In the wild, neighboring males do not appear to share any song units and, by this criterion, are as different as individuals from any where in the geographical range (recordings were obtained from as far apart as the west and east coasts of North America).

A very low mutation rate means that transmission is high, but also that varia tion will be slow to accumulate between populations. Although a low mutation rate does not seem likely in species with a strong learning component to song transmission, it is possible that most novel song types are quickly lost, so the cultural mutation rate is effectively low. I first consider factors affecting the per sistence of novel songs and then present some estimates of the effective muta tion rate.

Novel song types may be selected against, or neutral, or favored. Often, novel songs may be selected against because individuals do not easily recognize them. In the White-crowned Sparrow, the species-specific whistle at the beginning of the song is a trigger for young birds to learn components of song following (Soha and Marler 2000). Unusual songs without the whistle are unlikely to be copied. (Conversely unusual songs preceded by the whistle can be learned where they otherwise would not be; Soha and Marler were even able to get a White-crowned Sparrow to "sing" a ground squirrel alarm call by attaching the whistle to the front of the call).

Many unusual songs produced in the laboratory in the absence of adult tutors are sufficiently different that they are not readily recognized as belonging to con specifics. This has been shown by playback to males in nature for at least nine species (Thielcke 1973a; Shiovitz 1975; Becker 1978, 1982, p. 242; Lanyon 1979b; Thielcke and Wüstenberg 1985). Thielcke (1973a) showed that five out of six of the song types sung by hand-raised Coal Tits elicited no response from territorial males in the field. The sixth, which incorporated elements similar to call notes, was responded to at the same level as wild-recorded song (two songs tried). Thielcke (1970) also conducted an experiment in which he hand-reared a Short- toed Tree-Creeper in the laboratory and released it into the field; it responded to its own song, but not to those of wild birds. Finally, the Marsh Tit sings two kinds of songs. A simple repetitive song develops in isolation and is responded to by birds in the wild, whereas the more complex song does not develop in isolation and improvised versions are not responded to (Becker 1978).

Reduced recognition of a novel song could lead to a lower fitness for the singer through escalated interactions in disputes over territories (Chapter 13). If this is the case, males singing rare songs may lose their territories more readily, taking the songs with them. In addition, it seems likely that young males hear songs of many males, and they may more readily learn songs they hear repeatedly. Again, rare songs are rapidly lost from the population.

Songs are generally copied from neighbors, often highly accurately, which in itself implies a strong advantage to reproducing a song already present, as well as selection against novel types (Payne 1981, 1996; Morton 1987; Payne et al. 1988; Nordby et al. 1999; Todt and Naguib 2000). Two reasons for high copy fidelity have been suggested. First, shared songs may be attractive to females (Nottebohm 1969; Payne 1981). A young Indigo Bunting male that shares his song with a neighbor has a higher probability of gaining a mate than one that does not share (Payne et al. 1988). Female Song Sparrows pay more attention to a song that matches a song in her mate's repertoire than to one that does not match (O'Loghlen and Beecher 1999; Figure 10.8, *left*). Female Brown-headed Cowbirds respond less strongly to unusual songs, which are mostly sung by young males (O'Loghlen and Rothstein 2003).

The second reason that copying of neighbors may be favored is because of social interactions among males (Payne et al. 1988; Payne 1996). In the Song Sparrow, each male sings a repertoire of song types, and males usually reply to a neighbor with a song type that they share with that neighbor (Beecher et al. 1996). Early in the season, this is often the song type the neighbor has just sung, but, later on, it is more likely to be a different song type from the shared repertoire (Beecher et al. 2000). In the Banded Wren, aggressive interactions are accompanied by the singing of shared song types, and intention to retreat is indicated by singing a song type that is not in the other individual's repertoire (Figure 10.8, *right*; Molles and Vehrencamp 2001). The use of shared and unshared song types may thus signal different levels of aggressive intent (Krebs et al. 1981; Beecher et al. 2000; Vehrencamp 2001; Beecher and Brenowitz 2005).

The net result both of female preferences and of male interactions is that song types already in the population can have an advantage over rare, novel types. This is "frequency-dependent selection favoring the common type"; common songs are favored, and those that are rare are selected against. Such selection will tend to prevent the spread of any new cultural mutations; hence, the effective mutation rate is lower than the actual mutation rate.

Song variants that are present in a population can be used to estimate the effective rate of cultural mutation (i.e., the rate of mutation to song types that are not quickly lost). If one is prepared to make the assumption that the song

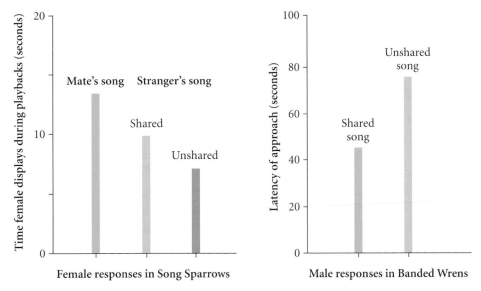

Figure 10.8 Experiments suggesting that the form of the song is less important than the context in which it is used. *Left:* Response of female Song Sparrows to a recording of a male singing a song that is found in the repertoire of their mates, as compared with a song that is not in their mates' repertoire (O'Loghlen and Beecher 1999). Sample size varied from 14–19 females tested. Females were captured and transported into the laboratory after their mates' repertoire had been determined. After being implanted with estradiol, they heard tapes of mate or stranger songs. Shown is total duration of copulation-solicitation displays (because the data are very skewed, no standard errors are shown; all differences are significant by nonparametric tests, $P < 0.01$). *Right:* Male Banded Wrens in Costa Rica sing ~22 different song types. They share about 80% of their song types with any given neighbor, and all song types sung by a male are typically sung by at least one neighbor. Molles and Vehrencamp (2001) placed a speaker in the middle of a bird's territory and one at the edge. For two minutes, three shared song types were repeatedly played from the central speaker and then either three shared or three unshared song types were played from the boundary speaker. Figure compares average response if a shared or unshared song type is played on the boundary ($N = 19$, $P < 0.05$). Latency of approach is the time taken after the treatment for the male to move at least 5 m toward the boundary speaker after it had been switched on. The unshared song type elicits a weaker response.

variants present in a population are selectively equivalent to each other and are evolving solely by drift, the number of variants present depends on a balance between input, through migration and mutation, and output, through drift. Drift is a function of population size; very small populations should consist entirely of individuals that sing the same songs, whereas large ones will have greater diversity. Based on formulae developed in population genetics, Lynch (1996) used estimates of population size and song-type diversity in different woodlands to obtain a combined mutation + immigration input of song types on the order of 10% per generation for Great Tits and Chaffinches. He used among-population differentiation (from other studies) to estimate the immigration rates of song types, and from this he suggested that mutation might account for about half the new input from each generation, or on the order of a few percent per generation. These estimates are based on many assumptions, but they are similar to those more directly observed from temporal changes in song traditions in a population of Indigo Buntings (Payne and Payne 1997; Figure 10.1; Appendix 10.1). Thus, in these species, the mutation rate of song types seems very high.

Because new song types originate by changes in the order and number of repetitions of syllable types, as well as in the syllable types themselves, the rate at which new syllable types appear in a population must be less than, or equal to, that of song types. Lynch and Baker (1993) estimated the cultural-mutation rate of syllable types in Chaffinches on the Canary Islands to be about one-fiftieth that of whole song types, and Baker and Jenkins (1987) thought that no new syllable types have arisen among Chatham Island Chaffinches since their colonization from New Zealand more than 100 years ago. On the other hand, the Eurasian Tree Sparrow in Illinois seems to have a high rate of production of new syllable types. The Illinois populations are descended from about 20 birds introduced from Germany in 1870. In 1985, only about 20% of the syllable types identified in Illinois were present in a German sample of 500 syllable types (Lang and Barlow 1997). In addition, only about 70% of syllable types are shared between populations separated by 20 to 40 km (Lang and Barlow 1987). Lang and Barlow (1997) argue that many new syllable types must have been produced since introduction. It is, of course, possible that the mutation rate increased as a direct result of the introduction, if the copying mechanism was disrupted or the use of unshared songs was favored.

In summary, song types may be subject to various forms of frequency-dependent selection. Selection favoring a common song type because it is readily recognized may lead to strong selection against new mutations and geographical invariance in songs, or at least in relevant song components, such as the White-crowned Sparrow's whistle. Presumably, in species with dialects, cultural mutations in those features of the song that define the dialect arise at a relatively slow rate, but, once they have become established, they are maintained by frequency-dependent selection favoring the common type.

POPULATION DIVERGENCE

Although many cultural mutations may be selected against, especially when rare, those that are not can spread to fixation either via selection because they have particularly favorable impacts on receivers, or through various drift-based processes. I consider each in turn.

Selection

Both theory and experimental evidence show that songs may plausibly evolve because new song types have favorable effects on listeners. Certainly many songs are attractive to humans. Computer simulations of neural nets, as discussed in the previous chapter, should apply equally to acoustic as well as to visual processing. Experimentally, Verzijden et al. (2007) showed that when individual Zebra Finches were trained to favor a shorter song over a longer one, they gave slightly stronger responses to even shorter songs, and, conversely, when trained to favor a slightly longer song over a slightly shorter song, stronger respones to even longer songs. Many experiments in psychology demonstrate that vocalizations amplify receiver responses to visual displays (Rowe 1999).

Experimental evidence for the stimulatory effects of song variants on females was given in Table 9.1 (p. 162). Particular components of a song, or different song types, may be more effective stimulants than others (Vallet et al. 1998; Todt and Naguib 2000; Forstmeier et al. 2002). Thus, in a Common Nightingale's repertoire of some 200 different songs, about 20% have a whistle tone, and these songs seem to incite singing by other males more easily (Todt and Naguib 2000). Hidden preferences may also exist, awaiting new mutations. A possible example comes from Balaban's (1988) atypical finding that male Swamp Sparrows singing one dialect respond more strongly to the playback of the song in the syntax of another dialect rather than their own.

In many species, females give more solicitations in response to larger song repertoires (Searcy 1992; Collins 1999). Okanoya (2004) showed that the Bengalese Finch, which is the domesticated variety of the White-rumped Munia, has, over the course of 240 years of domestication, developed a more complex song. (The number of syllable types is similar, but the song is sung in arrangements that are more varied.) G. Cardoso (*personal communication*) suggests that this variation represents relaxed selection for males to accurately reproduce species-specific songs (i.e., it resembles the plastic song that is a feature of song development). This is possible, but Okanoya (2004) did find that these songs stimulate females more than the wild-type song (Table 9.1). In this species, the breeder has selected for plumages and parental behavior and has deliberately paired individuals on the basis of those traits. Because of this, Okanoya (2004) suggests that song evolution was driven by female stimulation rather than by deliberate selection by the breeder or through sexual selection.

Cultural drift

Frequency-dependent selection favoring the common type may account for geographical invariance of songs even over large spatial scales. With respect to dialects, this kind of selection could account for the maintenance of dialect boundaries, but it is difficult to see how the dialects arise in the first place. It may be that, in some cases, a song type spreads to high frequency through chance (Lynch 1996), and it either overcomes any initial frequency-dependent selection (e.g., because the male singing the song type is a particularly dominant individual, whose songs get copied frequently) or is not subject to such selection. Conrads (1966) noted a very distinctive song type present in a population of Chaffinches that persisted locally for tens of generations. Because of the distinctiveness of the song, this population might be considered to have a discrete dialect, and it may be that the spread of this song type was a result of drift. Chilton and Lein (1996) thought such a process could lead to the formation of dialects among White-crowned Sparrows.

An alternative mechanism for dialect formation, which overcomes the problem of frequency-dependent selection favoring the common type, proposes that both the songs and the recognition of the songs are simultaneously and dramatically altered (Baker 1975). In Baker's model, dialects form as a result of the establishment in a new locality of a few males who had failed to learn songs in their natal area and so sing new song types. The population expands in space as a result of recruitment of additional immigrants, as well as offspring from the founders. The founder song types are learned, and they become the common song types in the population. These song types are subsequently maintained by frequency-dependent selection.

Baker's idea receives support from the work of Mundinger (1975), who studied the recent colonization of the House Finch in New York. The House Finch is a native of the western United States, but was often kept as a cage bird in the eastern states. In 1943, finches were found breeding on Long Island. They probably descended from birds released by a dealer when their sale was made illegal in 1940 (Elliott and Arbib 1953), although Hill (2002) suspects that escaped birds were likely to have been breeding in the wild before that. The species has subsequently spread through much of the eastern United States. In the early 1970s, Mundinger (1975) showed that House Finch populations along the coastline of Long Island were divided into dialects typically a few km on each side and separated by discrete boundaries spanning a few hundred meters. During the initial House Finch colonization, observed breeding colonies were few and separated by distances of 5–37 km (Elliott and Arbib 1953; Mundinger 1975). If the colonies were founded by the dispersal of a few insufficiently tutored young birds, different colonies should have varied in their songs, and, as the colony expanded, copying from neighbors would create small areas of distinctive songs. Dialects are not observed in the native range of the House Finch in California (Bitterbaum and Baptista 1979; Figure 10.3), where syllable-type diversity is much higher. It will be interesting to see if dialects disappear in New York, and, if so, over what timescale.

Thielcke (1970, 1973b, 1983) originally proposed that new songs arise through a failure to learn from adult tutors (the so-called "withdrawal of learning" hypothesis). In several examples, an isolate song from at least one hand-raised chick resembles the wild-type song of another subspecies or closely related species (Figures 10.2, 10.7). This includes the Common Chiffchaff (Thielcke 1983; Figures 10.6, 10.7), the Short-toed Tree-creeper (Thielcke 1970, 1973b; Thielcke and Wüstenberg 1985; Figure 10.7), the Willow Tit (Martens 1996), the Blue Chaffinch (Figure 10.2; Lynch and Baker 1991), the Brown-headed Cowbird (King et al. 1980), and perhaps the Dark-eyed Junco (Figure 10.7). In the case of the Chiffchaff, Thielcke (1983; see also Thielcke and Zimmer 1986) presents quantitative measurements of syllable structure showing that syllable types produced by hand-raised birds of the subspecies from Germany (*Phylloscopus collybita collybita*) resemble those of adult birds of the subspecies from Spain (*P. c. ibericus*) more than those of adult birds from Germany. For both the Chiffchaff and the Short-toed Tree-Creeper, Thielcke argued that a patchy habitat in Spain has led to many opportunities for founding events and hence, for song divergence.

Songs of birds raised in the absence of tutors can be very different from one isolate group to another. Only some of them resemble the songs of populations of wild birds. The value of the isolate bird studies is that they show the tremendous range of songs that can be produced in the absence of tutoring from adults. A great variety of the songs are produced in the laboratory, but hand-raised individuals housed together usually sing similar songs [Thorpe 1958a (see Figure 10.2); Thielcke 1973a; Kroodsma et al. 1995; Nordby et al. 2000], suggesting that small founding flocks would likewise converge in song and song recognition.

Selection vs drift

Small population size seems to be required if a new syllable or song type is to spread to fixation via cultural drift in a continuous population. To put numbers on this, I compared theoretical analyses of the rate of spread of a genetic mutation under drift and selection. Given several assumptions, Barton and Hewitt (1989) calculated that, in 10,000 generations, a mutation not subject to selection spreads, on average, 125 times the mean dispersal distance. (I am assuming here that dispersal distances from parent to offspring follow a normal distribution, as described further in Chapter 15.) By comparison, in 10,000 generations, a mutation that has a small (1%) selective advantage travels about 1700 times the mean dispersal distance (Fisher 1937), a 15-fold difference.

Consider the Coal Tit, which has a largely invariant song over 7000 km (Figure 10.5), and whose mean dispersal distance is estimated at 9.3 km in the U.K.

(Paradis et al. 1998). The spread of a neutral variant from one end of the species range to the other might take an average of about 60,000 generations, but a weakly selected one might take only 4,500 generations. The difference between rate of spread under neutrality and selection is actually larger than this because (1) occasional barriers to dispersal slow the rate of spread of alleles that are not subject to selection, but, provided there is any dispersal at all, they have little effect on the rate of spread of favored alleles (Barton 1979; Barton and Hewitt 1989), and (2) the favored mutation becomes fixed throughout the range as it spreads, whereas the neutral type coexists alongside the previous form. Fixation of a neutral song type may often take much more than N generations, where N is the population size (Hudson and Turelli 2003). In some birds, population sizes number many millions (Newton 1998, pp. 2–3).

Thus, neutral mutations are most likely to spread through a two-step process: first, the mutation is fixed in a small population, and then this population expands (Barton and Hewitt 1989). While small populations are needed for fixation by drift over reasonable timescales, large population sizes are more favorable for selection because a new favored mutation has a greater chance of appearing (Chapter 2).

Regiolects are characterized by song uniformity over large areas. Martens (1996) suggested that divergence through drift mechanisms has been important to their production. Conditions favorable to the process—small isolated populations—have probably arisen in response to climate fluctuations over the past 2–3 million years (Martens 1996). These should also be the conditions for "withdrawal of learning" to promote a burst of cultural mutation. In Chiffchaffs, for example, differences in regiolects (Figure 10.6) have been directly connected to differences generated during the rearing of isolated birds (Figure 10.7). Alternatively, some regiolect-specific features of songs could have spread as a consequence of their attractiveness. In such a case, a population would not have to be small. However, if the song is attractive, it might well spread rapidly throughout the range of the species, resulting in little population differentiation. Establishment of population differences may therefore require periods of geographical isolation so that different attractive mutations can arise in different populations, as discussed in Chapter 2.

SIMILARITIES WITH HUMAN LANGUAGE

In this short section, I briefly consider some of the similarities between bird song and human speech; it can be skipped without loss of continuity. Researchers in bird song often note the relevance of their studies to understanding human speech, at least in their grant proposals if not in their publications. The similarities between the development of bird song and the development of human speech (Marler 1970; Kuhl 2003) mean that studies of human language should be equally relevant to bird song. They are also relatively easy to read about, if only because linguists are better trained in writing than bird speciationists.

Parallels between bird song and speech development include learning, followed by rehearsal, and the gradual matching of vocalizations to the stored sounds (Kuhl 2003). Birds and humans both show a much greater propensity to learn when young, and, in many bird species, song learning ceases after the first year of life (Nordby et al. 2002). Social stimulation is essential. Writing about humans, Kegl et al. (1999, p. 223) state that "the source of language is within us but that the conditions for its emergence depend crucially upon community." Birds and humans alike develop very abnormal vocalizations when they are isolated or deaf, but they exhibit well-developed vocalizations when they are held together in peer groups. There are other more evolutionary parallels too. Human language and many bird songs are divided into dialects. When different dialects or members of closely related bird species overlap, individuals often sing songs of both groups (Chapter 14). Similarly, where human languages meet, many humans are bilingual. Languages and dialects promote assortative mating when populations mix (for birds, see MacDougall-Shackleton and Mac-Dougall-Shackleton 2001). Given these parallels, it is worth asking how human languages change (which has, of course, been studied in much more detail than bird song).

I have considered three ways in which bird song may diverge. Two are based in song learning—withdrawal of learning and drift of cultural mutations—and one is the selection of favorable mutations, which will apply whether the mutations are cultural or genetic. All have parallels in human languages. A remarkable example of withdrawal of learning in humans is the spontaneous development of Nicaraguan sign language in a single generation when deaf children were placed together in a school following the success of the Sandinistas. Deaf children had formerly been kept in isolation with their families, and most individuals had developed only rudimentary signs for communicating in the home (Kegl et al. 1999). The sign language is completely different from other sign languages and was not taught in the school, but rather, it developed among the children themselves.

Cultural drift also seems to be common among humans, and dialects often form within partially isolated social groups. A particular social group or dominant individual adopts a manner of speaking, often by incorporating ingredients present in a dialect elsewhere, and often as a mark of identification (Aitchison 1991, pp. 49–61). If the social group (or dominant individual) is respected and copied, new forms of speech can invade the population, not because of any inherent advantage of the new forms, but because of the prominence or prestige of the group (or individual) using them. This seems analogous to older birds being copied by younger ones.

While these mechanisms support the idea of cultural drift, a major mode of human language evolution is through selection on signaler and receiver, with new words and sounds frequently arising and occasionally spreading to fixation because of their utility or appeal (reviewed by Aitchison 1991). For example, new words are added for emphasis or politeness, but, after continued usage, they may lose their impact. Thus, in colloquial French, the phrase for *I don't know* has evolved from *je ne sais* (thought to have been prevalent from the ninth century through the thirteenth century), to *je ne sais pas* (in the seventeeth and eighteenth centuries), and now almost completely to *je sais pas* in some areas (Martineau and Mougeon 2003). The addition of the *pas* was probably added for emphasis (not *at all*). The loss of *ne* has occurred in association with other grammatical changes, and may reflect increased emphasis too, or simply obtaining the same result for less effort. This sort of thing is also seen in the quotation from Coryat at the beginning of the chapter, where the last syllable in *resonancie* has now been lost. Because word, grammar, and sound changes provide the environment for other changes to be preferred, language undergoes continuous change.

These results on language evolution suggest that contingency (i.e., the current form of the language), cultural mutation of words, grammar, and sounds, and general preferences for these novel features provide some of the main ingredients for a rapid rate of language change. Thus, human language evolves by mechanisms of cultural drift, in which words become popular because of who speaks them, as well as preferences for new cultural mutations because they are inherently attractive; perhaps it is reasonable to think that bird songs do too. "Withdrawal of learning," as in the example of the Nicaraguan sign language, may be much less likely an agent of change in humans than in birds, and selection of novel variants may be less common in birds than humans, where new words for new situations are constantly being invented. One important finding from human studies is the role of isolation; words very commonly spread into different languages. Pagel (2000) shows that the number of human languages in different regions of North America is associated with habitat diversity (classified into 23 major types, such as grassland, Mexican pine, and pine–oak). The explanation is not clear, but it may reflect reduced dispersal between groups exploiting different habitats.

GENETIC DIVERGENCE

Song differences often have a genetic underpinning. Young birds held together without tutors develop songs that can resemble those of closely related species, but the songs are more similar to conspecific song than to songs of distantly related species (e.g., Figure 10.7). When young are raised in isolation, species differ in the time when the young learn their songs, in the size of their repertoire (Kroodsma and Canady 1985), and in what they can learn from heterospecifics, presumably all a consequence of genetic differences (Beecher and Brenowitz 2005).

In some experiments, young birds preferentially learn conspecific song when they are given a choice of heterospecific or conspecific song (Nelson 2000a; Soha and Marler 2000). As noted earlier, in the White-crowned Sparrow, a species-specific whistle at the beginning of the song stimulates young birds to learn the notes that follow (Soha and Marler 2000). Likewise, if young male Swamp Sparrows and Song Sparrows are tutored with tapes containing songs of both species, they learn their own species songs preferentially, although they learn elements of the other species song when not given a choice (Marler and Peters 1977, 1988). Any preference to learn conspecific song is likely to have some genetic basis. The genetic influence need not be strong; it could be as simple as the young birds learning elements of a call note that they themselves produce innately, and this call note resembles elements in the adult's song. Such a mechanism has been proposed for some innate forms of species recognition (Chapter 13).

Genetic differences may arise subsequent to cultural evolution via genetic assimilation. In this model, individuals that learn songs most efficiently are selectively favored, leading to genetic evolution of song form, as in the Crusoe experiment (Chapter 7, p. 133). Assimilation cannot apply to species that do not learn their songs. In these species, changes must be driven by genetic evolution from the start. A way that songs evolve genetically in both species that do and do not learn songs is likely to be via selection favoring attractive variants. For example, some species that learn their songs have large song repertoires. Differences in song repertoire size across species may reflect benefits of large song repertoires in terms of attractiveness to females, benefits of small repertoires in terms of male competition, plus various costs to large repertoire size (Podos et al. 2004; Beecher and Brenowitz 2005). As a consequence of different selection pressures in different environments, repertoire size is likely to evolve genetically. Consistent with genetic differentiation, different subspecies of Marsh Wren retain differences in the size of their song repertoire when they are taken from the wild as nestlings and raised in captivity (Kroodsma and Canady 1985).

Even species that learn their songs have performance constraints; species differ in morphology and physiology, and learning takes place in the context of how easy it is for birds modify sound production (Slabbekoorn and Smith 2002; Podos et al. 2004; Suthers and Zollinger 2004). Because morphological and physiological differences represent responses to ecological selection pressures, this point is revisited when I consider the role of ecology in song evolution (Chapter 12).

Although most closely related species of birds may differ genetically in some aspects of song development, genetic differences are not essential to speciation. Any learning system that readily produces songs that are diagnosable as species-specific should suffice. Accurate song learning, together with accurate learning to recognize distinctive features of songs, should be sufficient for complete pre-mating reproductive isolation to develop, in the absence of any genetic differentiation (Chapter 13).

CONCLUSIONS

Songs may evolve because new components elicit favorable reactions from receivers, the mechanism discussed in detail in the previous chapter. The best-studied stimulatory effect is that multiple song types induce increased numbers of copulation-solicitation displays from captive females (Searcy 1992; Collins

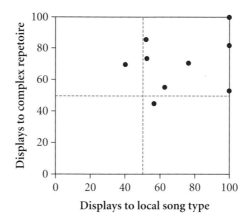

Figure 10.9 Nine female Yellowhammers (implanted with estradiol to elicit solicitation displays) were tested with recordings of local vs. foreign dialects (Norway vs. Denmark) and, in a separate experiment, recordings of three song types versus one song type from the local dialect (Baker et al. 1987). The preference for local vs. foreign dialect is scored as the fraction of copulation-solicitation displays directed towards that dialect. The preference for multiple songs is scored as the fraction of all displays given during the multiple-song playback sessions. Dashed lines divide preference for one type over the other.

1999). Songs may also evolve as a result of the drift of syllable types and "withdrawal of learning." In Yellowhammers, Baker et al. (1987) demonstrated female preferences for more song types and also a greater response to the local dialect (Figure 10.9). In this species, then, females may base their choice of mates on two criteria that have their roots in these two different mechanisms. Both mechanisms of song evolution (increased stimulation and cultural drift/withdrawal of learning) probably operate in species that learn songs. The two mechanisms are not mutually exclusive, and they interact with each other. For example, a novel song variant may have stimulatory effects against the background of one dialect, but not against the background of another.

Given that song learning seems to promote song divergence between populations, learning may also promote speciation, and this is suggested by the fact that nearly half of all bird species are songbirds (Fitzpatrick 1988). While the role for song learning in promoting speciation is difficult to assess, given that only three groups clearly have song learning, there is little evidence to support the idea, and some groups without learning also contain many species (Baptista and Trail 1992). Diversification rate (speciation minus extinction) in the early radiation of the lineages that led to the present-day species of songbirds was slower than the rate at a comparable stage in the diversification of the Ciconiiformes [gulls, storks, etc., 1027 species in the Sibley and Monroe (1990) compilation; Nee et al. 1992]. At least part of the recent diversification in South America is due to rapid speciation in the suboscines (tyrant flycatchers, antbirds, etc., 1151 species; Figure 6.9, p. 116), which do not have song learning. It appears, therefore, that song learning contributes to divergence in songs and provides a basis for premating isolation, but I suggest that this divergence is unlikely to be a major rate-limiting step in the production of new species.

SUMMARY

Geographical variation in songs occurs on all spatial scales, from rapid turnover of song types in the space of a few kilometers, to invariance over thousands of kilometers. In many species that learn their songs, different populations are characterized by song dialects with relatively sharp boundaries between them. In these species, males and females usually respond more strongly to songs from their own dialect, rather than to foreign songs. I consider three ways by which song learning could lead to song divergence between populations, all of which have been described for the evolution of human languages. First, some song variants may provide more stimulation than others, at least against the background of the songs currently being sung in the population. Second, more than half of all bird species learn their songs by copying, often from unrelated neighbors. Disruptions in the copying process lead to new song variants, and these variants may become established through drift. Finally, wholesale disruptions in the copying process, both in song copying and in song recognition, arise when insufficiently tutored young birds establish a new population. This "withdrawal of learning" mechanism may be an important cause underlying the formation of song dialects. Thus, song learning predisposes populations to evolve discrete song differences rapidly. Although species without song learning are not known to form dialects, they do commonly differ in their vocalizations. Mechanisms that do not require song learning, such as spread of attractive genetic mutations and ecological selection pressures, must be implicated in their divergence. While song is an important species-recognition mechanism, there is little evidence that song learning has accelerated speciation, and some groups without song learning have had high diversification rates.

In the following two appendices, I summarize studies on two species of songbirds that capture many of the features described in the text. These include the advantages of song sharing, the development of songs in isolation, and divergence from close relatives.

APPENDIX 10.1: SONG OF THE INDIGO BUNTING

The Indigo Bunting breeds in the eastern part of North America and spends the winter in Central America, the Caribbean, and northern South America. Based on a catalogue of songs collected from Iowa, Nebraska, Michigan, New York, and Kentucky, Thompson (1970), Shiovitz and Thompson (1970), and Baker and Boylan (1995) identified a total of ~147 syllable types in the Indigo Bunting's vocabulary. Each male sings a single song type, consisting of five to eight different syllable types (maximum 23) drawn from this pool (Shiovitz and Thompson 1970; Payne 1996; see Figure 10.1). A few percent of all males sing two song types (Payne and Payne 1997). Different males sing songs that differ in syllable type, syllable-type order, and/or repetition frequency of syllables. Although syllable types are not completely interchangeable in their position and some tend to occur at the beginning or end of the song (Baker and Boylan 1995), the potential diversity of songs is enormous.

In a study in Michigan, about 70% of young males learned their song from a single neighboring older male with whom they interacted frequently (Payne 1981, 1983; Payne and Payne 1997). Of the remainder, 20% retained a song they had learned before their arrival in the study area, and about 10% copied sequences and elements from more than one neighbor (Payne and Payne 1997). Young (first-year) males whose song matched that of an older male were more likely to obtain a mate than those whose songs did not match (87% of $N = 214$ matchers, cf. 66% of $N = 84$ nonmatchers; Payne et al. 1988). 7.5% of males ($N = 620$) changed their songs between years. These were mostly first-year birds changing to match a neighbor, and many had changed their song repeatedly in the previous season. Older males very rarely change their songs. When they do, they may switch to songs not recorded previously in the area (Payne et al. 1988).

Songs are often learned accurately and in their entirety, but, occasionally, syllable types are deleted or added, probably a result of copying from more than one tutor. In a Michigan study, Payne et al. (1988) required three syllable types in the same order for two songs to be considered a match. By this criterion, the half-life of a song type is only about 3.3 years (Figure 10.1), mostly because all birds singing a particular song type die. Because of copying, a song type persists about three times longer than a bird singing that song type.

In Michigan, no song types are shared between populations separated by 240 km, but many syllable types are shared. About 80% of the syllable types sung in Nebraska and Iowa are also found in Michigan (900 km distant; Baker and Boylan 1995), and more than 90% of the syllable types sung both in Kentucky and in New York (500–600 km distant) are also found in Michigan (Shiovitz and Thompson 1970; Emlen 1971). In other words, syllable types show little geographical variation across the species range, but song types vary considerably.

The Lazuli Bunting is a close relative of the Indigo Bunting. It breeds in the western United States and hybridizes with the Indigo Bunting in the Great Plains (described in Chapter 15). DNA evidence indicates that the species may be separated by more than 3 million years (Klicka and Zink 1997; Chapter 15). The singing behavior of the Lazuli Bunting is similar to that of the Indigo Bunting (Thompson 1976). A total of 122 syllable types have been identified from a survey across seven western states and Idaho and Nebraska (Thompson 1976; Baker and Boylan 1995). Thompson (1976) estimated that about 20% of the syllable types of each species were held in common between the species. However, quantitative analysis of two of these syllable types indicates that their morphology differs significantly between species (Baker and Boylan 1995). It appears that some syllable types may only diverge over millions of years, but it is possible that syllables leak through the shared border; individuals of one plumage type sometimes sing the song of the other species (Emlen et al. 1975; Baker and Boylan 1999).

APPENDIX 10.2: SONG OF THE CHAFFINCH

The Chaffinch is a common breeding bird across Europe and into Siberia. It has also been introduced into South Africa and New Zealand. Marler (1956) described 12 different vocalizations of the Chaffinch and the context in which they are used. These include 10 different calls, the plastic song [which is a soft warble that may start quite soon after the birds leave the nest and is particularly prominent in the bird's first year (Thorpe 1958b)], and the territorial song. Male Chaffinches have one to six different song types in their song repertoire [in England and Spain, each male sings two to three song types (Slater et al. 1980; Lynch and Baker 1993)]. Each song type is characterized by five to six different syllable types (occasionally up to 12; Lynch and Baker 1993). Within a song type, syllable structure and syllable order are stereotyped, with occasional variation in the number of times a particular syllable type is repeated (Slater and Ince 1979, 1982). A male typically sings each song type a few times in succession before switching to the next (Hinde 1958; Slater and Ince 1982). In a wood in southern England, a male shared a song type with an average of two out of three of his neighbors (Slater 1981). In tape-playback experiments, males appear to respond more strongly if played a song in their repertoire, and they show a tendency to match song types, that is, to sing the type that is played (Hinde 1958; Slater 1981).

In southern England, males sing from February to the beginning of July, with some irregular singing in September and October (Witherby et al. 1938). Song learning appears to start soon after fledging; a male tutored with a tape in his first two months of life reproduced the tutored song the following spring (Slater and Ince 1982). However, late-hatching birds may not hear any song until the fall or even until the following spring (Slater and Ince 1982), and Thorpe (1958a, b) showed that tutoring by tapes or live models can substantially influence song form up to about 13 months of age. After that time, little song learning takes place.

Across a 200-km transect in New Zealand, approximately 245 different syllable types were identified from 570 birds (>20 songs per bird were studied; Baker and Jenkins 1987; Lynch et al. 1989). Within each of these syllable types, syllable variants were also recognized (an average of 1.9 variants per type). Syllable variants are small differences that are repeatable among males and are therefore likely to be recognized by the birds as different. They are presumed to arise through copying errors, and they provide one way by which new types could evolve from old ones (Slater and Ince 1979). An alternative means of generating variation is by the invention of entirely new syllable types. Evidence for this is difficult to obtain in the field, because the alternative of immigration is almost impossible to reject (Slater and Ince 1979). However, in laboratory studies in which birds are not exposed to adult tutors, very odd syllable types are produced. Individuals hand-raised from the nestling stage and held in acoustic and visual isolation produce simple songs, although song length and number of notes are similar to those of wild Chaffinches (Figure 10.2). If similarly isolated individuals are held together in small groups, they produce more complex songs, resembling in structure those of wild Chaffinches but still containing syllable types different from those observed in nature (Thorpe 1958a, b; see Figure 10.2).

Different woodlands in a local area differ significantly in the frequencies of different song types (Slater and Ince 1979). Comparisons across wider geographic scales reveal a correlation between geographical isolation and the degree of differentiation (Baker and Jenkins 1987; Lynch et al. 1989; Lynch and Baker 1994). In the Azores and the Canary Islands, individuals sing more song types per bird (about four in the Azores, and more than five in the Canaries) than they do in Spain and England (on average fewer than three) (Lynch and Baker 1993). The high number of song types per bird on the Canary Islands is mainly attributable to the presence of several similar song types that differ from each other in minor rearrangements in the order of syllable types. However, more syllable types are found in the Azores than in Spain or England. The explanation for the difference between the islands and the mainland is not obvious. The density of Chaffinches on the islands is higher than it is on the mainland, especially in the Azores, and Lynch and Baker suggest that the greater syllable diversity reflects increased population size. Given the relatively small area, it is not clear that the population size is larger. It is possible that, instead, the higher density creates selection pressures favoring multiple song types (Chapter 12).

New Zealand birds that were introduced from England in the 1860s and 1870s have developed songs with shorter beginning segments and more elaborate end segments (Jenkins and Baker 1984). This may relate to the need for transmission through primarily coniferous habitat, if end phrases degrade less in this habitat. This is an ecological explanation for differences between songs, as considered further in Chapter 12. Birds on Chatham Island colonized from New Zealand more than 100 years ago. They have songs more closely resembling English birds, and they likewise occur in broadleaf trees (Baker and Jenkins 1987).

The closest relative to the Chaffinch is the Blue Chaffinch, which occurs on two islands in the Canary Islands, in the highland zone. The Chaffinch is found in the lowlands on these islands and on all the other islands of the Canary Islands. The Blue Chaffinch is thought to be the product of an earlier invasion of Chaffinch stock from the European mainland; molecular dating suggests that it has been separated from the Chaffinch for 0.3 million to 1 million years (Marshall and Baker 1998). The Blue Chaffinch's song is of lower frequency with shorter syllables than the Chaffinch's song (Figure 10.2). In some ways, it resembles the song of a Chaffinch that was raised in social isolation (Figure 10.2). Lynch and Baker (1991) suggest that the song of the Blue Chaffinch may have originated through a failure of the founding Chaffinch flock to be tutored by adults.

CHAPTER ELEVEN

Divergence in Response to Increased Sexual Selection

In a constant environment, the replacement of one socially selected trait by another may be difficult because the production of a new signal has costs to the signaler, and paying attention to a new signal may often have costs for the receiver. Sometimes, however, environments change to become more benign; for example, predation pressure may decline, or food levels may increase. Under such conditions, the costs of maintaining socially selected traits are reduced and, if the benefits remain the same, these traits should become more exaggerated. Populations may evolve more color, more complex displays, and/or more elaborate songs. Because there are many ways in which to be complex, but rather fewer ways in which to be simple, differences between populations will become magnified (Figure 11.1). In addition

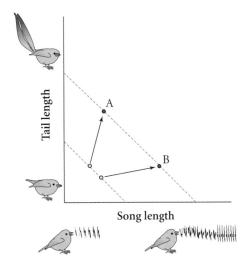

Figure 11.1 Two ancestral populations may differ slightly in two sexually selected traits (e.g., song and tail length) but be subject to approximately the same intensity of sexual selection, as indicated by the dashed line. An increase in the intensity of sexual selection leads to exaggeration of both characters, and also to an increase in the difference between the derived populations, *A* and *B*.

to exaggeration of those signals already present in populations, entirely new traits could become established in benign environments.

In this chapter, I investigate the possibility of population divergence in socially selected traits, when conditions favor an increased expression of such traits. In the extreme, selection pressures are identical in each population, so divergence is entirely due to any differences in the genetic make-up of the two populations. A great empirical system with which to evaluate this is the radiation of a single captive species into different breeds. Such breeds are often distinguished by what are considered to be sexually selected traits in wild species. Differences include colors, feather ornaments, vocalizations, and sometimes even displays. The agency of selection has been the artificial breeder and his or her appreciation of beauty (Darwin 1871). This has resulted in a huge diversity of breeds; more than 350 recognized breeds have descended from the ancestral Rock Dove (see Figure 11.2), with many variants within them. Darwin (1868) argued that the different breeds of the Rock Dove would probably be classified into at least six different genera if they were encountered in the wild, and there has been ongoing diversification since Darwin's time.

I compare differences among artificially selected breeds to differences among related species in order to ask the following questions:

| Aldrovandi 1610 Willughby 1676 | "Treatise" 1765 | Fulton 1876 | Lyell 1887 | Hill 1940 |

| Willughby 1676 | "Treatise" 1765 | Eaton 1858 | Lyell 1887 | Fulton and Lumley 1895 |

Figure 11.2 The accumulation of feather ornamentation in the Jacobin (*above*, described as "the chrysanthemum of the bird world" by Levi 1963), and Pouter breeds of the Rock Dove (from Levi 1963, pp. 97, 144).

1. How have breeds diversified over time? Is the pattern of diversification attributable to the appearance of attractive mutations? The alternative is that breeders have had specific goals and have attempted to harness genetic variation in the base population to achieve those goals (as an applied animal breeder does in trying to improve egg or meat production). Such deliberate selection has certainly been practiced on occasion by breeders, as exemplified by the great efforts made to make a red Canary, which was eventually achieved only via hybridization (Birkhead 2003).

2. How well does divergence among breeds act as a model for the process of divergence among species? The captive bird model is one in which conditions are far more benign than could be expected in the wild. A priori, we might expect diversification to be more limited in nature because many of the new mutations and variants produced in captivity have strongly deleterious effects, and the varieties are maintained only by careful husbandry (e.g., in the Jacobin breed of pigeon, feathers over the eyes completely obscure the bird's vision; Figure 11.2). In nature, the range of color patterns that can become established may be limited. However, in species that learn songs, different (cultural) song mutations arise very frequently. I conclude this chapter with an example in which song divergence between species may have been driven as a result of similar selection pressures for increased exaggeration.

DIVERGENCE AMONG DOMESTIC BREEDS

Many bird species have been domesticated. Some have been raised primarily for food, whereas others have been kept for show. The Rock Dove and the Red Jungle Fowl (pigeons and chickens) are the oldest domesticated species. Molecular analysis shows that chicken breeds are all descended from a single species (Fumihito et al. 1996), although domestication has happened independently at least nine times (Liu et al. 2006). The evidence that all pigeon breeds are descended from the Rock Dove is also very strong (Darwin 1868; Crawford 1990). On the other hand, some species, most notably the Canary, have been out-crossed to different species in order to introduce particular traits (Birkhead 2003).

The commoner domesticated species are listed in Figure 11.3 and Table 11.1. The following species are the main omissions: the Pheasant (partly feral), Helmeted Guinea-fowl, and Japanese Quail show relatively little diversification (Crawford 1990). The domestic goose, comprising at least 16 breeds, was originally domesticated more than 2000 years ago, and it may be descended entirely

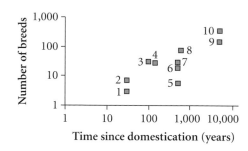

Figure 11.3 The number of breeds or varieties of domestic species plotted against time since domestication (from Price 2002). For the more recently domesticated species, the number of varieties is recorded; for the older ones (Canaries, Budgerigars, chickens, and pigeons), the number of breeds is noted. (1) Greenfinch: The first mutations appeared in British stocks in the 1950s (Lander and Partridge 1998). Four color varieties that affect loss or distribution of melanins (two are at the same genetic locus) are known. (2) Rosy-faced Lovebird: Seven color mutations have arisen since the 1960s in Australia (Shephard 1989). (3) Zebra Finch: The first color variety was recorded in 1921; 30 "mutations" are distinguished based on color variants and crests (Zann 1996, p. 288). (4) Budgerigar: Imported from Australia into Britain in 1840; first yellow and blue mutations in the 1880s (Bennett 1961; Rutgers and Norris 1972). Following Binks (1997), 28 breeds are listed (excluding composites; for example, the crest is counted only once). There are clearly many more variants of the Budgerigar than of the Zebra Finch. (5) Muscovy Duck: Six varieties (Roberts 1997). First domesticated at least 500 years ago when the color varieties became established (Crawford 1990). (6) Turkey: May have been first domesticated 2000 years ago. However, color variants were observed in the 1500s, and it is these stocks that may have given rise to present-day variants (Crawford 1990). Roberts (1997) lists 17 breeds, and there is an additional one in the American Poultry Association publication (Anon 1998). (7) Mallard: Has been domesticated repeatedly since at least 2000 years before the present (Crawford 1990). Modern breeds may be descended from a domestication event 500 years before present, or more recently (Crawford 1990). The American Poultry Association publication (Anon 1998) lists 27 varieties in 13 breeds. Roberts (1997) lists 25 breeds, of which ten are also on the American list. (8) Canary: Probably domesticated in Spain during the fifteenth century. In 1750, there are records of color varieties, and the crested form was first mentioned in 1793 (Bennett 1961; Vriends 1992). (9) Chicken: Domesticated chickens were in China 7500 years ago, but perhaps the source of modern varieties was 4500 years ago from the Indus (Crawford 1990). The American Poultry Association publication (Anon 1998) lists 382 varieties in 105 breeds. British Poultry Standards (Roberts 1997) lists 94 breeds without distinguishing varieties, and 53 of these breeds appear to be on the American list. The true number of breeds is much higher, because breeds from the rest of the world are not recorded. (10) Pigeon (Rock Dove): The first records were in fertility shrines about 6500 years ago, but may be older (Johnston and Janiga 1995). The first records of probable varieties date back 3000 years (Levi 1965; Hansell 1998). There are 350 breeds on a recent European list, but there are many more in other parts of the world (especially China) that are not recorded (Hansell 1998). Levi (1965) lists 329 breeds, some of which have many color varieties. More generally, species with many breeds have varieties within breeds that are unrecorded (e.g., color variants within breeds of pigeons), resulting in an underestimate of the disparity in diversity between those species with many breeds and those species with few breeds.

Table 11.1: Presence (X) of characters distinguishing breeds and varieties in domesticated species (from Price 2002)

	Number of breeds or varieties	Color and pattern	Crests	Shape[1]	Vocalizations	Abnormal feathers[2]	Exaggerated Tails	Flight displays[3]
Greenfinch	3	X						
Rosy-faced Lovebird	7	X						
Muscovy Duck	6	X						
Turkey	6	X						
Zebra Finch	30	X	X					
Budgerigar	28	X	X					
Mallard	28	X	X	X				
Canary	72	X	X	X	X	X		
Chicken	146	X	X	X	X	X	X	
Pigeon	350	X	X	X	X	X	X	X

For references and documentation of breed numbers, see the legend to Figure 11.3.

[1] Elongate body shape, as well as (in Mallard, chicken, and pigeon) modifications of bill and tarsus.
[2] Includes various feather modifications, such as that in the frizzle breed of poultry, in which each feather curls towards the head.
[3] Tumblers and rollers perform backward somersaults in the air. Other breeds have variants on the aerial display of the ancestral form (Baptista 1981).

from the Greylag Goose (Crawford 1990; Roberts 1997), but ancestry and time of domestication of the line that gave rise to present-day breeds seem more uncertain than in other species. Many species of captive finches are not included (e.g., Rutgers et al. 1977; Shephard 1989). I use the European Greenfinch, Zebra Finch, and Canary as representative. The other species are fairly recently domesticated, and they show results similar to those of the Zebra Finch and the Greenfinch reported in Table 11.1. For example, Shephard (1989) lists nine species of Australian finches, in addition to the Zebra Finch, that have been domesticated mostly in the twentieth century and that are "common" or "secure" in captivity in Australia. He describes from zero to eight varieties within these nine species (average 2.3), all due to color variants, and, in this way, these finches are similar to the Greenfinch (Figure 11.3).

I quantified measures of breed diversity for a given species as the number of breeds or varieties described in catalogs (Figure 11.3) or more simply by the presence of different kinds of traits (Table 11.1). Times of domestication of the various breeds are uncertain, but they clearly vary over two orders of magnitude (see legend to Figure 11.3). Time since domestication correlates with number of breeds (Figure 11.3; Spearman's rank correlation, $\rho = 0.75$, $N = 10$, $P = 0.03$). Breeds of chickens and pigeons, both of which were domesticated long ago, are distinguished by feather ornaments, beak and body shape, various knobs and wattles, vocalizations, and, in pigeons, unusual displays. Within each of these breeds there are many color varieties. However, only color varieties are present within species that have been more recently domesticated, and these varieties are often given the status of breeds. Thus the disparity between the older domesticated species and the more recently domesticated species is greater than it appears in Figure 11.3.

A main finding is a hierarchy of trait distributions among species; the more varied species differ in all the traits that distinguish the less varied species (Table 11.1). An exception to this hierarchy is the Bengalese Finch, which has evolved a more complex song (Chapter 10; Okanoya 2004), but not differences in shape or in crests, as far as I know. The changes in song were apparently not a result of selection by the breeder but, rather, appear to be a consequence of relaxed natural selection when birds are maintained in captivity (Okanoya 2004).

The patterns in Table 11.1 and Figure 11.3 suggest that all species are on a similar trajectory of diversification. This implies that recently domesticated species such as the Greenfinch could eventually become as diverse as pigeons. Despite this, not all traits diversify in the same way within each group, showing that the course of evolution is contingent on the form of the ancestor. For example, the song of the roller Canary has been selected for its melodies (Güttinger 1985). The oldest domesticated species, chickens and pigeons, which are nonpasserines, do not have the syrinx needed to develop such a varied song; it would presumably take many thousands of generations of selection to produce a singing chicken. The role of contingency is seen in more recent divergences too. As noted explicitly for Budgerigars by Binks (1997), some color mutations are recessive with respect to wild type but dominant with respect to other breeds. In other words, the trait is not visible in offspring from crosses between the breed and the wild progenitor, but it is visible in offspring from crosses between this breed and some other breeds. Such findings imply that the chances of one mutation becoming fixed depend on what other mutations have spread previously.

Mutation

Distinctive characters among breeds appear in an orderly manner, with color varieties always first (Table 11.1; Figure 11.3). A reasonable explanation for this is that attractive mutations are seized upon and propagated by the animal breeder, and that visible mutations in color arise the most frequently. In this section, I review the genetics of breed differences to ask if new mutations of large effect are the driving force behind the origin of new breeds. The alternative is that the breeder actively searches for individuals showing variation in the required direction and gradually builds up the character (e.g., in selecting for red coloration, he or she begins by searching for individuals that show some redness in their plumage). This alternative predicts a polygenic basis for breed differences. To examine this I review the genetic basis to the various characters.

With respect to color patterns, many differences between breeds are the result of single mutations, as has been shown by crossing among breeds and backcrossing to the wild type (Table 11.2). This strongly suggests that a new mutation was the initial step driving the development of the new variety. Mutations are often recessive (Sossinka 1982), meaning that they are hidden from selection when they first arise and are likely to be lost. Indeed, the bird-fancying literature contains graphic descriptions of attempts to save rare recessive mutations by inbreeding among close relatives (Binks 1997; Lander and Partridge 1998). This perhaps limits any analogy we wish to draw between domestic breeds and species in nature in which such inbreeding is unlikely, but it supports the idea that domestic breed divergence is driven by selection of attractive new variants.

Approximately 20% of the mutations that distinguish breeds of birds are on the Z (sex) chromosome (i.e., they are sex-linked; Table 11.2) and 80% are on the other chromosomes (termed the autosomes). The frequency of sex-linked genes is high, given the relative size of the Z chromosome. In the chicken, less than 4% of all the DNA and only about 2.7% of all the genes (a total of 369 genes) have been mapped to the Z chromosome (Hillier et al. 2004), although

Table 11.2: Inheritance patterns in crosses among breeds (from Price 2002)*

	Sex-linked			Autosomal			Reference
	Dominant	Intermediate	Recessive	Dominant	Intermediate	Recessive	
Pigeon	3	0	4	6	3	22	Levi (1963, 1965)
Chicken	3	1	2	6	10	12	Smyth (1990)
Muscovy Duck	0	0	1	1	2	6	Avanzi and Crawford (1990)
Mallard	0	0	2	4	3	4	Lancaster (1963)
Common Pheasant	1	0	4	0	2	11	Somes (1990)
Budgerigar	0	0	7	3	6	9	Binks (1997)
Zebra Finch	0	0	5	3	0	8	Zann (1996, p. 288)
Rosy-faced Lovebird	0	0	3	1	0	3	Shephard (1989)
Totals	7	1	28	24	26	75	
Totals (%)	4%	1%	17%	15%	16%	47%	

* Primarily color differences

this number may increase as the genetic map becomes increasingly refined (Dawson et al. 2006). One reason why so many traits causing breed differences are on the Z chromosome may simply be that many genes affecting sexually selected traits are on this chromosome. Alternatively, sex-linkage may be common because new mutations are more immediately visible or because they arise and become fixed more frequently. The reason they may be visible is that recessive mutations are exposed in the female who carries a single Z chromosome (Haldane 1924; Rice 1984). Males are ZZ and females are ZW, with the W chromosome degenerate (Chapter 16).

Sex-linked mutations may spread more frequently even if they are not recessive. Mutations occur relatively more frequently on the Z chromosome than the autosomes because, first, in any one generation, two-thirds of all copies of a Z-linked gene are in a male, whereas only one-half of all the copies of an autosomal gene are in a male, and, second, mutations are estimated to be five times more frequent in the male germ line than in the female germ line (Ellegren and Fridolfsson 1997). The higher mutation rate in males probably arises because there are many more cell divisions en route to producing a sperm rather than an egg. The much higher mutation rate in males leads to the expectation that favored Z-linked mutations will be established more frequently than autosomal mutations (Kirkpatrick and Hall 2004).

The preceding analysis applies to color patterns. Crests are also often determined by single major mutations, which may sometimes be dominant (e.g., in the Mallard; Lancaster 1963). In this case, new mutations are immediately visible to selection. Crests in the Budgerigar and Canary are circles of feathers growing out from a central point in the crown and may reflect a skull defect (Duncker 1931; Bennett 1961). In the Budgerigar, they are due to an incompletely dominant mutation, with greater expression in the homozygote. In the Canary, the crest is due to a dominant mutation that is lethal when homozygous (Bennett 1961). Similar mutations have also been reported in the Chestnut-breasted Munia (Mobbs 1992) and the Zebra Finch (Zann 1996, p. 288). In these species, the mutations have such deleterious effects that they are unlikely to spread in nature, but, again, they indicate the role of mutation in facilitating the generation of breed diversity: the appearance of a new mutation caught the breeder's eye and initiated the establishment of a new line. Once established, crests have been exaggerated by further selection, as a result of the breeder selecting among the continuous variation in the population. The increased development over three centuries of the feather hood in the Jacobin breed of pigeon is illustrated in Figure 11.2.

In the case of long or unusual tails, there is less evidence for the involvement of mutations of large effect. The tail of the fantail breed of Rock Dove has a large and variable number of feathers. Morgan (1918) crossed three homing pigeons that have a constant number (12) of feathers with fantails having between 29 and 32 tail feathers, and he then interbred the offspring. Based on these crosses, Morgan deduced that at least three genes must contribute to the difference in the number of tail feathers, as can be confirmed using Lande's (1981b) quantitative method (Price 2002). This is a minimum estimate, and it implies a polygenic basis. However, Darwin (1868, p. 223) did describe the appearance of minor visible tail variants in his pigeon stocks. Such variants may form the basis for initial selection by artificial breeders. Thus, the evidence is equivocal. Tail diversification may have been driven originally by the appearance of new readily visible mutations, or it may have been a result of breeders actively seeking out variants.

Courtship displays and vocalizations have evolved among breeds of captive pigeons, but it is not known how they originated. The ancestral Rock Dove performs an aerial display involving a few claps of the wings. Nicolai (1976) crossed breeds that perform more and louder claps with those resembling the wild type and found the loudness of the claps to be intermediate in the F_1 birds. With respect to vocalizations, a breed of pigeon differing in voice from the wild type was crossed with another resembling the wild type, and the voices of the F_1 birds were intermediate (Baptista and Abs 1983). Intermediacy of the F_1 birds may be consistent with a polygenic model, but it does not rule out the possibility that a single gene of large effect is of importance. The rarity of differences in displays and vocalizations seems to make it likely that breeders seized on a new variant rather than selecting on the base population. The somersaulting habit of the Tumbler breed of pigeons, which is not a courtship display, appears to be largely attributable to a single recessive gene affected by additional genetic or environmental influences (Entrikin and Erway 1972).

In conclusion, the appearance of new mutations clearly plays a role in determining patterns of diversification among breeds. Color variants arise first and seem to be most often attributable to mutations of sufficiently large effect that they are readily visible to the breeder. Other traits, such as crests and unusual tails, may have a lower mutation rate, or the mutations may be of small effect, or they may be less appreciated by the breeder. There is good evidence that these feather ornaments have become greatly exaggerated by careful and prolonged selection of relatively minor variants (Darwin 1868, p. 224; see Figure 11.2), but it is not known if the appearance of the variants was what drove the initial attempt at selection or if breeders specifically selected for increased feather length among the normal variants in the population.

SPECIES DIVERGENCE AND BREED DIVERGENCE

Both single-gene and polygenic differences are likely to distinguish related species in colors and feather ornaments (Price 2002), and in this way resemble differences among breeds of a single species. However, species crosses have rarely been taken through to the second generation, which, until recently, was the only way to distinguish alternative modes of inheritance. New genetic techniques, such as the identification of specific genes involved in adaptation, are now being used to study the genetics of species differences. In one example, Mundy et al. (2004) showed that mutations in the same gene (*MC1R*) affect melanin patterns in a passerine, a goose, and at least two species of skuas. Mutations in this gene also cause breed differences in chicken and quail (Nadeau et al. 2006). Over the coming years, molecular studies should lead to a good understanding of the genetics of species differences. It is beginning to look as if single-gene differences of large effect are more important than previously appreciated. If so, perhaps the analogy from domestic breeds to species is not far-fetched.

Species diversification resembles breed diversification in other ways. Darwin (1868, p. 243) noted: "…the varieties of one species frequently mock distinct but allied species." This is especially obvious for the domestic pigeon (Baptista et al. 1997, pp. 105–106; Table 11.3). Baptista studied the 322 extant species of the Columbiformes (pigeons and doves) and showed that many of the characteristic features that distinguish species also distinguish breeds of the domestic pigeon. These include the color of bare parts (exposed skin, eye color), feather ornaments (tails, crests, leggings), and feather color patterns (Table 11.3). Baptista's result gives some credence to the use of artificial breeds as a model for the divergence of sexually selected traits among species.

A second similarity between species diversification and breed diversification comes from the observation that crosses among species sometimes result in color patterns that resemble a third related species (Table 11.4). In two cases in ducks, a display not present in either parental species, but present in other duck species, is present in the hybrid (Kaltenhäuser 1971; Price 2002). Likewise, crosses among pigeon breeds and among chicken breeds sometimes produce offspring closely resembling the wild type (Darwin 1868, pp. 197–201, 242). The among-breed crosses regenerate the wild type presumably because recessive mutations with different phenotypic effects have become established in each breed. Because they are recessive, their expression will be masked in the hybrid, exposing the ancestral pattern. In the case of the crosses between species, it is possible that the resemblance of the hybrid to a third species is not because the third species carries ancestral traits, but instead that the third species is the derived form: it was formed in nature as result of hybridization of the other two (Ghigi, described in Phillips 1915; Scherer and Hilsberg 1982; Chapter 15). However, the results from the domestic breed studies are consistent with the idea that hybrids in species crosses reveal an ancestral form that has been masked by the fixation of recessive mutations.

To the extent that patterns in domesticated breeds of birds can be applied to those in nature, they support a model of population divergence whereby different mutations accumulate in different lines or populations, and divergence is driven

Table 11.3: Traits present in domestic breeds of the Rock Dove, as well as various wild species of pigeon and dove (Baptista et al. 1997; L. Baptista, *personal communication*)

Character	Breed	Species
Swollen eye cere	Carrier, Steinheimer Bagdad	Diamond Dove, Bare-eyed Pigeon
Swollen nasal cere	Barb, Dragoon	Seychelles Blue-Pigeon, Carunculated Fruit-Dove
Feathered leggings	English Tumbler, Lahore	Fruit-Doves (*Ptilinopus*), Cloven-feathered Dove
Frontal crest	Trumpeters	Topknot Pigeon
Occipital crest	American Crest, Archangel	Crested Quail-Dove, Green Imperial-Pigeon
Extra tail feathers	Fantail	Crowned Pigeons (*Goura*), Pheasant Pigeon
Elongated tail and wings	Swift	Passenger Pigeon
Sexual dichromatism	Reehani Dewlap, and several other breeds	Ground Doves (*Claravis*), Orange Dove
Absence of oil gland	Fantail, White Carneau	Fruit-Doves (*Ptilinopus*), Crowned Pigeons (*Goura*)
Orbital skin blue	Rock Dove	Jambu Fruit-Dove
Orbital skin yellow	Roller	Bare-faced Ground-Dove
Orbital skin red	Barb	Diamond Dove
Orbital skin white	Homer	White-crowned Pigeon
Orbital skin grey	Rock Dove	Luzon Bleeding-heart
Iris color white	Baldhead Tumbler	Speckled Wood-Pigeon
Iris color red	Rock Dove	European Turtle-Dove
Iris color blue	Tung Koon Pak	Luzon Bleeding-heart
Iris color yellow	Squabbing Homer	Grey-chested Dove
Black forehead patch	Spot	Stephan's Dove
Penguin color pattern	Lahore	Tambourine Dove
Green body, white tail	White-tail	Nicobar Pigeon
Black head, white neck and breast	Gazzi Modena	Snow Pigeon

Table 11.4: Color patterns appearing in hybrid crosses that resemble a third extant species

Parent 1	Parent 2	Color pattern	Present in:	
Red Shoveler	Northern Shoveler	Pale cheeks / grey sides of head	Australian Shoveler	Harrison and Harrison (1963a)
Red Shoveler	Northern Shoveler	White neck ring[1]	Mallard, other ducks	Harrison and Harrison (1963a)
Northern Shoveler	Hybrid*	Pale cheeks	Australian Shoveler	Harrison and Harrison (1971)
Mallard[2]	Gadwall, Speckled Teal, Spot-billed Duck	Green crown stripe	Eight other Anas species	Harrison (1963a)
Tufted Duck	Ferruginous Pochard	Green gloss on crown	Other Aythya species	Harrison (1963a)
Common Teal[3]	Northern Shoveler	Bridled face pattern	Baikal Teal	Harrison and Harrison (1963b)
Tufted Duck	Common Pochard	Grey back, black head	Lesser Scaup	Scherer and Hilsberg (1982)
Northern Bobwhite[4]	Scaled Quail	Crest form	Crested Bobwhite	Johnsgard (1971)
Northern Bobwhite	Gambel's Quail	Spotted flank feathers	Elegant Quail	Johnsgard (1971)
Chestnut-breasted Munia	Diamond Firetail	Scalloped breast	Scaly-breasted Munia	Harrison (1963b)

* Hybrid between the Cinnamon Teal and the Northern Shoveler.
[1] The white neck ring is present in young Northern Shovelers, but not in adults.
[2] The Mallard was crossed separately with each of the three species listed.
[3] Seven other pairs of species also produce a bridled face pattern in the hybrid. The cross that is listed produced an individual that could have been mistaken for a Baikal Teal.
[4] See Figure 14.8, p. 318, third line.

because these different mutations have favorable effects on the viewer of the mutation (i.e., in captivity, the breeder; in nature, conspecifics). Because divergence among breeds has happened over such short time spans, new mutations clearly arise sufficiently frequently that a similar process of diversification could happen in nature. However, differences between natural populations and artificial breeds make divergence among breeds an imperfect model. In particular, the same variant may be much more likely to become established in nature than in domestic breeds. This will lead to parallel evolution, that is, the repeated evolution of similar features, rather than to divergence. First, the artificial breeder is specifically on the lookout for traits that distinguish his or her breed from others, whereas the receiver of a signal in nature has no regard for what is going on elsewhere. Second, as emphasized by Wallace (1858, p. 59), breeds are removed from the rigors of natural selection. For example, crests in the Canary are due to a dominant mutation that is lethal when homozygous. The feathers in the Jacobin pigeon so greatly obscure the bird's vision (Figure 11.2) that its chances of survival to the age of reproduction in nature are zero. Because many mutations have multiple effects, some of which are likely to be deleterious, a much smaller subset of all possible mutations should be established in nature than in captivity. Third, the circumstances of breed propagation, in small semi-isolated populations with mating often between relatives, are favorable for the accumulation of loss-of-function recessives that make up many of the mutations that distinguish domestic breeds (Sossinka 1982; Table 11.2). In nature, such circumstances are likely to be much rarer, making it more difficult for recessive mutations to increase from a low frequency.

These considerations suggest that, in nature, a smaller spectrum of mutations is likely to become established than in domestic breeds. This reduces the possibilities for divergence between populations, and it increases the chances of parallel changes, at least when compared with domestic breeds. Thus, mutations in the *MC1R* gene are repeatedly involved in the generation of melanic plumages in birds, despite the identification of 120 genes (in the mouse) that affect melanism (Mundy 2005). Mundy suggested that this is because mutations in the *MC1R* gene have fewer deleterious side-effects than mutations in other genes affecting melanism, a reasonable proposition that needs to be tested. In addition, the *MC1R* mutation to melanism is a dominant mutation, which has a much higher chance of becoming established (because it is immediately visible and subject to selection) than does a recessive mutation.

Parallel evolution in the wild

Parallel evolution in nature is common and widely recognized (Wake 1991; West-Eberhard 2003). For color patterns, this can be qualitatively shown by a study of the common names of species. Sibley and Monroe (1990) list no fewer than 41 species (in 40 different genera) that are named black-throated something, eight that are black-hooded, eight that are blue-capped, seven that are blue-fronted, and many other cases of multiple species named for the same

Table 11.5: Examples of parallel evolution within genera, based on phylogenetic reconstructions

Genus	Number of species	Trait	*No. species trait present (no. times originated)	
Shanks (*Tringa*)	16	Red legs	2 (2)	Pereira and Baker 2005
Woodpeckers (*Veniliornis*/ New World *Picoides*)	23	Pied back pattern	10 (2)	Moore et al. 2006
Australo-Papuan scrubwrens (*Sericornis*) [1]	14	White above eye	4 (3)	Christidis et al. 1988
Indian Ocean bulbuls (*Hypsipetes*) [2]	7	Green coloration	5 (2)	Warren et al. 2005
Asian leaf-warblers (*Phylloscopus*) [1]	42	Pale crown stripe	14 (2)	Price and Pavelka 1996
Orioles (*Icterus*) [1]	25	Black pigment on lesser coverts	6 (6)	Omland and Lanyon 2000
Grackles and allies	40	Colorful "shoulder" patch	10 (4)	Johnson and Lanyon 2000
Grackles and allies [1]	40	High UV reflectance on male breast	8 (6)	Eaton 2006

NOTE: In all cases, phylogenetic inference suggests that the ancestor did not have the patch, except for the woodpeckers. In the woodpeckers, the ancestor is inferred to have been pied (with two losses and one subsequent gain). *The number of originations is inferred from phylogenetic reconstructions under the principle of parsimony, which minimizes the number of times the trait has evolved, e.g., if two sister species share the trait, the trait is assumed to have originated in their ancestor.

[1] In these studies, other traits showing multiple evolution were also recorded.
[2] Count is actually of five species, two of which contain two subspecies.

color patch. Some quantitative assessments of parallel evolution in color based on phylogenetic analysis are listed in Table 11.5. The most detailed is of the 25 species of New World orioles that constitute the genus *Icterus*. Omland and Lanyon (2000) studied the coloration of 44 different feather tracts that vary among species and/or subspecies in this genus. Forty-two of these patches showed evidence for multiple gains and/or losses of color. For example, the presence of black pigment on the lesser wing coverts is present in six different species, and each of these species probably gained the pigment independently (Table 11.5). In this group, parallel evolution is striking, and it even occurs at

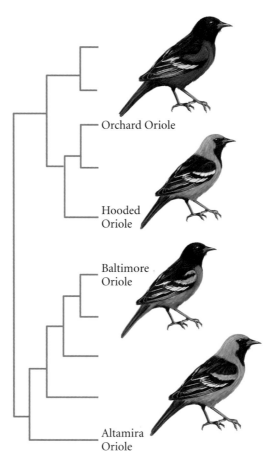

Orchard Oriole

Hooded Oriole

Baltimore Oriole

Altamira Oriole

Figure 11.4 A phylogeny for some New World Orioles (redrawn from Hoekstra and Price 2004; the phylogeny is from Allen and Omland 2003). Two distinctive color patterns (the Altamira and Baltimore types) have arisen independently in two different clades. Each of these patterns has also arisen a third time within the group as a whole.

the level of whole plumages. Two conspicuous patterns in the genus are the so-called Altamira and Baltimore types (Figure 11.4). Each has evolved in three different lineages, apparently being built up piecemeal through the addition and subtraction of patches (Omland and Lanyon 2000).

Many examples of geographical variation within species are consistent with parallel evolution. Remsen (1984) described 28 zoogeographical species along the eastern slopes of the Andes that show a leapfrog pattern, in which three taxa replace each other from north to south; the northern and southern taxa have similar color patterns, but the one in the middle is different. Examples of similar within-superspecies variation in Asia include the grey-mantled Hooded Crow, whose distribution to the east and west is bounded by the all-black Carrion Crow (pp. 354–355), and the Coal Tit (Figure 10.5, p. 195; Martens et al. 2006). By themselves, such examples are not strong evidence for parallel evolution because the different color pattern in the middle might be the derived one (Remsen 1984). However, sometimes each color form occupies two or more separate regions, making it more likely that at least one form has originated more than once; this occurs in brush finches (*Atlapetes*) along the Andes (Remsen and Graves 1995b). Molecular studies have also been used within species to demonstrate parallel evolution (Omland and Lanyon 2000; Ödeen and Björklund 2003). In the Yellow Wagtail complex in Asia, some color patches are present in populations of each of three phylogenetic species, and absent from other populations of these species (Ödeen and Björklund 2003; Pavlova et al. 2003), but, in this case, introgression by hybridization is a reasonable alternative to recurrent mutation (Pavlova et al. 2003).

Examples of parallel evolution in color show that, although divergence can potentially occur through the accumulation of different color patterns, it is likely to occur at a slower rate than would be inferred from the diversity produced among domestic breeds, because the same patterns repeatedly spread.

Song evolution has also been analysed in a phylogenetic context. In the New World orioles, song shows many examples of parallelism and convergence, and indeed song evolution seems very similar to that of color (J. Price et al. 2007). Some whole songs are very similar between quite distantly related species. J. Price et al. (2007) suggest that this reflects a constraint—a limited number of song variants that can be produced—but in some cases it could be a result of learning from heterospecifics. A few phylogenetic studies of song have been conducted on other groups, and these have found much less convergence (Päckert et al. 2003; J. Price and Lanyon 2004). In song, cultural mutations should often produce a great variety of syllable and song types, and, in the next section, I describe a case in which it does appear that song divergence has occurred as a result of fixation of different "arbitrary" variants.

Catchpole and Slater (1995, p. 190) suggest that communication between males favors short songs, so each male can listen for replies from other males (described further in Chapter 12). Finally, the more closed habitat in the north (mature forest vs. shrubs at tree line) may lead to a greater emphasis on acoustic rather than visual cues.

The analysis suggests the importance of an interaction between geography allowing some differentiation, and parallel selection pressures accentuating the differences. Variation in song form seems to have arisen across the southern part of the range, and may reflect processes of cultural mutation and drift. The differences have been magnified by parallel selection pressures as the species expanded its range northward. These kinds of interactions between ecological selection pressures and sexual selection were detectable only because of the special geographical arrangement inherent to a ring species. They may be a common way of producing diversity in socially selected traits.

PARALLEL EVOLUTION AND DIVERGENCE

The facts that color patterns carried by a species (1) are occasionally observed appearing de novo in domestic breeds (Table 11.3), (2) can appear in hybrids between two species that do not have the trait (Table 11.4), (3) are often found to have repeatedly evolved in related lineages (Table 11.5), and (4) can sometimes be mimicked by environmental (e.g., hormonal) manipulations during development in another species (Price 2006) are some of the strongest pieces of evidence that available genetic variation, rather than specific selection pressures, contributes to directions of evolution. Why should a stripe on the crown repeatedly evolve, whereas a cross on the crown is never observed? One could argue that this is because receivers of signals favor crown stripes and ignore crown crosses, but a more reasonable explanation is that the stripe is developmentally easier to produce and serves the purpose just as well. In terms of the peak-shift jargon (Chapter 5), there are alternative adaptive peaks, but some are more easily reached by available genetic variation, and other peaks may be out of reach altogether.

Because parallel evolution due to similar underlying genetics is most likely among close relatives, it places a constraint on the rate at which populations and related species diverge under parallel selection pressures. In environments in which new ornaments are generally favored, the same ornament may repeatedly spread. In many cases, populations diverge because they are in different environments and these environments favor different traits, the subject of the next chapter. But, if populations experience parallel selection pressures in similar environments, they may still diverge if patterns of genetic variation differ between the populations, or if different mutations arise. The extraordinarily high cultural mutation to different song types makes songs particularly prone to divergence, as considered in the ring species example.

In this chapter, I have continued to emphasize the stimulatory value of new traits as the main reason why they become established in a population. If different traits arise in different populations, populations diverge. An alternative mechanism of population divergence by sexual selection in uniform environments is through the so-called runaway process. As modeled by Lande (1981a) and Kirkpatrick (1982), populations occupying identical environments diverge as a result of genetic drift in female preferences, and this can trigger a rapid escalation in the strength of female preference and male trait (e.g., review in Arnqvist and Rowe 2005). At first sight, this is an attractive model to explain the great diversity of sexually selected traits, because populations may run away in many different directions. I have not discussed it so far because it seems unlikely to me to be of great importance. The original model of the runaway process relies heavily on the absence of costs and benefits to the female of choosing different kinds of males (Day 2000). Extensions do show how some costs can be maintained when there are deleterious mutations (Pomiankowski et al. 1991) or spatial variation in the strength of selection acting on the trait (Day 2000), but, even in these cases, moderate costs should severely reduce the potential for divergence. The runaway process relies on genes that determine the female preference being nonrandomly associated with the genes that determine the male trait (i.e., a genetic correlation, Lande 1981a; Kirkpatrick 1982; Day 2000). The genetic correlation implies that males carrying genes for an exaggerated version of the trait also carry genes from their mother that determine strong preferences. As the male trait evolves, the female preferences get pulled along with it (i.e., female preferences genetically evolve). This has the effect of changing the strength of the female preference, thereby causing additional evolution in the male trait. Whenever the male trait and female preferences are genetically variable, such a genetic correlation will inevitably be present, but there is little reason to believe it will be large, and there is some empirical evidence that it is small (Qvarnström et al. 2006a). Besides the lack of support for the runaway model, other modes of sexual selection certainly operate and are important, as illustrated by the diversification of domestic breeds. In this case, the breeder is the "female" and the breed is the "male," and diversification cannot be driven by a runaway model, which is based on the premise that female preferences genetically evolve. Indeed, one can state with some certainty that there has been no genetic evolution of the preferences of the artificial breeder, even if fashions have changed.

Divergence due to the spread of new mutations is most likely to occur if populations and species are repeatedly gaining and losing sexually selected traits. As one trait is lost and another evolves, divergence can proceed. Phylogenetic analysis can be used to reconstruct ancestors and thereby test for directions and rates of evolution (Price and Birch 1996; Burns 1998; Omland and Lanyon 2000; J. Price et al. 2007). At least in some groups, color patterns do evolve frequently (Omland and Lanyon 2000), as do some features of songs (Päckert et al. 2003; J. Price et al. 2007). Most often, the character "sexual dimorphism in color" has been examined, and this can evolve either by females becoming brighter to resemble males, or by males becoming duller; the former is estimated to be surprisingly common (Irwin 1994; Burns 1998; Badyaev and Hill 2003). In a survey across all passerine birds, Price and Birch (1996) found that conspicuous sexual dimorphism evolves frequently, and we obtained a crude estimate that a more monomorphic lineage has a probability of about 1% per million years of evolving into a more dimorphic lineage. Figuerola and Green (2000) estimate that the transition between apparent sexual monomorphism and sexual dimorphism has occurred at least 50 times in the Anseriformes (ducks and geese) alone. Thus, phylogenetic analysis suggests that repeated evolution of sexually selected traits is common. Patterns are even more striking within species. Peterson (1996) identified 158 examples of conspicuous geographical variation in sexual dimorphism in color or color pattern (usually one population is dimorphic and another more or less monomorphic). In a comparison of migratory and resident subspecies across 72 different species, Fitzpatrick (1998b) found that 64 pairs (89%) showed observable differences in coloration. In summary, repeated gain and loss of socially selected traits create opportunities for these traits to diverge, as a result of different mutations, by chance, becoming established at different times.

CONCLUSIONS

The idea that divergence may be produced even under parallel selection pressures was quite common in the last century. Huxley (1914, p. 516) stated "…as always selection of accidental variations has led to very diverse results" and Armstrong (1947, p. 317) developed an analogy of animal signals to signaling with flags by ships at sea. Flags need to be bright against their background to be visible, but Armstrong emphasized that the differences between them have nothing to do with the environment but instead reflect arbitrary conventions. While the evidence summarized in this chapter supports the possibility that animal signals are similarly arbitrary conventions, evolutionists have more recently emphasized the power of selection to fine-tune sexually selected traits to what may be small environmental differences between populations (Endler 1992). The influence of ecological factors on the evolution of socially selected traits has cropped up in all of the last three chapters, and it is now time to deal with it directly.

SUMMARY

New signals may be favored because of the reactions they elicit in receivers. If different signals arise in different populations, divergence will occur even if the populations occupy similar environments, and this is especially likely if conditions change so that increased expression of sexually selected traits is generally favored. Diversification among breeds of domesticated birds provides a model for this sort of process. Species that have been domesticated for several thousand years are divided into many breeds, distinguished on the basis of color, feather ornaments, morphology, and vocalizations. Diversification is often due to the animal breeder seizing on attractive traits and, subsequently, refining the trait by selection. Thus, the process of breed divergence is analogous to a model of divergence in nature due to the accumulation of "arbitrary" signals in different populations. However, at least two features of breed divergence lead one to expect that this is an imperfect model, and new mutations in nature will more often lead to parallel, rather than divergent, evolution. First, the animal breeder is often specifically looking for new traits. Second, he or she is able to select from among a great variety of new mutations, many of which are deleterious in the wild, or are recessive and easily lost without special inbreeding programs. In nature, divergence is most likely if traits frequently mutate to different forms. This applies to cultural mutation of songs. I describe one example in which song divergence between a pair of closely related species seems to have resulted from parallel selection pressures for increased complexity, with the different form of the complexity attributable to the accumulation of different cultural mutations.

CHAPTER TWELVE

Social Selection and Ecology

Social signals are used because they elicit favorable responses from conspecifics. As the environment varies, so different signals should vary both in their costs and in their effectiveness. Environmental influences include types of predators encountered, effects of the light and acoustic environment on signal transmission, effects of population density on the frequency of social interactions, and the influence of patchy resources on the distribution of conspecifics. In this chapter, I ask how these factors affect the intensity and targets of social selection. First, some environments favor greater expression of socially selected traits than others. Second, different ecological factors often result in different socially selected traits being favored, so that even populations experiencing similar intensities of selection may diverge in such traits. Throughout this chapter, I focus on sexual selection, including sexual selection by female choice and by male competition for resources that females then choose. Other forms of social selection are clearly affected by the environment. They have been little discussed, but they are likely to follow principles that are similar to those that that govern sexual selection.

After six chapters away from the subject, I am finally returning to ecological speciation: How does reproductive isolation between populations arise as a consequence of divergent selection pressures (Schluter 2000, 2001; Rundle and Nosil 2005)? In Chapter 5, I considered how divergence in such traits as body size and habitat occupancy could cause both premating and postmating isolation. That chapter emphasized the role of natural selection; here, I consider sexual selection. I restrict myself to asking how sexually selected traits diverge between populations in different environments. The following chapters focus on the question of how this divergence is linked to reproductive isolation.

A research program aimed at uncovering environmental correlates of sexually selected traits really dates back only to Crook (1964), who studied weaverbirds in Africa (reviewed and updated in Lack 1968). The strongest pattern Crook identified was for insectivorous forest species to be solitary, monogamous, and monomorphic (although both sexes are often colorful), whereas granivorous savanna species are colonial, polygynous, and generally cryptically colored (but with males acquiring an ornate breeding plumage). Crook (1964) argued that a massive flush of food for the granivorous species meant that females could raise young alone. This, in itself, would not drive polygyny, but

Lack (1968), following Verner (1964), suggested that there would then be advantages to a female that enters into a polygynous relationship because the number of safe nesting sites is limited, and one male could control several sites. Crook (1964) identified other differences in social behavior among weaverbirds that are apparently driven by ecological factors. These included the types of threat displays that males make to other males and how nest position (e.g., in a tree or in grass) can affect display posture when a male advertises his nest to a female.

My goal in this chapter is to identify environmental correlates of the expression of sexually selected traits, as far as is currently known. I address three main questions:

1. In continuation from the previous chapter, I ask: what environmental features drive variation in the intensity of sexual selection? Does a higher intensity of selection translate into a greater exaggeration of sexually selected traits?
2. How do differences among environments generate a shift in the target of sexual selection? I review correlations between various ecological features of a species (e.g., habitat occupied, body size) and plumage patterns and songs.
3. What is the role of ecological variation in driving divergence in color and song among island populations? Many island species are assigned species status based on their colors and, more rarely, their songs.

BACKGROUND

First, I review some theory on the way ecology drives the evolution of sexually selected traits. It is easiest to develop the argument through sexual selection by female choice. Consider a female searching for a mate, and assume that she prefers a male that has an extreme value for a particular secondary sexual trait (e.g., a long tail or bright plumage); this is the common pattern (Ryan and Keddy-Hector 1992). The *strength* of an individual female's preference can be thought of as the number of males she will visit before making a mating decision. If she mates with the first male she encounters, every male has the same expected mating success. There will be no sexual selection. On the other hand, if she visits and then chooses among many males, she is likely to mate with a male that exhibits an extreme expression of the trait, and sexual selection will be strong.

Whenever sexual selection is present, males with more extreme values of the trait have a mating advantage, and the trait evolves to become more exaggerated until it is balanced by various costs, such as attraction of predators. The result is a continuum of possible equilibrium points corresponding to different strengths of female preference. At each point, the sexual-selection advantage to a male is balanced by the potential costs incurred in expressing the trait (Figure 12.1).

Figure 12.1 can be used to illustrate some of the diverse ways in which sexually selected traits may evolve as environments change. Male traits can become more or less exaggerated even if the strength of female preference remains constant (consider a vertical line above the arrow on Figure 12.1). Two ways in which this can happen are illustrated. First, if the costs of the sexually selected trait are reduced (such as in a predator-free environment), the trait will become more exaggerated. Second, if some males obtain many mates and others none, sexual selection will be stronger than it would be if all males have similar mating success. In a polygynous mating system, it is theoretically possible for one (exhausted) male to mate with every female and for all the other males to fail to mate, a potentially huge fitness advantage to an attractive individual. Under true monogamy, each male obtains just one mate, and the only advantage to attractive males accrues because some females are of higher fecundity than others. The high-fecundity females come into breeding condition early and pair with the most attractive males in the population; females in lower condition choose among the other males (Kirkpatrick et al. 1990).

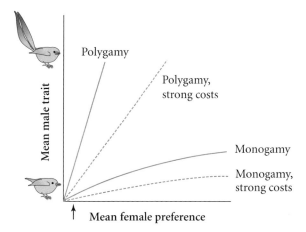

Figure 12.1 The average degree of exaggeration of a sexually selected trait in a population plotted against average strength of female preference in that population (after Kirkpatrick et al. 1990). In the model of monogamy, all males obtain exactly one mate, and the only advantage accruing to a male is that the first females to choose are in higher condition and invest more in reproduction. In the lines for *strong costs,* males suffer relatively higher mortality for a given increment in tail length.

In the preceding examples, male traits evolve even if the strength of the female preference does not change. In addition, an increase in the strength of female preferences—movement from left to right along the x-axis in Figure 12.1—always results in greater exaggeration of a male trait. Just as male traits evolve in response to costs (e.g., predation risk) and benefits (e.g., mating success), so a female's willingness to search for an extreme male should depend on the costs of the search balanced by any benefits she receives.

One important cost to females in those species in which males contribute to raising the brood is that ornate males may be worse parents than less ornate males. For example, ornate males may attract predators to the nest, or their ornament may hinder the gathering of food to provision the young. In European populations of the Barn Swallow (described in Chapter 9), long-tailed males have higher mating success, but they are less efficient at collecting food for the young. Such costs to females are absent from systems without male care, such as lekking species (also described in Chapter 9). Thus, lekking species should be more strongly sexually selected than monogamous species for two distinct reasons. First, a given strength of female preference translates into a greater strength of sexual selection on males (Figure 12.1), and, second, female preferences evolve to become stronger because the costs to the female of mating with particularly exaggerated males are low.

The benefits a female receives from prolonging her search for a male with a particularly exaggerated trait are assumed to result because he is of high quality, say, if he provides more parental care, or is free of parasites (Chapter 9). These benefits can be decomposed into three factors (Schluter and Price 1993):

1. The correlation between a female's preference and the trait of the male she mates with (Figure 12.2). Females with stronger preferences (i.e., females that search longer) should end up mating with the more exaggerated males, but the extent to which they will actually do so depends on how well they can distinguish among the males. Because of this, we (Schluter and Price 1993) referred to the correlation between the female preference and the male trait as the "detectability" of the trait.
2. The correlation between the male's "quality" and the expression of the male trait (Figure 9.13, p. 181). We termed this correlation the "honesty" of the trait. If a female mates with a male with a large value of an honest trait, she should get a male in high quality; but, if she mates with a male with a large value of a dishonest trait, she may often be disappointed.
3. The benefits that a male of given quality imparts to the female. (In addition to direct contributions, such as help with raising young, benefits may include effects on the offspring, such as transmission of genes for parasite resistance, although we did not explicitly model this.)

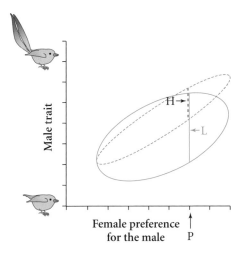

Figure 12.2 Comparison of mating success in an environment in which males are difficult to distinguish (*solid blue*) and in one where they are easy to distinguish (*dashed black*). The ellipses describe the distribution of female preferences and the trait of the male she actually mates with; because the system is here assumed polygynous, some males can be counted more than once. Consider a female with preference P. In a low-detectability environment, she, on average, mates with a male with characteristics that are close to average (L). In a high-detectability environment, she will, on average, mate with a more extreme male (H). Note also that, in a high detectability environment, the ellipse is displaced upwards; the intensity of sexual selection on males is increased even for the same average female preference (Schluter and Price 1993, compare with Figure 9.13, p. 181).

These factors are multiplicative and balanced by costs of searching for males, so the strength of female preferences is expected to evolve to an equilibrium given by:

$$\frac{\text{Detectability} \times \text{Honesty} \times \text{Benefit from male}}{\text{Costs of exerting choice}}$$

The logic of the result can be inferred by considering extremes. Suppose all males appear the same to the female in the prevailing environment (Detectability = 0), or the male trait is not correlated with quality (Honesty = 0), or differences in male quality have no effect on female fitness (Benefit = 0). In any one of these situations, either the females would be unable to choose among the different males, or they would gain no fitness benefits if they did, and the best thing to do would be to mate with the first male encountered. If all three of these factors are greater than zero, then females should search among at least a few males, resulting in some sexual selection on the male trait.

Increasing search costs may be a main determinant of how long a female will search. In this case, and if search costs remain roughly similar from one environment to the next, the average number of males surveyed by a female remains roughly constant. By definition, the intensity of sexual selection does not differ between environments. However, females are under selection to shift their attention to those traits that give them the most benefits in the prevailing environment (Schluter and Price 1993). Changes in a novel environment that affect either detectability or honesty could lead to rapid turnover in male traits, by altering the benefits to females. For example, a movement from dark to light habitats may mean that visual differences are more readily detected than sound differences, which might cause the female to focus on plumage rather than on song. Likewise, if a high abundance of a blood parasite in one environment becomes a major cause of variation in offspring fitness, and if infection by that parasite is particularly well revealed by a bare patch of male skin (Zuk et al. 1990), females should direct their attention to variation in this trait.

In the following sections, I ask how first, the strength and second, the target of sexual selection varies between environments.

RESOURCE DISPERSION

The more that individual males differ in the number of mates they attract, the stronger the sexual selection should be. This leads to two questions: (1) How do environmental differences affect the degree of polygyny? (2) Does an increase in the disparity among males in their mating success lead to an increase in the exaggeration of sexually selected traits, as predicted from theory?

Mating system

The degree to which resources are clumped in space influences the mating system. In particular, the ability of a single male to monopolize a high-quality resource is often associated with polygyny (Verner 1964; Emlen and Oring 1977; Davies 1991). This cannot be true of leks, which are the most extreme form of polygyny. As described for the Greater Sage Grouse (Chapter 9), males gather at display sites and are visited by females solely for mating. Females receive no material benefits, so resource defense is immaterial. Nevertheless, ecology probably explains the evolution of leks. The so-called hot-spot model of lek formation begins with the observation that leks are present in species where females have extremely large and overlapping home ranges for various possible ecological reasons, such as the need to search for rare foods (Bradbury 1981). Because resources controlled by males do not affect female reproductive success, female choice becomes focused more on male characteristics (Brad-bury 1981). Given this, females could prefer clustered males for several reasons. For example, it makes it easier for them to distinguish among the males, or they may simply be more stimulated by the presence of multiple males (Bradbury 1981; Endler and Théry 1996). Some other models of lek formation have fewer ecological underpinnings (e.g., females are stimulated more greatly by the presence of multiple males, without the large female home range requirement), but these seem less compelling (Davies 1991).

True monogamy is rare because, even among socially monogamous species, the presence of extrapair copulations results in some males having higher mating success than others. Factors driving differences among species in the average amount of extrapair copulation are the focus of much current research (Griffith et al. 2002). Some of the strongest correlates are ecological. Extrapair copulations are more common in systems in which males contribute less to raising the young (Møller and Cuervo 2000), in which male assistance is less critical to the successful raising of a brood (Møller 2000), and in which annual male mortality is high (reviewed by Griffith et al. 2002). The usual explanation for these associations is that females that engage in extrapair copulations are at risk of losing parental care from their social partner (Arnqvist and Kirkpatrick 2005), and this cost is less important in some environments than it is in others. In environments in which male mortality is high, males should continue to provide costly parental care even if they are cuckolded, for their chances of surviving to breed again are relatively low.

Sexual selection and mating system

Increased variance in mating success among males should increase the strength of sexual selection (Figure 12.1). This appears to be the case. Across species, the degree of sexual dimorphism in color or color pattern correlates with polygynous mating (Dunn et al. 2001, Figure 12.3), with the frequency of extrapair paternity (Møller and Birkhead 1994; Owens and Hartley 1998; however, this association needs to be reevaluated as more data appear on extrapair copulation rates, see Griffith et al. 2002), and with a presumed surrogate of this, testis size (Dunn et al. 2001). Extravagant, sexually dimorphic feather ornaments are associated with polygynous mating (Cuervo and Møller 1999). With respect to song, Read and Weary (1992) found that, in nine out of eleven passerine genera, syllable type-repertoire size was positively correlated with the tendency to be polygynous, a significant deviation from chance.

There are many exceptions to the rule that polygyny is associated with extravagant sexually selected traits, and these exceptions are often attributed to special ecological factors. Out of 150 lekking taxa (species and a few sub-

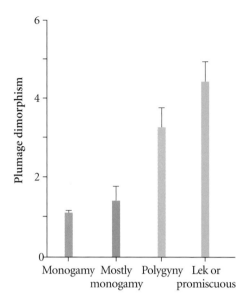

Figure 12.3 Association of social mating system with sexual dimorphism in color (from Dunn et al. 2001). Average scores (+ standard error) for all species in each category are plotted (number of species is, from left to right, 764, 33, 46, 53 respectively). Plumage dimorphism was obtained by comparing males and females for differences in shade and in color on five areas of the body, resulting in a 10-point scale. The *x* axis represents grades of polygyny: "Monogamy" implies that < 5% of the reproductive males are polygynous; "Mostly monogamy" implies that 5–15% of the males are polygynous (i.e., are paired with more than one female); and "Polygyny" implies that > 15% of the males are polygynous.

species), approximately 10% are dull-colored and monomorphic (Bleiweiss 1997). Among species of hummingbirds that lek, those that are dull-colored and monomorphic forage over widely dispersed resources (as predicted from the hot-spot model of lek formation), but both the males and the females forage among dispersed feeding sites. Males rarely engage in intraspecific aggression at food sources, so signals that would be displayed in aggressive interactions should not evolve. Bleiweiss (1997) suggested that, by being dull-colored, these males reduce the risk of aggressive attacks by dominant hummingbird species. A lekking species of hummingbird that has a moderately bright-colored male is intraspecifically aggressive at nectar sources, and two even more strongly dimorphic lekking hummingbird species frequently engage in intraspecific aggression and probably have high social status in interspecific interactions as well. Thus, in lekking hummingbirds, the evolution of bright male plumage may be driven less by mate choice than by social encounters over food resources (Bleiweiss 1997).

When polygyny arises through resource defense, females may base their choices on the resources a male controls and not what he looks or sounds like. This means that competition among males for those resources should be the critical factor driving the evolution of secondary sexual traits. Among New World blackbirds, bright "shoulder" patches are associated with marsh-nesting habit, and marsh-nesting is associated with polygyny (Johnson and Lanyon 2000). Experiments have shown the importance of the bright shoulder patches in male competition. Male Red-winged Blackbirds with their patches painted out tend to lose their territories, but those that were able to retain territories attracted mates at the same rate as unmodified birds (Smith 1972). Male aggression may be particularly high on marshes, not only because of the premium on defending high-quality territories and thereby attracting multiple females, but also because of the high density of birds on these marshes (Johnson and Lanyon 2000).

In addition to plumage, resource-based polygyny affects the evolution of song. Catchpole (1980, 2000) studied the migratory *Acrocephalus* warblers of the Old World. These are marsh-nesting birds with varying degrees of polygyny. They occupy a niche that is similar to that of the marsh-nesting blackbirds of the New World, but they have drab plumages. Song is a main target of sexual selection. Catchpole relates differences in song length across species to differences in the mating system:

1. In the Great Reed-Warbler in Europe, 40% or more of the territorial males may be polygynous (i.e., have more than one mate on their territory), and 20% or more have no mate. Males sing relatively short songs of two kinds. The longer song (~4 seconds) appears to be involved more in mate attraction, and the shorter song (~1 second), in territorial defense.

2. In the Sedge Warbler, fewer of the males are polygynous (up to 20%). Male song almost ceases after pairing up, suggesting that it is important for mate attraction, and the song does not seem to be used in territorial defense (Catchpole 2000). By contrast with the Great Reed Warbler, the song is elaborate. One song lasts about 15 seconds and is immediately followed by another in long sequences. Catchpole (1980, 2000) calls the Sedge Warbler song the acoustic equivalent of the Indian Peafowl's tail (perhaps it should be called the warbler's tale).

3. The Marsh Warbler seems to have the lowest rate of polygyny (~7% of males polygynous in one population). The male has a small territory, the female forages mostly off the territory, and territory does not seem to be a target of sexual selection. The male sings an extraordinarily varied song prior to attracting a mate, perhaps consisting entirely of syllables drawn from other species and sometimes continuing almost unbroken for four

hours. Syllables from 212 different species have been recorded in Marsh Warbler songs (Dowsett-Lemaire 1979). Each individual has a repertoire of, on average, 76 identified species plus probably some unidentified ones (Dowsett-Lemaire 1979).

Among these *Acrocephalus* warblers, the most polygynous species, the Great Reed-Warbler, sings the shortest song. In the Great Reed-Warbler, female choice is likely to be based at least partly on the quality of the male's territory. Catchpole (1980, 2000) notes that, when territorial defense is important and songs are addressed to males in territorial disputes, gaps are needed during which the singer can listen for replies, and hence, rather short songs may be favored. If songs are addressed to females rather than to males, then we might expect a more continuous signal to be used, because females listen for songs of males, but do not sing themselves; the songs should also become exaggerated and attractive (Catchpole and Slater 1995, p. 190). Thus, variation in song form among *Acrocephalus* warblers may have less to do with the intensity of sexual selection and more to do with a changing target of sexual selection, from a focus on the territory to a focus on the male. These differences in song match the kinds of differences inferred by Huxley (1938) for plumage traits. Huxley argued that those involved in female attraction (of which he used the Indian Peafowl as the prime example) should be beautiful, whereas those traits used in male competition should be bold and strikingly patterned (see Chapter 9).

ADAPTIVE RADIATION AND THE EVOLUTION OF SOCIALLY SELECTED TRAITS

Resource abundance and resource dispersion affect both the intensity and the target of social selection. But variation in resource abundance and dispersion has rarely been related to species diversification during adaptive radiation. Instead, segregation along various ecological axes, such as prey size (associated with body size), feeding method, and habitat occupied, creates ecological differences that facilitate coexistence. Such ecological differences can impact the evolution of sexually selected traits, perhaps with little overall change in the intensity of sexual selection. To assess this, in the following sections, I review correlates of songs and plumages with habitat, body size, feeding method, and diet.

Songs

Song frequency is strongly correlated with (log) body size (e.g., $r = -0.77$ for 90 European passerines, Wallschläger 1980; $r = -0.95$ among populations of the

Dusky-capped Flycatcher, Lanyon 1978, p. 475; see Table 12.1 for other comparisons). Small-bodied species sing at higher frequencies. This is thought to reflect selection for efficient vocal transmission. Efficient transmission depends both on song frequency (the lower the frequency, the better the transmission) and on volume. Large vibrating structures produce low frequencies more efficiently than small vibrating structures, so the term [(volume at source)/frequency] is maximally efficient at different frequency values for different body sizes (Ryan and Brenowitz 1985). The strong correlate of song frequency with body size means that, in an adaptive radiation, diversification along the body-size axis results in song evolution. Clayton (1990b) studied the songs of two subspecies of Zebra Finch that differ in body size and associated song frequency; hybrids are intermediate in both traits. As described further in Chapter 13, Clayton used a cross-fostering study to show that rearing environment did not alter several aspects of songs, including frequency. She also showed females responded to the innate (rather than learned) features of the songs, although specifically which features are not known.

An individual bird's physiology and morphology should affect the ease with which it can produce certain kinds of songs (Slabbekorn and Smith 2002; Podos et al. 2004; Suthers and Zollinger 2004). In Darwin's finches, syllable repetition rates, as well as frequency bandwidth (i.e., the difference between the maximum and minimum frequency) of the syllable may be partly mediated by beak size (Podos 2001). Within one species (the Medium Ground-Finch), individuals with the largest beaks tended to sing syllables with relatively low bandwidths and low repetition rates, whereas individuals with smaller beaks had, on average, higher bandwidths and repetition rates. This suggests that altered beak size, which is a fundamental feature of the Darwin's finch radiation, could lead to evolution of song as an associated change. In support of this, the warbler finches, which have the smallest beaks of all the finches, have the highest syllable bandwidths and some of the highest trill rates in the group.

With respect to the habitat axis, low-frequency sounds are optimal for transmission except when they are very close to the ground (Marten et al. 1977). However it is possible that, in some environments, the effects of ambient noise (e.g., low-frequency wind and high-frequency insect noise) can create an intermediate sound window of optimal frequency in which to sing (Ryan and Brenowitz 1985). In fact, when body-size effects are accounted for, correlations of song frequency with habitat (forest vs. open country) are weak but sometimes significant (e.g., Badyaev and Leaf 1997; Ryan and Brenowitz 1985; see Table 12.1). Hunter and Krebs (1979) found that Great Tits in dense forest tend to sing in a narrower range of frequencies than those in open woodland (the forest birds sing at the low end of the frequency spectrum of the woodland birds). Hunter and Krebs suggest that low frequencies reflect the greater need for efficient trans-

Table 12.1: Comparative studies of ecological correlates with song or call variables

Comparison group	Song variable	Ecological correlate	Reference
44 species of New World doves	Higher frequency	Smaller body size, closed habitat	Tubaro and Mahler 1998
41 species of tinamous	Higher frequency	Smaller body size	Bertelli and Tubaro 2002
41 species of tinamous	Narrow frequency bandwidths	Closed habitats	Bertelli and Tubaro 2002
16 populations of the Satin Bowerbird	Lower frequency, less modulation	Denser habitats	Nicholls and Goldizen 2006
30 taxa of Old World warblers	Higher maximum frequency	Smaller body size, open habitat	Badyaev and Leaf 1997
117 antbird species	Higher frequency	Smaller body size	Seddon 2005
159 Panamanian species	Higher frequency*	Smaller body size, open habitat	Ryan and Brenowitz 1985
11 species of *Sylvia* warblers	Higher frequency	Smaller body size	Bergmann 1976
90 European passerines	Higher frequency	Smaller body size	Wallschläger 1980
120 North American species	Higher maximum, minimum, and dominant frequencies	Smaller body size	Wiley 1991
14 species of Darwin's finches[1]	Higher modal frequency	Smaller body size	Bowman 1979
9 passerine genera[2]	Larger song repertoire	Migratory habit	Read and Weary 1992
96 antbird species[3]	Lower frequency	Closer to ground	Seddon 2005
30 taxa of Old World warblers	Fewer, longer syllables, separated by longer intervals	Closed habitats	Badyaev and Leaf 1997
66 North American warblers and buntings	Longer intervals between repeat units, more pure tones	Closed habitats	Wiley 1991

177 Panamanian species	Longer intervals between repeat units, more pure tones	Closed habitats	Morton 1975
41 cardueline finches	Fewer, more widely spaced notes	Dense vegetation	Badyaev et al. 2002
13 Himalayan passerine species	Higher frequency in narrower range, less modulation	Water torrents	Martens and Geduldig 1990

* This is the "emphasized" frequency, the frequency to the nearest 500 Hz that has longest trace on the sonogram (Morton 1975).

[1] The correlation with body size is weak, but there is a much stronger correlation with the size of the internal membrane of the syrinx.

[2] Within the nine genera, seven showed a positive association of migratory habit and large repertoire, which is not significant by a binomial test. However, a comparative method using the individual species resulted in a significant association.

[3] After morphology is controlled for in a multiple regression.

mission in dense forest; singing males are also farther apart in these forests. By contrast, Great Tits in noisy urban environments filled with low-frequency sounds tend to restrict their songs to the higher-frequency part of the spectrum (Slabbekoorn and Peet 2003; Slabbekoorn and den Boer-Visser 2006). In woodlands, without either the forest or the urban constraints on transmission, social selection pressures favoring complexity may come into play.

Besides frequency, a second, and stronger, correlate of habitat with song is that species more often sing songs with rapidly repeated notes in open country than they do in forests (Chappuis 1971; Morton 1975; Wiley 1991; Badyaev and Leaf 1997). This high trill rate in open country is attributed to: (1) influences of reflecting boundaries in forests, which blur intervals between notes and (2) the presence of occasional air turbulence in open country, which affects only a portion of the repeat so that some of the message gets through (Wiley and Richards 1982, pp. 152–156; Wiley 1991; Brown and Handford 1996).

THE RUFOUS-COLLARED SPARROW Song dialects often do not correlate with obvious habitat boundaries, and it has usually been proposed that they have originated by some form of cultural drift (Chapter 10). The White-crowned Sparrow is the classic example (Chapter 10), and a congener of the White-crowned Sparrow, the Rufous-collared Sparrow, or Chingolo, forms the classic exception. This species is the sole representative of the genus *Zonotrichia* in South America (four additional species of *Zonotrichia* occur in North America). The Rufous-collared Sparrow occurs across a massive range of habitats, from Mexico to the southern tip of South America, and from sea level to more than 4000 meters altitude. It is currently divided into 25 subspecies. Songs vary substantially within populations in Ecuador (Handford 2005), but farther south they are organized into dialects, with relatively little variation within dialects (Nottebohm 1975; Handford 1988, 2005). Unlike the White-crowned Sparrow and other dialect species, these regional-scale dialects are often strikingly associated with different habitats (Figure 12.4). Dialects can extend over long and narrow areas associated with particular habitat types (Handford 2005) and even track the habitat as it goes up narrow valleys, forming fingers into another dialect (Handford 1988). The boundary between a dialect associated with thorn woodland (*blue circle* in Figure 12.4) and other dialects associated with other habitats, parallels the vegetation transition for more than 2000 km (Handford 2005).

Across dialects, the association of a higher trill rate with more open habitat, as predicted from the acoustic adaptation hypothesis, is shown nicely on the eastern side of the Nevado de Aconquija in Argentina (Figure 12.4). The pattern is destroyed on the western side of the same mountain, where a high-altitude

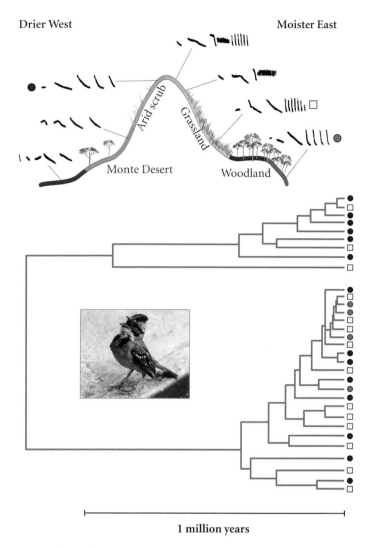

Figure 12.4 *Top:* Dialects of the Rufous-collared Sparrow on a west-to-east transect in northern Argentina, over an altitudinal range of 1600 m to 3000 m (redrawn from Nottebohm 1975; photographs of some of these habitats are in Handford 2005). Each dialect is associated with a different vegetation type, and extends over a large geographical range. *Below:* an estimate of genetic distances between individuals belonging to three of the dialects, as indicated by the symbols. The tree has been redrawn and edited from Lougheed et al. (1993). It is based on the percent difference in mitochondrial DNA sequence estimated from restriction-site polymorphism analysis. Scale bar represents 1 million years under the standard calibration of 2% mitochondrial DNA divergence per million years. Photograph of the bird by Juan Pedro Paz-Soldán.

open area (● in Figure 12.4) has one of the slowest trill repetition rates (Notte-bohm 1975; Handford 1988), but this and an example near the Atlantic coast (Tubaro and Segura 1994) are two of the few outliers in a generally strong cor-relation between trill rate and habitat structure (Handford 2005, P. Handford, *personal communication*). Handford (2005) notes that this area may formerly have been more forested, and cleared by European settlers. Elsewhere, patterns of dialectical variation are also explained by more recent historical patterns of vegetation, rather than vegetation in the current human-altered environment (Handford 1988; Handford and Lougheed 1991; Handford 2005). Sparrows in areas where woodland has been cleared for more than 100 years retain their characteristic dialect (Handford 1988).

A mitochondrial-DNA analysis shows that different individuals from the same dialect can carry haplotypes that have perhaps been divergent for more than a million years (Figure 12.4), and genetic distances are comparable with distances between the most recently separated North American species in the genus (Zink et al. 1991). Lougheed et al. (1993) suggest that such a deep mito-chondrial split may have been preserved because each mitochondrial lineage was originally carried by a different subspecies, one belonging to the coastal plains and the other to the mountains. Thus over long periods of time (perhaps in geographically separated locations) dialects have become associated with habitat in a manner predicted by the acoustic adaptation hypothesis. Overlay-ing this, it appears that, as in other dialect species, social pressures act to main-tain dialects. The dialects have remained stable despite recent changes in vegeta-tion, and the mitochondrial DNA analysis implies that migration between the dialects regularly occurs, with migrants adopting the songs of the dialect.

Plumage

Analyses of plumage brightness and sexual dimorphism have mostly been based on subjective assessment of brightness and/or degree of sexual dimor-phism (reviewed by Badyaev and Hill 2003), although these are now being replaced by quantitative measurements of plumage reflectance (Endler and Théry 1996; McNaught and Owens 2002; Gomez and Théry 2004; Eaton 2005). Based on a sample of passerine birds, Eaton (2005) concluded that most species are sexually dimorphic in color to at least some degree, although the average difference between the sexes may be small and the distributions overlap. Most analyses based on sexual dimorphism so far are thus comparisons between what are judged to be more and less dimorphic species.

Bright color is correlated with darkness of habitat, canopy nesting, migratory habit and diet (Table 12.2). Explanations for these correlations include: (1) the direct availability of pigments in the diet results in their being more easily uti-lized in the plumage (Ryan et al. 1994; Olson and Owens 2005)—for example, carotenoid pigments are more abundant in fruits than seeds (Olson and Owens 2005); (2) changing detectability—the conspicuousness of colors and color pat-terns depends on the background against which displays of these colors take place (Endler 1992; Endler and Théry 1996; Heindl and Winkler 2003); (3) increased aggression resulting from high density favors the use of various color-ful badges, e.g., at fruiting sources (Orenstein 1973); and (4) various impacts of predation (Baker and Parker 1979; Endler 1992; Martin and Badyaev 1996; Badyaev and Hill 2003; Gomez and Théry 2004). Some studies have found cor-relates of parasite infection with brightness (Read 1987; Scott and Clutton-Brock 1989; Pruett-Jones et al. 1991). Bright plumages may have evolved to indicate parasite loads (Hamilton and Zuk 1982), but this begs the question of why different species differ in parasite infection. In two studies in which fru-givory and/or foraging in the canopy were controlled, the association with para-site infection disappeared (Pruett-Jones et al. 1991; Garvin and Remsen 1997).

Song and plumage

If the intensity of sexual selection is constrained, plumage should be negatively associated with song, whereas, if most of the difference between environments is in the intensity of sexual selection, song and plumage should be positively associated. Qualitatively, there is evidence for both effects. At one extreme, the Superb Lyrebird in Australia has conspicuous displays and amazing feather plumes, and Mayr (1963, p. 97) described it as "perhaps the most gifted vocalist in the world." Other species are dull, essentially monomorphic, and have lim-ited vocalizations (e.g., *Acrocephalus* warblers on remote islands, see p. 264). On the other hand, within closely related groups, Shutler and Weatherhead (1990) and Badyaev et al. (2002) found a negative association between color dimor-phism and song length. In these groups, species with longer songs tend to be dull and less dimorphic (Figure 12.5). One possibility is that such negative asso-ciations represent arbitrary, alternative solutions to a similar intensity of sexual selection, but such an explanation is perhaps unlikely. Songs and plumage col-ors are quite different from each other (for example, songs can be turned on and off at will), so it seems more likely that environmental features affect which traits are favored. Darwin (1871, p. 56) suggested: ". . . if bright colors were dan-gerous to the species, other means would have to be employed to charm the females; and the voice being more melodious would offer one such means." If this sort of thing is happening, it has yet to be demonstrated. More the opposite, really: Badyaev et al. (2002) found that, after controlling for altitude, body mass,

Table 12.2: Comparative studies of ecological correlates with plumage coloration

Comparison group	Trait	Ecological variable	Reference
Waterfowl	Male brightness, dimorphism	Breeding latitude	Scott and Clutton-Brock 1989
154 North American passerines	Male brightness	Arboreal nests	Johnson 1991
Finches, North American warblers	Male brightness, female brightness	Height of nest above ground	Martin and Badyaev 1996
Louisiana woodland community	Brightness	Height of nest above ground	Garvin and Remsen 1997
Peruvian rain forest	Brightness	Foraging height	Garvin and Remsen 1997 Walther et al. 1999
New Guinea avifauna	Brightness	Foraging height	Pruett-Jones et al. 1991
20 pairwise comparisons, Australia	Male brightness and short-wavelength colors	Open habitat	McNaught and Owens 2002
40 species, French Guiana rainforest	Male brightness and long-wavelength colors	Foraging height	Gomez and Théry 2004
37 pairwise comparisons	Male brightness	Migratory habit	Fitzpatrick 1994
516 western Palearctic species[1]	Brightness	Southern point nonbreeding range	Baker and Parker 1979
126 cardueline finches	Male brightness, sexual dimorphism	Low elevations	Badyaev 1997
1010 species[2]	Sexual dimorphism	Bush cf. ground and tree nesting	Dunn et al. 2001
New Guinea avifauna	Brightness	Frugivory	Pruett-Jones et al. 1991
516 western Palearctic species[1]	Brightness	Feeding in flocks	Baker and Parker 1979
Leaf-warblers, *Phylloscopus*	Brightness	Darkness of habitat	Marchetti 1993
Blackbirds, Icteridae	Presence of carotenoid color patches	Marsh nesting habit	Johnson and Lanyon 2000
140 families[3]	Carotenoid color in plumage	Carotenoids in diet	Olson and Owens 2005

[1] Taken from Baker and Parker's tables of multiple regression analyses in their Appendices. These are the only two variables to be consistently and strongly associated with coloration.
[2] Testis mass and social mating system are both weakly but highly significantly correlated with plumage dimorphism, and this ecological factor remains significant when they are included in a multiple regression.
[3] The variable tested was proportion of species in the family showing carotenoid coloration in their plumage; carotenoid coloration in bare parts was also tested, and no association with diet was demonstrable. Diet was divided into nine categories ranging from low carotenoids (seeds and nuts) to high carotenoids (diatoms and algae).

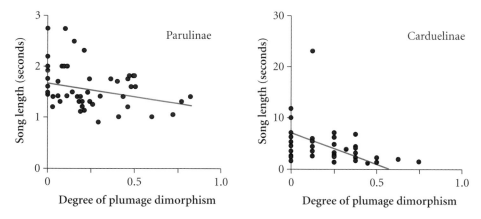

Figure 12.5 Negative associations of song length and plumage dimorphism. Plumage dimorphism is scored on an arbitrary scale by subtracting male-brightness scores from female-brightness scores. Data for the paruline warblers are from Shutler and Weatherhead (1990) ($r =$ –0.34, $N = 56$) and for the cardueline finches from Badyaev et al. (2002) ($r = $–0.36, $N = 41$. When the Canary, which is the outlier, is omitted, then $r = $–0.48, $N = 40$). Lines are the reduced major axes (omitting the Canary).

and vegetation density, the strength of the negative relationship, if anything, increased.

ISLAND PATTERNS

Color patterns and feather ornaments vary greatly among islands within archipelagoes, particularly in the tropics (e.g., Figure 3.5, p. 60; Wallace 1891; Mayr and Vaurie 1948; Mayr and Diamond 2001). Similar variation in vocalizations is less well documented, but is certainly present in the Galápagos Islands (Bowman 1979; Grant et al. 2000), the Canary Islands (Lynch and Baker 1994; Helbig et al. 1996; Martens 1996), and the Azores (Päckert and Martens 2004). Variation across islands has provided a breeding ground for many theories of population differentiation and speciation. As noted in Chapter 3, one idea is that, in small populations, such as those that are found on islands, drift causes random genetic differentiation from the source. This has little theoretical or empirical support for morphology, mainly because of the overwhelming power of selection, but cultural drift could be important in the evolution of songs. In addition, two other features of islands are: (1) restricted dispersal between them and (2) likely variation in the biotic environment, creating different natural selection pressures (Chapter 8). These two features are the main factors behind the "time-in-isolation" and "ecological" contributions to speciation (Figure 2.3, p. 17). Plumages and songs provide much of the basis for distinguishing among closely related species. It is worth having a look at island patterns in some detail in order to ask if special environmental features of islands drive social selection pressures.

Unfortunately, potential environmental causes of variation that might affect the evolution of sexually selected traits within archipelagoes have been little studied. Bowman (1979) compared the sound-transmission properties of several of the Galápagos Islands, finding differences in both absolute transmission efficiency and the relative rates at which sound attenuates across different frequencies, probably due to differences in vegetation. Bowman showed that, on one island, sound attenuated relatively uniformly across frequencies, but, on another, higher frequencies attenuated much more quickly than lower frequencies. A population of the Large Cactus-Finch on the island with relatively uniform rates of sound attenuation sang across a wider frequency bandwidth than a population on the other island. Bowman suggested a causal link between the different vegetation on the two islands and the differences in this feature of the two populations' songs.

More studies have compared islands with mainlands. Differences in plumage are summarized in Table 12.3. Plumages are often duller on islands, but a stronger pattern is that sexual dimorphism is often modified and often reduced. Peterson (1996) found that, within species, dimorphism is reduced on the

Table 12.3: Comparative studies of island–mainland differences within species in plumage coloration

Comparison group	Trait	Description	Reference
71 pairwise comparisons of subspecies (all non-migratory)*	Dull males	55% are duller on the island, 33% brighter, and 12% show no recorded difference	Fitzpatrick 1998b
Variation in sexual dimorphism, 36 comparisons	Dull males	80% of males became duller on island	Peterson 1996
Variation in sexual dimorphism, 14 comparisons	Bright females	80% of females became brighter on island	Peterson 1996
Geographical variation in 19 species[1]	Dull plumage	52% duller, 5% brighter, 43% no change	Grant 1965b

* Mostly western Palearctic passerines.
[1] Comparison of Mexico with the Tres Marias islands.

island in 83% of 30 cases where sexual dimorphism differs between an island and its presumed mainland source (from the Appendix in Peterson 1996; see Figure 12.6). When sexual dimorphism decreases, it is sometimes the result of the female becoming brighter (Peterson 1996; Badyaev and Hill 2003; Table 12.3). The studies reported in Table 12.3 (and Figure 12.6) do not control for the number of times monomorphic and dimorphic species have colonized islands. If monomorphic species colonize islands more frequently than dimorphic ones (as might be expected, because there are more monomorphic species; Price and Birch 1996), then the tendency for populations to lose rather than to gain dimorphism on islands, as estimated in Peterson's study, is even higher.

Plumage reduction is striking in the ducks. In four of six comparisons between a mainland subspecies and an insular subspecies, the male of the insular subspecies has substantially duller plumage and resembles the female (Lack 1970; Fitzpatrick 1998b). In all six insular subspecies, the clutch size is smaller

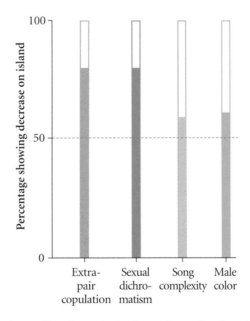

Figure 12.6 Comparative studies of sexual selection and sexually selected traits on mainlands and islands. Horizontal line indicates increases and decreases are equally likely. *Extrapair copulations* compares populations of the same species [$N = 12$, four (33%) are migratory; Griffith 2000]. *Sexual dichromatism* compares subspecies or allospecies, one of which is dimorphic and the other more-or-less monomorphic ($N = 50$, Peterson 1996). *Song complexity* is the average number of syllable types in a male's song ($N = 13$, from Table 12.4). *Male color* is based on descriptions in handbooks, mainly from the western Palearctic, and compares island and continental subspecies ($N = 77$, Fitzpatrick 1998b); only those subspecies showing a difference in one direction or the other are tabulated [15 (19%) are seasonal migrants and the rest are residents].

(although the eggs are larger) than that of their mainland relative (Lack 1968, p. 228). A difference in clutch size between the mainland and islands suggests that environments differ in ecologically meaningful ways. Females are apparently putting less into reproductive effort, and it seems reasonable to think that males are too, only they do so by not displaying as much. Among the ducks, pair-bonds may be more stable on islands, meaning that, in any given breeding season, fewer males are involved in competition for mates, and across ducks as a whole, stability of the pair-bond is associated with monomorphism (Scott and Clutton-Brock 1989; Figuerola and Green 2000). The fact that, among some species on islands, dimorphism is sometimes reduced because females become brighter rather than the males becoming duller may also be related to long-term pair-bonds. Bright female plumage seems to be associated with female aggressiveness, especially towards other females (Hinde 1952; West-Eberhard 1983; Irwin 1994).

Patterns of song variation in island–mainland comparisons are not as well documented as variation in plumages. Among *Acrocephalus* reed warblers, those singing the least complex songs are island residents, such as the Seychelles Warbler (Catchpole and Komdeur 1993). The *Acrocephalus* species on Henderson Island was at one time famous for apparently not singing at all, but Graves (1992) found that males do utter a series of long, thin notes that resemble a song. I compiled pairwise comparisons of closely related taxa between mainland and island relatives (Table 12.4) and between introduced and source populations (Table 12.5). In the island-mainland comparisons, island males have songs with fewer syllable types per song in 7/12 (58%) of the cases in which this can be quantified (Figure 12.6). Changes in song complexity seem to be common, but, as far as can be assessed, increases in complexity on islands are about as frequent as decreases.

One explanation for changes in song complexity on islands is cultural drift. Small population sizes during founder events may have caused reduced syllable-type diversity on islands (e.g., Lynch and Baker 1994; Baker 1996; Baker et al. 2006). Increases in song complexity on islands have also been attributed to cultural mechanisms, notably a failure to learn from tutors (Chapter 10; Päckert and Martens 2004; reviewed by Baker et al. 2006). Although drift and withdrawal of learning could apply to some cases, environmental factors explain many differences in songs between islands and mainlands. The mainland subspecies of the Japanese Bush-Warbler is highly polygynous with no paternal care. An island subspecies occurs at higher densities, has paternal care, produces smaller clutch sizes, and probably has much less polygyny (Hamao and Ueda 2000). Island males have more song types than the mainland subspecies, but each song is simpler. This is expected if male competition for territories becomes more important in the high-density island situation and female choice becomes less important.

Typically, when reduced song complexity is present on islands, the reduction is not only at the population level but also at the level of what an individual male sings. Males on islands often have fewer syllable types in their repertoire than males on the mainland (Table 12.4). This would not be expected on a sampling hypothesis but is predicted if sexual selection favoring elaborate songs is weaker on these islands. Chaffinch populations on Chatham Island have fewer syllable types than the ancestral populations on New Zealand (Baker and Jenkins 1987; Table 12.5). Most of the differences between Chatham and New Zealand Chaffinch songs are in the relatively fewer different syllable types at the end of the song, which is associated with a lack of an elaborate song ending. This has in turn been related to different transmission properties of the habitat on Chatham (Baker and Jenkins 1987; see Appendix 10.2, p. 216). Selection to reduce the length of the song ending was likely accompanied by loss of syllable types in the population as a whole.

More direct evidence indicates that the intensity of sexual selection is often reduced on islands. Among socially monogamous passerines, the frequency of extrapair paternity is lower in island populations when compared with mainland populations (Griffith 2000; Figure 12.6). Winter Wrens are polygynous in most parts of Europe, but apparently not on islands off the northern parts of Britain (Armstrong 1955). Verner (1964) argues that this is because harsh island conditions result in low breeding densities. This means that all surviving males can occupy relatively high-quality areas, lowering the variation in resource quality among territories needed for polygyny to evolve.

I suggest that differences in plumage dimorphism and song complexity between islands and the mainland commonly reflect differences in the intensity of social selection, as well as shifting targets of selection, and it is likely that these are driven by ecological conditions peculiar to islands.

GENERAL CONCLUSIONS ON SOCIAL SELECTION

Nonecological and ecological processes

Divergence in socially selected traits between populations may be achieved through appearance of new variants that have favorable effects on receivers, whenever populations become geographically separated (Chapters 9 and 10). Divergence should be aided in environments that favor generally higher levels of social selection (Chapter 11), and I have shown in this Chapter that different environments often specifically favor different kinds of socially selected traits. If

Table 12.4: Comparisons of songs between mainland and island populations

Species	Location	Distance (km)	Taxonomy	Syllable types*	
				Mainland	Island
Singing Honeyeater	Australia	20	Subspecies	33	8
Red-capped Robin*	Australia	20	Population	5	4
European Robin*,[7]	Europe–Tenerife	1000	Subspecies	8	4
Brown Creeper[1]	San Francisco Bay	1	Population	13	13
Bewick's Wren	California	30	Subspecies	88	56
House Wren*	California–Guadalupe	500	Subspecies	13	12
Black-capped Chickadee*,[7]	Massachusetts	10	Population	–	–
European Blue Tit*,[8]	Morocco–Tenerife	100	Subspecies	6	7
Goldcrest*,[3]	Iberia–Azores	1500	Subspecies	5	7
Firecrest*	Europe–Madeira	1000	Subspecies	2	8
Japanese Bush-Warbler*	Japan	1000	Subspecies	2	3
Chaffinch	Morocco–Madeira	500	Population	20[6]	26[6]
House Finch*	California	28	Subspecies	13	15
Fox Sparrow[4]	Newfoundland–Nova Scotia	600	Population	42	15
Dark-eyed Junco*,[5]	California–Guadalupe	500	Species	1	5

* Comparisons are of populations of approximately equal size, except for row entries indicated by *, which are of individual birds. Averages of counts have been rounded to the nearest integer.

[1] The similarity in number of syllable types may be because of regular immigration. One bird sang a song identical with the mainland song and is not included in the syllable count.

[2] Males of the mainland form sing a single song that may be sung at a continuum of different frequencies. Males of the island form sing two discrete songs at different frequencies, but some males may sing more songs.

[3] This is based on variation in the initial part of the song. On both the mainland and in the Azores, individual males sing a repertoire of songs derived from variation in the end part [repertoire size is ~23 both on the mainland (Table 8.1) and on the Azores ($N = 3$ birds)].

Song types[*]		Ratio of mainland to island parameters			
Mainland	Island	Song length	Syllable types per song	Syllables per song	Reference
47	11	–	1.4	1.3	Baker 1996
–	–	1.5	1.4	1.7	Baker 2006
–	–	1.9	2.2	1.9	Stock and Bergmann 1988
–	–	0.9	1.4	1.1	Baptista and Johnson 1982
20	18	1.0	1.8		Kroodsma 1985
36	–	0.6	1.0	1.4	Barlow 1978; Platt and Ficken 1987
1	2	1.0	–	–	Kroodsma et al. 1999
1	1	1.0	0.8	1.0	Becker et al. 1980
1	3	0.9	0.7	0.6	Becker 1977; Päckert and Martens 2004
3	1	0.3	0.3	0.8	Päckert et al. 2001
4	2	1.0	1.3	1.3	Hamao and Ueda 2000
–	–	–	–	–	Lynch and Baker 1993
4	5	0.9	0.9	–	Bitterbaum and Baptista 1979, Table 1
–	–	–	1.4	1.6	Naugler and Smith 1991
3	2?	1.0	0.2	1.3	Mirsky 1976

[4] The comparison is between the average of two islands in Newfoundland (16 males, 19 males) with one island off Nova Scotia (11 males).

[5] Island data estimated from sonograms in Mirsky (1976); mainland data from sonograms in Marler et al. (1962). Guadalupe Junco is considered a separate species (Sibley and Monroe 1990).

[6] This is the effective number of syllable types, which is a measure of diversity rather than absolute numbers. It is less than the observed number in the sample, but it has the advantage that comparisons among populations are less affected by sample size.

[7] Based on 163 songs from 18 Teneriffe birds and 92 songs from 13 German birds. On Teneriffe, mimicry of five other species has been noted.

[8] See Salzburger et al. (2002) for taxonomy.

Table 12.5: Comparisons of songs between source and introduced populations

Species	Source	Introduction	Time (years)	Syllable types*	
				Source	Introduced
Western Gerygone[1]	Australia	Island, 20 km	50	–	–
Eurasian Tree Sparrow	Germany	Illinois, U.S.	120	150[2]	120[2,3]
Chaffinch	New Zealand	Chatham Island	100	40	20
House Finch	California	Eastern U.S.	20	69–141	28–46

* Comparisons are of populations in source region and in an introduced region of approximately equal size. The Western Gerygone and Chaffinch were natural colonizations; the Eurasian Tree Sparrow and House Finch were human-aided introductions.

[1] On the island, a completely different song type is present in the repertoire of about 55% of the males (Baker 2006).

[2] This is the "effective number" of syllable types (see Table 12.4, note 6).

[3] The difference is not significant by a two-tailed t-test $(P = 0.18)$, where population is replicate (four populations in Germany and eight in Illinois).

the prime causes of population divergence in socially selected traits could be disentangled, we would be a lot closer to an understanding of the roles of geographical isolation (i.e., limited gene flow) and ecology in speciation. This disentanglement is difficult, and, given the current state of knowledge, perhaps the strongest conclusion we can make is that both processes must normally be involved.

Given sufficient time in geographical isolation, populations should diverge in socially selected traits without guidance from selection arising out of environmental differences. In the model I have emphasized, divergence arises because a new mutation stimulates receivers favorably, and different mutations arise in different isolated populations. Genetic drift in traits and in female preferences for those traits may also be involved in driving population divergence in uniform environments (e.g., Lande 1981a), but, for traits subject to strong selection pressures, as might be expected when they affect mating success, the role of genetic drift may be of relatively minor importance (see p. 49).

Identifying the importance of nonecological processes to population divergence is not easy. No clear methodology can be applied because, in their purest form, such models predict no correlations of environments with traits. Part of the difficulty is that, if a new socially selected trait does become established solely because of its improved effect on receivers, it will be modified by selection in other contexts. Endler (1992) notes the considerable value of being cryptic to

Song types*		Ratio of source to introduced parameters			
Source	Introduced	Song length	Syllable types per song	Syllables per song	
1+	2+	1	–	–	Baker et al. 2003
–	–	–	–	–	Lang and Barlow 1997
–	–	0.94	1.46	0.78	Baker and Jenkins 1987
13–71	2–6	0.87	0.86	0.94	Mundinger 1975; Bitterbaum and Baptista 1979

predators when a conspicuous trait is not actually being used in intraspecific displays. Thus, correlations with different predators and predation pressures arise secondarily, and the origin of the trait becomes obscured. On the other hand, a model of adaptation to different environments does predict strong correlations of environmental variables with trait expression. It can suffer from the drawback that, if one searches through enough environmental variables, one is bound to find a correlate (Endler 1986). Additional functional analyses are needed to verify any interpretations.

For songs, parallels with the evolution of human language suggest a component of arbitrary divergence. Few would suggest that environmental differences have anything to do with the rapid divergence of French from English, although some "lunatic fringe" theories (Aitchison 1991, p.106) once posited impacts of environmental variables (such as living at high altitudes) as agents of language change. Birds are different, and both within-species and among-species components of songs can clearly be related to habitat and to other ecological variables, most notably the remarkably strong association of song frequency with body size. Slabbekoorn and Smith (2002) suggested that we should focus on these aspects of songs, for they may be the ones critical to species recognition in adaptive radiation. However, correlations of song traits with habitat are often weak. Geographic separation also results in song divergence through (cultural) mutational turnover. If different isolated populations occupy different habitats, the result will be an association between song characteristics of a population and the habitat it occupies, even if the song traits have evolved by nonecological processes. This might contribute to the strong association of dialect with habitat in the Rufous-collared Sparrow in addition to any acoustic adaptation in the song, and for the association of habitat diversity with human language diversity described by Pagel (2000; see Chapter 10).

Color patterns have less of a cultural component than the songs of many species, so we might expect them to evolve more deterministically in response to environmental differences. Indeed, Endler and Théry (1996, p. 450) suggested that, given knowledge of an ancestral form, it might be possible to predict accurately the color patterns that will arise in a new environment. This is too optimistic. Even if an environment favors only a subset of all possible sexually selected traits, it is unlikely to favor just one. The particular form that actually becomes established must depend on appropriate genetic variants being present, and, because of this, some random differentiation between populations placed in similar environments is to be expected. Snow (1976) suggested that the South American bellbirds (Figure 9.1, p. 159) have a large random component underlying their divergence. This seems likely.

Sexual selection and speciation

Sexual selection might promote speciation by accelerating divergence among populations in sexually selected traits. This is an attractive hypothesis because of the intimate connection between sexual selection and the recognition of conspecifics. As sexually selected traits diverge, recognition of those traits may also diverge, resulting in the generation of premating reproductive isolation between populations (Lande 1981a; West-Eberhard 1983; Price 1998). In support of a role for sexual selection in accelerating speciation, several studies found that high numbers of species or subspecies are found in taxa that are particularly sexually dimorphic in coloration or feather ornaments (Barraclough et al. 1995; Møller and Cuervo 1998; Owens et al. 1999). However, Morrow et al. (2003) and Phillimore et al. (2006) used larger data sets and more refined analyses than those used previously, and found no correlation of species diversification rate with color dimorphism. Both sexual selection and the number of species in a taxon may be incorrectly measured when sexual dimorphism in color is the trait studied. First, it is quite difficult to see why dichromatism alone should capture the intensity of sexual selection; other sexually selected traits, such as song complexity, are not included in these tests. In some taxa, song length is negatively associated with the degree of sexual dichromatism (Figure 12.5). Second, as noted in the introduction to this book, there has been a recent explosion in the number of recognized species, most of it due to an increasing study of vocalizations among taxa that are very similar in their plumages (Isler et al. 1998; Irwin et al. 2001a; Alström and Ranft 2003; Remsen 2005), and these remain to be incorporated into the comparative tests.

If speciation rate is correlated with sexual dichromatism in at least some taxa, this may reflect the rapid formation of allospecies. Such allospecies are not usually recognized on the criterion of reproductive isolation because they are geographically separated from each other, but, most often, on the degree of difference in coloration (Figure 1.1, p. 5). For example, Barraclough et al. (1995) found a marginally significant association of sexual dimorphism in color with the number of species in a taxon, but the association disappears when reanalyzed to ask if dimorphism correlates with the maximum number of species in the taxon that occur sympatrically (Price 1998). Diversification in color patterns among populations, and hence production of allospecies, may be especially likely to happen in a depauperate, fragmented region, such as along a mountain range (e.g., the hummingbirds in Figure 6.11, p. 118) or across islands in an archipelago (e.g., the Golden Whistler in Figure 3.5, p. 60).

Sexual selection may be generally important in promoting reproductive isolation when species come into sympatry. Species that have diverged in sexually selected traits should also diverge in mating preferences for those traits (Chapter 13). In addition, any hybrids between such species may have significantly reduced mating success, a form of postmating isolation (Chapter 14). Despite these expectations, it is not clear if groups suffering intense sexual selection establish sympatry more quickly than those subject to weaker sexual selection. (As argued in Chapters 2 and 6, the ease of range expansion into sympatry is likely to be an important limit on speciation rates). For example, sympatric *Dendroica* warblers of North America are colorful and sexually dimorphic, suggesting a role for sexual selection in their production. They are also ecologically similar (MacArthur 1958; Price et al. 2000). It is possible that the coexistence of these warblers in the breeding season is facilitated by reproductive isolation, connected to the large divergence in plumage patterns. However, the warblers do differ in feeding ecology, albeit in relatively small ways (MacArthur 1958; Price et al. 2000). Different species have different migratory schedules, and most sympatric species diverged millions of years ago (Lovette and Bermingham 1998; Figure 6.3), so they are likely to produce unfit hybrids (Chapter 15). Such ecological and genetic differences may play a key role in promoting coexistence and reproductive isolation among species in this group.

In summary, allospecies might be more likely to be recognized in groups in which color patterns are an important target of sexual selection. However, the evidence that more strongly sexually selected groups have higher speciation rates than less strongly sexually selected groups is weak. The lack of any strong effect on speciation is not because sexual selection is unimportant, but more because it is so widespread. Indeed, the main conclusion from this chapter is that populations of most species readily diverge in socially selected traits, especially if the ecological conditions they experience differ and if they are geographically isolated. This divergence provides many cues that could be used by birds in identifying individuals that belong to their own species. The achievement of premating reproductive isolation depends critically on how females choose mates, and a large degree of divergence between populations in socially selected traits will be of little consequence if females do not utilize these differences in making decisions about choosing mates. The underlying factors driving recognition of conspecifics are the subject of the next chapter.

SUMMARY

Many ecological factors affect the intensity, as well as the target, of social selection. In some species, males differ greatly in the number of offspring they sire. High variance in male mating success is often correlated with resource dispersion; some males control good resources, and several females may pair with those males to gain access to those resources. Polygyny (i.e., one male mating with multiple females, other males with none) is correlated both with increased song complexity and with increased plumage dimorphism, but there are many exceptions to the pattern. Often, the exceptions have been given compelling ecological explanations. Even when the strength of sexual selection stays approximately constant, characteristics of songs and plumages vary with environments. Across species, various song and plumage traits are correlated with habitat features (such as the tendency for canopy birds to be more brightly colored than understory birds), and song frequency is very strongly correlated with body size. With respect to the mainland, plumage dimorphism is more often reduced than increased on islands, and this reduction is best attributed to ecological conditions on those islands. Songs may increase or decrease in complexity on islands; these changes have been related both to environmental conditions and to cultural influences. I conclude that ecological factors are a major cause of the evolution of socially selected traits. It is difficult to determine the importance of alternative nonecological models of population divergence, the subject of previous chapters. Overall, however, we expect differences in socially selected traits between populations to arise rather easily. Evidence that divergence in such traits provides a rate-limiting step on the development of reproductive isolation is weak.

CHAPTER THIRTEEN

Species Recognition

Acritical step in speciation is the development of assortative mating—the tendency to mate with members of one's own population. Traits such as plumage patterns, songs, morphology, and displays are used to identify individuals as suitable mates (Clayton 1990a; Price 1998). As populations diverge in these traits, if assortative mating is to occur, then the recognition mechanism must also diverge. It is not immediately obvious how the trait and the recognition of the trait evolve together, rather than interfere with each other (Pfennig 1998). Consider the appearance of a new color variant. Individuals carrying this variant may not be recognized as conspecifics; a species recognition mechanism imposes selection against novel types (Seiger 1967; Laland 1994). Alternatively, the novel color variant may be sufficiently attractive that it overrides the own-species preference. Then, if all individuals in all populations have a similar bias, populations will not diverge in the recognition mechanism, even if the trait remains restricted to one population (Jones and Hunter 1998; Pfennig 1998; Collins and Luddem 2002).

In birds, at least part of the resolution of the conflict between sexual selection and species recognition depends on the finding that species recognition results from learning of species characteristics, usually beginning with one's parents. "Social recognition" is defined as the process whereby animals become familiar with conspecifics and remember them (Mateo 2004). Social recognition has many consequences for the origin of premating isolation. In this chapter, I consider the learning process, how it leads to species recognition, and how it might be modified genetically. I start with an example that illustrates most of the points, and then I consider the generality of the results.

SEXUAL IMPRINTING

The example comes from a comparison of the two subspecies of Zebra Finch (*Taeniopygia guttata guttata* and *T. g. castanotis:* Clayton 1990a–d; reviewed by Clayton 1990a). The subspecies *T. g. castanotis* (a male is illustrated in Figure 9.10) is widely distributed throughout Australia, whereas *T. g. guttata* occurs in the Lesser Sunda Islands of Indonesia; the two subspecies are separated by about 600 km of ocean (Zann 1996, p. 256). When flocks are placed in aviaries, the two subspecies pair assortatively but if they are held in isolated mixed pairs, they readily breed together and produce fertile offspring. *T. g. castanotis* is up to 10% larger in linear dimensions than *T. g. guttata* (Clayton et al. 1991). Among traits expressed only in males, *T. g. castanotis* males have a larger black breast band (Clayton et al. 1991) but sing shorter songs at a lower frequency (Clayton 1990b) than *T. g. guttata* males. Displays differ in that *T. g. castanotis* males court in a less upright position than do *T. g. guttata* males, and, during display, a *T. g. castanotis* male will raise the feathers at the back, rather than at the front, of his head (Clayton 1990d). Cross-fostering between subspecies shows that, although young birds copy syllable types from their foster father, song frequency and song length have genetic components. Cross-fostered *T. g. castanotis* sang slightly longer songs with more syllables than normally raised young, and cross-fostered *T. g. guttata* sang slightly shorter songs with fewer syllables than did normally raised individuals, but these differences were not significant. (Note that songs do have to be learned from some tutor, and, in the absence of any tutor, songs of young birds are abnormal.) Body size and plumage development are not affected by cross-fostering (Clayton 1990a).

Cross-fostering had a large effect on mate preferences in pairwise choice tests, as well as in pair formation when placed in aviaries (Clayton 1990c):

1. When both males and females of both species were cross-fostered, they all ended up in hybrid pairs (19/19; 100%).
2. When only the males were cross-fostered, they showed a significant preference for females of the subspecies that raised them, if those females were presented behind a one-way mirror. However, the preference was reduced when the females were allowed to see the males and thereby to interact with them. No hybrid pairs formed in the aviary (0/18; 0%).
3. When only the females were cross-fostered, results were similar to those in the male cross-fostering experiments, but a few of the pairs formed were hybrids (3/16; 19%)

Clayton showed that plumage, song, and male behavior appear to be used in determining female preferences. First, she used paint to enlarge the breast band of several *T. g. guttata* males, thereby making the size of the breast band similar to that of a *T. g. castanotis* male (Clayton 1990d). She placed an individual male in a central cage surrounded by ten compartments (the "finkodrome"). Five of these compartments contained females of one subspecies, and five contained females of the other. Unpainted males directed most of their attention to their own subspecies, whereas painted males showed no such association (Figure 13.1). Presumably, the difference was in response to the reactions of the females. When painted and unpainted *T. g. guttata* males were introduced together into

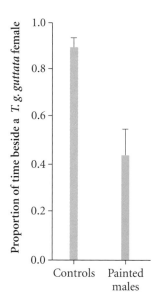

Figure 13.1 A Zebra Finch *T. g. guttata* male, either painted to resemble *T. g. castanotis* or an unpainted control, was placed in a central cage surrounded by ten females in separate cages (five *T. g. guttata* and five *T. g. castanotis*) for ten hours. Bars (+ standard error) give the proportion of time the male perched in front of females of his own subspecies (*N* = 10 in each group). From Clayton (1990d).

an aviary, *T. g. castanotis* females tended to pair with the painted males and *T. g. guttata* females tended to pair with the unpainted males, although the difference was not significant.

With respect to song, females react strongly to the songs from the subspecies of the father that raised them and weakly to songs of the other subspecies, irrespective of whether these females were cross-fostered or normally raised (Figure 13.2). Thus, the rearing environment of the female has a strong influence on her preferences for male song.

Male zebra finches display more interest in females who look like their mothers (Vos et al. 1993; ten Cate et al. 2006; see also Figure 9.11, p. 175). In turn, this interest likely affects female preferences, and, in general, mutual interactions between the two sexes are likely to be involved in many pairing decisions (Servedio and Lande 2006). Clayton (1990a) showed that normally raised females prefer consubspecific males that were normally raised, rather than cross-fostered males. This is not because these two classes of males sing differently; Clayton showed that females respond similarly to tape playback of the songs. Although other unmeasured differences between normally raised and cross-fostered males (e.g., differences in courtship displays) cannot be ruled out

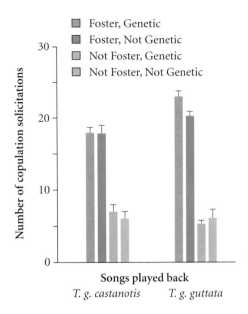

Figure 13.2 Mean number of copulation-solicitation displays given by Zebra Finch females of two subspecies in response to song playbacks (+ standard deviation: $N = 8–12$ in each group). The females were raised either by their own subspecies or by the other subspecies, and they were tested when fully grown after they had been given a dose of estradiol. The first four columns are responses of females to *T. g. castanotis* songs, and the second four are responses to *T. g. guttata* songs. There is a strong influence of foster father on female preference and no evidence for genetic effects. From Clayton (1990b).

as a cause for the mating preference, the result probably reflects males showing more interest in females who look or sound like the females that raised them.

Clayton's study shows that cross-fostering has a large effect on mate choice, and it leads to complete assortative mating when both sexes are cross-fostered. Such results are general. Young birds form a strong sexual preference based on learning about the individuals that rear them during a short sensitive period; a process termed *sexual imprinting* (Lorenz 1937; Clayton 1993; ten Cate and Vos 1999). Sexual imprinting has been demonstrated in more than 100 bird species, including taxa as diverse as the Ostrich and passerines, mostly by cross-foster-ing of captive species among different, sometimes distantly related, species, including humans (ten Cate and Vos 1999). In nature, the tendency of individu-als to pair heterospecifically as a consequence of being raised by heterospecifics is difficult to document by observation, but several cross-fostering experiments have led to the formation of mixed pairs (Cheke 1969; Harris 1970; Fabricius 1991; Ma and Lambert 1997; Slagsvold et al. 2002). Imprinting on color variants also affects mate preferences within species, as shown in Zebra Finches (ten

Cate et al. 2006), pigeons (Warriner et al. 1963), *Coturnix* quail (ten Cate and Bateson 1989), and the Mallard (Lorenz 1937; Cheng et al. 1979).

An early example of imprinting was described by Lorenz (1937, p. 263):

> Portielje of the Amsterdam Zoological gardens raised a male of the South American Bittern [possibly the Fasciated Tiger-Heron] who when mature courted human beings. When a female was procured, he first refused to have anything to do with her but accepted her later when left alone with her for a considerable time. The birds then successfully reared a number of broods, but even then Portielje had to refrain from visiting the birds too often, because the male would, on the appearance of the former foster father, instantly rush at the female drive her roughly away from the nest and, turning to his keeper, perform the ceremony of nest relief, inviting Portielje to step into the nest and incubate!

This learned preference for humans by hand-reared birds creates problems in animal conservation and animal husbandry. Among hand-reared Ostriches on English farms, males give courtship displays to humans, and females solicit copulations from humans, although it is not clear if these behaviors adversely affect the reproductive success of the birds (Bubier et al. 1998). The California Condor recently declined to near extinction, and, between 1987 and 1992, the entire population was held in captivity (in a "condor-minium"). Recognizing problems with imprinting, handlers used puppets resembling condor heads and beaks to feed chicks that were to be released into the wild (Meretsky et al. 2000). The strategy has been only partially successful. Some reintroduced birds have approached bystanders for food handouts and have vandalized tents (Meretsky et al. 2000, 2001).

The Zebra Finch, as well as Portielje's bittern, illustrate that sexual preferences established as a result of early learning may be difficult to alter (note that these are "preferences" and do not imply a refusal to mate with other types). However, social experience can modify preferences to some extent. Within species, associations with siblings or flock mates of a different color can alter sexual preferences away from the foster parent (Cooke and MacNally 1976; Kruijt et al. 1983; Kruijt and Meeuwissen 1993).

Oetting et al. (1995) made an elegant study of the impact of the later social environment on the development of sexual preferences in the Zebra Finch. They allowed Male Zebra Finches to be raised by Bengalese Finches and then placed the males in isolation from 40 days of age. At age 99 days, the male Zebra Finches were presented with a nest. The following day, they were exposed to a Zebra Finch female for one hour. Despite exposure to the Zebra Finch, when given pairwise choice tests four days later, all of the Zebra Finch males sang exclusively at a Bengalese Finch female and ignored a Zebra Finch female ($N = 10$); thus sexual imprinting had led to strong preferences for the heterospecific. However, when the experiment was repeated without the presentation of a nest on day 99, five of the ten males directed some song to the Zebra Finch female.

The interpretation of this is that excitement brought on by the nest served to consolidate the imprinting on the foster parent, which was then difficult to reverse. Without the presentation of the nest, several males did modify their preferences. Further experiments with 20 males in the no-nest situation showed that males that modified their preferences most on exposure to a Zebra Finch female were: (1) the ones most aroused (in terms of both general activity and measured corticosterone levels) when exposed to the female, and (2) those that were fed least by the foster Bengalese Finches when young (Figure 13.3).

Genetic contributions to species recognition

The overwhelming evidence for learning, with cross-fostering sometimes resulting in strong preferences to mate with species that look very different, raises the question of whether there is any genetically based tendency towards preferentially learning about one's own species. In many sexually dimorphic species young birds are raised entirely by the mother. Males may look very different from the female, yet females do not see or interact with them when young. It had been thought that in these species, females have innate preferences for conspecifics (e.g., in the Mallard, Schutz 1965). Subsequent experi-

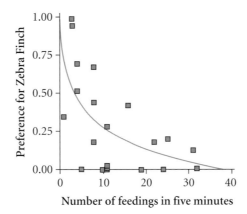

Figure 13.3 Preference of Zebra Finch males for Zebra Finch females, after they had been raised by Bengalese Finches. Males were given pairwise choice tests between female Zebra Finches and Bengalese Finches at age 104 days, after exposure to a Zebra Finch female at day 100. Preference is the fraction of songs sung to the Zebra Finch in one hour of observation. The x axis is the number of feedings given by an adult Bengalese Finch to the Zebra Finch males when they were young and soliciting food, in five-minute focal watches conducted every two days, during the 20-day period after fledging. A logarithmic curve is fit to the data ($P < 0.05$). Zebra Finch males that receive more attention from Bengalese Finches when young maintain stronger preferences for their foster species. From Oetting et al. (1995). Bischof and Clayton (1991) obtained a similar result.

ments cast doubt on this interpretation. In the Mallard, it now appears more likely that males imprint on the characteristics of their mother, direct their attention to similar females, and females mate with males that show the most interest in them (Cheng et al. 1979; Kruijt et al. 1982). In other sexually dimorphic species without male parental care, such as lekking grouse, young females may learn about suitable mates by following older females (e.g., Höglund et al. 1995).

In nature, experimentally cross-fostered individuals often end up pairing with members of their own species (Fabricius 1991; Slagsvold 2004). Although this could reflect an own-species bias, it is perhaps more likely to result from repeated social interactions with normally imprinted conspecifics. Such conspecifics are expected to respond to a cross-fostered individual throughout its life, based on its plumage and morphology, plus any innate component to its vocalizations and behavior. The result of repeated social interactions should be a redirecting of preferences of cross-fostered birds back in the direction of the conspecific, further strengthened because normally raised heterospecifics will show little interest. Slagsvold (2004) cross-fostered whole broods of Pied Flycatchers into Great Tit and Blue Tit nests in Norway and found no cross-species pairings in the subsequent breeding season. The Pied Flycatcher is migratory, spending the winter in Africa, whereas the tits are resident in Norway. It is possible that flycatchers have an innate genetic recognition mechanism [Sæther et al. (2007) suggest that genes for recognition lie on the Z chromosome]. However, an alternative is that conspecific pairing results from extensive interactions with normally raised conspecifics. This alternative seems especially likely because the reciprocal cross-fostering experiment could not be done, so there were no cross-fostered tits that might have shown an interest in Pied Flycatchers.

The obvious place to search for an innate tendency to mate with one's own species is among brood-parasitic species. About 1% of all bird species are brood parasites. They lay their eggs in the nests of other species, and in most parasitic species the young are raised by the foster species, so sexual imprinting on the parents cannot work. The North American Brown-headed Cowbird has been recorded laying its eggs in the nests of more than 220 different species, of which more than 140 have been shown to have reared the cowbird young to fledging (Lowther 1993). King and West (1977) kept female cowbirds isolated from males from the time they were undeveloped eggs. They were held in pairs in auditory and visual isolation from other birds. Several of these females gave copulation-solicitation responses to male cowbird song, while not responding to control songs of other species, implying innate recognition. Hauber et al. (2001) suggested that a "chatter call," which is made by both sexes and is relatively invariant across the whole species range, is the key. Nestlings beg in response to the chatter call, and fledglings and juveniles approach the chatter call. How this innate response arises is unknown, but the form of the chatter call shows some similarities to the begging call. Thus, one possibility suggested by Hauber et al. (2001) is that the recognition results from young birds learning their own calls.

Although the chatter call seems to provide a trigger, it only sets the stage for a great deal of social learning in the cowbird, which is initiated as young cowbirds gather in flocks once they are independent. One vocalization, the male flight-whistle, is divided into local dialects and is socially learned (Rothstein and Fleischer 1987). Demonstrated female preferences for the local dialect (O'Loghlen and Rothstein 1995) are also clearly learned, because of high juvenile dispersal among dialects (O'Loghlen and Rothstein 1995; Anderson et al. 2005). The Brown-headed Cowbird is divided into three subspecies, and females respond most strongly to the song of their own subspecies (King et al. 1980). When young birds were housed with adults of two different subspecies for more than one year, both a male's song and female mate preferences for song were modified toward the subspecies with which they were kept (Freeberg et al. 1999, 2001). These birds were subsequently housed together, and they mated assortatively by culture conditions (Freeberg 1996). Indeed, young males that were housed with Canaries actively court female Canaries instead of female cowbirds, although such preferences can be reversed the following year if the birds are housed with conspecifics (Freeberg et al. 1995). Other brood parasites also seem to require social learning to develop species-recognition cues. Soler and Soler (1999) studied the Great Spotted Cuckoo, whose young are reared by the Black-billed Magpie. Once fledged, young chicks are visited by adult cuckoos and normally gather in small groups. Soler and Soler moved chicks outside the normal range of the species. Young raised alone did not associate with each other and did not appear to recognize each other as conspecifics.

In a few species that are not brood parasites, young birds do not see either parent. A female Australian Brush-Turkey deposits her eggs in a mound of decaying vegetable material, which acts as an incubator. The young hatch asynchronously, dig themselves out, and often do not see any conspecifics for some time. Göth and Evans (2004) used robots in an aviary to infer that behaviors (such as pecking) and UV reflectance, perhaps especially of the legs, were innate cues that the young used to socialize with conspecifics. Chicks will respond to another chick's call (a soft "grunt") by reducing their level of activity and by scanning, but they show no response to adult vocalizations (Barry and Göth 2006), which they presumably learn later.

Thus the evidence implies that, in species in which the young are not raised by the parents, weak innate preferences are strengthened by much social learning. Despite the lack of more direct evidence, it is possible that in those majority of species in which young are raised by one or both parents, a similar process operates. Although there is no strong evidence for an innate component, three lines of reasoning support this possibility. First, Davies et al. (2004) used cross-fostering experiments to show that young of two passerine species showed innate fear responses to their own species' alarm calls. This could form the basis for an own-species bias (in this study, responses to alarm calls were further strengthened as a result of learning). Second, in Clayton's experiments on the Zebra Finch, when only one cross-fostered sex was introduced into an aviary, most (90%) of the pairs that formed were with its own subspecies rather than its foster subspecies. This suggests normally raised birds more strongly determine mating patterns than cross-fostered birds, which might result from an own subspecies bias. However, in this case, cross-fostered individuals were held in sibling groups which likely influenced social learning. Third, the tendency of males to preferentially learn their own species songs is fairly well established (Chapter 10; Nelson 2000a; Soha and Marler 2000; Beecher and Brenowitz 2005), and song-learning shows many similarities to sexual imprinting (Baptista et al. 1993; ten Cate et al. 1993).

In summary, learning contributes strongly to species recognition, and sexual imprinting drives species recognition in most species. There may be some genetically determined preference to learn features of one's own species, which may be as simple as learning features of one's own innately produced call note. However, own-species biases do not seem to be very important. In the vast majority of species, the first individuals that a young bird encounters are one or both of its parents. Learning about features of the parent makes an innate recognition mechanism redundant. Thus, it is not obvious that innate recognition should be maintained in systems where parents provide care to their young. By contrast, in such systems, a learning system is essential if chicks are to distinguish their parents from other conspecifics, as described below.

In the following sections, I consider the learning process throughout the life-history, from hatching to pair formation—that is, how individual recognition of the parents by the young is eventually translated into the choice of a conspecific as a mate. I then ask how the presence of heterospecifics may lead to genetic modification of what is learned later in life.

FILIAL IMPRINTING

The first manifestation of learning in young birds is filial imprinting, which is narrowly defined as the learning process accompanying the following response of precocial birds (Hinde 1962; Bolhuis 1991). (Precocial birds are species in which the chicks leave the nest soon after hatching, whereas altricial ones are blind and helpless upon hatching.) Filial imprinting is adaptive in that it usually results in a chick following its mother. A similar process is also recognized in altricial birds as a begging response (Junco 1988). In many wild bird species, the chick recognizes the parent's call and will respond to it, but not to the call of other conspecifics (Halpin 1991, pp. 235–240; Marler 2004, pp. 159–162).

Chicken and quail provide the model systems for the study of filial imprinting. In these species, the chicks follow, and develop a strong attachment to, the first moving object they see. They will then not follow other distinct objects but will instead develop a fearful reaction to them. Some salient features of imprinting are *predisposition, generalization,* and *discrimination.*

PREDISPOSITION Predispositions are weak. Chicks have naïve biases to approach certain objects more than others, but these biases are not very specific. For example, it is easier to imprint chicks on moving objects and on some colored lights than it is to imprint them on stationary objects or on lights of other colors, but Bateson (1966, p. 185) suggested the latter result simply reflects the conspicuousness of the color against the background. Chicks have a tendency to approach a rotating stuffed fowl over a rotating red box. Experiments have shown that it is the image of the head and neck of the fowl that leads to this bias. Chicks were as likely to approach a stuffed polecat as a stuffed hen (Johnson and Horn 1988). Naïve ducklings do show predispositions for the maternal species-specific call, but much less so if they have been devocalized (i.e., prevented from calling themselves) when still in the egg and held in isolation from other developing eggs so that they cannot hear calls (Gottlieb 1978). The weakness of specific predispositions mirrors the findings from sexual imprinting.

GENERALIZATION After an animal has learned about something, it is likely to respond in the same way to a similar, but not identical, stimulus, to learn any differences, and to update them as the same representation. This is termed stimulus generalization. Hinde (1961) noted that the parent appears in many shapes and sizes and against many backgrounds, and therefore the young bird should not be very selective in the cues it uses. Precise individual recognition of the mother under a variety of circumstances becomes possible by learning the ensemble of characteristics that identify her (e.g., her shape, color, and call). To learn these characteristics, chicks may actively work to expose themselves to different aspects of the mother (Jackson and Bateson 1974), and it has been shown that chicks more readily imprint both on the call of the mother and on her appearance when both are presented simultaneously than when they are presented separately (Bolhuis 1991, p. 321; Bolhuis and van Kampen 1992), a principle that applies to learning more generally (e.g., Rowe 1999; see Chapter 9). These sorts of updates and the use of multiple cues may be important in generalizing from single individuals to conspecifics during the development of sexual preferences (Figure 13.4).

DISCRIMINATION Part of the nature of the imprinting process is that, once the young bird has formed an attachment to a particular object, it avoids discretely different objects (Bolhuis 1991, p. 310). Hinde (1961) noted the strong selective advantages to avoiding potential predators as well as unfamiliar adults. The tendency to discriminate rather than to generalize depends both on the characteristics of the different objects and how close together in time they are observed. Objects differing in form alone are most easily discriminated in subsequent tests if they are presented simultaneously; this presumably reflects the tendency to update similar objects presented sequentially as the same representation. Objects differing both in form and in color are more easily distinguished than objects differing in only one feature, and these objects are easily distinguished when presented sequentially (Honey et al. 1994).

Filial imprinting may be widespread because of the need for individual recognition and, in particular, recognition of the parent. It would be impossible for an innate mechanism of parental recognition to evolve. This would require precise genetic transmission of the parents' characteristics to the offspring, and transmission of the ability of the offspring to recognize and respond to those characteristics. Grafen (1990) discussed similar problems with respect to recognition of relatives who are not the parents. Sharp et al. (2005) show that, in the Long-tailed Tit, recognition of relatives as an adult is achieved by learning calls from the parents and siblings while in the nest.

INDIVIDUAL RECOGNITION

Individual recognition requires learning, and it does not stop with parent-offspring relationships. Characteristics of individuals are learned throughout the life of a bird. Many advantages accrue both to recognizing individuals and to being recognized on the basis of past encounters. In the case of threat, the combination of individual recognition and the memory of past interactions is one of the most honest indicators of fighting ability (van Rhijn and Vodegel 1980). It is likely to be much harder to bluff individual identity than other indications of fighting ability, such as intensity of a threat display (van Rhijn and Vodegel 1980). Thus, the frequency and relative proportion of different threat displays change as flockmates become familiar with one another (Douglis 1948; Balph 1977; Whitfield 1988; Wilson 1992; Figure 9.6, p. 168): contests settled by display or fights are now resolved with much less interaction, to the benefit of both birds. In this way dominance hierarchies based on individual recognition become established (Guhl and Ortman 1953; van Rhijn and Vodegel 1980). Douglis (1948) showed that 27 hens in a flock were able to recognize each other individually, based on staged pairwise interactions, although Guhl and Ortman (1953) stated that individual recognition appears to be lost within two weeks of separation. Douglis (1948) reported more anecdotal evidence that 97 flock-mates individually recognize each other.

Individual recognition is important in many other contexts. Mates may recognize each other even after a separation of more than seven months (Herring Gulls, Tinbergen 1953; Zebra Finches, Immelmann 1959; Ring Doves, Morris and Erickson 1971). Re-forming pairs go through early courtship more quickly and have higher reproductive success than do unfamiliar pairs, even in laboratory experiments (Yamamoto et al. 1989).

The main function of some variable signals seems to be to enable other birds to rapidly learn differences between individuals (Beecher 1982; Whitfield 1987; Dale et al. 2001). In many territorial species, neighbors recognize each other based on song (Temeles 1994; Stoddard 1996). This has been demonstrated not only by comparing a male's reaction to playback of the song of his neighbor with his reaction to the song of a stranger, but also by playing a neighbor's song from both a familiar location and an unfamiliar location. Godard (1991) showed that, in Hooded Warblers, males remember their neighbors' songs after a period of seven months away in the winter quarters. Similarly, females recognize mates based on their songs or calls (Brooke 1978; Lampe and Slagsvold 1998). A sophisticated mechanism of individual recognition of mate, offspring, and parent by call has been demonstrated in six species of penguins (Aubin and Jouventin 2002).

Plumages are also used in individual recognition. The Ruddy Turnstone is a small wader that uses a silent head-bob aggressive display, exhibiting a white throat and a black breast prominently. Whitfield (1986) carefully painted head patterns on models to look like a neighbor or a stranger, and then placed them on the territory boundaries of ten males. He found much stronger aggressive responses of the territorial bird if the face patterns resembled those of a stranger rather than those of a neighbor. In winter, turnstones form flocks consisting of some tens of birds. Painting the face patterns of individual birds in a captive flock of ten resulted in the manipulated birds receiving more attacks for about 30–60 minutes, after which they regained their former status (Whitfield 1988).

SPECIES RECOGNITION

Individual recognition easily leads to species recognition. Given that a variety of different views can be recognized as the same individual (Hinde 1961; Bolhuis 1991, pp. 316–318), it is fairly straightforward to see how individual recognition

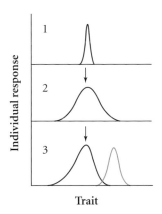

Figure 13.4 Model of development of a mating-preference curve through an individual's lifetime (after Irwin and Price 1999). (1) The distribution of signals that a chick associates with its parent. (2) The young bird generalizes its response curve outward from the distribution of the parents' traits to recognize other conspecifics. (3) In the presence of a similar heterospecific (*blue curve*), as a result of experience with this heterospecific, the preference curve may contract on one side.

can be generalized to other similar individuals (Figure 13.4). The problem comes at the other end of generalization: why do individuals not generalize so much that they include members of other species?

The problem is difficult, because only a fraction of the species-specific cues are needed to identify conspecifics, and presumably any one cue that resembles that of a heterospecific could lead to mistaken identifications. Thus, call notes, songs, and plumages are each sufficient to identify a conspecific. In Pied Wagtails, strangers flying over a territory give a typical "chis-ick" call; if this call is heard by a resident, the resident responds with a special "chee-wee" call, and the intruder leaves (Davies 1981). In song-playback experiments, only a tape-recorded conspecific song is needed to elicit strong responses from males. Likewise, the presentation of stuffed mounts without songs leads to male reactions (e.g., Ratcliffe and Grant 1983). Females also respond to conspecific song or plumage when presented in isolation (Table 9.1, p. 162; Figure 13.5). The use of relatively few cues is essential for an efficient recognition system. Between modalities, birds that are seen do not always vocalize, and vocalizing birds cannot always be seen. Within modalities, transmission of song components varies with environmental conditions, and nonvocalizing individuals will not always be viewed in the same aspect or against the same background. This point was already made for filial imprinting and individual recognition, but it applies equally to species recognition.

The use of different cues to identify conspecifics, either in isolation or together, must usually arise because they become associated with one another through learning, as noted in the section on filial imprinting. Most fieldwork on

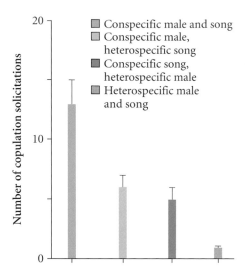

Figure 13.5 Average number of copulation solicitation displays (+ standard error) given by female buntings in response to the presence of males, with or without conspecific songs and calls (Baker and Baker 1990). All birds came from allopatric populations and were held in the laboratory for 10 months prior to the experiment. Nineteen Indigo Buntings and 20 Lazuli Buntings were tested, and their responses were combined. A similar result had been obtained in a previous year.

associative learning has been based on reactions of males to other males. This has shown that learning can sometimes be across wide phenotypic gaps (e.g., very unusual songs can become associated with a particular plumage type). Richards (1979) discovered three towhees that were singing Carolina Wren songs, as well as the songs typical of their species. Other towhees that had one of these three towhees as neighbors responded aggressively to the playback of the mimic Carolina Wren songs, whereas towhees from elsewhere did not respond to these songs. Thus, associative learning of the mimic song with a towhee's plumage, morphology, and possibly other vocalizations, resulted in the mimic song being recognized as coming from a conspecific. Generalization at this level suggests that quite different heterospecifics could sometimes be embraced as conspecific. I first consider some examples of generalization between closely related species, and then how generalization may be limited, so that heterospecifics are not treated as conspecifics.

At least seven studies have inferred that learning regularly leads to increased responses of males to songs from related species. The result is that heterospecific males are reacted to in a similar way as a conspecific, and males of the two species defend mutually exclusive territories (Rohwer 1973; Sorjonen 1986; Gil 1997; Irwin and Price 1999). The Blackcap and the Garden Warbler are interspecifically territorial, and they respond aggressively to each other's quite similar songs. Matyjasiak (2005) showed that associative learning has led to both song and plumage being used by Blackcaps to identify Garden Warblers. In Poland, Blackcaps arrive back from their winter quarters three to four weeks before Garden Warblers. During this time, Matyjasiak played Garden Warbler songs on Blackcap territories, accompanied by stuffed mounts of each species. In 27 of 31 times, the Blackcap on the territory attacked the Garden Warbler mount but not the Blackcap mount. When Matyjasiak played Blackcap song on 39 other territories, the owner always attacked the Blackcap mount. Apparently, Blackcaps had learned to associate Garden Warbler song with plumage, and they remembered this association over eight months.

In the Western Meadowlark and Eastern Meadowlark, males show stronger aggressive responses to heterospecific song where both species co-occur than where one species is found alone (Rohwer 1973). It is clear that this is a result of learning, rather than an innate response, because the compared males can be less than one field apart. The two species look very similar (Figure 2.6, p. 27). In this case, it seems that males are learning that two different songs are associated with essentially the same plumage pattern, just as the towhees were doing with their wren-mimic neighbors.

These examples suggest that members of other species can come to be treated as conspecifics through learning, especially when they are similar in at least one cue. Although the examples are of situations in which males are competing for territories, they raise the question: how are most heterospecifics recognized as different? If the phenotypic gap between species is sufficiently large, and if a sufficient number of cues are always available, this should not be a problem. Nelson (1989) measured 14 variables on the songs of 12 species coexisting with Field Sparrows in the eastern United States. He found that each measurement differed somewhat among species, and, once he had incorporated seven different measurements, every species song was correctly classified 98% of the time. This means that, provided individuals are always able to use an ensemble of characteristics to identify conspecific songs (and plumage and other traits are also sufficiently different that whole songs are not generalized), they are unlikely to generalize sufficiently to recognize a song as belonging to another species. Field Sparrows show reduced responses to modifications of (1) frequency, (2) number of phrases, (3) duration of a trill, (4) internote interval, and (5) note shape when the other song features are held constant (Nelson 1988). In the first four, decreased responses occurred when the trait was modified two to three standard deviations beyond the mean (i.e., just at the limit of natural variation). Thus, associative learning would be required to recognize a rare variant along a single dimension, but even this amount of generalization would be insufficient to mistakenly include heterospecifics, because they differ along multiple dimensions.

Nelson's study was in a community of quite distantly related species, and problems of recognition are likely to accrue if taxa are more similar—that is, during the early stages of speciation. As noted above, associative learning by males sometimes results in heterospecifics being embraced as conspecifics. However, at other times, learning can result in reduced responses so that heterospecifics become increasingly ignored (Figure 13.4). For example, Gill and Murray (1972) found a lower response of Blue-winged Warblers to the song of Golden-winged Warblers where they occurred in sympatry. Gill and Murray (1972) suggested that this results from "learning that a particular song represents a particular plumage type." These species do hybridize, but they have overlapping territories and heterospecific males learn to tolerate each other. Irwin and Price (1999) give three other examples in which aggressive responses are lower in sympatry than allopatry. The boundary between learning that results in simply ignoring other heterospecific males and in learning that results in increased responses to heterospecifics seems critical to understanding the origins of premating isolation when similar, closely related species come into sympatry, but the factors leading to one or other outcome are little studied (Chapter 14).

The observations described above are all based on males competing with each other. While they show that associative learning can lead to increased or decreased responses to heterospecifics, results based on male competition do not necessarily translate to mating decisions by females. A typical male may interact frequently with the same neighbor, giving him time to learn the characteristics of that individual. On the other hand, a typical female may visit several different males during her search for a mate (Bensch and Hasselquist 1992), and she is likely to respond most strongly to those males that resemble other conspecifics and her parents. She is also likely to integrate multiple cues, after observing potential partners for a prolonged period, rather than making decisions based on a single cue. Baker and Baker (1990) examined the response to heterospecific cues to males of female Indigo Buntings and Lazuli Buntings drawn from allopatric populations. They found that, when they either (1) exposed females to heterospecifics but played conspecific song or (2) exposed females to conspecifics but played heterospecific song, they got moderate responses (Figure 13.5). Females gave strong responses to conspecific males with conspecific songs and weak ones to heterospecific males and heterospecific songs. Patten et al. (2004) found similar results for two subspecies of Song Sparrows. These results suggest that females are stimulated by multiple cues, presumably learned through sexual imprinting, so, given that heterospecifics differ in several ways, they may infrequently be chosen as mates. However, especially when conspecifics are rare, stimulation by similar-looking heterospecifics may indeed be sufficient to result in mating and/or pair formation (Chapter 14).

Genetic influences on species recognition

The need to learn about heterospecifics must come with costs. In his study of male interactions in meadowlarks, Rohwer (1973) showed a possible advantage to a male to rapidly recognize his neighbors. In areas where the species overlap, males seem to fight more often than they do in pure populations of one or the other species. Rohwer (1973) found that, where the two species occur together, 30% of 358 males had wound marks around the beak, whereas, in areas where only one species occurs, only 15% of 124 males had wound marks. He argued that the unfamiliar song of a heterospecific makes it more difficult for males to learn and remember individuals. The result is that a neighbor is not immediately recognized as the rightful owner of an adjacent territory, leading to escalated contests.

The time it takes to learn about other species implies the presence of selection to modify the extent to which individuals need to learn, especially if these species are frequently encountered. Thus we might expect that a relatively narrow generalization curve (Figure 13.4) should be genetically assimilated, resulting in the more rapid recognition of other similar species. One test of this would be to compare the reactions of individuals imprinted on identical conspecific parents that have never encountered heterospecifics. The prediction is that individuals from sympatric areas would learn to recognize heterospecifics more quickly than would individuals from allopatric areas. This has not been

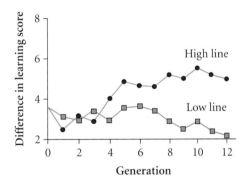

Figure 13.6 Evolution of learning demonstrated in an artificial selection experiment. Kovach (1990) examined the preferences of one-day-old quail chicks for a red or a blue light in a maze with a total of 14 choices. A score of 14 indicates that blue was always chosen, and a score of 0 means that red was always chosen. Kovach then imprinted the chicks by holding them in a box for 12 hours (in the early generations of the study) to 30 hours (in later generations) and presenting a light at one end of the box for 10 minutes and then the other end for 10 minutes. He imprinted those chicks that showed preferences for red (a score of 7 of less) onto a blue light. Conversely, those chicks that showed a preference for blue light were imprinted onto a red one. He then repeated the preference test. The selection experiment consisted of breeding from individuals who had the highest change in learning score (indicating high imprintability) or, in a separate line, the lowest change in learning score (low imprintability). For each selected pair, one individual had been imprinted to blue and one to red, and the genetic evolution of a response to color was minimal. [Kovach (1980) had previously demonstrated that one-day-old chick preferences to each color could be rapidly modified by selection.] The figure is from Kovach (1993).

done. However, an artificial selection experiment on filial imprinting by Kovach (1990, 1993) shows that genetic variance affecting generalization is present (Figure 13.6). Kovach (1990) selected on imprinting of one-day-old Japanese Quail chicks, by breeding from individuals that imprinted the most and least efficiently, and discarding those that were intermediate. After measuring the chicks' preferences for a blue or red light in a maze, he attempted to alter their preferences by imprinting the chicks on one or other color over a period of 12 hours or more (see Figure 13.6 for details). Those chicks that preferred blue were imprinted on red, and those that preferred red were imprinted on blue. He then measured preferences again. He selected a line in which preferences could be easily modified by imprinting and one in which they could not. He found that he could select for both an increased and decreased modifiability of the imprinting process (Figure 13.6).

These results indicate that the degree to which preferences can be modified by learning rapidly evolves given the correct selection pressures. Thus, if quite similar looking and sounding heterospecifics are a regular feature of the conspecific's environment, conspecifics should come to ignore them with little learning, or without the need for learning at all. This contributes to the process of reinforcement of mate preferences, considered more fully in the next chapter.

SPECIES RECOGNITION, SEXUAL SELECTION, AND SPECIATION

Given the finding that species recognition is largely attributable to learning, we can return to the questions posed at the beginning of this chapter. First, sexual imprinting should result in individuals of different appearance not being recognized as conspecific. This therefore imposes selection against novel traits, so how do such traits invade and spread in the population? Second, how can a trait and recognition of the trait change together? This is needed if individuals in one population are to recognize individuals in another population as different. I consider these questions in turn, beginning with the conflict between sexual selection and species recognition.

One feature of the learning process is that individuals that are not too different are updated as the same representation. In this way, a slightly deviant individual, or one carrying a novel trait, can be recognized as conspecific. If the trait is quite different from the norm, then this may take repeated encounters, with some cost and, hence, reduced fitness. In other examples, however, there may be little cost to bearing a new trait.

One way by which a new trait may become established with little cost has been demonstrated in a study of the Javan Munia, a relatively dull-colored, sexually monomorphic finch. The species has chocolate-brown upperparts and hood, and it is pale below (Figure 13.7). Witte et al. (2000) and Plenge et al. (2000) attached a single red feather to the forehead of each of several males (Figure 13.7). They found that females showed no preference for unadorned males over adorned males in choice tests (Figure 13.7), implying no cost of mating with males carrying the novel trait. When males with the red feather raised their young, the young became imprinted on the red feather and then preferred

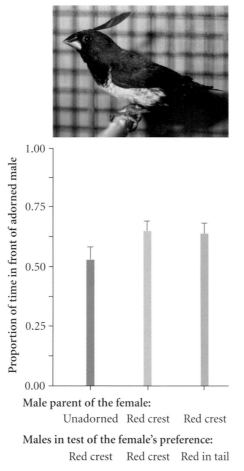

Figure 13.7 Preferences induced by imprinting (from Plenge et al. 2000). Columns give the fraction of time females directed their attention to males in pairwise choice tests of an adorned male or an unadorned male (+ standard error). Female Javan Munias were raised by unadorned parents or by parents distinguished by the fact that the male of the pair either had a red feather attached to his forehead or had red paper stripes added to his tail. (All female parents were unadorned.) Females raised by pairs in which no adornment had been added to the male did not discriminate between adorned and unadorned males. Females raised by pairs in which the males had added adornment preferred males with red color in their plumage. Photograph provided by Klaudia Witte.

males carrying the red feather in subsequent choice tests (Figure 13.7; Witte et al. 2000). In this way, a conspicuous trait could become established, first because it was not discriminated against by individuals imprinted on the ancestral type, and, second, because it was favored by individuals imprinted on the novel type. Plenge et al. (2000) carried the study further and showed that females from adorned parents preferred males with four red paper stripes artificially introduced among their tail feathers (Figure 13.7). This implies that the induced preference for red was quite general and could lead to selection favoring the appearance of red patches elsewhere in the plumage. Similarly, in a field experiment, Qvarnström et al. (2004) showed that female Collared Flycatchers paired with males that were adorned with a red patch preferred red-patched males the following year, whereas other females had no preference. In this case, learning as an adult seems to have favored the spread of a novel trait, in a mechanism rather similar to that shown for the Javan Munia.

These experiments provide a model for the spread of new traits through imprinting and learning, but a dual recognition process is still implied: first, a conspecific is chosen and, second, a certain conspecific is preferred. This dual process is observed in many experimental studies. For example, females prefer male songs from their own region, as well as more complex songs within each region (Baker et al. 1987, Figure 10.9, p. 212; Clayton 1990a, b). Female Japanese Quail imprinted on a white strain preferred that strain, but, within the strain, there was a preference for black spots painted on the breast of the male (ten Cate and Bateson 1989). Exactly how this dual selection process is resolved in nature is unclear (Pfennig 1998), but often heterospecifics in the vicinity are so different that they are ignored by females. In other cases, resolution of species recognition and sexual selection may be through the different roles of the sexes, with males choosing conspecifics to court and females choosing among courting males (Williams 1983, p. 305).

The second question raised at the beginning of this section was how do the trait and recognition of the trait evolve together? This is more easily understood. Sexual imprinting automatically results in the evolution of recognition along with the trait, and it provides a powerful mechanism for assortative mating across populations (Laland 1994). As described above, subspecies of Zebra Finch pair assortatively when given the choice in an aviary, but, when cross-fostered, they pair assortatively by foster parent (Clayton 1990a, c). Although sexual imprinting results in strong mating preferences for individuals resembling the parents, such preferences can be modified by social experience. In this case, the ability to learn about and to develop preferences for new partners reduces the possibility of population differentiation. Factors limiting cross-species mating are considered further in the next chapter.

CULTURAL SPECIATION

I end this chapter with an example of complete speciation apparently attributable entirely to learning. The brood-parasitic *Vidua* finches of Africa are divided into two main groups, the indigobirds (ten species) and the paradise and waxbill whydahs (nine species). All these species parasitize other finch species. They lay their eggs in the nests of these other species, and the foster parents incubate the eggs and rear the young. Seventeen of the 19 *Vidua* species each largely parasitize just a single host species. Robert Payne has studied them in depth, initially by driving around Africa in an old van. He coined the term "cultural species" to describe their formation.

Within the parasitic finches, species recognition appears to be based largely on song, which is learned from the foster parent. Males sing songs that resemble those of their host species, and females learn the song of the host species when they are young and use this as a cue in choosing males with whom to mate. The Village Indigobird parasitizes the Red-billed Firefinch. Male Red-billed Firefinches sing a single song type. Male Village Indigobirds have a repertoire of three or four mimic song types as well as on the order of 20 or more nonmimic song types (Payne 1979; Payne et al. 1998), which they probably learn from other indigobirds, having first been primed by the song of their foster parent (Payne et al. 1998). In the Village Indigobird, song learning from foster fathers has been demonstrated in the laboratory using both Red-billed Firefinches and Bengalese Finches as hosts; the young males then learn additional songs from companions raised by the same host species (Payne et al. 1998).

Females use song to recognize conspecifics. In aviary experiments with wild-caught birds, both female indigobirds and female whydahs were preferentially attracted to their host's song (whether sung by the host, or by a male of their own species), rather than to the song of a species that is not the host (Payne 1973a, b). Payne et al. (2000) raised female Village Indigobirds in captivity using either Red-billed Firefinches or Bengalese Finches as hosts. They found that when adult, the female preferentially approaches a speaker playing a tape of a male song that resembles the song of the female's foster parent. Field research also supports the idea that song is a main cue used by females to identify conspecifics. In regions where two or more indigobirds co-occur, about 2% of all visits by female indigobirds to singing male indigobirds were to a heterospecific and not to their own species (Payne et al. 2000). In half these cases, the heterospecific male was singing the song characteristic of the female's species; Payne et al. (2000) suggest that these males may have been raised by the "wrong" host. In the other cases, females visited a heterospecific male singing songs characteristic of the male's normal host. Payne et al. (2000) suggest that, in these cases, the females themselves may have been raised by the male's host and so developed preferences for heterospecific male song.

In addition to recognizing conspecific males as mates, females have to discover nests of their foster species in which to lay their eggs, and this also seems to develop by an imprinting like process; perhaps a female follows a female that resembles her foster mother. Payne et al. (2000) showed that, after rearing female Village Indigobirds in captivity under different hosts, in a mixed aviary, the females visit and lay eggs in the nests of the species that raised them; six females reared by Bengalese Finches laid only in Bengalese Finch nests, and eight females reared by Red-billed Firefinches laid only in firefinch nests.

Although females can distinguish among different species of indigobirds based on mimic songs, the cues by which the females prefer to associate with their own species rather than their foster species are unknown. Male indigobirds include a number of other vocalizations in their repertoire besides the mimic songs, including a chatter call and begging calls (of both their own and the host species, Payne 1979; Payne et al. 1998). Own-species begging calls seem to be innate and different from those of the host (Payne et al. 1998, 2001). Possibly, the female recognizes the begging calls in a male song having learned her own calls in the nest, but this has not been tested.

These results suggest that a new parasitic species may rapidly originate when one or more females lay in the nest of a new host species. Young males learn the songs of the new foster species. Young females learn and respond to the song of their foster fathers, and they learn where to lay their eggs. In this way, premating isolation is rapidly established between finches that parasitize different host species.

Much recent work on the phylogenetic relationships of the parasitic finches has shown that the indigobirds are very closely related to one another (Klein and Payne 1998; Sorenson et al. 2003, 2004; Sefc et al. 2005; Figure 13.8). They have become the classic example of likely sympatric speciation in birds (Sorenson et al. 2003; Chapter 2). Comparing Cameroon with Malawi and Zimbabwe, indigobird species are, with one exception, more closely related to other species in the region than to the species using the same host in the other region, thereby providing evidence that they have speciated sympatrically (Klein and Payne 1998; Sorenson et al. 2003; Sefc et al. 2005). Payne and Payne (1994) found populations of indigobirds exploiting unusual hosts. One population is clearly a recent colonist in that it looks identical to the form that parasitizes the typical host (Payne et al. 2002). The recent divergence among the parasitic species, their genetic similarity within geographical regions, the observed recent colonization events, and the fact that females are occasionally observed visiting males of different species suggest a dynamic, ongoing process of colonization of hosts.

In many *Vidua* species, the mouth coloration and the pattern of the nestlings match the colors and patterns of the nestlings of the host species (Nicolai 1974;

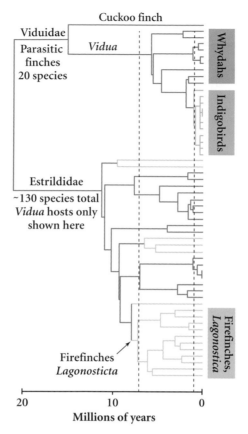

Figure 13.8 Phylogenetic tree based on mitochondrial DNA sequences depicting relationships among parasitic finches of Africa and their hosts (redrawn from Sorenson et al. 2003). The Cuckoo Finch parasitizes warblers, which are not on the tree. Indigobirds parasitize those finch species indicated by the *orange lines*, and whydahs parasitize the other finches on the tree (*blue lines*, mostly belonging to the genera *Estrilda* and *Pytilia*). Dashed lines represent estimated ages of divergence among firefinch hosts and the indigobirds that parasitize these hosts, respectively. Note the shallowness of the parasite tree.

Payne et al. 2001; Payne 2005b; Figure 13.9). In captivity, Payne et al. (2001) found that Red-billed Firefinches were more likely to raise Village Indigobird young successfully than the young of finch species that do not mimic the firefinch nestlings, but nonmimetic young are occasionally reared to independence. In a field experiment, Schuetz (2005) found that painted Common Waxbill chicks had only slightly reduced growth rates, and no reduction in survival to fledging, when compared with controls. These results imply that, although exploitation of a new host may initially result in reduced fitness, such hosts can be colonized. Then, the reduced fitness of nonmimetic young results in selection to match the host's mouthpart coloration. In this way, genetic changes in mouth-

Hosts   Parasites

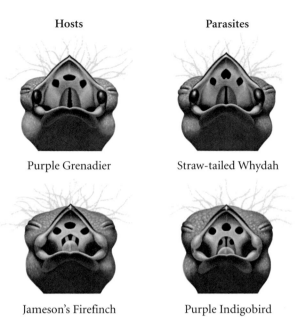

Purple Grenadier   Straw-tailed Whydah

Jameson's Firefinch   Purple Indigobird

Figure 13.9 Mouthparts of *Vidua* finches and their hosts (from Nicolai 1974). One whydah and one indigobird are shown. Within the indigobirds, the differences are much smaller (Payne 2005b).

part coloration follow after the colonization of a new host. Once mouthpart coloration has evolved to become different, hybrids between species may have intermediate coloration, resulting in some reduced fledging success and, hence, some postmating isolation. Modifications in mouthpart coloration should also make it more difficult for a well-adapted parasite to reinvade the ancestral host. Differences among the indigobird species in mouthpart coloration are quite small, but differences between the indigobirds and whydahs are large (Figure 13.9), and perhaps colonization of each other's hosts is difficult.

CONCLUSIONS

The examples of this chapter show why a system of learning is likely to be widely used. First, many of the traits are used both in individual recognition and in species recognition; a learning mechanism is essential for individual recognition. Given that an individual-recognition mechanism is in place, it is easy to see why this mechanism leads to species recognition. Second, learning means that an individual can use a subset of many cues to identify species and to update representations. This imparts flexibility, but it also means that individuals may embrace heterospecifics as one of their own species. Such a flexible system slows the completion of speciation, for it can lead to mating with heterospecifics. How this gets resolved so that mating becomes restricted to conspecifics is the topic of the next chapter.

SUMMARY

In most birds, recognition of members of one's own species as suitable mates is achieved almost entirely through learning. "Sexual imprinting" is defined as the learning of species-specific characteristics by young birds, usually from the parents during a sensitive period, which results in a sexual preference for one's own species later in life. Once established, such a preference can be difficult to reverse. Because of sexual imprinting, divergence in such features as song, plumage, and morphology results in the species-recognition mechanism evolving in parallel, thereby rather easily creating the basis for premating isolation. Learned preferences are widespread probably because of the value of recognizing parents and other individuals during ongoing social interactions; individual recognition is impossible to achieve via genetic mechanisms. The learning process begins with chicks recognizing their parents, continues through individuals recognizing each other in social encounters, and ends with adults distinguishing between conspecifics and heterospecifics during territorial interactions and pair formation. A learning system imparts flexibility to species recognition so that unusual conspecifics, as well as similar heterospecifics, can eventually be recognized through a process of generalizing from a few cues. Any species-recognition mechanism, including sexual imprinting, imposes a constraint on the appearance of novel features because individuals carrying those features may be discriminated against. However, one consequence of learning and generalization is that new traits can sometimes become established. In one experimental study on the Javan Munia, some individuals were adorned with a red feather. They were not discriminated against by individuals raised by normal parents, but they were preferred by individuals imprinted on parents that themselves carried the feather.

CHAPTER FOURTEEN

Mate Choice
at the End of Speciation

In the vast majority of sympatric species, individuals from one species rarely mate with those from another. As described in the previous chapter, the tendency to mate with one's own kind develops as populations diverge in such characteristics as morphology, vocalizations, and plumages, as well as in the habitats they occupy. Preferences for these traits diverge in parallel (e.g., Table 10.1, p. 190; Chapter 13; and for habitat, Davis and Stamps 2004; Beltman and Haccou 2005). Divergence may be sufficient that individuals from one population never mate with individuals from another and speciation is complete. An isolated species on an oceanic island (e.g., a Dodo) would probably never recognize its closest relative (i.e., a pigeon) as a suitable mate, provided it does not accidentally become misimprinted.

On the other hand, when taxa have not diverged greatly, they should merge when they come together. A recent example of this is the blue and white morphs of the Snow Goose in North America, which, until the 1920s, were essentially allopatric and were sometimes considered to be separate species. As a result of changed agricultural practices, the two morphs came into contact, and they now freely interbreed, albeit with some assortative mating, to produce fully fit offspring (Cooke et al. 1988). Recent human-induced habitat change seems to have led to increased levels of hybridization of many other taxa besides the Snow Goose, by bringing individuals of related taxa into more frequent contact (Mayr 1963, p. 128; Rhymer and Simberloff 1996), and, in the next chapter, I describe several wide zones of overlap and hybridization that appear to be transient effects of habitat alteration. In addition, human introductions have created hybrid populations. Rhymer and Simberloff (1996) list three bird species and four subspecies that are now threatened as a result of hybridization with related introduced taxa.

Lying between the extremes of a complete failure to interbreed and a rapid merging of taxa is the situation in which individuals from each taxon sometimes mate with individuals from the other, but the taxa remain distinct. In this case, additional evolutionary processes can lead to a reduction in cross-taxon matings, that is, a strengthening of premating isolation. In this chapter, I investigate various conditions under which this happens. I envisage a situation in which two differentiated taxa spread into each other's range and hybridize at the level of a few percent, producing offspring with low fitness. I consider factors that favor a reduction or an increase in reproductive isolation and use empirical studies of mate choice in nature to ask how likely these conditions are to be met.

Many closely related species have been recorded as occasionally hybridizing (McCarthy 2006). From his examination of museum skins, Mayr (1963, p. 114) estimated that one out of 60,000 wild birds is a hybrid. He also suggested that perhaps one in 20,000 birds of paradise might be a hybrid (Mayr 1945); both these estimates are very much guesses. Grant and Grant (1992), based on an update of Panov (1989), found that just over 9% of all bird species have been recorded as hybridizing in nature. In many cases, hybridization between a pair of species has been recorded just once (e.g., Short 1969; Rohwer et al. 2000). When hybridization is as rare as this, the strength of selection against hybridization is so weak that it is likely to be difficult to reduce further. However, given that hybridization regularly occurs, albeit infrequently, it seems plausible that hybridization was more common for some taxa when they first came together, and that it became reduced to low levels as a result of various interactions in sympatry. In the theory of reinforcement, selection against individuals that pair with heterospecifics and produce offspring of low fitness strengthens the tendency to mate with members of one's own taxon (Dobzhansky 1940; Howard 1993; Servedio and Noor 2003; Coyne and Orr 2004, their Chapter 10). Assessment of this theory is the main goal of this chapter.

Mate choice and sexual selection can influence the completion of reproductive isolation in several ways besides reinforcement. First, traits involved in male competition, such as songs and plumage patterns, can either diverge or converge as a result of interactions between males from similar taxa. If these traits are also the subject of mate choice, the chances of females mating with heterospecifics are correspondingly altered. Second, the strengthening of mate choice may be influenced by other costs and benefits of choosing heterospecifics besides those related to the production of hybrids. The classic example is a cost of search. When conspecifics are rare, it can take a long time to find a conspecific mate, and the cost of mating with a common heterospecific is reduced. A final role for mate choice in the completion of reproductive isolation is through *postmating* isolation. This happens if any hybrids that are produced have low mating success because they look or sound different from conspecifics.

I first consider the theoretical background to reinforcement. I then review empirical evidence on male signal evolution in sympatry and the costs and benefits of female choice of heterospecifics. I conclude that a refinement of mate choice through reinforcement may well be important in completing the speciation process, but only between taxa that have already established substantial premating and postmating isolation prior to contact. Sexual selection against hybrids may also contribute to reduced gene flow, as hybrid males do sometimes seem to be discriminated against.

BACKGROUND

In the previous chapter, I noted the dual processes involved in mate choice: sexual selection and species recognition. Sexual selection often favors extreme values of male traits, whereas preferences for one's own species are a stabilizing force (see also Figure 10.9, p. 212). In principle, either the strength of sexual selection, or the strength of stabilizing preferences for one's own species, or both, could be modified in response to selection that favors mating with conspecifics. First, reinforcement could lead to an increase in the strength of sexual selection for extremes. In this model, females who mate with intermediate males sometimes mate with members of the other species, whereas females who mate with extremes always mate with their own species (Figure 14.1). If hybrids are unfit, then females with extreme preferences produce more surviving offspring, on average. As a result of this selection, female preferences diverge. Divergent female preferences in turn drive divergence in the male traits. Eventually, male distributions of the two species do not overlap, and assortment is complete.

Reinforcement can also operate through increased female discrimination of males. In the discrimination model, a male trait, or combination of traits, is already nonoverlapping between the two taxa. Some females are assumed to be

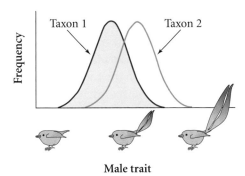

Figure 14.1 Sexual-selection model of reinforcement (Liou and Price 1994). Females in taxon 1 that choose males with relatively long tails and females in taxon 2 that choose males with relatively short tails sometimes mate with members of the other taxon. If hybrids have low fitness, females that choose males in the other extremes of the distribution have higher fitness.

less discriminating than others. These females end up mating with heterospecifics more often, and they have relatively low fitness. In this case, selection operates to narrow the window of recognition of the trait, so females eventually respond only to conspecific males (Figure 13.4, p. 285). Discrimination evolves without any necessary evolution of male traits.

Intuitively, reinforcement should always work if hybridization is costly. However, several factors work against the process. First, and probably most critically, if hybrid fitness is not too low, then sufficient gene flow between the species results in both female preferences and male traits getting across the species boundary; they then collapse back into one species. Second, if hybrid fitness is very low, heterospecific matings essentially produce no offspring. In this case, one population may decline and go extinct, especially if it is initially rare, because many individuals tend to mate with the other species (Liou and Price 1994). Third, selection will be strong only when a large number of individuals are engaged in heterospecific matings. If two species meet in a narrow hybrid zone, the majority of individuals of one species lie outside the zone and do not encounter the other species. Thus, reinforcement is most expected among incipient species that show extensive geographical overlap.

Liou and Price (1994) studied the generation of complete premating reproductive isolation between incipient species in the sexual-selection model (Figure 14.2, *left*). Kelly and Noor (1996) and Servedio (2000) studied models of increased discrimination. They did not investigate whether reproductive isolation can be completed but, rather, asked if the tendency to mate with conspecifics increases, as opposed to the alternative of the two taxa collapsing into a hybrid swarm. Servedio (2000) assumed that a large population is receiving immigrants at a low rate from another population. She asked what happens to an allele present at low frequency that favors increased discrimination of a trait that distinguishes the two populations (Figure 14.2, *right*). Both the sexual selection and discrimination models illustrated in Figure 14.2 are consistent in indicating that complete mixing of the two populations is often the expected outcome, even when hybrid fitness is moderately low. However, reinforcement regularly occurs when hybrid fitness is low and strong assortative mating is present before the taxa spread into sympatry. Liou and Price (1994) showed that reinforcement is possible when only a small fraction of each species encounter each other, but it is less likely than when the two species completely overlap.

These theoretical findings have been largely confirmed by laboratory experiments. In particular, very low hybrid fitness is needed before reinforcement can be experimentally demonstrated, but then it occurs regularly (Rice and Hostert 1993). If hybrid fitness is zero, the process of a strengthening of premating isolation is sometimes referred to as reproductive character displacement rather than

reinforcement (Butlin 1987; Goldberg and Lande 2006). This is because reproductive isolation between the taxa is already complete, so, under the biological
species concept, the two taxa should be considered good species. Butlin's argument has been accepted by others, including me (Liou and Price 1994). However,
the distinction between extremely low levels of gene flow and no gene flow may

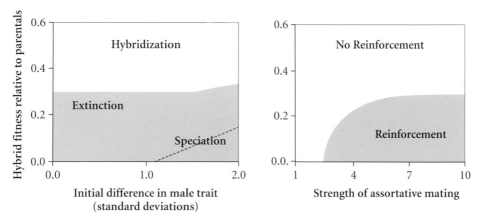

Figure 14.2 Conditions for reinforcement under different genetic and behavioral assumptions.
Two populations are assumed to have diverged in postmating isolation *(y* axis) and premating
isolation *(x* axis). Interior of the plots show conditions under which premating isolation
increases. *Left:* This simulation investigates the completion of speciation when females prefer
extreme males, as in Figure 14.1. The average initial difference between the two populations in
the male trait (in standard deviation units, see Figure 14.1) is given on the *x* axis; a large
difference implies little overlap in the male distributions and substantial premating isolation
prior to contact. In this model, a parameter for the rate of population growth is included. When
population growth dips due to low hybrid survival, extinction of one or the other population
can occur. Speciation occurs if the male traits and female preferences diverge sufficiently from
the initial conditions so that there is no cross-mating. The increasingly blue-shaded area
indicates the parameter space in which this sometimes happens (because of chance effects, either
extinction or speciation is possible); below the dashed line speciation always happens (from
Liou and Price 1994). *Right:* This simulation investigates the spread of a gene for assortative
mating based on some trait that can take on two values (here called "red" and "orange"; i.e., red
individuals tend to mate with reds and oranges tend to mate with oranges). The question is
whether a gene causing assortment would increase when introduced at low frequency, the
alternative being a merging of the two populations (from Servedio 2000). The *x* axis gives the
strength of assortative mating imparted by the gene. A value of 1 = random mating, and a value
of 10 means that, if the two types of males were in equal frequency, a female is 10 times more
likely to mate with a male carrying the same trait value as she does than with a male who differs
from her. In each generation, 5% of the population is made up of immigrants of the rarer
species. The immigrants are red. Individuals of the resident (commoner) species are mostly
orange, and orange is assumed to have a natural selection advantage of 1% over the red morph
(so frequencies of the two morphs are maintained by a balance between selection and
migration). Hybrid breakdown is due to interactions at two other gene loci. The shaded area
indicates the region over which a gene for assortment increases.

not be critical to the progress of reinforcement. In this book, I use the term "reinforcement" to indicate a strengthening of premating isolation resulting from low hybrid fitness, whether or not hybrid fitness is zero. The term "reproductive character displacement" is used in its original meaning, which describes a pattern of increased premating isolation in sympatry compared with allopatric populations (Howard 1993), whatever the process that produced it.

Reinforcement leads not only to greater premating isolation, but also usually to an increase in *postmating* isolation, because hybrids have low mating success (Kirkpatrick and Servedio 1999). At the end of speciation, few hybrids are produced, and selection against hybrid matings is weak, simply because few individuals are actually involved in such matings. Countering this effect, as females develop stronger preferences for their own species, increased discrimination against hybrids reduces hybrid mating success. Thus, sexual selection against hybrids is a form of postmating isolation that is increasing at the same time as the number of hybrids is decreasing, and hence could be important to the final cutoff of gene flow between species.

Search Costs

The reinforcement models assume a fixed cost to a female of mating with a heterospecific male, namely, low hybrid fitness. This cost may be ameliorated if heterospecific males provide superior benefits to females, such as access to a good territory or help with raising young (Pierotti and Annett 1993). Costs and benefits also vary depending on the abundance of conspecifics. A strong prediction to emerge from models of mate search is that individuals should get less choosy the longer they search—females broaden the range of male phenotypes that they find acceptable (Real 1990). This rule evolves solely in the context of conspecific mate choice. But it implies that females are more likely to be sufficiently stimulated to mate with heterospecifics if conspecific males are difficult to find. The rule predicts not only that rarity should be correlated with tendency to hybridize, but also that the more similar heterospecific male signals are to conspecific male signals, the more likely hybridization will be.

Wilson and Hedrick (1982) took search costs further and showed how it could actually pay for a female to mate with a heterospecific male if she finds him first, rather than continuing to search for a conspecific. First, females may suffer a cost whenever they encounter a male, whether conspecific or heterospecific. For example, a male's display may induce various hormonal responses (Chapter 9). Second, females may pay a cost in terms of time spent in continued search. Wilson and Hedrick (1982) showed that these costs lead to an adaptive solution whereby females should switch from the rule "always mate conspecifically" to "accept the first male encountered" as either the relative abundance and/or the absolute abundance of conspecifics declines (Figure 14.3). When conspecifics are rare, a female may benefit from mating with the first male she finds, because of the costs she expects to pay in the process of continuing her search. If encounter costs and other forms of search cost are incurred, when one species is rare, hybridization may not be disfavored and could actually be under selection to increase in the population.

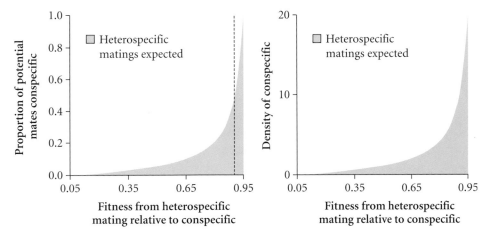

Figure 14.3 Females of a rare species searching for mates may choose heterospecifics with no loss in fitness, even if the hybrids they produce have relatively low fitness. Only conspecific mating is expected in the unshaded area. In the shaded area, heterospecifics will be chosen if they are encountered first. In both examples, it is assumed that males are encountered in proportion to their abundance. *Left:* The cost is assumed to be entirely a result of encountering males [from equation (3) in Wilson and Hedrick 1982]. Here, the encounter cost = 0.95. The switch from "accept the first individual encountered" to "accept only conspecifics" depends on the relative abundance of the two species because, if heterospecifics are relatively common, a female may encounter many heterospecifics before she finds a conspecific, each time paying a cost. *Right:* The cost depends only on the time it takes to find a conspecific. The y axis is scaled by setting search cost at the point $y = 1.0$ to equal to the benefits of a conspecific mating; the horizontal line from the point $y = 1.0$ intersects the curve where hybrid fitness is half that of conspecific fitness (calculated from equations in MacArthur and Pianka 1966). The switch from "accept the first individual encountered" to "accept only conspecifics" depends on the abundance of the conspecific only. This is because a decision to continue searching is made every time a female encounters a heterospecific, and it depends only on how long she expects to take to find a conspecific from then on. In both of these examples, a rare species may be selected to mate randomly and a common species to mate assortatively (termed the "speciation conflict" by Wilson and Hedrick 1982). This is illustrated in the left graph: to the left of the dashed line, if one species lies in the shaded area, the other must lie in the unshaded area; to the right of the dashed line, if one species lies in the unshaded area, the other must lie in the shaded area. The speciation conflict may be resolved by one sex, e.g., the female having greater control over matings (Wilson and Hedrick 1982).

In summary, theory predicts that a precondition for reinforcement between taxa is that levels of premating and postmating isolation are already high when the taxa come into contact. In addition to these conditions for reinforcement, divergence rather than merging of taxa is most likely if (1) traits used as mating cues differ strongly between taxa, (2) any direct benefits to females from mating with heterospecifics are relatively low, and (3) the costs of searching for conspecifics are low. Before discussing reinforcement, I first consider the empirical evidence for each of these three factors.

SIGNAL SIMILARITY AS A CAUSE OF HYBRIDIZATION

Some hybridization events result from errors inherent in the learning system of mate recognition, notably the misimprinting that arises when young are raised by parents of a different species (Grant and Grant 1997a; Payne et al. 2000; Slagsvold et al. 2002; Chapter 13). However, many examples of hybridization are likely to result from individuals (usually females) being sufficiently stimulated to mate by heterospecifics that have displays and signals similar to those of their own species. This has been shown for songbirds, in which males have learned the song of another species. For example, in the indigobirds described in Chapter 13, some females visited heterospecific males that were singing the song normally sung by the female's species, probably because these males had developed from eggs placed in nests of the female's host and had learned that host's song (Payne et al. 2000). In Darwin's finches, out of three of six hybrid pairs between the Medium Ground-Finch and the Common Cactus-Finch on Daphne Island, the male sang a heterospecific song (Grant and Grant 1997a). Grant and Grant (1997a) thought that one Medium Ground-Finch male sang a Common Cactus-Finch song because a pair of Common Cactus-Finches had raised it. The Common Cactus-Finches had apparently taken over a nest of a female Medium Ground-Finch after she had laid at least one egg. I have included this example because it illustrates how unusual events may sometimes result in the formation of hybrid pairs. However, the examples also show that, if species regularly converge in their signals when in sympatry, the chances of hybridization should increase. On the other hand, if signals regularly diverge, chances of hybridization should decrease. Signals may sometimes converge and sometimes diverge as a result of social interactions between males.

Signal convergence

Rohwer (1973) suggested that, in areas where the distribution of the Western Meadowlark overlaps with that of the Eastern Meadowlarks, the two species converge in plumage. He argued that this enables individuals to be more readily recognized by the other species and to defend territories without escalating fights. Whether this is a genuine case of plumage convergence needs to be evaluated, but there are many examples of increases in song similarity in sympatry, which, in songbirds, is relatively easily achieved by learning. In hybridizing species, males of one species often sing songs or phrases belonging to the other species as well as to their own species (Dobkin 1979; Helb et al. 1985; Haavie et al. 2004; some examples are given in Table 14.1). Mixed singers are presumably copying from adult conspecifics and similar-looking heterospecifics defending territories against each other, as well as from other mixed singers. Because mixed singing is associated with interspecific territoriality, copying of heterospecifics likely results in easier communication in aggressive interactions (Emlen et al. 1975; Dobkin 1979; Haavie et al. 2004). No special adaptive reason for heterospecific copying needs to be invoked, because young birds normally copy unrelated territorial adults of their own species (Chapter 10). Indeed, mixed and bilingual singers are found at dialect boundaries within species (Baptista 1975; Trainer 1983; Tubaro and Segura 1994; Nelson 2000b).

Table 14.1: Examples of mixed singers in hybrid zones

Species pair		Location	% Mixed singers (sample size)		% Hetero-specific pairs with a mixed singer (sample size)[*]	
First species	Second species		First species	Second species		
Collared Flycatcher[1]	Pied Flycatcher	Öland, off Sweden	0% (> 160)	65% (49)	100% (10)	Alatalo et al. 1990, Haavie et al. 2004, Qvarnström et al. 2006b
Common Nightingale	Thrush Nightingale	Northern Germany	0.05% (200)	28% (239)	100% (6)	Lille 1988
Indigo Bunting[2]	Lazuli Bunting	Wyoming	0% (13)	13% (46)	0% (< 15)	Baker and Boylan 1999
Common Chiffchaff[3]	Iberian Chiffchaff	Pyrenees	8% (52)	10% (46)	< 40% (71)	Helbig et al. 2001

NOTE: In all cases, males are interspecifically territorial.
* This tallies heterospecific pairs for males of the species that most commonly sing the mixed song. In the Collared Flycatcher and the Pied Flycatcher, the opposite direction of the cross occurs, even in the absence of mixed singing by the Collared Flycatcher.
[1] Known to be dominant in interspecific interactions.
[2] A song consisting of two or more Lazuli Bunting phrases and two or more Indigo Bunting phrases was classified a hybrid song. In addition to these records in the table, some individuals of both species (based on plumage) sang the other's song. No pure females (based on plumage) were paired with males of the other plumage type singing mixed songs, but many heterospecifically mated males were singing the other species song.
[3] Forty-two heterospecific pairs did not involve mixed singers. Twenty-nine heterospecific pairs were between a mixed singer and one or other species. Mixed singers were identified to species based on mitochondrial DNA, and some were likely hybrids.

Mixed singing is widespread in some hybrid zones (Table 14.1), and, in other zones, hybrids are common and sing intermediate songs (e.g., Robbins et al. 1986). On the islands of Gotland and Öland in the Baltic Sea, the Collared Flycatcher hybridizes regularly with the Pied Flycatcher. Pied Flycatcher males that occur near Collared Flycatchers often sing songs comprising elements of both species, whereas ones that occur farther away do not (Alatalo et al. 1990; Haavie et al. 2004; Qvarnström et al. 2006b). In all of the hybridization events that involved male Pied Flycatchers, and in which male song has been recorded, the male Pied Flycatchers sang mixed songs (Qvarnström et al. 2006b).

Song similarity can result in increased chances of hybridization in two ways. First, it may be that the mixed singers are able to occupy territories preferred by heterospecific females. Second, mixed singing is likely to stimulate heterospecific females directly. Neither of these effects has been conclusively demonstrated. However, in Darwin's finches, males singing the other species' song as a result of learning mistakes are far more likely to be involved in heterospecific matings than those singing conspecific song, and this is unlikely to have anything to do with territory quality (Grant and Grant 1997a).

Divergence of signals

Male competition for territories may sometimes favor divergence in signals, rather than convergence, especially if it is advantageous for a subordinate male to signal his subordinate status. How this might happen has been shown for a single species, the Lazuli Bunting (Greene et al. 2000). In Montana, relatively dull-colored males settle near relatively bright-colored ones. Greene et al. (2000) found a likely benefit to bright males in allowing dull males to settle nearby because the bright males are likely to obtain extrapair copulations with the dull male's mate, and the dull male may provide a buffer restricting a female's access to other bright males. Dull males are cuckolded through extrapair copulations at twice the rate of bright males [Greene et al. (2000) found that on average 1.5 chicks in a nest of a dull male were sired by other males, and on average 0.76 chicks in a nest of a bright male were sired by other males]. There is also a benefit to dull males in being allowed to settle. Intermediate males are restricted to suboptimal habitat, presumably as a result of aggression from the bright males, and they consequently have lower reproductive success than dull males. In this way both dominants and subordinates gain by signaling their status; the former by gaining in paternity and the latter by gaining access to good territories.

Likewise, divergence as a result of male competition may be favored between species, if they compete for territories and one species is dominant to the other. This is illustrated by the Collared Flycatcher and Pied Flycatcher. Male Collared Flycatchers are black and white, but male Pied Flycatchers vary from black and white to brown and white (the fully black forms are illustrated in Figure 14.4). On Gotland, Alatalo et al. (1994) showed Collared Flycatchers occupy preferred

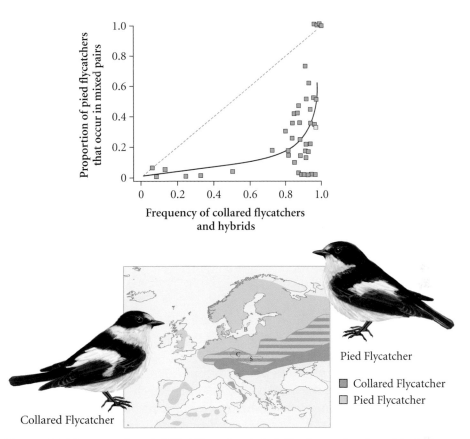

Figure 14.4 The proportion of all Pied Flycatchers in hybrid pairs as a function of the frequency of Collared Flycatchers, for 47 localities in the Czech Republic and Slovakia, indicated by the letters C and S, respectively, on the accompanying map (redrawn from Sætre et al. 1999). The yellow square is the island of Gotland in the Baltic. Gotland, and Öland to the west, where both species also breed, are indicated in yellow and by the letter B in the map (from Veen et al. 2001). The curved line is a polynomial regression line fit to the mainland data. The dashed line gives the expectation under random assortment. Note that the drawing of the Pied Flycatcher is from the zone of allopatry; it is much browner in the Czech Republic and Slovakia, where the two species overlap.

deciduous habitats. In these preferred habitats, the Pied Flycatchers are browner than they are in coniferous habitats, so brown color is positively associated with the presence of the Collared Flycatcher. As on Gotland, across central and eastern Europe, the Pied Flycatcher is, on average, browner where it co-occurs with the Collared Flycatcher than it is in areas outside the zone of sympatry (Sætre et al. 1999). In cage bird experiments, Sætre et al. (1993) found the Collared Flycatchers were more aggressive to darker Pied Flycatchers than to browner, female-like Pied Flycatchers. In the field, Alatalo et al. (1994) moved males of the two species towards each other by moving their nest boxes and found that

Collared Flycatchers evicted blacker Pied Flycatcher males from their nest box at a greater distance than they evicted browner males. The reduced aggression elicited by browner Pied Flycatchers may have contributed to divergence between the two species in zones of sympatry (Sætre et al. 1993; Alatalo et al. 1994), and also resulted in a reduction in cross-species matings. Sætre et al. (1997) show that divergence in male plumage is associated with divergent mate preferences, and female Pied Flycatchers from different locations prefer males of the local phenotype.

In summary, interactions between heterospecific males competing for territories can lead both to convergence and divergence in social signals. The conditions favoring convergence or divergence are not well understood, but it is likely that they result from the same kinds of selection pressures that favor convergence or divergence among males of different social status within species. Limited evidence indicates that convergence increases, and divergence decreases, the probability of hybridization.

BENEFITS TO MATING HETEROSPECIFICALLY AS A CAUSE OF HYBRIDIZATION

The time a female is prepared to devote to continuing her search for a mate is assumed to depend on the costs of the ongoing search balanced against the potential benefits of finding a mate superior to those encountered so far (Chapter 12, p. 247). When search costs are high, females may mate with a heterospecific male when they cannot find a conspecific, even if they are applying a rule that has evolved solely in the context of conspecific mate choice.

Empirically, the rarity of conspecifics is a strong correlate of the propensity to hybridize (Grant and Grant 1997a; Nuechterlein and Buitron 1998; Randler 2002). For example, Nuechterlein and Buitron (1998) used playback experiments to show that male Western Grebes were more than twice as likely to respond to a female Clark's Grebe advertising call late in the season than they were early on. The number of unmated female Western Grebes is likely to become relatively scarce as the season progresses, and males become less choosy as their chances of acquiring a mate are reduced. Some studies of natural invasions of one species into another's range have found hybridization that disappears with time (Table 14.2), with the transient effect attributed to the rarity of new colonists (Short 1969). Across space too, one consequence of recent habitat alteration is the creation of small patches of one habitat surrounded by another, simulating a rarity effect. Chapin (1948) studied three species of paradise flycatchers in Africa; two occur in rainforest and one occurs in more disturbed habitat (woodlands). He found three areas with hybrid populations, and he

Table 14.2: Examples of reduced hybridization subsequent to a range expansion

Species pair		Details	
Great Spotted Woodpecker	Syrian Woodpecker	Hybridization occurred only on a narrow front as the Syrian Woodpecker expanded its range.	Short 1969[1], p. 90
European Blue Tit	Azure Tit	Expansion of the Azure Tit into eastern Europe in the 1870s and 1880s was accompanied by hybridization. Hybridization was less frequent by the 1950s (but the range had retracted too).	Vaurie 1957
Slender-billed White-eye	Silvereye[2]	The Silvereye invaded Norfolk Island in 1904 and rapidly increased in numbers. Three out of 75 specimens collected in 1912–1913 were hybrids; none have been recorded since.	Gill 1970
Sooty Myzomela	Cardinal Myzomela	The Cardinal Myzomela invaded San Cristobal in the Solomons > 100 years ago. Hybrids were reported in 1908 and 1927, but none were reported in 1953.	Mayr and Diamond 2001, p.180

[1] Secondary source: the original could not be located.
[2] It is unclear if rarity could have been involved in this example, because hybridization events apparently occurred after the Silvereye had increased to large numbers (Gill 1970).

suggested that hybridization arose when pieces of rainforest containing very few individuals of the rainforest species became surrounded by habitat containing the woodland species. Although rarity is a strong correlate of propensity to hybridize, it turns out that males of a rare species are about as equally likely to be involved in heterospecific matings as females (Randler 2002). This implies that, in cases where the male of the rare species is involved in hybridization, factors other than search costs are driving female choice, or that male choice of females is an important factor in matings (Servedio and Lande 2006).

Costs and benefits to heterospecific pairings have been most thoroughly evaluated for the Pied Flycatcher and the Collared Flycatcher. Sætre et al. (1999) surveyed different populations in a hybrid zone between these two species and found the proportion of hybrid matings to be high only where the Pied Flycatcher is uncommon (Figure 14.4). On Gotland, Pied Flycatchers form about 4% of all flycatchers, and about one-third of all Pied Flycatchers pair with Collared Flycatchers (Veen et al. 2001; Figure 14.4). In accord with Randler's (2002) general findings, the two kinds of crosses (female Collared × male Pied and female Pied × male Collared) are approximately equally frequent.

Female Collared Flycatchers pair with male Pied Flycatchers, particularly late in the breeding season. Male Collareds are common, so rarity is not likely to explain why female Collareds fail to pair with their own species (Veen et al. 2001). However, late in the season, female Collareds paired with Pieds fledge about 40% more young than do female Collareds paired with male Collareds (Veen et al. 2001). It appears that male Pied Flycatchers hold better territories at this time (Wiley et al. 2007). In fact, the costs to female Collareds of pairing with male Pieds are quite low anyway because females paired with Pieds copulate with nearby conspecifics. A remarkable 60% of all chicks of female Collareds paired with male Pieds are Collared Flycatchers and not hybrids (Veen et al. 2001). Note that this is very similar to the inference that female Lazuli Buntings mated with dull males seek out bright males for copulations (Greene et al. 2000).

Female Pied Flycatchers also pair with male Collared Flycatchers. On Gotland, Pied Flycatchers are rare, so female Pied Flycatchers may find it difficult to locate unpaired males belonging to their own species. The alternatives to continuing to search for unpaired males are to enter into a bigamous relationship or to pair with a male Collared Flycatcher. On the Swedish mainland, Alatalo et al. (1981) found that the secondary female in a bigamous relationship receives almost no help from the male in raising the brood, and she achieves a fledging success less than half that of a monogamous female. Given these costs, pairing with a male Collared Flycatcher may not be so detrimental. A second factor is the quality of territory maintained by heterospecific males [experimental manipulations indicate that females are likely to use territory quality in mate-choice decisions (Alatalo et al. 1986b)]. On Gotland, the Pied Flycatcher is rare in preferred deciduous habitat. So female Pied Flycatchers may be pairing with heterospecifics, not just because of the difficulty of finding unpaired males of their own species but also because they are choosing preferred habitats in which to breed.

Benefits of pairing with heterospecifics have been inferred in some other hybrid zones. The Black-capped Chickadee and the Carolina Chickadee meet in a hybrid zone in eastern North America (Chapter 15). Females choose dominant males in choice tests in the lab, even if the males are heterospecific (Bronson et al. 2003b). Within species of chickadees, females benefit by choosing dominant males, because they gain superior territories and may have higher survival (Bronson et al. 2003b). Male Mallards are socially dominant over other species, and this leads to hybridization (Brodsky et al. 1988). Females that pair with a dominant male may get better protection from harassment by other males (Brodsky et al. 1988). In a hybrid zone between California Quail and Gambel's Quail, both species co-occur in mixed coveys, and mating between members of the same covey is associated with earlier reproduction and higher reproductive success than mating with an individual from another covey, whether or not it is between heterospecifics, hybrids, or conspecifics (Gee 2003).

In summary, several factors seem to mitigate the costs of hybridization in places where hybridization occurs reasonably frequently. However, any benefits that accrue to a heterospecific mating are qualified by hybrid fitness. When hybrids have zero fitness, hybridization can never be favored, and reinforcement always is.

REINFORCEMENT

Hybridization likely arises as an epiphenomenon of what is going on within species. Females are strongly selected to mate, so, in the absence of conspecifics, they may be stimulated by signals resembling those of conspecifics. Any costs to mating with a heterospecific are ameliorated if heterospecifics are more easily found than conspecifics, or if they provide superior material benefits. Various costs and benefits of hybridization result in selection to alter the frequency at which it occurs. Whether the level of hybridization has increased in response to the direct benefits of heterospecific mating is not clear, but may apply to the fly-catchers on Gotland. The level of hybridization should decrease whenever costs of hybridization are high, as in the theory of reinforcement.

In taxa other than birds, some examples exist in which species differ more in sympatry than in allopatry in traits that are associated with mating (Howard 1993). There are several examples in birds, including the Collared Flycatchers and Pied Flycatchers (Sætre et al. 1997). Seddon (2005) found that in antbirds, sympatric pairs of species are more different in their songs than allopatric pairs. Patterns such as these may be a result of reinforcement (i.e., selection against hybridization in the zone of overlap), but other explanations have been proposed (e.g., Noor 1999; Servedio 2001; Albert and Schluter 2004). First, divergence in sympatry may be driven by ecological character displacement, with associated evolution of mate preferences and other traits involved in mating (Grant and Grant 1997b; Dieckmann and Doebeli 1999; see also Chapter 5). Ecological character displacement arises if individuals from different taxa that are ecologically the most similar compete for resources and consequently have relatively low survival or fecundity (Figure 5.3, p. 81). This causes divergence in many traits that affect premating and postmating isolation (Chapter 5), and, in Darwin's ground finches, ecological character displacement is thought to contribute to reduced cross-species matings (Grant and Grant 1997b; Chapter 5). Second, divergence in sympatry may result from competitive interactions among males to achieve any sort of mating, rather than as a result of selection to avoid heterospecific matings. This may be one reason why male Pied Flycatchers are browner where they are in sympatry with Collared Flycatcher than they are in allopatry (Alatalo et al. 1994), although, in this example, reinforcement may also be involved (Sætre et al. 1997). Third, Galbraith (1956) noted that two very similar-looking species in the Golden Whistler complex co-occur without hybridizing on New Guinea. He suggested that "the predevelopment of isolating mechanisms may be related to the intensity of selection against interbreeding with other species" (Galbraith 1956, p. 180). In other words, selection for discrimination against congeners in different zones of allopatry had narrowed the window of recognition sufficiently that, when the two forms came together in sympatry, they already showed substantial assortative mating.

Models of reinforcement show that a high level of both premating and postmating isolation is required if taxa are not to merge into one. These conditions may sometimes be difficult to achieve. In the case of the Pied Flycatcher and Collared Flycatcher, hybrid females are infertile, and many of the hybrid males are likely to have low fertility (Sætre et al. 2003). This implies a substantial cost to mating heterospecifically. Despite such costs, hybridization continues. In Eastern Europe, hybridization may reflect the dynamics of a hybrid zone (Chapter 15), with most members of each species simply not encountering the other species, and gene flow into the zone opposes selection to complete the assortative mating. On Gotland and Öland, ongoing hybridization may partly be a consequence of the inferred recent invasion of the Collared Flycatcher (perhaps 200 years ago, Alatalo et al. 1990). In addition, on Gotland the costs of heterospecific pairing are mitigated via benefits from being on a relatively high quality territory (for both female Collared and Pied Fycatchers), plus the ability to undertake extrapair copulations (for female Collared Flycatchers) and reduced search costs (for female Pied Flycatchers).

Hybridization may not always be as costly as it seems. However, no countervailing advantage to females from mating heterospecifically exists if hybrid fitness is zero. In addition, in this case, it is impossible for preferences and male traits to leak across the species border, and reinforcement becomes a likely outcome. When hybrid fitness is close to zero and species overlap extensively, I suggest reinforcement is an important step in completing premating isolation. The alternative is that mate preferences diverge sufficiently in allopatry so that species never recognize each other in sympatry.

If divergence in premating isolating mechanisms in allopatry is sufficient to drive premating isolation in sympatry, the extent to which taxa show premating isolation should be correlated with the degree of divergence in traits that form the basis of mate recognition. Although this idea has not been quantitatively tested, many examples do not fit the pattern, and a common sentiment in the literature is that "acquisition of reproductive isolation and morphological divergence are not closely correlated" (Mayr and Gilliard 1952, p. 334; here Mayr and Gilliard are using morphology more generally than I have in this book, and they include plumage variation as well as size and shape). First, good sympatric species may differ only slightly and, second, variation within a single randomly mating population of one species may be much larger than differences between species. I describe examples of each.

With respect to slight differences among similar species, a recently discovered species complex of somewhere between five and seven flycatcher-warblers in China and the Himalayas sing similar songs and look very similar despite millions of years of separation (Alström and Olsson 1999, 2000; Martens et al. 1999; Päckert et al. 2004). Until a few years ago, all of them were thought to belong to a single species. Two species that breed sympatrically over a large region of the Himalayas are illustrated in Figure 14.5.

In other cases, individuals within species can differ greatly, yet still mate randomly. Song differences within one randomly mating population of Darwin's finches are greater than those seen between other species of Darwin's finches (e.g., Millington and Price 1985; Figure 14.6). Presumably, young birds are generalizing from the morphological characteristics of their parents to learn to recognize different songs sung by similar-looking individuals. Color-pattern polymorphisms are also common (Buckley 1982; Fowlie and Kruger 2003). In African bush-shrikes, polymorphic variation within a single species is similar to fixed color differences seen among other related species (Hall et al. 1966).

In Chapter 5, I suggested that random mating among the discrete size morphs of the Black-bellied Seedcracker arose because all offspring are similar to one or another morph and hence "hybrids" have the same fitness as parentals

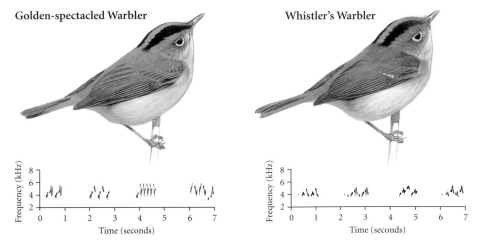

Figure 14.5 Two species of flycatcher-warbler that breed sympatrically in the Himalayas despite having similar songs and plumages. They are segregated altitudinally but have considerable overlap. They were distinguished only in 1999 (Alström and Olsson 1999, 2000; Martens et al. 1999). They may be separated by as much as 6.5 million years (!) according to mitochondrial-DNA sequences [note that, as in other examples in this book, this is based on the HKY-Γ distance metric ($\Gamma = 0.2$) with 2% per million years as the estimated rate of mitochondrial DNA divergence (Appendix 2.1); from sequences in Olsson et al. 2004]. The figure is reproduced from Alström and Olsson (1999), painted by Ian Lewington. Songs differ in that Whistler's Warbler has the less varied phrases.

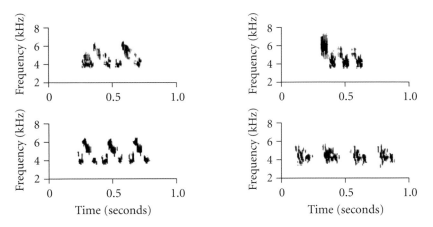

Figure 14.6 Four distinctive song types of the Medium Ground-Finch on Daphne Island in the Galápagos (sonograms redrawn from Gibbs 1990). These types are fairly discrete and can be classified to type by naïve observers, although there is also quantitative variation within them (Grant and Grant 1996c).

(p. 85). Similar reasoning may apply to color polymorphisms and variable songs within populations of a single species. The converse is that, when different species do show strong conspecific mate preferences, it is because hybrids between them have reduced fitness. This, in turn suggests that reinforcement of mate preferences may be common. Indeed, many behaviorists have interpreted sexual imprinting, including evolution of a narrow recognition window in response to the presence of heterospecifics, as having evolved to avoid heterospecific matings (Lorenz 1937, p. 265; Immelmann 1972, p. 166; Clayton 1993). Given that very similar-looking species coexist, reinforcement may operate more through increased discrimination of traits rather than through any evolution of the traits themselves.

SEXUAL SELECTION AGAINST HYBRIDS

I finish this chapter with a consideration of the role of mate choice in postmating isolation, that is, in reducing the fitness of hybrids. Evaluating the extent to which hybrids experience low mating success is not easy, because hybrids are often rare, hence difficult to study. In some hybrid zones, it is thought that hybrids are not sexually selected against, even though the parental types mate assortatively (Grant and Grant 1992, 1998). Random mating is found in other zones (Ingolfsson 1970; Moore 1987; Good et al. 2000) with no detectable selection by female choice. However, hybrids do usually look, sound, or display differently from either of the parental species (Price 2002, and see Table 11.4, p. 231). Some examples are shown in Figures 14.7 and 14.8. Hybrid displays may

Costa's Hummingbird   Hybrid   Anna's Hummingbird

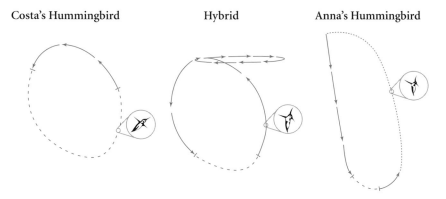

Figure 14.7 Aerial displays of Costa's Hummingbird, Anna's Hummingbird, and hybrids between them. Five different hybrid males were observed making the circle at the top of the display, although they sometimes displayed in a manner similar to the display of Costa's Hummingbird. The long-dashed lines indicate the portion of the flight over which sound is produced. Redrawn from Wells et al. (1978).

be quite unusual (e.g., the hummingbirds illustrated in Figure 14.7) and it is difficult to believe that this has no effect on mate choice. In the case of North American quail (Figure 14.8), Gee (2003) showed that female Gambel's Quail and California Quail prefer to associate with their own species in laboratory studies, but she found that, in a hybrid zone, they pair randomly. However, the majority of individuals in the hybrid zone were hybrids.

Some laboratory and field studies have identified likely sexual selection pressures against hybrids. First, Clayton (1990b) measured female responses to hybrids between the two subspecies of Zebra Finch. Male hybrids between these subspecies sing songs that are different from those of either parental species, and females of both parental types respond poorly to those songs (Figure 14.9).

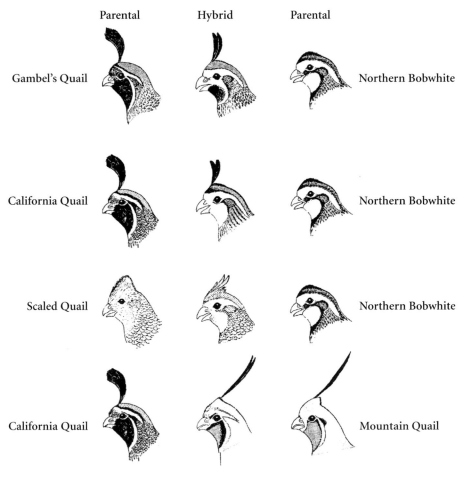

Figure 14.8 Face markings of hybrids in four quail crosses (from Johnsgard 1971; reproduced with permission of the American Ornithologist's Union and the *Auk*).

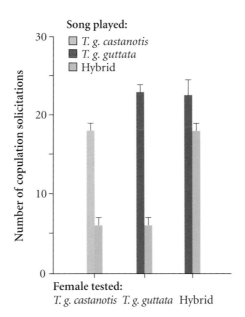

Figure 14.9 Mean number of copulation-solicitation displays (+ standard deviation) by estradiol-implanted female Zebra Finches in response to playback of male songs. Twelve females from each subspecies (*Taeniopygia guttata castanotis* and *T. g. guttata*) were tested, as were 12 hybrid females that had been raised by *T. g. castanotis* females and *T. g. guttata* males (data from Clayton 1990b).

Hybrid females do respond to songs produced by hybrid males, but they respond more strongly to songs resembling the male that reared them (Figure 14.9). Second, in experiments with Indigo Buntings and Lazuli Buntings, females gave lower sexual responses to males that sing the song of one species but had the plumage of another (Baker and Baker 1990; Figure 13.5, p. 286). This pair of species has also been studied in the field. Baker and Boylan (1999) showed that pure plumage types mate assortatively, but female hybrids mate randomly, implying sexual selection against male hybrids.

A few field studies in addition to the one on the buntings have found or inferred some sexual selection against hybrids. Female Greater Prairie-Chickens actively avoided a displaying hybrid Prairie Chicken × Sharp-tailed Grouse (Sparling 1980). Blue-winged Warbler × Golden-winged Warbler hybrids had about 60% the pairing success of the parental species (Ficken and Ficken 1968). In Western Grebes and Clark's Grebes, the early-season pairing success of hybrids was about 30% that of the commoner species, the Western Grebe (Nuechterlein and Buitron 1998). Moore and Koenig (1986) found that male hybrids between the western and eastern subspecies of the Northern Flicker had smaller broods than the parental species. They raised the interesting possibility that females paired with hybrids were less stimulated by these males and hence put less effort into reproduction. Based on experimental manipulations of color patterns by Nobel (1936), Moore (1987) suggested that an important component of selection against hybrid flickers is male competition for territories, with male hybrids outside the hybrid zone being less able to establish themselves because their unusual color patterns act as a less efficient signal. Finally, Hermit Warbler × Townsend's Warbler hybrid males in the State of Washington were as successful in keeping their territories as were male Hermit Warblers, but they had only about 75% the success of the Townsend's Warbler males (Pearson 2000).

In Chapter 5, I considered the possibility that ecological factors reduce hybrid fitness, thus creating some postmating isolation between taxa. This may be particularly important when divergence has occurred along multiple ecological dimensions (e.g., prey size, habitat, predator-avoidance strategies) because several dimensions are cumulatively likely to lead to higher hybrid unfitness, and they may not all break down simultaneously if the environment changes. Sexual selection against hybrids differs from ecological causes of hybrid unfitness in several ways. First, environmental fluctuations should have less effect on hybrid fitness. Second, it is easy to imagine differences among species continuing to accumulate in sexually selected traits (i.e., along multiple dimensions, such as different patches of color, feather ornaments, vocalizations, etc.), even if environmental differences do not change much, so the hybrids may become increasingly different from the parental species. Third, hybrid fitness decreases as preferences for conspecifics strengthen: the stronger the preferences, the less likely the hybrids are to be chosen. Thus, as the amount of premating isolation increases, so does the amount of postmating isolation. Although these features make sexual selection against hybrids a potentially strong source of postmating isolation leading to reinforcement, in many species, the lowered fitness is likely to fall mainly on males. However, especially in species with long-term pair bonds, female hybrids are probably also discriminated against by males.

CONCLUSIONS

Reinforcement may be a common feature of the end stages of speciation, provided that species overlap extensively. Models and data reviewed in this chapter imply that the evolution of premating isolation via reinforcement requires substantial loss of hybrid fitness—that is, postmating isolation must be essentially complete. Hybrids may have low fitness from ecological causes, because of sexual selection against them, or because they have intrinsically low viability or fecundity. These alternative causes of low hybrid fitness reappear in the next two chapters.

SUMMARY

Premating reproductive isolation may increase as a result of interactions between taxa in sympatry. In particular, if individuals that hybridize are selected against because hybrids are unfit, then selection favors a strengthening of preferences for conspecifics (reinforcement). Other interactions in sympatry can lead either to an increase or to a decrease in premating isolation. First, ecological character displacement causes divergence in traits involved in premating isolation, but ecological selection pressures can also lead to convergence (Chapter 5). Second, competition among males of different species for territories may sometimes result in divergence of male traits, although it can also cause convergence. Third, females that mate with heterospecifics may gain advantages that compensate for low hybrid fitness, such as access to better territories. Theory suggests that reinforcement is likely only when both premating and postmating isolation are strong prior to populations coming into contact. Although direct evidence for reinforcement is difficult to obtain, I suggest that strong postmating isolation may generally be present among widespread sympatric species that do not hybridize, and that reinforcement when postmating isolation is strong is often involved in completing the tendency to mate with conspecifics.

CHAPTER FIFTEEN

Hybrid Zones

Hybrid zones are areas where genetically distinct populations meet, mate, and produce hybrids (Barton and Hewitt 1985, 1989). They represent a snapshot in the progress towards speciation; individuals from one population mate with individuals from another, but crossmating does not lead to a collapse of the two populations back into one. The presence of viable and often fertile hybrids in the hybrid zone means that some genes can get from one population to the other. On the other hand, the maintenance of qualitatively different phenotypes on each side of the hybrid zone, which may often be classified as separate species, implies that many genes fail to cross the zone. In this chapter, I review patterns in hybrid zones to ask how some genes become restricted to one species, whereas others pass through the zone.

One reason for conducting this survey is that hybridization may not only retard divergence, but it may also play a creative role in speciation. The introgression of genes from one species to another increases the chances that new and favorable genetic combinations will arise (Grant and Grant 1994, 1996b; Arnold 1997; Dowling and Secor 1997; Barton 2001). First, novel genetic variation may result in a population being able to respond to unusual selection pressures. Second, a population that consists largely of hybrids may become separated from the rest of the species. Subsequent establishment of a unique combination of traits (e.g., because they are favored under sexual selection) results in the population becoming diagnostically different and, perhaps, eventually, a different species (Panov 1992).

Hybridization is likely to occur regularly during the early stages of adaptive radiation (Seehausen 2004). This is certainly the case in Darwin's finches. Grant et al. (2005) studied 42 populations from nine species of Darwin's finches. They compared populations of a pair of species that came from the same island, with populations from the same pair of species drawn from different islands. They showed that populations from the same island tend to be genetically more similar to each other than to populations from different islands. This result is most easily explained if co-occurring species occasionally hybridize. On Daphne Island, gene flow between finch species has been traced by observing hybridizing pairs and their descendants. Introgression affects the mean, increases the variance, and alters correlations between traits (Grant and

Grant 1994; Grant et al. 2005). For example, over the past 15 years, hybridization between the Common Cactus-Finch and the Medium Ground-Finch has occurred at a rate of about 1% per generation. This has led to morphological and genetic convergence of the two species (Grant and Grant 2002b; Grant et al. 2004, 2005). About half of the Common Cactus-Finches on the island now carry some recently introgressed genetic material from the Medium Ground-Finch. Grant and Grant (1994, 1996b) have argued that introgression elevates the potential for evolution to occur in novel directions.

Young adaptive radiations, of which Darwin's finches are the prime example in birds, are characterized by ecological differentiation and extensive sympatry. Ecological differences and assortative mating, rather than geographical separation, are the main factors maintaining the species as distinct entities. On the other hand, as noted elsewhere in this book, many closely related species are allospecies, characterized by geographically separated ranges, sometimes with small areas of overlap, and ecological similarity. Where such allospecies meet, they may hybridize, forming a hybrid zone. Unlike Darwin's finches, which have similar color patterns but differ strongly in morphology (size and shape), most allospecies are characterized by similarity in morphology but differ in song and plumage. It is mainly plumage variation that has been used to indicate the presence of hybrids.

At least 200 bird hybrid zones are known. In the Palearctic region, Aliabadian et al. (2005) list 52 species pairs with narrow zones of overlap, of which 19 (37%) are known to hybridize at least in some locations (see also Haffer 1992; Newton 2003, his Chapter 10). Again in the Palearctic, Haffer (1992) lists 23 pairs of what are classified as well-marked subspecies with wide contact zones and extensive hybridization. In North America, Mayr and Short (1970) describe 44 superspecies that contain more than one species, of which five show extensive overlap and hybridization, and an additional 12 have overlap with limited hybridization. Mayr and Short (1970) also list 19 examples of hybridization in North America between what they classified as subspecies. In Australia, Ford (1987) identified approximately 80 hybrid zones, most of them between currently classified well-marked subspecies. The number of hybrid zones is certainly underestimated, both because many areas of the world are still poorly known and because nearly all hybrid zones that have been identified are based on conspicuous color-pattern differences between taxa. More subtle features, such as songs, chromosomes, and morphology, have been less studied. In other groups, many of the best-studied hybrid zones involve chromosomal differences (Barton and Hewitt 1985).

Following Barton and Hewitt (1985), I assume that hybrid zones result from contact between formerly allopatric taxa and have not formed in situ within one continuously distributed population. Endler (1977) showed that it was theoretically difficult to distinguish between these alternatives, based on present-day patterns, but, as I argued in Chapter 4, there is good evidence that many species with currently parapatric distributions have initially diverged in allopatry, and there appears to be no good evidence that any have formed without a period of geographical isolation. This includes many currently hybridizing taxa. In particular, the time of initial divergence between hybridizing taxa can sometimes be correlated with absolute geographical barriers such as ice sheets (e.g., Bensch et al. 1999; Rohwer et al. 2001; Weir and Schluter 2004; see Figure 2.7, p. 28).

In this chapter, I describe bird hybrid zones and consider how they may be selectively leaky to certain genes. To do this, I first describe relevant theoretical results on the causes of the width of hybrid zones. I then consider the characteristics of hybrid zones. I ask why different traits and genes might move across a hybrid zone to different extents and conclude by considering how differential introgression might lead to the formation of distinct populations, subspecies, and species. Throughout, I refer to hybridization as occurring between species, but this should be taken as a form of shorthand, and I include both what are recognized as subspecies and species in the analyses. (Likewise, gene flow between two species is often defined as introgression, and that between subspecies, as intergradation; here I lump them together as introgression). In the Appendix to this chapter, I provide information on each bird hybrid zone that has been studied to the extent that some reasonable estimate of its width—the distance over which hybrid phenotypes are found—has been made. This information is summarized in Table 15.1, and it is referenced at various places in the chapter.

BACKGROUND

In order to characterize a hybrid zone, the geographical distribution of phenotypes needs to be mapped (e.g., Figure 15.1). To do this, every individual in the sample is first assigned a *hybrid index* based on its phenotype. An individual is given a value of 0 if it looks like one species and a value of 1 if it looks like the other, with intermediate values for intermediate phenotypes. Usually this is done by scoring several different features of the phenotype (e.g., separate patches of color) and then taking the average. The hybrid index thus varies from 0, for an individual that appears to belong to one species, to 1.0, for individuals that appear to belong to the other. The index is directly comparable to the frequency of an allele, *a* or *A*, which is fixed for one form in one species (i.e., all individuals have the genotype *aa* and have a value of 0) and fixed for the other form in the other species (all individuals have the genotype *AA* and have a value 1.0).

Table 15.1: Characteristics of bird hybrid zones*

Species[1]		Location	Ecological differences	Zone width (km)	Zone length (km)	Age[2] of taxa (my)	% Hybrids	Assortative mating	Movement (km/yr)[3]
Rock Partridge	Red-legged Partridge	Southwestern Alps	Altitudinal gradient	15	150	2.4	86	–	–
California Quail	Gambel's Quail	Southern California	Wetter vs. drier	20	400	1.1	88	1	–
Red-breasted Sapsucker	Red-naped Sapsucker	Western U.S.	Generally wetter vs. drier	20	160	–	10	1.9	0
Green-backed Woodpecker		South-central Africa	Woodland vs. open habitats	300	1500	–	40	–	–
Northern Flicker		Great Plains of U.S.	Generally wetter vs. drier	300	4000	0.0	84	1	0
Red-billed Hornbill		Northern Namibia	None obvious in overlap	126	>50	–	27	1.3	–
Vinaceous Dove	Ring-necked Dove	Northwestern Uganda	None obvious	6	–	1.5	–	–	–
†Western Gull	Glaucous-winged Gull	Pacific coast of U.S.	Sand vs. vegetation nesting	800	–	0.6	70	1	Y
White-collared Manakin	Golden-collared Manakin	Panama	None obvious	10	10	–	80	–	0.04
Variable Antshrike		Bolivia	Altitudinal gradient	212	100	0.9	–	–	–
Variegated Fairywren		Australia		200	800	–	95	–	–

Species	Species (2)	Location	Habitat					
Varied Sittella		Australia		300	1000	–	86	–	–
Carrion Crow	Hooded Crow	Europe	Meadows vs. stubble	20	2000	0.0	12	1.8	1.5
†Australian Magpie		Southeastern Australia	Thicker vs. open forest	165	600	0.0	60	1	0
Black-faced Woodswallow		Australia	Generally wetter vs. drier	350	–	–	71	–	–
Timor Figbird		Australia	None obvious	370	–	–	81	–	–
Collared Flycatcher	Pied Flycatcher	Northern Europe	Deciduous vs. mixed	20	200	2.0	3	1.2	0
Common Nightingale	Thrush Nightingale	Eastern Germany	–	–	–	–	2.5	1.3	Y
Black-eared Wheatear	Pied Wheatear	Iran	–	400	–	–	65	–	–
Rufous-naped Wren		Chiapas, southern Mexico	Generally wetter vs. drier	30	10	–	90	–	0
Black-capped Chickadee	Carolina Chickadee	Eastern U.S.	Small: forests vs. woodlands	25	2500	4.0	58	1	1
Tufted Titmouse	Black-crested Titmouse	Eastern U.S.	None obvious	125	500	0.2	85	–	0
Red-eyed Bulbul	Garden Bulbul	South Africa	Ecotone	15	80	2.5	70	–	1.3
White-eared Bulbul	Red-vented Bulbul	Northern Pakistan	Drier vs. wetter	200	–	–	–	–	–
Melodious Warbler	Icterine Warbler	Northeastern France	Open canopy vs. dense bushes	–	400	1.9	3.5	1.7	2
Willow Warbler		Sweden	None	350	300	0.0	–	–	–
Common Chiffchaff	Iberian Chiffchaff	Pyrenees	None obvious	20	70	3.1	10	1.4	2

(continued)

	Species[1]	Location	Ecological differences	Zone width (km)	Zone length (km)	Age[2] of taxa (my)	% Hybrids	Assortative mating	Movement (km/yr)[3]
House Sparrow	Willow Sparrow	Alps, northern Italy	None	30	600	4.0	40	–	0
Black-throated Finch		Australia		70	120	–	63	–	–
†Yellowhammer	Pine Bunting	Siberia	Scrub vs. light forest	1000	2000	–	50	–	Y
Black-headed Bunting	Red-headed Bunting	Iran	–	100	100	–	35	–	–
Collared Towhee	Spotted Towhee	Mexico	None obvious	30	450	4.5	100	–	–
Eastern Towhee	Spotted Towhee	Great Plains of U.S	None obvious	150	1000	–	80	–	–
†Blue-winged Warbler	Golden-winged Warbler	Northeastern U.S.	Different successional stages	600	1000	1.7	13	1.6	Y
Yellow-rumped Warbler		Western Canada	Boreal cordilleran forest	150	1000	0.01	100	–	–
Hermit Warbler	Townsend's Warbler	Western U.S.	None	125	125	0.5	26	1.7	Y
†Rose-breasted Grosbeak	Black-headed Grosbeak	Great Plains of U.S.	None obvious	250	900	2.2	50	–	0
†Indigo Bunting	Lazuli Bunting	Great Plains of U.S.	Generally wetter vs. drier	400	800	3.3	30	1.4	13
†Baltimore Oriole[4]	Bullock's Oriole	Great Plains of U.S.	Generally wetter vs. drier	150	1800	3.2	–	–	10
Common Grackle		Eastern U.S.	Pine marshes	160	1300	–	90	–	1

* See Appendix 15.1 for further explanation of column headings. Where width varies at different locations, a midpoint value has been used.
† The taxa in these seven pairs have historically (< 100 years) spread into each other's range, resulting in recent historical increases in zone width.
[1] When the two hybridizing taxa are considered to be conspecific, the second column has been left blank.
[2] Age in millions of years is based on divergence in mitochondrial DNA, assuming the standard rate of 2% per million years (Appendix 2.1, p. 36).
[3] "Y" indicates movement recorded, without quantitative information. Column is left blank if the hybrid zone has not been studied in multiple years.
[4] Among fully adult-plumaged birds only; assortative mating is lower among juvenile-plumaged birds (Edinger 1985).

A smooth curve can be fit to the individual indices plotted along a line transect through the zone (e.g., Figure 15.1). Note that this index does not measure the degree of hybridization but, instead, measures the extent of overlap between taxa. (If no hybridization were occurring, in the center of the zone, the average hybrid index would be 0.5, being composed of individuals with indices of 0 and 1.0 in equal proportions.) In theory, the width of the zone is defined as the inverse of the tangent to the steepest part of this curve (see Figure 15.1), and with random mating among the overlapping taxa, this roughly corresponds to the distance over which the average index changes from approximately 0.1 to 0.9 (Moore and Price 1993). In practice, the width of a hybrid zone has often been reported as the distance over which signs of hybridization are regularly observed (i.e., individuals carrying features of both the parental species are recorded).

Some sophisticated models have been built to try to work out what features determine the width of a stable hybrid zone (Barton and Hewitt 1985, 1989). The simplest model assumes a balance between dispersal into the zone and decreased fitness of hybrids. A hybrid zone set in this way is termed a "tension zone." High dispersal distances between place of hatch and place of reproduction result in a wide zone. High hybrid fitness leads to a wide zone, and low hybrid fitness to a narrow zone. The zone is stable because individuals of one species become increasingly rare as one moves through the hybrid zone, so they end up mating with the other species more frequently and thereby more often produce hybrids with low fitness. In other words, if one were suddenly to mix the two species together, either the two species would collapse into a hybrid swarm or one species or the other would go extinct. It is only because of the spatial segregation of the two species that the zone remains stable (Bazykin 1969).

In this model, suppose one species is fixed for an allele a (i.e., all individuals are aa homozygotes), and the other for an allele A (i.e., all individuals are AA homozygotes). Assume that the hybrids, which have the genotype Aa, have a lower fitness than either homozygote because of their genetic makeup, and that individuals mate randomly everywhere (i.e., there is no preference to mate with conspecifics). Zone width is approximately given as $w \approx \frac{3.75\mu}{\sqrt{s}}$, where μ is the average dispersal distance between offspring and parent and s is the selective disadvantage of the heterozygotes (Bazykin 1969). (The theory is actually based on the square root of the mean of the squared values of the dispersal distances: the "root mean square distance." This theory assumes that dispersal distances are normally distributed, and, in this case, the mean dispersal distance is about 80% of the root mean square value. I use the 80% correction and discuss the mean dispersal distance throughout this book because it is intuitively easier to contemplate.)

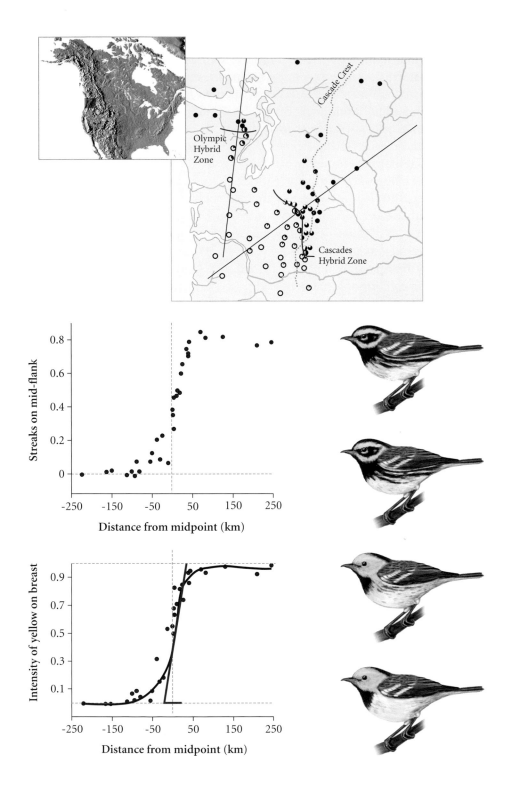

To put numbers on this, if the heterozygote, *Aa*, has only 90% the chance of surviving than either homozygote does, then the selection coefficient, $s = 0.1$. Substitution in Bazykin's equation gives a zone width of about twelve times the average dispersal distance. As a result of selection against hybrids, the two species may remain homogeneous for either *AA* or *aa* over much of their respective ranges, with a short zone of introgression where dispersal is balanced by selection.

When several genes affect fitness, a similar result pertains (Barton and Hewitt 1985, 1989). However, in this case, selection on one gene affects the other gene because genes from one species arrive in the zone from outside as a block (they are all brought in by an individual). Following successive generations of hybridization, recombination between the genes removes the association between them, so selection acts increasingly independently on each locus. Consequently, the shape of the curve describing gene frequencies as a function of distance differs from the shape when a single gene affects fitness; it is relatively steeper near the center of zone (Barton and Hewitt 1985, 1989; see Kruuk et al. 1999). This difference may be difficult to detect empirically.

Ecological causes of selection

In the tension-zone model described above, the parental species suffer no loss of fitness wherever they are found in the zone (i.e., their survival, their mating success, and the number of offspring they produce are all unaffected by geographical location). All the selection falls on the hybrids. In two other forms of selection, both parental and hybrid fitness are affected, and fitness also depends on where the individuals are in the zone. The two kinds of selection are environment-dependent selection (Chapter 5) and sexual selection (Chapter 14).

Most theory has been done on the way in which ecological, rather than social, factors influence zone width (Barton and Hewitt 1989; Kruuk et al. 1999). In one such model, Kruuk et al. (1999) assumed random mating and two

Figure 15.1 Two hybrid zones between the Hermit Warbler and Townsend's Warbler in the state of Washington, U.S. (Rohwer et al. 2001). Each point shows the mean hybrid index at the locality based on eight characters and typically about 10 males (Rohwer and Wood 1998). Each individual was scored for each trait on a scale of 0 to 7, which was then scaled to 1.0 by dividing by 7. On the right are the scores for the character "mid-flank" and "intensity of yellow on the breast" across the Cascades transect. "Mid-flank" is a measure of the amount of black streaking on the sides. On the lower graph, a smooth curve has been fit by eye to the points *(blue line)*. This is to illustrate zone width *(thick horizontal red bar)*, which is given by the point of intersection of the tangent to the curve *(red line)* with the lines $y = 0$ and $y = 1$. Pure Townsend's Warblers *(top)*, pure Hermit Warblers *(bottom)*, and two hybrids are illustrated (redrawn from Rohwer et al. 2001).

adjacent environments. Suppose that one species is fixed for an allele *a*, and the other is fixed for an allele *A*. In the environment occupied by the first species, *aa* individuals have a selective advantage when compared with the *AA* individuals. In the other environment, the advantage is exactly reversed. The hybrid *Aa* has intermediate fitness in each environment (Kruuk et al. 1999). For comparable selection intensities, the hybrid zone is of comparable width, and the plot of gene frequencies over space is of a shape that is similar to zones produced by genetic incompatibilities (Barton and Hewitt 1989; Moore and Price 1993; Kruuk et al. 1999).

When ecological factors are involved, the parental species suffer if they are in the wrong environment. Thus, the width of the hybrid zone is set partly by its effects on parental fitness. This means that overlap between species is limited even in the absence of hybridization. Indeed, many pairs of species do have narrow zones of overlap despite the apparent absence of hybridization (Mayr and Short 1970; Prigogine 1980; Haffer 1992; Aliabadian et al. 2005). It is therefore possible that the width of many hybrid zones is largely independent of hybrid fitness but, instead, is set by interactions among the parental species.

In Chapter 6, I presented a model of species range limits based on ecological interactions, assuming no hybridization (Case and Taper 2000; Figure 6.7, p. 111). In this model, competition between the parental species, coupled with the adaptation of each species to the environment in which it is found, limits penetration of one species into the other's range. A relevant result from the range-limit models is that, if the environment changes sharply at one point, range boundaries accumulate at this point (Figure 15.2). This is because individuals of one species are especially poorly adapted to the conditions on the other side of the steep transition and are more easily outcompeted. Penetration of one species into the other's range is also greatly reduced over what might be expected when the environment shows a more gradual change (Figure 15.2). Thus, zones of overlap between ecologically similar species should accumulate and be relatively narrow when environmental conditions change rapidly, whether or not hybrids are being produced.

If environments occupied by two species show relatively gradual change from one type to the other, the ecological model predicts a relatively wide zone of overlap. In such cases, it is possible that hybrids, which are often morphologically and ecologically intermediate to the parental species and may suffer no intrinsic defects, have higher fitness in transitional habitats than either of the parental species (Moore 1977). This form of selection results in a hybrid zone in which hybrids are maintained by selection in the center of the zone and persistence of the zone does not depend on dispersal of individuals from outside. In distinction from a tension zone, this kind of hybrid zone is termed "bounded hybrid superiority" (Moore 1977).

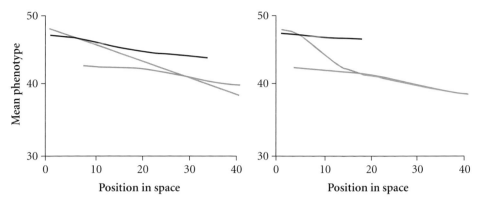

Figure 15.2 A model of range limits set by ecological competition (from Case and Taper 2000; see also Chapter 6, p. 111). The kinds of resources vary in space so that the optimal phenotype for a single species would follow the black line. The left-hand figure is similar to Figure 6.7 (p. 111) and shows a large overlap of the two species whose distributions are given by the *colored lines*. In the right-hand figure, there is an abrupt break in the resources, and the range boundaries accumulate at the break with less overlap. In these figures, the lines representing the distribution of species, taken from Case and Taper (2000), are terminated when population density is 10% of the maximum.

Social causes of selection

Assortative mating is common in hybrid zones. This mostly reflects the tendency of parentals to mate with conspecifics. Short (1969) restricted the term "hybrid zones" to zones of overlap with essentially random mating, and he used the term "zones of overlap and hybridization" where the parental types are at higher frequency than expected under random mating throughout the zone. These are both now called hybrid zones. Jiggins and Mallet (2000) redefined Short's terms as unimodal and bimodal hybrid zones, respectively.

Very little theory has been done on how sexual selection and assortative mating affect the width of a hybrid zone. Sexual selection against hybrids reduces their fitness, so should narrow the zone. If few hybrids are being produced because most members of the parental species mate assortatively, intuitively one might expect hybrid zones to be relatively broad. This is because many individuals in the center of the hybrid zone mate conspecifically, producing high fitness offspring, which can disperse further through the zone. However, Goldberg and Lande (2006) made a model of strong assortative mating along an ecological gradient, setting hybrid fitness to be zero, and showed that zone width is substantially narrower than it would be without hybridization (more complex models where hybrid fitness is not zero have yet to be made). The reason is that, even if assortative mating is strong, when one species is rare, individuals still end up pairing heterospecifically, and produce low-fitness offspring.

PATTERNS

A plot of age of hybridizing taxa vs. zone width (based on divergence in mitochondrial-DNA sequences) is shown in Figure 15.3. Note that the age of the hybridizing taxa says nothing about the age of the hybrid zone itself; many zones have probably become established relatively recently (i.e., since the Pleistocene epoch; Rohwer et al. 2001), and the taxa that currently hybridize may have initially become separated millions of years previously. The plot in Figure 15.3 shows considerable scatter. However, I suggest that roughly three types of hybrid zones can be identified. They are: (1) wide transitional zones, (2) narrow zones in which hybrid fitness is apparently low, and (3) wider zones in which hybrid fitness may be lower than that of the parental types, but not as low as in the narrow zones. In addition, at least one hybrid zone (between a pair of gull species) shows elements of bounded hybrid superiority, with hybrids having higher fitness in the center of the zone than either parental type.

First, if hybrid fitness is low, theory predicts that, when two species overlap extensively and also hybridize extensively, the hybridization is temporary. The result is a collapse to one or other species, but the surviving species may carry some genes from the other. Because these zones represent a temporary state, it may seem unlikely that many would be around at the present day. However, sev-

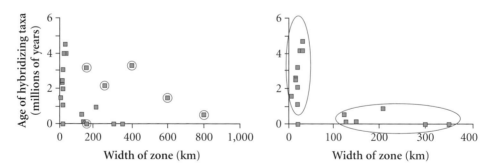

Figure 15.3 *Left:* Estimated times of divergence between parental taxa plotted against hybrid-zone width (from Table 15.1). The encircled points are of taxa that have recently spread extensively into each other's range (marked † in Table 15.1). *Right:* The same plot as on the left, but with the encircled taxa removed (note the scale change on the *x* axis). Two types of hybrid zones are recognized in this plot, as indicated by the ellipses. In one type, zones are narrow and the hybridizing taxa are old, and, in the other, they are wide and the hybridizing taxa are young.

eral hybrid zones, including some of the widest, seem to be associated with the recent spread of the range of one of the taxa into that of the other, associated with anthropogenic change. For example, three hybridizing pairs of species in the Great Plains of the United States have been recorded as moving into each other's range over the past 50 years, attributed to habitat modification (tree planting; Rhymer and Simberloff 1996). Zones that are associated with historical expansion of one species into the range of the other are indicated by a circle over the points in the left-hand plot of Figure 15.3 and by a dagger symbol (†) in Table 15.1. The zones are often wide, and some show the postulated rapid collapse to one or another species. For example, the Blue-winged Warbler is spreading into the range of the Golden-winged Warbler, and the former replaces the latter in any one locality within 4 to 50 years after arrival (Gill 2004).

If the recent wide zones are excluded, the age of the hybridizing taxa is negatively correlated with zone width (Figure 15.3, *right*). For simplicity I divide these remaining zones into two categories. First, some hybrid zones are between long-diverged taxa, and are fairly narrow (see Figure 15.3, *right*). These hybrid zones appear, more or less, to fit the classic conception of a tension zone—a balance between dispersal and low hybrid fitness. Where tested, hybrids show intrinsic forms of fitness loss (i.e., fitness is independent of the environment in which the hybrids live). Bronson et al. (2003a) studied the narrow hybrid zone (15–30 kilometers) between the Black-capped Chickadee and the Carolina Chickadee, which diverged perhaps 3–4 million years ago (they are not sister species; Gill et al. 2005). Bronson et al. (2003a) moved pairs of each species into the hybrid zone and showed that they had equivalent fitness, but pairings between individuals normally present in the hybrid zone resulted in approximately half the fledging success of the pure pairs. In the center of the zone, Bronson et al. (2005) found reduced fledging success for male parents that were genetically intermediate, and also reduced fledging success for pairings between individuals that were genetically dissimilar.

The final category of hybrid zone consists of those zones that are relatively wide (>100 km) and the hybridizing taxa are fairly recently diverged (Figure 15.3, *right*). It might be objected that the youth of these taxa (as evaluated by mitochondrial DNA distances) is misleading, but, instead, mitochondrial DNA from one taxon has introgressed into the other; this seems to be a common occurrence in many groups (Chan and Levin 2005). However, arguments based in historical biogeography (e.g., Bensch et al. 1999; Rohwer et al. 2001; Weir and Schluter 2004) and the fact that several of the taxa, although only separated by short distances, remain distinctly different in their mitochondrial DNA, indicate that most of these hybridizing pairs of taxa are in fact young. Where studied, hybrids in these zones do not appear to have major intrinsic defects, being both viable and fertile (e.g., Bensch et al. 1999; Rohwer et al. 2001).

Assortative mating is common in bimodal hybrid zones, where hybrids are relatively rare, and these zones are relatively narrow (Figure 15.4). One explanation for the narrowness of these zones is that hybrids have very low fitness. The age of hybridizing taxa is generally old (Table 15.1), so they might be expected to exhibit a relatively high level of both premating and postmating isolation.

If we exclude the apparently transient zones and concentrate on hybrid-zone classes (2) and (3), theory predicts that the width of the zone should result from a balance between dispersal and selection against hybrids, as well as any selection on the parentals. With respect to any influence of dispersal, when widely different taxa (including insects, plants, and mammals) have been compared, average dispersal distance (from birth to reproduction) is indeed positively correlated with zone width (Barton and Hewitt 1985, p. 129). Data on birds are not available to test this, but, if any correlation exists, it is not likely to be very strong. In England, the mean dispersal distance for the Willow Warbler (20.8 km) is almost identical to that for the Pied Flycatcher (20.6 km) (Paradis et al. 1998), yet the hybrid zones in which they are involved (in Sweden and the Czech Republic respectively) differ greatly in width (350 km vs. 20 km). Thus, it is difficult to demonstrate that dispersal has important influences on zone width. Instead, zones may be wider between younger than between older pairs

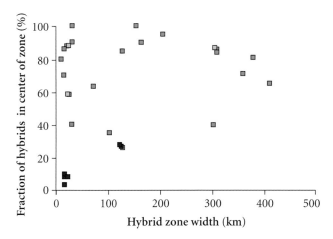

Figure 15.4 The fraction of hybrid phenotypes in the center of the zone plotted against zone width (from Table 15.1, excluding recently spread taxa, marked † in Table 15.1). In some hybrid zones, mated pairs have been observed, allowing assortative mating to be directly measured, as indicated by the color of the symbols: (□) Random mating, (■) Nonrandom mating. Zones with few hybrids are typified by non-random mating and are relatively narrow. The point at 300 km and 40% hybrids is the Green-backed Woodpecker, which lives in a patchy environment and has been relatively poorly surveyed.

of species because hybrids have higher fitness when hybridizing species are younger. Social selection and ecological factors may also contribute to the setting of zone width and differ among zones. I consider social and ecological causes in turn.

Social selection

An individual using a signal that is locally common seems to have advantages both in territory defense and in mate attraction (Chapter 13). This has been best demonstrated for songs: males and females respond strongly to their own dialect and less strongly to other dialects (Chapter 10; Table 10.1, p. 190). It likely applies to color patterns too. Moore (1987) and Moore and Price (1993) suggested that a failure to recognize different plumage patterns among the hybridizing flickers of the Great Plains makes it difficult for individuals of one type to establish territories in the range of the other: "Flickers that cross the hybrid zone have difficulty defending territories because individuals of the resident subspecies are not intimidated by the signals communicated by the immigrant flicker" (Moore and Price 1993, p. 221). Moore and Price had no direct evidence that this was the case in the flicker system, but Stein and Uy (2006) found that in a manakin hybrid zone in Panama (Figure 15.5), male manakins with yellow collars had a mating advantage over those with white collars, but only on those leks where the yellow collar phenotype is at high frequency. Selection favoring the locally common phenotype will limit the width of the hybrid zone, because individuals that disperse into the range of the other species have low fitness (Moore and Price 1993).

Ecological factors

The position of hybrid zones is often correlated with ecological differences (noted for about two-thirds of the cases in Table 15.1). In particular, Moore and Price (1993) presented a strong case for ecology in determining not only the position but also the width of hybrid zones in the Great Plains of the United States. They thoroughly studied the hybrid zone between the Yellow-shafted Flicker and the Red-shafted Flicker (these are usually considered subspecies of the Northern Flicker). The zone lies in the rain shadow of the Rocky Mountains and hence is associated with a steep transition in vegetation type. It coincides with multiple range boundaries in other bird and mammal species. Most convincingly, the width of the zone varies with the steepness of inferred ecological gradients (Figure 15.6).

The extreme form of ecological influence is in the bounded hybrid-superiority model, where hybrids actually have higher fitness than parentals in the cen-

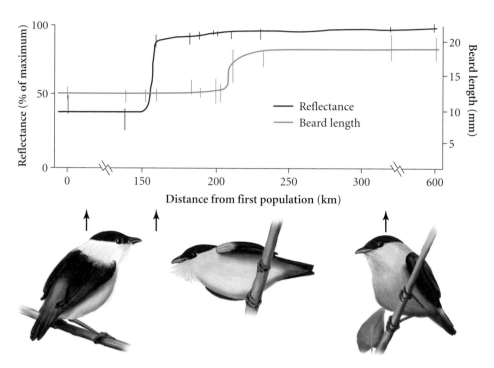

Figure 15.5 Hybrid zone of the White-collared Manakin and the Golden-collared Manakin in Panama (after Brumfield et al. 2001). The cline for morphology, molecules, and beard length is congruent, and it is displaced about 50 km from that for collar color (Parsons et al. 1993; Brumfield et al. 2001). Vertical lines are twice the standard deviation of the individual values in a population, and they show the positions at which populations were sampled; a smooth line was fit to the mean of these points. The hybrid male (*center*) came from the population indicated, where both pure parentals can also be observed (C. Stein and A. Uy, *pers comm.*). In this region, collar color has introgressed from the Golden-collared Manakin, but not belly color.

ter of the zone. The best example of this is the hybrid zone between the Glaucous-winged Gull and the Western Gull along the Pacific coast of the United States. Good et al. (2000) showed that hybrids, which are ecologically and morphologically an admixture of the two parental species, do well in environments that are an admixture of those experienced by the parental species (Chapter 5).

In many hybrid zones, social and ecological selection pressures, and the intrinsically low fitness of hybrids, may all be involved in setting zone width and position. In the hybrid zone between the Carrion Crow and the Hooded Crow in Italy, behavioral observations suggest the possibility of sexual selection against hybrids, as well as ecological forms of selection via habitat choice and social

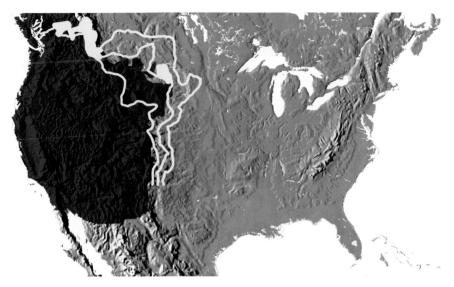

Figure 15.6 Hybrid zone of the Northern Flicker [where the two subspecies, named the Red-shafted Flicker (*red*), and the Yellow-shafted Flicker (*green*), meet]. The contour lines mark the approximate positions where the average hybrid index is 0.2, 0.5, and 0.8, respectively. The yellow areas indicate places where both parental types can be found. The hybrid zone twists and broadens in association with changing environmental conditions along the rain shadow of the Rockies (Moore and Price 1993).

dominance in competition for food (the Carrion Crow is dominant; Saino and Scatizzi 1991). Intrinsic loss of hybrid fitness is suggested by the finding that egg volume, chick survival, and number of chicks fledged per nest are all lower for hybrid females than females of either the Carrion Crow or the Hooded Crow (Saino and Bolzern 1992; Saino and Villa 1992) and that some hybrid females lay abnormally thin-shelled eggs (Saino and Villa 1992). The Collared Flycatcher/Pied Flycatcher hybrid zone was extensively discussed in the last chapter. These species have been separated from each other for much longer than the Carrion Crow and the Hooded Crow, and intrinsic loss of fitness is substantial; hybrid females are infertile and hybrid males have low fertility (Veen et al. 2001; Sætre et al. 2003). But assortative mating within the flycatcher hybrid zone (Sætre et al. 1999), as well as direct competition between males of the two species (Alatalo et al. 1994), raises the possibility of sexual selection against hybrids. Habitat differences may impose some ecological selection. However, Veen et al. (2007) rule out the possibility that migratory differences are important, as hybrids appear to show similar migration routes to the Pied Flycatcher.

It is likely that all three forms of selection usually operate in hybrid zones, with intrinsic loss of hybrid fitness becoming increasingly important the older the taxa involved. Complete loss of hybrid fitness can take millions of years (Chapter 16), and, in very young zones, intrinsic causes of the loss of hybrid fitness may be small. Ecological and social selection should be the important determinants of width in these hybrid zones.

ZONE MOVEMENT

A large number of hybrid zones have been recorded as moving historically, even with little change in their width. Excluding the cases in which one taxon is spreading into the range of the other (i.e., the circled points in Figure 15.3), 60% (nine out of 15) of all the hybrid zones that have been censused at different times have shown detectable movement of at least part of the zone. A typical rate of movement is on the order of 1 km per year (Table 15.1). Causes of zone movement can be used to assess the kinds of selection pressures operating to set zone position and width, and also to ask why it is that some traits may introgress from one species to other, whereas others do not. Sexual selection seems often to be involved, as well as various ecological forms of selection.

Social dominance has been invoked as a factor in zone movement for several hybrid zones (Pearson and Rohwer 2000; Rohwer et al. 2001; McDonald et al. 2001; Bronson et al. 2003b). In the Townsend's Warbler/Hermit Warbler hybrid zone (Figure 15.1), evidence from cline shape implies that the hybrid zone is moving south, with Townsend's Warblers replacing Hermit Warblers (Rohwer et al. 2001). Pearson and Rohwer (2000) showed that, in areas where one or the other species occurs, Townsend's Warblers are more than four times as aggressive in responding to mounts of Hermit Warblers than Hermit Warblers are in responding to mounts of Townsend's Warblers (Figure 15.7, *left*). Probably for this reason, within the zone, Townsend's Warblers are more successful at holding territories than Hermit Warblers, and they have higher pairing success (Pearson 2000). Part of the Carolina Chickadee/Black-capped Chickadee hybrid zone has moved north about 100 km over the past 100 years, with the Carolina Chickadee replacing the Black-capped Chickadee (Curry 2005). Male Carolina Chickadees are dominant over male Black-capped Chickadees when matched for size, and, in aviary experiments, females associate with dominant males (Bronson et al. 2003b; Chapter 14).

Finally, in the hybrid zone between the Golden-collared Manakin and the White-collared Manakin in Panama, two male color traits have introgressed from the Golden-collared Manakin into the White-collared Manakin popula-

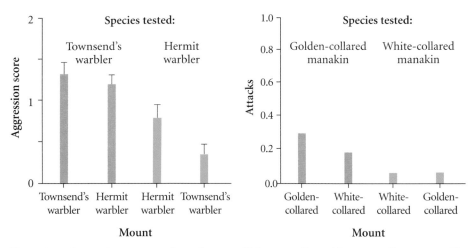

Figure 15.7 Aggressive responses by males to stuffed conspecific and heterospecific mounts. All experiments were done in allopatric areas, where only one or the other species breeds. The top line in each figure gives the test species and the bottom line gives the species of the stuffed mount. *Left:* Interspecifically territorial Townsend's Warblers and Hermit Warblers (+ SE) in Washington state. The rankings are based on a composite score of several aggressive behaviors, including number of pecks and hits of the stuffed mount and number of wing flicks. $N = 39$ individuals in each experiment. Townsend's Warblers are more aggressive than Hermit Warblers by t-tests (from Pearson and Rohwer 2000). *Right:* Lek-mating *Manacus* manakins in Panama (from McDonald et al. 2001). Each experiment was conducted with 17 males. The fraction of all males that attacked the mount is shown. Golden-collared Manakins are more aggressive than White-collared Manakins, by a Mann-Whitney U test comparing the average aggressive score of a male to a heterospecific vs. conspecific mount.

tion (Figure 15.5). Measurements of size and shape, some molecular markers, and some secondary sexual traits, most notably the length of the male's "beard," vary concordantly across a hybrid zone whose width is about 10 km (Parsons et al. 1993; Brumfield et al. 2001). The hybrid zone for collar and belly color is displaced some 50 km further north. Trait separation may be ongoing. Based on museum specimens, over the past 100 years the main hybrid zone has shifted eastward a few kilometers and hybrids with yellow collars have shifted a few kilometers to the northwest (Brumfield et al. 2001). In any case, in the area of overlap, many individuals that are genetically White-collared Manakins carry yellow collars. Experiments with stuffed mounts have shown that the male Golden-collared Manakin is aggressive (Figure 15.7, *right*). Hybrids (i.e., yellow-collared individuals) are also aggressive (McDonald et al. 2001). So the differential introgression of the yellow collars may be partly due to aggression-associated differences. This is not the whole story, however. At mixed leks within the zone, Stein and Uy (2006) found that females preferred to mate with yellow-collared males, but they were unable to detect any differences in aggression.

Perhaps associated with social dominance, altered ecological conditions are likely to contribute to the movement of hybrid zones. The three examples listed above are all of the dominant form moving into the range of the subordinate form. McDonald et al. (2001) and Rohwer et al. (2001) infer that aggression provides a widespread advantage. However, aggression can be costly (e.g., Duckworth 2007), and advantages will accrue only in favorable environments. In some nonhybridizing pairs of species that have limited range overlap, it is thought that the dominant species occupies the more productive areas, excluding the subordinate species, which must make do in the poorer areas of the resource gradient, where the dominant species cannot easily subsist (e.g., Robinson and Terborgh 1995; Gross and Price 2000). This implies that, as total productivity fluctuates, range boundaries should move. In Italy, the Carrion Crow is dominant to the Hooded Crow (Saino and Scatizzi 1991), but the hybrid zone between them has not moved substantially. However, in Scotland and Denmark, the hybrid zone between these two species runs from east to west, with the Hooded Crow to the north of the zone and the Carrion Crow to the south. It is only in these locations that substantial movement of the hybrid zone has been detected in historical times. In both of these cases, the zone has moved to the north (Cook 1975; Haas and Brodin 2005). Many southern species in Britain that do not have a closely related congener to the north of them are showing northward movement of their northern range limit, which has been attributed to the pervasive warming trend (Thomas and Lennon 1999). Thus, the movement of the Carrion Crow to the north may be a consequence of improved conditions for this species, pushing the Hooded Crow in front of it.

DIFFERENTIAL INTROGRESSION

In the previous sections, I considered how dispersal and various forms of selection affect zone width and zone movement. I now turn to ask how zone width or zone movement may differ among traits. In this case, some traits introgress further into one or another species than other traits do, producing novel phenotypes: a creative role for hybridization.

Differential introgression can happen in at least four ways:

1. Dispersal may differ among traits.
2. Different traits are subject to different patterns of selection in the hybrids.

3. The mechanism of inheritance differs among traits.
4. Some genes may be tightly associated with others that are under selection, whereas others are less tightly associated.

I consider each of these in turn.

Dispersal

It may seem odd that dispersal can differ among genes, as they are all simultaneously carried when an individual moves. Differences can arise, for example, when a gene is present in one sex more frequently than it is in the other (i.e., on the mitochondrial or sex chromosomes). The simplest example of differential introgression is then if one sex (e.g., the male) disperses further than the other (i.e., the female). In nonhybridizing species, males dispersing beyond the normal range limits of the species would have little reproductive success, but, in hybridizing species, these males can mate with females of the other species. In the Hermit and Townsend's Warblers, mitochondrial DNA from Hermit Warblers can be found in populations of pure Townsend's Warblers (according to their color patterns) more than 2000 km north of the zone of hybridization (Rohwer et al. 2001). Rohwer et al. (2001) suggest that the hybrid zone has moved south following contact between the two species perhaps 5000 years ago. An explanation for the high frequency of Hermit Warbler mitochondrial DNA in the north is that zone movement was largely due to aggression and southward movement of Townsend's Warbler males, with females settling more at random. The mitochondrial DNA, which is inherited only through the female line, would not move but would get left behind as the zone moved south. An example of the opposite pattern is at one location in the Blue-winged Warbler / Golden-winged Warbler hybrid zone, where the mitochondrial DNA of the Blue-winged Warbler rapidly replaces that of the Golden-winged Warbler, and this may be explained by successful dispersal of females beyond the leading edge (Gill 1997); elsewhere, introgression is more symmetrical (Shapiro et al. 2004).

Dispersal may also differ among genes when a gene itself affects the chances of an individual dispersing. A possible example of this is considered in the next section.

Selection

First-generation (F_1) hybrids should be fairly uniform (all individuals are Aa and, at a second locus, Bb, and so on). However, F_2 hybrids that result from crosses among the F_1 hybrids, as well as backcrosses of the F_1 hybrids to the parental species, exhibit a range of phenotypes due to recombination and segregation (e.g., the genotype *AABb* is produced; this genotype never occurs in the parental species or in the F_1 hybrids). Some of these hybrid genotypes may confer high fitness on their bearers, and such combinations will then show deeper introgression into the other species than those combinations that reduce fitness. This has been suggested for the Hermit Warbler/Townsend's Warbler hybrid zone (Figure 15.1). At the southern edge of this zone individual birds show more signs of introgression of the melanin-based traits on the belly from the Townsend's Warbler into the Hermit Warbler (second picture down in Figure 15.1), than the carotenoid-based ones (third picture down in Figure 15.1; Rohwer et al. 2001). Possibly the melanin-based traits are more strongly correlated with aggression. It is also possible that melanin-based traits are associated with a greater tendency to disperse (and aggression and dispersal tendency may themselves be correlated, as has been shown in a study of the Western Bluebird; Duckworth 2007).

Inheritance

A third way in which different traits show different patterns across a hybrid zone arises out of differences in inheritance patterns. In the models described earlier, the *Aa* hybrid differs from both the *AA* and the *aa* genotypes. If *A* is dominant to *a*, then the F_1 hybrid phenotype looks the same as that of the *AA* homozygote. Assume that, at the center of the hybrid zone, the frequency of the *A* allele is 0.5 and there is random mating. The frequency of individuals showing the phenotype of the *A* allele is the frequency of *AA* homozygotes (0.25) plus *Aa* heterozygotes (0.5) = 0.75. On the other hand, if *A* is recessive to *a*, the frequency of individuals showing the phenotype of the *A* allele would be only 0.25. Different inheritance patterns account for the displacement of the face pattern across the Hermit Warbler/Townsend's Warbler hybrid zone (Figure 15.1; Rohwer and Wood 1998). The yellow face of the Hermit Warbler appears to be genetically dominant to the black-patterned face of the Townsend's Warbler, so hybrids that are phenotypically intermediate in other characters most often have Hermit Warbler face patterns (third picture down in Figure 15.1).

Associations between genes

Mitochondrial DNA is transmitted only through the female line. Hybrid females usually have lower fitness than hybrid males (Haldane's rule, Chapter 16). One result of this is that mitochondrial DNA should introgress less than autosomal DNA. Limited introgression of mitochondrial DNA is seen in the Collared Flycatcher and the Pied Flycatcher (Sætre et al. 2003), in which the female hybrids are known to be infertile (Veen et al. 2001), and is also present in a chiffchaff hybrid zone (Helbig et al. 2001; Bensch et al. 2002) and a hybrid zone in gulls (Crochet et al. 2003).

If different genes are nearby on the same chromosome, they are said to be closely linked. Close linkage means that the spread of one gene will be affected by patterns of selection on the other, because it will take more generations for recombination to break down the association (Borge et al. 2005). In the Pied Flycatcher/Collared Flycatcher hybrid zone, genetic markers on the autosomes have introgressed into the Collared Flycatcher from the Pied Flycatcher, but genes on the sex chromosomes have not, possibly because F_2 and backcross males that carry one sex chromosome from each species have especially low fertility (Sætre et al. 2003). A pattern of higher introgression of autosomal markers rather than sex-linked ones is also present in the Carolina Chickadee / Black-capped Chickadee hybrid zone (Sattler and Braun 2000; Bronson et al. 2005). These are studies of presumed neutral genetic markers whose spread may be greatly delayed by selection on other loci. A favored allele should still manage to become dissociated from the deleterious ones and then spread, and even if the favored and deleterious genes show little recombination, the time delay is relatively small (Barton 1979). However, following Rieseberg (2001), Navarro and Barton (2003) developed a model that showed how chromosomal inversions can create regions of the genome that are essentially completely linked. In such a case, favored alleles may become trapped in the inversions. There is no direct evidence of such a role for inversions in birds as yet; the model is considered further in the next chapter.

DOES HYBRIDIZATION HAVE A CREATIVE ROLE?

Hybridization likely occurs throughout the different stages of an adaptive radiation, with few incipient species immune from it. First, early in adaptive radiations, ecological differences between closely related, often young, sister species lead to extensive sympatry. Frequent encounters result in some hybridization. Again in the early stages, ecological conditions may occasionally change, resulting in increased cross-taxon matings, increased hybrid fitness, and an increased rate of introgression (Chapter 5). Late in adaptive radiation, related ecologically similar species remain geographically separated, but geographical barriers come and go, leading to contact and to possible hybridization events. Large-range changes associated with Pleistocene climate fluctuations over the past 2.4 million years must have continually isolated many populations and then brought them back into contact (Hewitt 2000; Newton 2003, his chapters 9–11). Such changes are not a peculiar consequence of the Pleistocene, and they have occurred frequently in earlier geologic times as well (Haffer 1997, 2002). For example, sea levels were higher from 5.5 to 4.0 million years ago than they are now, turning mountains into islands (Nores 1999; Woodruff 2003). It may take millions of years for complete hybrid infertility or inviability to develop (Chapter 16), and, throughout this time, gene exchange can potentially take place.

Although populations may come into contact and hybridize, it is not necessarily the case that all parts of the genome introgress to the same extent (Wu 2001; Mallet 2005). Some alleles, such as those favored in the environment of one population but not in that of another, or those that work poorly when heterozygous (i.e. when one allele comes from one taxon and the other from the other taxon), remain restricted to their respective taxa. On the other hand sexually selected traits that are stimulating to receivers may spread across the hybrid zone and right through the other species. In their analysis of the manakin hybrid zone, McDonald et al. (2001) argued that this was happening, and the yellow collar might eventually spread throughout the range of the White-collared Manakin. While this is a plausible example of the spread of an unconditionally favored trait, it is also possible that the pattern will not spread completely throughout the range of the species. The yellow-collared males have now reached a river; although some individuals of this phenotype are found on the other side of the river, their spread may be delayed or even prevented by frequency-dependent selection favoring the common form (Brumfield et al. 2001; Stein and Uy 2006). More generally, introgression may be related to recent changes in the environment, in which case spread will be limited to regions over which the ecological conditions are appropriate.

Large numbers of hybrids in a hybrid zone increase the chances of a creative role for hybridization in speciation. Even in zones in which hybrids have low fitness, the occasional novel recombinant may be thrown up that has high fitness (Barton 2001). A difficulty in fixing such a recombinant is that the favorable combination of genes will be broken down again immediately as a result of mating with another individual. It may therefore rarely be able to spread (Barton 2001). This difficulty may be circumvented in at least two ways. First, hybrid populations may differ in predictable ways from the parental types. Different hybrid phenotypes may be recurrent and in moderately high frequency because different genotypes and phenotypes extend to variable distances from the center of the hybrid zone, due to different patterns of dispersal, selection, or inheritance. A second means of differentiation is that a hybrid population becomes separated from the other species, and different genes from each species become fixed in this population, as suggested for some species of wheatears by Panov (1992).

Several subspecies do seem to have arisen via hybridization of other subspecies or species, although molecular evidence that this is the case is lacking. The Adelaide Rosella is likely a hybrid population resulting from the intermixing of two species (the Crimson Rosella and the Yellow Rosella, which are often themselves considered to be subspecies), and wild Adelaide Rosellas are indistinguishable from hybrids of these species produced in captivity (Forshaw 1969, pp. 184, 189). Additional examples include subspecies of the African Paradise Flycatcher (Chapin 1948), the Dark-eyed Junco (Miller 1941), the Golden Whistler (Galbraith 1956), the Variable Wheatear (Panov 1992), and the Pine Bunting (Panov et al. 2003). In addition, the Italian Sparrow (which is sometimes given species status and sometimes considered to be a subspecies of the Willow Sparrow) seems to have arisen via hybridization between the House Sparrow and the Willow Sparrow, perhaps 3600 years ago (Johnston 1969). Other former subspecies have been demoted when it was determined that they are, in fact, the product of introgression between a pair of other taxa (Marantz 1997; Brumfield et al. 2001; Isler et al. 2005). At the species level, in Table 11.4 on p. 231, I gave several examples of hybrids between two species that resemble a third species. Some have suggested that these resemblances indicate the way in which new species could have been formed, although they may instead represent the recurrent appearance of a trait present in a common ancestor (see Chapter 11).

In summary, hybrid zones could play a role in speciation by being selectively leaky to certain genes. The importance of this process is difficult to assess. Plausibly, divergence to the level of full species would proceed along roughly similar lines if all populations were held in continuous geographical isolation. This has certainly happened often enough, such as when speciation occurs on isolated oceanic islands, where opportunities for hybridization seem small. Surveys of distinctive subspecies for signs of introgression are likely to be the best way to determine the importance of a creative role for hybridization in speciation.

CONCLUSIONS

Recent interest in hybrid zones has led to a few controversies and the development of some quite advanced theory. Moore (1987) suggested that many zones were examples of bounded hybrid superiority where, in the center of such a zone, hybrids have higher fitness than the parental species. By contrast, Barton and Hewitt (1981, 1989) argued that a balance between dispersal and the intrinsic loss of hybrid fitness sets the width of most zones. Subsequently, Moore and Price (1993) inferred that the ratio of dispersal distance to zone width was so high in their main study species (the Northern Flicker) that low hybrid fitness is likely to be involved in setting width, and Barton and colleagues accepted that environmental gradients play an important role in setting width. Models of ecological selection and intrinsic loss of hybrid fitness show that cline shape cannot be used to distinguish the importance of these two factors (Kruuk et al. 1999). Studies on hybrid zones instead demonstrate a role for both. Ecological factors are implicated by the finding of environmental correlates with zone position and zone width, and direct evidence implicates ecological, intrinsic, and social factors in affecting the fitness of hybrids in hybrid zones.

Probably the greatest value of studies in hybrid zones is the insight they give into the build-up of reproductive isolation between taxa. Investigations into mate choice demonstrate how premating isolation develops (Chapter 14). Measurements of hybrid fitness are starting to provide the crucial information on the complementary roles of social, ecological, and intrinsic causes of post-mating isolation. Results from these population-level studies can be tested, because they lead to differing predictions about patterns of introgression and cline width (e.g., Szymura and Barton 1991; Rohwer et al. 2001). Thus, studies on hybrid zones are likely to be increasingly employed as speciation research continues.

SUMMARY

Hybridization between recently diverged taxa leads to the opportunity for gene exchange, which slows divergence in some traits but also creates novel genotypes that could facilitate evolution in novel directions. In birds, more than 200 hybrid zones are known in which different species or subspecies meet in zones of overlap and hybridize to produce a variety of different phenotypes. Most hybrid zones have been identified based on variation in color patterns. I identify three kinds of hybrid zones. Zones of the first kind reflect the recent expansion of one species into the range of the other, usually attributed to anthropogenic disturbance. These zones are often very wide, but this is likely to represent a temporary state. The second kind are narrow, and the hybridizing taxa are typically separated by more than 1 million years. The narrowness of these zones is likely due to strong selection against hybrids balanced by dispersal into the zone. Zones of the third kind are wider, and they are found between relatively young taxa. In all kinds, the position of the hybrid zone may move from one year to the next. Both zone width and zone movement are likely to be influenced both by sexual selection and by environmental conditions, which affect not only the fitness of the hybrids but also the fitness of the parental species. Because inheritance, dispersal, and selection can vary among traits, a single hybrid zone may have a different width for different traits, and not all traits need move at the same rate when a hybrid zone moves. Examples include differential introgression of mitochondrial DNA from one species into the other, the differential spread of traits associated with social dominance, and the failure of sex-linked genes to spread across the zone, perhaps because the sex chromosomes carry more infertility factors. This means that distinct populations of hybrid individuals can form, and these could give rise to new taxa. Several subspecies seem to have been produced as a result of hybridization, but whether hybridization plays an important role in the creation of bird species remains uncertain.

APPENDIX 15.1: LIST OF HYBRID ZONES

The list below is of bird hybrid zones where (with one exception) there has been at least an attempt to measure zone width. The list is intended to indicate the sources and quality of data used in Table 15.1, in Figure 15.3, and in Figure 15.4. Numbers refer to the Sibley and Monroe (1990) list. Methods for quantifying observations are described below.

Habitat differences. This indicates whether the parental types were noted as being in different habitats in the vicinity of the hybrid zone.

Zone width. Based on formal mathematical models, the width of the zone is the inverse of the maximum slope of the cline, but it is approximated as the distance over which gene frequencies of one species change from about 10% to about 90% (Moore and Price 1993). Here, it is usually the distance over which hybrids were observed. Unless stated, width is based on color patterns, and the zone itself was discovered based on color-pattern variation.

Mitochondrial-DNA distances. These are between the parental forms, and they are sometimes taken from the original literature, so the distance metric used varies. When sequences were taken from Genbank and calculated here, I use the HKY-Γ ($\Gamma = 0.2$), as indicated by "HKY" (see Appendix 2.2, p. 38). The effect of different corrections is small at short distances (<4%), but some of the longer distances may be underestimated if the distances are taken from the original literature.

Assortative mating. This is measured in two different ways, as indicated by superscripts.[1,2] (1) In zones where parentals are common, this is calculated by ignoring hybrids altogether. Observed and expected frequencies of the four possible pairs (two homospecific and either the male or the female of the first species in the pair) are calculated and tested in a 2×2 chi-square contingency table. (2) In zones where hybrids predominate, this is the Pearson correlation coefficient between male and female phenotype scores. Significant assortative mating is indicated whenever $P < 0.05$.

Fraction of birds with hybrid phenotypes. Based on a hybrid index, extremes are assigned to the pure form and the rest to hybrids. Superscripts[1,2,3] refer to method of assessment of the fraction of hybrid phenotypes. (1) Only localities in which at least one hybrid was recorded are included. The measure presented is ratio of the total number of hybrids to the total number of individuals summed across all these localities. (2) The measure is the maximum number in one locality inferred from a figure or table in a relevant publication. (3) The measure is among observed mated pairs at one or two localities.

Movement. Indicates any change in the position of the hybrid zone, when the zone has been surveyed in different years.

137. Rock Partridge and Red-legged Partridge. Zone runs ~150 km along southwestern French Alps to the coast (Randi and Bernard-Laurent 1999).

Habitat differences. Zone is associated with altitude.

Width. ~15 km according to plumage, but 70–160 km according to allozymes.

Genetic differentiation. 4.8% mitochondrial DNA (Randi 1996).

Assortative mating. Not known.

Fraction of birds with hybrid phenotypes. [2]86% based on plumage.

Movement. Not known.

315. California Quail and Gambel's Quail. Crest (also known as the topknot), genetic markers, morphology, and color are roughly concordant with each other, i.e., show similar clines (Gee 2004). Hybridization occurs along >400 km in southern California.

Habitat differences. Hybrid zone is across a sharp climatic gradient from drier and hotter to wetter and cooler. However, there are no differences in nesting ecology or habitat use where individuals of the two species co-occur.

Width. ~20 km.

Genetic differentiation. 2.2% mitochondrial DNA (Zink and Blackwell 1998).

Assortative mating. Random mating (Gee 2003).

Fraction of birds with hybrid phenotypes.[2]88% (Gee 2003).

Movement. Not observed in five years, but frequencies of plumage characteristics of the more mesic species (California Quail) in the center of the zone were positively correlated with winter rainfall (Gee 2004).

590. Red-breasted Sapsucker and Red-naped Sapsucker. Main zone is ~160 km long in northern California and southern Oregon.

Habitat differences. Western wetter form meets eastern drier form along an ecotone (Johnson and Johnson 1985).

Width. ~20 km. This is a zone of sympatry and hybridization set by patches of suitable habitat colonized by both species, and it is not known to be clinal within a patch.

Genetic differentiation. 0.1–0.7% mitochondrial-DNA restriction fragments depending on individuals compared (Cicero and Johnson 1995).

Assortative mating. In the area of hybridization where pure types are in approximately equal frequency, the ratio of the observed number to the expected number of pairings is 1.9 and significant (Johnson and Johnson 1985). Migratory differences result in displacement in timing of breeding.

Fraction of birds with hybrid phenotypes. [3]10%.

Movement. None known, but a mountain range was colonized by one species between 1910 and 1960, resulting in establishment of an additional hybrid zone. Little movement was apparent in 23 years in British Columbia (Scott et al. 1976).

602. Green-backed Woodpecker (two subspecies). Hybrid zone runs ~1500 km in Zaire, Rwanda, Burundi, and Angola (Prigogine 1987).

> *Habitat differences.* Second growth for one subspecies, more open habitats for the other.
>
> *Width.* ~300 km, based on limited samples in patchy habitat.
>
> *Genetic differentiation.* Not studied.
>
> *Assortative mating.* Not studied.
>
> *Fraction of birds with hybrid phenotypes.* [1]40% based on plumage in the Katombe region.
>
> *Movement.* Not studied.

681. Northern Flicker. Zone runs 4000 km from Texas to Alaska, shows curvature.

> *Habitat differences.* None noted in center of zone.
>
> *Width.* ~300 km in south (Moore and Price 1993). In northern U.S., may be more than twice as broad, and then narrows < 50 km as it turns westward [Figure 15.6; based on maps in Short (1965) and Moore and Price (1993)]. Width is correlated with the breadth of a vegetational ecotone and the width of the area in rain shadow from the Rockies (Moore and Price 1993). Morphological differences of about 8–10% in structural characters match color cline (Short 1965).
>
> *Genetic differentiation.* ~ 0%. Mitochondrial-DNA haplotypes show little association with geography in the region of the cline (Moore et al. 1991).
>
> *Assortative mating.* None in the central U.S. (Moore 1987). Some assortative mating ($r = 0.35$) between color indices in British Columbia, probably due to separation of breeding times (Wiebe 2000).
>
> *Fraction of birds with hybrid phenotypes.* 84% among mated pairs in center of zone (Moore 1987).
>
> *Movement.* Little, if any, movement over more than 100 years in central U.S. (Moore and Buchanan 1985).

929. Red-billed Hornbill. Currently considered subspecies but hybrids and heterosubspecific pairings have relatively low fitness; the length of the zone is unclear but > 50 km. Based on plumage and eye-color differences; however, morphomoetrics, calls and courtship displays also differ (Delport et al. 2004).

> *Habitat differences.* Not noted in center of zone, but differences present in allopatry.
>
> *Width.* 126 km based on eye-color and plumage.
>
> *Genetic differentiation.* Not known.
>
> *Assortative mating.* [1]Ratio of observed to expected number of pure pairings is 1.3 and significant; suggested to result from call differences.
>
> *Fraction of birds with hybrid phenotypes.* [3]27%.
>
> *Movement.* Not known.

2493. Vinaceous Dove and Ring-necked Dove. Based on song (plumage differences are slight). Zone runs about 10 km, perpendicular to a lake (P. M. den Hartog, *personal communication*).

> *Habitat differences.* None.
>
> *Width.* ~6 km according to vocalizations.
>
> *Genetic differentiation.* 3% mitochondrial DNA (HKY).
>
> *Assortative mating.* Not known.
>
> *Fraction of birds with hybrid phenotypes.* [2]100%.
>
> *Movement.* Not known.

3205. Glaucous-winged Gull and Western Gull. Zone runs along the Pacific coast from Oregon to British Columbia (Bell 1996); asymmetric, with introgression more towards the north than the south.

> *Habitat differences.* Vegetation in which nests are placed differs, and there are also foraging differences; these differences are also reflected in differences between the hybrids and parentals (Good et al. 2000).
>
> *Width.* 800 km, based on color patterns; 700 km based on microsatellite variation (Gay et al. 2006). Morphology and allozymes show shallower clines, but differences between parentals are small (Bell 1996).
>
> *Genetic differentiation.* 1.1% mitochondrial DNA (Gay et al. 2005).
>
> *Assortative mating.* Random mating in center of zone (Good et al. 2000).
>
> *Fraction of birds with hybrid phenotypes.* [2]Up to 77% (Bell 1996). Hybrids do as well or better in middle of zone, suggesting bounded hybrid superiority (Good et al. 2000).
>
> *Movement.* No directional movement, but width has greatly increased, and there may have been little hybridization 100 years ago.

4532. White-collared Manakin and Golden-collared Manakin. In Panama. Length of zone is only about 10 km, as it runs up against mountains and coast (Brumfield et al. 2001).

> *Habitat differences.* None noted. Forest composition is superficially similar along the transect.
>
> *Width.* For several molecular and morphological markers, as well as a secondary sexual trait (beard length), width is about 10 km. However, one allozyme locus has a width > 100 km. The color of the collar and belly has a width of about 4 km, but this cline is displaced 50 km west of the other markers (Figure 15.5; Brumfield et al. 2001).
>
> *Genetic differentiation.* Not known.
>
> *Assortative mating.* Mating is polygamous; at a lek, one male may obtain > 70% of all matings (Parsons et al. 1993), so assortative mating is unlikely (but difficult to con-

firm because females do not have the secondary sexual traits that form the basis of study).

Fraction of birds with hybrid phenotypes. [1]About 80%, based on genetic markers; at least 50% appear to be second generation (Brumfield et al. 2001). The fraction may be higher, as the plumage of most birds appears to be intermediate.

Movement. Apparently shifted eastward several km in 100 years, except for collar color, which has shifted northwestward (Brumfield et al. 2001).

4589. Variable Antshrike. In Bolivia. Although not completely sampled, there are two probable contact zones between three subspecies, each about 100 km (Brumfield 2005). Originally identified based on color patterns, shown also for song (Isler et al. 2005) and DNA (Brumfield 2005).

Habitat differences. Not studied. However, the hybridizing taxa differ in elevation.

Width. Based on DNA, widths are 95 km and 329 km, respectively, with large confidence limits (Brumfield 2005).

Genetic differentiation. 1.5–2.2% mitochondrial DNA (Brumfield 2005).

Assortative mating. Not studied.

Fraction of birds with hybrid phenotypes. Probably 100%, at least in the center of the broader zone. Populations in this location were originally classified as a separate subspecies.

Movement. Not studied.

5163. Variegated Fairywren. Length of zone >800 km in northwestern Australia (Ford and Johnstone 1991). Only females differ substantially between the subspecies, and width is based on female color characters.

Width. ~200 km.

Fraction of birds with hybrid phenotypes. [1]95% (based on crown and back color). Other characteristics not studied.

5590. Varied Sittella. Length of zone >1000 km. Queensland, Australia (Ford 1980; Short et al. 1983a, b). Several hybrid zones between different subspecies; one studied by Short et al. (1983b) ranges from coast to central Queensland ~250 km wide and ~600 km long. At its northwestern tip in central Queensland, this zone meets and intergrades with three other subspecies to produce great diversity (Ford 1980; Short et al. 1983a). *Fraction of birds with hybrid phenotypes.* [1]86% (Ford 1980). Other characteristics not studied.

5749. Hooded Crow and Carrion Crow. The range of the grey-backed Hooded Crow extends from eastern Europe through central Russia and is nested within the Carrion Crow, with whom it forms a hybrid zone to both the west and the east (Meise 1928). In the west, the zone extends from northern Italy north about 1200 km to the Baltic (Meise

1928; Haas and Brodin 2005) and is also present in Scotland. In the east, the zone extends about 3500 km north from east of the Caspian Sea almost to the Siberian coast.

Habitat differences. In winter in Italy, during pair formation, Carrion Crows selected meadows and Hooded Crows selected maize stubble; this corresponds with predominant habitats in allopatry (Saino 1992).

Width. ~20 km in Italy (Saino et al. 1992), 10–50 km in Scotland (Cook 1975). Appears wider in east (Meise 1928). Width is narrower in mountainous areas, i.e., northern Italy, western Scotland. In Italy, a cline in allozyme variants seems to be shallower than the cline in morphology (Saino et al. 1992), although differences between parental populations are small.

Genetic differentiation. 0% mitochondrial DNA. 336 bp of mitochondrial cytochrome *b* is virtually identical from Carrion Crows both to the west and to the east of the Hooded Crow, as well as from the Hooded Crow (Kryukov and Suzuki 2000).

Assortative mating. Ratio of observed to expected number of pure pairings based on random expectations was between 1.25 and 2.0 in an Alpine study in Italy, and significant (Saino and Villa 1992). Assortative mating is attributed to habitat choice along an altitudinal gradient and aggressive interactions in winter flocks, which tend to sort by subspecies even when they are found foraging in the same pasture.

Fraction of birds with hybrid phenotypes. In one locality where Carrion Crows made up 51% of the parentals, [3]12% (Italy; Saino and Villa 1992).

Movement. Moved north in eastern Scotland (approximately 80 km in 55 years), but not in western Scotland (Cook 1975). Moved north in eastern Denmark (19 km in 80 years), but less so in western Denmark (Haas and Brodin 2005), elsewhere not clear. Movement in Scotland attributed to generally lower altitudes in east, enabling the more aggressive Carrion Crow to push north as climate changed (Cook 1975), but this explanation deemed unlikely by Haas and Brodin (2005).

5819. Australian Magpie. In southeastern Australia, length ~600 km connecting a white-backed and black-backed form (Burton and Martin 1976).

Habitat differences. None (Hughes et al. 2002). Outside the zone, habitat differences were suggested to be related to visibility (white-backed form historically in more heavily forested habitat; Kallioinen et al. 1995).

Width. Two transects: 110 km and 220 km.

Genetic differentiation. 0% (Hughes et al. 2001).

Assortative mating. Random, although extragroup paternity may be high, and its possible contribution is yet to be assessed (Hughes et al. 2003).

Fraction of birds with hybrid phenotypes. [2]50–70% (Burton and Martin 1976).

Movement. None apparent over 20 years (comparison of Burton and Martin 1976; Hughes et al. 2002), but the degree of introgression of intermediates is increasing

(~60 km in ~60 years; Kallioinen et al. 1995), apparently in association with human habitat alteration.

5831. Black-faced Woodswallow. Length 1000 km in northern Queensland, Australia (Ford 1978). *Width* ~350 km. *Fraction of birds with hybrid phenotypes.* [1]63%, [2]80%. In allopatry, one is the dry-country form and one is the wet-country form.

5865. Timor Figbird. Queensland, Australia (Ford 1982). Length not recorded. *Width.* ~370 km along the coast. The hybrid zone is in dry country, and contact was originally perhaps in gallery forests along creeks. Morphology is roughly concordant with color pattern. *Fraction of birds with hybrid phenotypes* [1]81%. Thought to be random mating, but no direct evidence.

6476. Pied Flycatcher and Collared Flycatcher. Runs east–west through the Czech Republic for ~200 km; hybridization occurs outside this zone also (Figure 14.4, p. 309). In addition, the much-studied isolated populations of the Collared Flycatcher on the islands of Gotland and Öland in the Baltic Sea hybridize with the Pied Fycatcher (Alatalo et al. 1990, 1994; Veen et al. 2001). These hybridizing populations are thought to be a result of recent (~200 years) colonizations by the Collared Flycatcher (Alatalo et al. 1990).

Habitat differences. Collared Flycatchers are found in deciduous lowland and in more southerly regions; Pied Flycatchers are in mixed forests at higher elevations, extending to higher latitudes. In sympatry, Collared Flycatchers are socially dominant to Pied Flycatchers and may exclude them from preferred deciduous habitats (Alatalo et al. 1994).

Width. 20–25 km, based on map in Sætre et al. (1999).

Genetic differentiation. 3.9% DNA mitochondrial cytochrome *b*. Female hybrids are sterile, so there is no introgression of mitochondrial DNA. Sex chromosome (Z) also shows no introgression, probably because they carry infertility factors (three out of 11 males heterozygous at the sex chromosomes were probably sterile; Sætre et al. 2003). In a DNA analysis of 30 genes, 0.8% of sympatric Pied Flycatchers and 36% of Collared Flycatchers had at least one allele from the sympatric congener on hotland (Sætre et al. 2003), but introgression lower in central Europe (Borge et al. 2005).

Assortative mating. In the Czech Republic, where Collared Flycatchers make up 85% of the birds, the ratio of the observed to the expected number of pure-type pairings is 1.2 and significant (Veen et al. 2001). Thought to be a result of demonstrated female preferences based on color pattern (Sætre et al. 1997) and song, but habitat differences may also be involved.

Fraction of birds with hybrid phenotypes. 3% among mated pairs in one study site in the Czech Republic (Veen et al. 2001).

Movement. No evidence for large-scale movement. At various sites in the Czech Republic, the generally low frequencies of one or the other species have mostly fluctuated erratically in studies lasting several years (Sætre et al. 1999).

6555. Thrush Nightingale and Common Nightingale. In eastern Europe. Dimensions have been little studied.

Habitat differences. Habitats are more similar in sympatry than in allopatry, especially for the Common Nightingale (Sorjonen 1986). In sympatry in Poland, Common Nightingales occupied hillier areas.

Width. No study found.

Genetic differentiation. Not known.

Assortative mating. Ratio of observed to expected number of pure pairings based on random expectations is 1.3 and significant (Becker 1995). In eastern Germany, 89% of broods were between conspecifics, 6% were heterospecific matings, and 5% were backcrosses.

Fraction of birds with hybrid phenotypes. In eastern Germany, where the frequency of the Common Nightingale is 86.5%, 2.5% (five birds) were hybrids. All were males (Becker 1995).

Movement. In the last 40 years, the Thrush Nightingale's range has moved 100 km southwest. The Common Nightingale has also withdrawn (Sorjonen 1986), but to a lesser extent, so the zone of overlap has probably increased.

6666. Black-eared Wheatear and Pied Wheatear. Zone determined by pied plumages of males, particularly of back, throat, and sides of neck, which appear to segregate independently. They meet in four contact zones, the longest between the Black Sea and the Caspian Sea, and hybridization is known in all zones (Haffer 1977).

Habitat differences. None noted.

Width. About 400 km in northern Iran. One character, the presence of a black throat in the eastern species, occurs throughout the western species, where it is found in frequencies of >50%, although the patch maybe smaller, and Haffer (1977) attributes this to differential introgression of this character. The white throat has also penetrated into the range of the eastern species, but at a low frequency.

Genetic differentiation. Not known.

Assortative mating. Not studied. A few cases of experimental removal of one member of a pair resulted in replacement by the opposite species [original observations by Panov and Ivanitzky (1975); this account is based on Haffer (1977)].

Fraction of birds with hybrid phenotypes. [2]65%. No locality where both parental types coexist was studied, i.e., the remaining 35% belongs to just one of the parental species.

Movement. Not studied.

6882. Rufous-naped Wren. Two subspecies meet along a narrow band of coastal habitat ~10 km wide (Selander 1964, 1965) in Chiapas, southern Mexico. Morphology varies concordantly with color patterns; subspecies differ by about 10% in characters such as wing length. Song variability is higher in the center of the zone.

Habitat differences. Drier areas are occupied by one subspecies. However, limited habitat diversity in the center of the zone leads to no obvious habitat differences where they co-occur.

Width. ~30 km for color patterns.

Genetic differentiation. Not studied.

Assortative mating. Thought to be random in the center of the zone.

Fraction of birds with hybrid phenotypes. [2]90% in the center of the zone.

Movement. Little evidence for change over 25 years (Selander 1964, 1965).

6980. Carolina Chickadee and Black-capped Chickadee. The hybrid zone was originally identified based on songs, many of which are mixed in the hybrid zone. (Color and morphological differences between the species are small but measurable.) Ranges are more or less contiguous along ~2500 km from southern Kansas to the east coast in the U.S. (Black-capped Chickadee to the north), but they are often separated by geographical gaps, so the two species do not meet. Hybrids are formed where the ranges overlap. The two species are not each other's closest relatives.

Habitat differences. Very little (Brewer 1963). The zone runs parallels to the interface between forested plateau and plains in Missouri, but the center of the zone is displaced from the ecotone by about 5 km.

Width. ~30 km in Illinois, based on song (Brewer 1963). 15 km along a stretch 120 km long in SW Missouri, based on song (Robbins et al. 1986). 15–30 km across one zone in the Appalachians based on a cline of diagnostic nuclear markers (Sattler and Braun 2000; Bronson et al. 2005). Width is correlated with the steepness of the altitudinal gradient. Cline in morphology is generally coincident with these clines. One autosomal locus tested has introgressed much further than the two sex-linked loci and the mitochondrial locus (Sattler and Braun 2000).

Genetic differentiation. 8% mitochondrial DNA (HKY). For mitochondrial restriction sites, 3.4% is the estimate.

Assortative mating. Based on morphology, it is thought that mating is random in Missouri (Robbins et al. 1986).

Fraction of birds with hybrid phenotypes. In one area in the Appalachians with about 66% Black-capped alleles, > 16% of the birds were potentially F_1, and 58% carried at least one allele from both species based on diagnostic markers at four loci. In the center of the zone, hybrids have about 50% the fledging success of the parental forms. Reproductive success is positively correlated with the genetic similarity of the pair, and lower if the male is genetically intermediate (Bronson et al. 2005). In a controlled experiment in which parental forms were moved into the zone (Bronson et al. 2003a), pairs in the hybrid zone had 50% the reproductive success of the parental forms.

Movement. In Ohio, where the zone runs east to west, the southern boundary of the Black-capped Chickadee range moved 100 km north in 100 years, where it has been replaced by the Carolina Chickadee (Brewer 1963; Bronson et al. 2003a, b).

7026. Tufted Titmouse and Black-crested Titmouse. These were merged into one species in 1983 (Grubb and Pravosudov 1994), but they are now considered as two species. Zone ~500 km long through mid-Texas from near the coast (Braun et al. 1984), but contact is limited to five areas where woodlands come together (Dixon 1955).

Habitat differences. None obvious (Dixon 1955), but there is no known locality at the present time where both parental species co-occur (Dixon 1990). Dixon suggests climatic differences distinguish the forms, and suggests bounded hybrid superiority.

Width. 100–150 km (Dixon 1990).

Genetic differentiation. 0.4%, based on mitochrondrial-DNA restriction sites (Klicka and Zink 1997).

Assortative mating. Thought unlikely.

Fraction of birds with hybrid phenotypes. >85% in some localities (Dixon 1955, p. 177).

Movement. Apparently no movement or change in width over 100 years (Dixon 1990), although there may have been small fluctuations (Dixon 1955).

7149. Garden Bulbul and Red-eyed Bulbul. Based on color of eye-wattle (bare skin patch round the eye). Zone is at least 80 km long, but it may be more in South Africa (Lloyd et al. 1997). Weight and beak measures vary concordantly with the color cline.

Habitat differences. The species occupy different habitats and meet across an ecotone associated with a mountain barrier.

Width. 10–20 km.

Genetic differentiation. ~5%, based on mitochondrial-DNA restriction sites.

Assortative mating. Not studied.

Fraction of birds with hybrid phenotypes. 70% in the center of the zone.

Movement. The position of the zone shifted 5 km between censuses in a wet and a dry year (separated by 4 years). All hybrids ($N = 8$) had mitochondrial-DNA from one parental type, suggesting one-way introgression.

7156. White-eared Bulbul and Red-vented Bulbul. These two species overlap over 1000 km in northwestern India and Pakistan with occasional hybridization, but, in northern Pakistan, near the range limits of both species, there is a hybrid zone extending over perhaps 200 km (Roberts 1992, p. 85). A second species, the Himalayan Bulbul, is often considered a subspecies of the White-eared Bulbul. It extends throughout the Himalayas, where it has an overlapping distribution with the Red-vented Bulbul (2500 km), and it probably also hybridizes in northern Pakistan.

Habitat differences. Little studied. Hybrids may be in wetter areas than the White-eared Bulbul (the Red-vented Bulbul has spread in association with irrigation and cultivation, replacing the more dry-adapted White-eared Bulbul in many areas) and more insectivorous than the Red-vented Bulbul. Other features not studied.

7590. Melodious Warbler and Icterine Warbler. Based on song and morphology (Icterine Warbler has a wing about 15% longer); plumages are very similar. Runs ~400 km through northeastern France to the English Channel (Faivre et al. 1999; Secondi et al. 2003).

Habitat differences. Icterine Warbler in dense high bushes, Melodious Warbler in canopy in more open areas (Secondi et al. 2003).

Width. The zone of overlap is about 130 km, but the frequency of hybrids along the transect is unclear (Secondi et al. 2003).

Genetic differentiation. 3.7% mitochondrial DNA (HKY; Helbig and Seibold 1999).

Assortative mating. Hybridization has increased as the Icterine Warbler has declined and the Melodious Warbler has spread northeast. In the 1990s, in eastern Burgundy, 12% ($N = 66$) of pairs were heterospecific. The ratio of observed to expected among conspecific pairs is 1.7 and significant.

Fraction of birds with hybrid phenotypes. In eastern Burgundy, among 384 birds in which the ratio of Melodious Warbler to Icterine Warbler was 5.5, 13 birds (3.5%) were intermediate and likely hybrids.

Movement. The zone has moved northeast about 110 km in 55 years (Secondi et al. 2003). In eastern Burgundy, the ratio has shifted from 1:1 to 5.5:1 in favor of the Melodious Warbler over 20 years (Faivre et al. 1999). Although there are signs of morphological introgression from Melodious Warbler into Icterine Warbler, there is no Icterine Warbler mitochondrial DNA found in Melodious Warbler populations in allopatry (Secondi et al. 2003).

7656. Willow Warbler. Based on morphology, especially measures of size (Bensch et al. 1999). Plumage color (brown-green) varies concordantly with size. Length ~300 km across central Sweden and Norway. The hybrid zone may be set partly by selection against hybrids, as birds from the north migrate south and ones from the south migrate southwest.

Habitat differences. Not studied, but if there are differences they are not large.

Width. 350 km.

Genetic differentiation. 0% in mitochondrial DNA and microsatellites.

Assortative mating. Not known. However, southern birds may arrive on their breeding grounds two weeks earlier than northern birds. There is an association between color and morphology within the hybrid zone that may result from assortative mating and/or dispersal into the zone.

Fraction of birds with hybrid phenotypes. Not known. In a measure of size, the average difference between parentals is only about 0.8 standard deviations of the populations in the center, implying that most phenotypes would be consistent either with coming from one or other parent or with being hybrids.

Movement. Not studied.

7657. Common Chiffchaff and Iberian Chiffchaff. Based on song and call notes, which are distinct in the two species; plumage patterns are very similar. Hybrid zone stretches from the coast of the western Pyrenees ~70 km inland (Salomon 1987; Salomon et al. 1997).

Habitat differences. Not noted.

Width. ~20 km based on song (Salomon 1987; Salomon et al. 1997). Among 94 birds in the hybrid zone, only five had a mismatch of DNA with song or call, implying low gene flow through the female line. Nuclear DNA (microsatellites and AFLP markers) is more similar across the hybrid zone, suggesting gene flow through the male line (Helbig et al. 2001; Bensch et al. 2002).

Genetic differentiation. 6.2% mitochondrial DNA [HKY, and hence differs from Helbig et al. (2001)]. There is little variation within each species (Helbig et al. 2001).

Assortative mating. In the center of the zone, where pure types are in equal frequency, the ratio of the observed to the expected number of pairings is 1.4 and significant (Helbig et al. 2001).

Fraction of birds with hybrid phenotypes. Based on genetic (AFLP) markers, about 10% of the individuals are hybrids (F_1 and first-generation backcrosses) (Bensch et al. 2002). A crude estimate of F_1 hybrid fitness (based on adult frequencies) is about 60% the fitness of the parentals (Helbig et al. 2001).

Movement. Moved southwest about 20 km in 10 years (comparison of maps in Salomon 1987; Salomon et al. 1997). Zone also may have broadened slightly (Helbig et al. 2001; Salomon et al. 1997), but this is not clear from the maps.

8321. House Sparrow and Italian (Willow) Sparrow. Zone wraps around the Alps from the Mediterranean to the Adriatic, ~600 km, meeting in the high-altitude passes (Summers-Smith 1988, pp. 121–126).

Habitat differences. None.

Width. ~30 km.

Genetic differentiation. 8.0% mitochondrial DNA (HKY).

Assortative mating. Not studied.

Fraction of birds with hybrid phenotypes. [1]40% (Lockley 1992).

Movement. No major movement is thought to have occurred over 50+ years. Fluctuations of about 10 km over seven years in one study in the west (Lockley 1996) is attributed to human development. The Italian Sparrow is probably itself a result of hybridization between the House Sparrow and the Willow Sparrow, suggested to date to perhaps 3600 years ago (Johnston 1969).

8636. Black-throated Finch. Length ~120 km, Queensland, Australia (Ford 1986; several other hybrid zones are described in lesser detail in this paper).

Width ~70 km.

Fraction of birds with hybrid phenotypes. [1]63%.

Other categories not studied.

8882. Yellowhammer and Pine Bunting. Broad area of overlap and hybridization (Panov et al. 2003), runs east–west ~2000 km and north–south ~1000 km in central Siberia.

Habitat differences. Yellowhammer more in scrub and steppe and Pine Bunting more in light forest with conifers. There are slight morphological differences.

Genetic differentiation. 0% in mitochondrial DNA, but, because they are phenotypically so different, this may be a result of introgression (U. Olsson and A. Roubtsov, *personal communication*). Allopatric populations differ significantly in nuclear markers (A. Roubtsov, *personal communication*).

Assortative mating. Not studied.

Fraction of birds with hybrid phenotypes. About 50% in some populations. No populations with both parental phenotypes are known, but some nearby populations can contain one or the other.

Movement. The Yellowhammer is moving east in association with human-altered landscape and, to a lesser extent, the Pine Bunting is moving west. Hybridization is increasing.

8981. Collared Towhee and Spotted Towhee. The cline described here is along a transect from west to east across several more-or-less separated plateaus in southern Mexico (Sibley 1950, p. 173). Mean "length" is 50 km (width of a plateau). Sibley (1950) terms this an "elongate hybrid zone" because introgression is along a series of plateaus, which are relatively narrow.

Habitat differences in sympatry. None.

Width. Intergradation is over 450 km, but there are gaps separating different plateaus.

Genetic differentiation. 8.9% mitochondrial DNA (HKY).

Assortative mating. Not studied.

Fraction of birds with hybrid phenotypes. [2]100%. Elsewhere in Mexico, two isolated mountaintop populations are hybrid swarms. At other localities, they are altitudinally segregated and either overlap without interbreeding or overlap with very limited hybridization.

Movement.   Not studied.

8982. Eastern and Spotted Towhee. Based on plumage. Considered subspecies in Sibley and Monroe (1990), but now given species status. The two taxa meet in the Great Plains of the U.S. They probably come into contact, or are at least close to each other, over 1000 km., from Kansas to Manitoba, but substantial hybridization has only been demonstrated along the Platte River in Nebraska (Sibley and West 1959).

Habitat differences. Drier–wetter.

Width. Along the Platte river, Nebraska, zone width is ~150 km.

Genetic differentiation. Not studied.

Assortative mating. Not studied.

Fraction of birds with hybrid phenotypes. [2]80%.

Movement. Not studied.

9037. Blue-winged Warbler and Golden-winged Warbler. The species overlap and hybridize where they meet over a large area of the eastern U.S.

Habitat differences. Golden-winged Warblers occupy earlier successional habitat than Blue-winged Warblers, but territories often overlap.

Width. Overlap and hybridization may be 500 km in some directions.

Genetic differentiation. 3.5 % mitochondrial DNA (HKY; I. Lovette, *personal communication*).

Assortative mating. Frequency of hybridization is 5–10% where both species overlap. For pure types, the ratio of observed to expected was 1.6 and significant in northern New York state, where Golden-winged Warblers are about three times as common as Blue-winged Warblers (Confer and Larkin 1998).

Fraction of birds with hybrid phenotypes. [3]13% at one locality (Confer and Larkin 1998); below 10% at others (Gill et al. 2001).

Movement. Over the past 100+ years, Blue-winged Warblers have been extending their range northward and westward. They seem to replace Golden-winged Warblers completely in any one locality within 50 years after they first arrive (Gill 1980), but the time may be as short as four to five years (Gill 2004). This may be partly a response to different stages of vegetational succession. Direct evidence for competitive interference is weak (Confer and Larkin 1998), but Golden-winged Warblers have reduced reproductive success in the presence of unmated male Blue-winged Warblers (Confer et al. 2003; Gill 2004).

9054. Yellow-rumped Warbler. Zone runs from Alberta to southeastern Alaska, ~1000 km (Hubbard 1969). Two well-differentiated forms meet along a line where there is much unforested habitat and contact is limited to about 15% of the total length. Morphology (wing length) and allozymes are roughly concordant with plumage patterns (Hubbard 1969). Call notes differ between the taxa; in the area of overlap, some birds appear to have intermediate calls.

Habitat differences. Not studied; however, the species occupy northern boreal and western cordilleran forest and they co-occur where these meet.

Width. Based on allozymes, 150 km (Barrowclough 1980). Some aspects of the color pattern introgress further into the parental populations than others.

Genetic differentiation. 0.2% mitochondrial-DNA (Milá et al. 2007b).

Assortative mating. Not studied.

Fraction of birds with hybrid phenotypes. 100% in center of zone.

Movement. Not studied.

9056. Hermit Warbler and Townsend's Warbler. Three zones in separate mountain ranges of the Pacific Northwest of the U.S., where Townsend's Warbler lies to the north and east of the Hermit Warbler. The two best studied are ~100 km and 150 km long (Rohwer and Wood 1998; see Figure 15.1).

Habitat differences. None noted.

Width. ~125 km for color patterns. The species differ in both black (melanin) and yellow (carotenoid) characters, and the black characters have a slightly wider cline than the yellow ones (Rohwer et al. 2001). The mitochondrial DNA cline is much broader. Hermit Warbler DNA is found at high frequency in Townsend's Warblers up to 2000 km north of the hybrid zone, although Townsend's Warbler DNA does not penetrate very far south.

Genetic differentiation. 0.9% mitochondrial DNA.

Assortative mating. For parental types, the ratio of the observed number to the expected number of pairings is 1.7 and significant (Pearson 2000).

Fraction of birds with hybrid phenotypes. [3]26% (Pearson 2000).

Movement. Movement not directly observed, but inferred to be extensive since the last glaciation, with Townsend's Warblers displacing Hermit Warblers as they move south (Rohwer et al. 2001). This is based on: (1) the asymmetry of the shape of the cline, with few Hermit Warbler plumage characteristics extending north or east of the hybrid zone, whereas Townsend's Warbler plumage characters extend >100 km south or west; (2) the presence of Hermit Warbler DNA far to the north (movement south is attributed to male aggression, and males do not carry the mitochondrial DNA).

9569. Rose-breasted Grosbeak and Black-headed Grosbeak. About 900 km from Kansas to southern Canada in the Great Plains.

Habitat differences. Thought to be slight (West 1962).

Width. 200–300 km in South Dakota and Nebraska. Hybridization rare in North Dakota.

Genetic differentiation. 4.4% (uncorrected DNA; Klicka and Zink 1997).

Assortative mating. Based on five pairs, West (1962) thought there might be assortment.

Fraction of birds with hybrid phenotypes. [1]40%, [2]58%. (West 1962).

Movement. Position roughly stable over 20 years, although the zone may have broadened.

9600. Indigo Bunting and Lazuli Bunting Songs and morphology vary roughly concordantly with color (Kroodsma 1975; Emlen et al. 1975). ~800 km long in North Dakota, South Dakota, and Nebraska.

Habitat differences. Small, if any. The two species defend interspecific territories (Emlen et al. 1975).

Width. ~400 km in North Dakota (Kroodsma 1975).

Genetic differentiation. 6.6% mitochondrial DNA. Allozymes differ eight times more between allopatric populations ~1000 km apart than between sympatric ones (Baker and Johnson 1998).

Assortative mating. Interspecific crosses about 10% in Wyoming. Considering the pure types only (and the fraction of Lazuli is 70%), ratio of observed Lazulis mated conspecifically to expected under random mating = 1.4 and significant (Baker and Boylan 1999).

Fraction of birds with hybrid phenotypes. In three areas in which the Lazuli Bunting was two to three times as common as the Indigo Bunting, the frequency of males with hybrid plumage was 13% (in Nebraska; Emlen et al. 1975), 33% (in Wyoming; Baker and Boylan 1999), and 44% (in North Dakota; Kroodsma 1975).

Movement. Indigo Buntings spread west 200 km in 15 years in northern Nebraska and appear to be replacing the Lazuli Bunting (Emlen et al. 1975). The cline appears to be asymmetric, with more introgression from the Indigo Bunting into the Lazuli Bunting than the converse (Kroodsma 1975).

9642. Baltimore Oriole and Bullock's Oriole. Length ~1800 km (Oklahoma to Alberta, Canada).

Habitat differences. None noted at one site in Colorado (Edinger 1985).

Width. ~150 km in Kansas (Rising 1983; Allen 2002). Cline similar for mitochondrial DNA, much shallower for some nuclear markers (Allen 2002).

Genetic differentiation. 6.4% mitochondrial DNA (HKY).

Assortative mating. Among adult-plumaged birds, there was strong assortative mating in a Colorado study (Edinger 1985). Following Allen (2002, her chapter 4), index scores of 0–0.15 and 0.85–1 are considered parental, and among these, the ratio of observed to expected pairings is [1]1.8 and highly significant ($N = 40$ pairs). Spearman's rank correlation of index scores is $r = 0.54$ ($N = 80$ pairs). In subadults, assortative mating appears less strong than in birds in full adult plumage, and 18 mixed pairs were all Baltimore Oriole males × Bullock's Oriole females. Assortment is not attributable either to habitat selection or to arrival date (which is similar for the two species, Edinger 1985).

Fraction of birds with hybrid phenotypes. In the center of the zone in Kansas, [3]70% (see Rising 1996). In a Colorado study, [3]52% (Edinger 1985). Despite a large number of birds with intermediate hybrid scores, distribution is strongly bimodal, and only 12% of all birds lie in the middle third of the distribution.

Movement. In Nebraska, there is some evidence that the zone moved west 200 km in 20 years (Rising 1983). In southern Kansas, the zone has probably remained stable for more than 30 years (Rising 1996).

9691. Common Grackle. Two subspecies. The zone between them extends from Louisiana to New England in U.S., about 1300 km (Huntington 1952).

Habitat differences. The parental forms differ in Louisiana. One subspecies occupies drier (pine) areas and the other marshes (cypress–tupelo gum). The marsh subspecies has a larger bill.

Width. 25–60 km in Louisiana, depending on interdigitation of preferred habitats. Width may be more than 300 km in north. In Louisiana a cline in morphology is broader, associated with latitude, and changes in a more or less continuous manner both inside and outside the hybrid zone (Huntington 1952; Yang and Selander 1968).

Genetic differentiation. Not known.

Assortative mating. Not known. Random mating is suspected within mixed colonies (Huntingdon 1952), but it is possible that habitat assortment occurs even over small spatial scales (Yang and Selander 1968).

Fraction of birds with hybrid phenotypes. [2] >90% (Yang and Selander 1968).

Movement. In Louisiana, moved northwards ~30 km in 30 years, without change in width. Attributed to ameliorating climate.

CHAPTER SIXTEEN

Genetic Incompatibility

Take its jeans down.

—M. Ashburner, when asked how to identify the sex of a fly

Postmating isolation is an essential part of many speciation events. Without it, hybridization leads to the merging of two species back into one. Natural and sexual selection against hybrids contribute to postmating isolation, but these causes are susceptible to breakdown when environments change. A final, absolute barrier to gene flow comes either when hybrids are not produced even though crossmating occurs, or, if hybrids are produced, they die before they mate, or they are sterile. In this chapter, I assess the rate at which such intrinsic loss of fitness develops. My main goal is to ask if intrinsic loss of fitness is associated with speciation itself or builds up after speciation is completed. In the latter case, it serves as an additional barrier to gene flow, but is not involved directly in the production of new species. A subsidiary goal is to assess possible contributions of chromosomal evolution to the development of genetic incompatibilities.

The loss of hybrid viability and fertility is somewhat easier to study than ecological and social causes of lowered hybrid fitness because crosses can be conducted in captivity. Rigorous, controlled experiments have been done in fruit flies (*Drosophila*, Coyne and Orr 1989, 1997; Turelli and Orr 2000). In birds, amateur breeders have conducted most hybridization experiments (Gray 1958; McCarthy 2006). Because the data for birds are limited and their quality variable, I highlight relevant results from *Drosophila*. The chapter is in three parts. First, I consider the theoretical background, then the empirical literature on the rate at which hybrid fitness is lost, and, finally, chromosomal evolution.

BACKGROUND

Postmating isolation can arise either because sperm from one species fail to fertilize the egg of the other (postmating, prezygotic isolation), or because hybrids have low fitness (postzygotic isolation). The low fitness of hybrids forms the containing the mother's W chromosome will develop into daughters (having received the Z from the father), and all of the eggs containing the mother's Z chromosome will develop into sons (again having received a Z from the father). This means that any mutation on the W chromosome that causes the mother to invest most in those eggs that will develop into daughters is favored, even if this causes mothers to invest less in sons; the W chromosome from the mother is never present in a male, so the fitness of the W chromosome depends only on the fitness of females. Miller et al. (2006) show that any mutation on the Z chromosome that causes the mother to invest more in sons (even at the expense of daughters) will also be favored. Such mutations result in a sex ratio skew, which may be countered by mutations arising elsewhere in the genome.

These various forms of sexual and genetic conflicts can potentially lead to rapid genetic divergence between populations, especially on the sex chromosomes. Interactions between genes that have evolved through such mechanisms have often been postulated to cause a reduction in hybrid fitness (Frank 1991; Hurst and Pomiankowski 1991; Hurst et al. 1996; Rice 1998; Tao and Hartl 2003).

Genetic incompatibilities may arise following multiple amino acid substitutions, but most substitutions appear to have little effect on hybrid fitness. In *Drosophila,* Orr and Turelli (2001) estimated a probability of only about one in ten million that two amino acid differences between a pair of species, when brought together in an individual, cause complete sterility of that individual. Orr and Turelli's estimate comes with a number of assumptions, especially that every interaction causes complete rather than partial sterility, and that the interaction is due to recessive effects, that is, it is not expressed in hybrids of the form *AaBb*, where any deleterious effects of the *a*–*B* interaction are masked by the *A* and *b* alleles respectively, but only in later generations in genotypes of the form *aaBB* (as described later in this chapter, evidence from *Drosophila* indicates that such recessive interactions are the most common, and the first to arise during population divergence). The implication is that several thousand mutations need to be substituted in each line before there is a reasonable chance that a genetic interaction leads to sterility. Presgraves (2003) estimated that an even smaller fraction of all substitutions produce hybrid inviability. These results imply that a prolonged period of geographical isolation should be the most favorable for the development of incompatibilities, because it gives the opportunity for divergence at many loci. Whenever incipient species come together and hybridize, gene flow across the species border reduces the possibility of incompatibilities developing.

The spread of unconditionally favored alleles throughout a species range can be delayed or prevented even in the presence of hybridization. This happens if an unconditionally favored allele is closely linked (i.e., nearby on the same chromosome) to another allele that is favored only in part of the species range (e.g., because of special features of the environment in this part of the range). Anything that prevents or reduces recombination between the two genes will assist in restricting the unconditionally favored allele to the same part of the species range as the locally favored allele, and hence will accelerate the build-up of genetic incompatibilities. Recombination is often suppressed by chromosomal rearrangements, especially inversions (an inversion results from a removal of a chromosome segment, its rotation through 180°, and insertion back in the same position). Such rearrangements may play an important part in facilitating the development of genetic incompatibilities in the face of hybridization (Noor et al. 2001; Rieseberg 2001; Navarro and Barton 2003). Noor et al. (2001) showed that genes affecting hybrid sterility between two closely related species of *Drosophila* mapped to the chromosomal-inversion differences between them.

Given this background, I now review the rate at which genetic incompatibilities accumulate in birds, and then the potential role of chromosomal rearrangements. Nothing is known about the genes involved, so we are confined to documenting timescales over which various forms of hybrid unfitness and chromosomal rearrangements arise.

ACCUMULATION OF INCOMPATIBILITIES IN BIRDS

Postmating prezygotic isolation

Most of this chapter is concerned with loss of fitness of hybrids produced as a result of matings between species, but it is probable that genetic differences between species also reduce the chances of successful fertilization. T. R. Birkhead has drawn attention to research in poultry breeding, which shows that, in many cases, less sperm reaches the egg after a heterospecific mating than after a conspecific one (reviewed by Birkhead and Brillard 2007). Following fertilization, sperm travels through the vagina; even in conspecific matings much of the sperm does not make it, but most of that which does enters the so-called sperm storage tubules, from which it is released over a period of days and transported passively to the point of fertilization. Because the vagina is a major barrier in conspecific crosses, this barrier seems likely to be generally important when reduced numbers of sperm reach the egg in heterospecific crosses (Birkhead and Brillard 2007). Steele and Wishart (1992) found that artificial insemination of turkey sperm into the vagina of female chickens results in no fertilization, but insemination at the level of the sperm storage tubules was successful, confirming that the difficulty of traversing the vagina is a critical barrier in this particular cross.

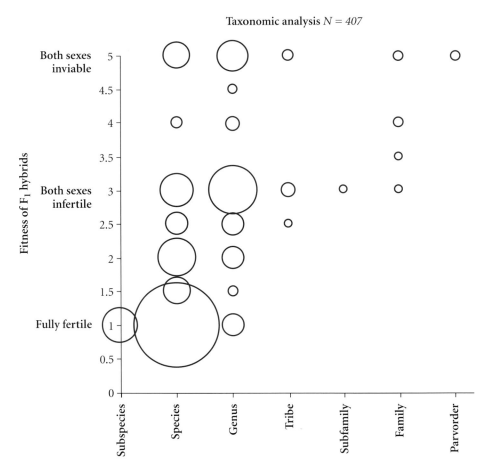

Figure 16.2 Plot of fitness index against taxonomic categorization of cross (from Price and Bouvier 2002). The area of the circle reflects the number of observations. The smallest circle represents a sample size of one. Fitness index is as follows: 1–fertile; 1.5–one sex fertile, the other sex some individuals recorded as fertile; 2– one sex fertile, the other sex viable but infertile; 2.5–one sex sometimes fertile, the other sex viable but infertile; 3–both sexes viable but infertile; 3.5–one sex viable and infertile, the other sex sometimes present in reduced numbers; 4–one sex viable and infertile, the other sex missing; 4.5–some individuals of one sex present, the other sex missing; 5–hybrids not present, both sexes missing.

contributing more to the initial build-up of postmating isolation and speciation, and other components develop only after postmating isolation is more or less complete. The rate of development of postmating, prezygotic isolation has not been assessed, but the rate of loss of different forms of hybrid fitness has been, as described in the following sections.

Fitness loss of hybrids can occur in four ways. Either the male hybrid or the female hybrid may have either reduced fertility or reduced viability. In fact, the order of loss of these four components seems to be fairly predictable:

1. Hybrids of both sexes are produced, but the females are infertile.
2. Hybrids are produced, but both sexes are infertile.
3. Hybrids are only males and these males are infertile.
4. Crosses produce no offspring at all.

There seem to be few exceptions to this pattern. First, Haldane's rule states that, among hybrids, if one sex has lower fitness than the other, it is the heterogametic sex. The rule is perhaps the strongest pattern in speciation (Laurie 1997; Coyne and Orr 2004, pp. 284–299). In birds, the female is the heterogametic sex and Haldane's rule is strongly supported. Thus, with respect to fertility, in 72 crosses in which both hybrid sexes were present, male hybrids had higher fertility than the female hybrids, and in just three crosses (with very low sample sizes), the female had higher fertility than the male. With respect to viability, in 15 crosses, the male had higher viability than the female with no exceptions. One possible complication is that these hybrids have not all been checked by dissection. At least in the Galliformes (chickens and pheasants), females with degenerate ovaries sometimes take on male plumage (Smith and Thomas 1913), but the converse is very rare. This probably only applies to a small fraction of females. For example, in a study of hybridization between Reeves' and Common Pheasants in which the gonads were checked, Haldane's rule was still strongly supported (Smith and Thomas 1913).

The analysis also suggests that, within females, hybrid fertility is lost before hybrid viability. This is not so easy to assess, because it is impossible to ask whether inviable offspring would be fertile if they grew up. However, Wu (1992) suggested that if fertility and viability are lost at the same rate, and we restrict our attention to crosses that produce fertile males but low-fitness females, we should see about the same number of crosses that produce infertile females as crosses that produce inviable females. In fact, the ratio is 72:3. This is a large difference, but it comes with three caveats. First, it is not always easy to distinguish infertile females from a failure of females and males to mate, or the sperm to inseminate the egg. Second, it is not easy to tell if the apparent infertility of hybrids is a result of early inviability of their offspring rather than true infertility. The vast majority of the infertility records listed by Gray (1958) are accompanied by simple statements, such as "the hybrids are sterile," without further comment. Third, a low viability of females may not be recorded. By this I mean that the presence of relatively few hybrid females is not remarked upon, but a low fertility of females relative to that of males is noted. Despite these qualifiers, the very large difference in the number of infertilily rather than inviability cases suggests that fertility is indeed commonly lost before viability, as has been found in *Drosophila* and other animals (Wu and Davis 1993; Rice and Chippindale 2002).

Finally, it appears that hybrid male fertility is lost more quickly than hybrid female viability. We found 100 cases in which both the male hybrids and the female hybrids were present but with apparently low fertility, but only three cases in which male hybrids were fertile and female hybrids were missing (Price and Bouvier 2002). This is a large difference, but the conclusion that hybrid male fertility is lost more quickly than hybrid female viability remains somewhat controversial (Lijtmaer et al. 2003). First, even if both sexes are present and both have reduced fertility, females may be rare, even though this is not recorded. If this is the case, both female viability and female fertility have been reduced (Smith and Thomas 1913; Gray 1958). Second, as noted in the previous paragraph, it is difficult to distinguish infertility from other causes that lead to an absence of offspring in crosses (i.e., failure to mate, failure of offspring to survive); as described in the next section, Lijtmaer et al. (2003) attributed presumed cases of low "fertility" in F_1 individuals to low viability of their offspring. However, sometimes the gonads have been checked, confirming low fertility of both sexes. For example, among hybrids between the Muscovy Duck and the Mallard, 17 females and 28 males were all confirmed to be sterile (Mott et al. 1968).

The time to loss of hybrid fitness can be roughly calibrated using mitochondrial-DNA distances (Figure 16.3). One of the more interesting results is the appearance of a rough "speciation clock" (Turelli and Orr 2000): hybrid unfitness accumulates steadily with time. [The speciation clock has been renamed the "intrinsic loss of hybrid fitness clock" by Bolnick and Near (2005), which, while closer to the truth, is not such a catchy phrase.] The clock holds both across and within groups. For example, it is seen within the ducks (Tubaro and Lijtmaer 2002). It is also seen when ducks in the genus *Anas* are compared with the pheasants (Figure 16.3); the *Anas* ducks in the dataset are young and have relatively high hybrid fitness, whereas the pheasants are older and have lower hybrid fitness (Price and Bouvier 2002).

Species crosses that result in more or less complete hybrid infertility (scores of 2.5 and 3 in Figure 16.3) have an average sequence divergence of 14% ($N = $ 25). This suggests that species that produce completely infertile hybrids last shared a common ancestor on average about 7 million years ago. The lowest value is at 1 million years (for the Gadwall crossed with the Falcated Duck; this needs further verification) and the next lowest value is at 2.5 million years. Times to loss of fertility are, of course, upper bounds (fertility may have been

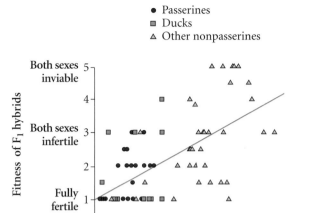

Figure 16.3 Plot of fitness index against time a pair of species have been separated (from Price and Bouvier 2002, with a few updates). Points are based on HKY-Γ mitochondrial distances (Γ = 0.2, see Appendix 2.2, p. 38), converted to time using the standard divergence rate of 2% per million years. For the fitness-index classification, see Figure 16.2. Regression line (forced through the point 0,1) is $y = 0.465x$ ($r = 0.69$, $N = 107$). "Ducks" refers to the genus *Anas* and close relatives only (Johnson and Sorenson 1999). Most nonpasserines, apart from ducks, are galliform birds (sequence data from Kimball et al. 1999), lovebirds (*Agapornis* spp.; sequence data from Eberhard 1998), and doves in the genus *Streptopelia* (sequence data from Johnson et al. 2001).

lost earlier), but they are often much longer than the times to speciation documented in Chapter 2 (i.e., from 0.6 million years upwards for ecologically similar, geographically separated species, and even less for young species in adaptive radiations).

The average distance among species that produce no hybrids is 25% ($N = 7$). This is about 12.5 million years ago, based on the molecular-clock calibration, and more distant than the loss of fertility. The shortest distance is 21%, more than 10 times longer than the shortest distance to loss of fertility. The relatively large distance between species that produce inviable offspring fits the earlier conclusion that hybrid fertility is lost before hybrid viability. Viability is lost extremely slowly in some cases (Prager and Wilson 1975). The most distant crosses resulting in any offspring are those between the Helmeted Guineafowl (from Africa) and the domestic chicken (originally from Southeast Asia) and the Helmeted Guineafowl and the Indian Peafowl. The Helmeted Guineafowl is estimated to have last shared a common ancestor with the chicken at least 30 million years ago (Mayr 2005), and perhaps more than 40 million years ago (Cooper and Penny 1997; Dimcheff et al. 2002). When guineafowl hens are artificially inseminated with chicken sperm, up to 70% of the eggs may hatch. The majority of hybrids that survive to adulthood are males, which are infertile (McCarthy 2006, pp. 51–52). Apparently, natural hybridization occurs in Africa, where introduced chickens have gone feral.

A shortcoming of the previous analysis is that it is restricted to F_1 hybrids. Lijtmaer et al. (2003) present a quantitative study of pigeons and doves based on careful experiments by Whitman (reported in Riddle 1919, pp. 31–40). This dataset only partially overlaps that analyzed by Price and Bouvier (2002), and it is important because Lijtmaer et al. (2003) investigated crosses among the hybrids, as well as backcrosses to the parental species. They studied the proportion of unhatched eggs in species crosses and assumed that this was a direct measure of hybrid inviability. As in other datasets, F_1 hybrid inviability increases with the length of time the species in the cross have been separated (Figure 16.4). Backcrosses show a (nonsignificant) tendency towards having lower hatching success than the F_1 hybrids (Lijtmaer et al. 2003). Crosses among the F_1 hybrids have significantly lower hatching success than the crosses

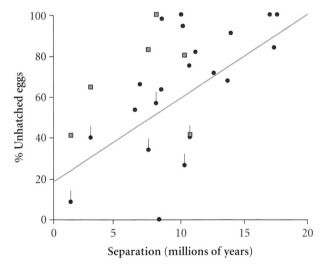

Figure 16.4 Plot of percentage of unhatched eggs in crosses of various species of doves against the genetic distance between them, after Lijtmaer et al. (2003). For this plot, I recomputed the genetic distances using sequences from GENBANK and the HKY-Γ metric (using sequences described in Lijtmaer et al. 2003), and then cast the distances into millions of years of separation, using the standard 2% figure. ● Results of crosses between the parental species. Line is fit to these points only: $y = 17.9 + 2.3x$, $r = 0.64$, $N = 21$. ■ Hatchability of F_2 individuals: these are the results of crosses within the F_1 generation from the six crosses indicated with the vertical tick mark.

that produced the corresponding F_1's (Wilcoxon signed rank test, $P < 0.05$, $N = 6$; see Figure 16.4). Lijtmaer et al. (2003) interpreted this to be a result of lower F_2 viability than F_1 viability, but, given the results described in the previous sections, it may instead reflect low F_1 fertility. Three studies that directly measured F_2 fitness have been reported (Table 16.1). Two of those studies showed no reduction in fitness in either the F_1 or the F_2 generation, but the third study, which shows some reduced fitness in the F_1, has greatly lowered fitness in the F_2.

Two clear results from these analyses are in need of explanation: (1) the more rapid loss of fertility than of viability among species crosses and (2) Haldane's rule.

FERTILITY LOSS It does not appear that fertility is inherently more susceptible to genetic disruption than viability; in *Drosophila,* recessive lethals are commoner than recessive steriles (e.g., Wu and Davis 1993). Instead, the more rapid loss of fertility than viability has received two explanations. The first, in an extension of the Dobzhansky–Muller reasoning, asks why genes affecting fertility within species should be evolutionary labile. Rapid evolution of genes affecting fertility is expected under a number of sexual-selection, sexual-conflict and meiotic drive models (Wu and Davis 1993; Rice 1998; Rice and Chippindale 2002; Tao and Hartl 2003).

Hybrid infertility may also develop relatively quickly if species differ in chromosome structure and/or chromosome numbers. Chromosomal rearrange-

Table 16.1. Examples of F_2 fertility

Species in cross		F_2 generation	Geography	Millions of years	Reference
Common Quail	Japanese Quail	Fully fertile	Allopatric	1.25	Deregnaucourt et al. (2002)
Red Junglefowl	Grey Junglefowl	Embryo viability of F_2 generation is 3% that of F_1	Parapatric	0.75	Morejohn (1968)
Mallard	Northern Pintail	Fertile	Sympatric	4.2	Sharpe and Johnsgard (1966)

NOTE: In the junglefowl (*Gallus* spp.) cross, the F_1 individuals have relatively low fertility, but, in the other crosses, the F_1 individuals are fully fertile. Ages are based on mitochondrial-DNA distances, using the cytochrome *b* gene and sequences in GENBANK (and the HKY-Γ distance metric, see Appendix 2.2), except for the quail (*Coturnix* spp.) cross, which was taken from the reference. Distances are converted into time using the standard calibration of 2% divergence for every million years of separation.

ments can affect normal segregation during meiosis, resulting in F_1 hybrids with low fertility (King 1993). This should affect fertility of the F_1 hybrids but not their viability. However, the idea has theoretical problems; if chromosomal rearrangements themselves cause low fertility, any newly arising rearrangements will be selected against, so they are unlikely to become established in the first place. Several mechanisms to circumvent this have been proposed (Lai et al. 2005). For example, it is possible that multiple rearrangements could build up in different populations, each not deleterious on the background on which it arises, but, in sum, the multiple rearrangements could produce hybrids that do experience difficulties at meiosis, as in the genetic model for the accumulation of Dobzhansky–Muller incompatibilities (Figure 16.1). Baker and Bickham (1986) presented such a model. Consider three chromosomes (A, B, C). In one population, A and B fuse to make a chromosome AB; this new chromosome can pair normally against the original pair of chromosomes so it is not selected against. In another population, B and C fuse to make BC, and, again this new chromosome is not selected against. However, pairing and segregation of the new chromosome combinations is difficult, causing a breakdown in fertility. This may apply to several mammalian groups (Baker and Bickham 1986), but fusions of this kind do not seem to be very common in birds (Christidis 1990).

Causes of infertility have been studied in some hybrids (Table 16.2). Three sterile hybrid males from a cross between two species of *Columba* pigeons had normal testes (Smith 1912). Infertility probably results from sperm abnormalities. About 50% of the spermatozoa were beaded and twisted, and an additional 40% of the sperm were twice the size of normal sperm (Smith 1912). Smith attributed this to: (1) a failure of chromosomes to segregate normally at the first meiotic division and/or (2) a failure to go through a second meiotic division at all, so sperm carry twice the chromosomal complement. These failures seem to support the chromosomal model, but there is no direct evidence that they result from chromosomal differences in structure and number, rather than genetic incompatibilities that affect meiosis.

Female hybrids were also studied in pheasants and ducks. Their infertility arises because the ovaries are degenerate (Poll in Smith and Thomas 1913; Mott et al. 1968). Smith and Thomas (1913) attributed this to problems of meiosis similar to those that the males experience. They argued that, because meiosis is initiated in females when the female is herself still an embryo, a failure to go through meiosis might lead to gonad loss rather than to gamete degeneration. However, normal meiosis is arrested in the female embryo before chromosomal segregation, and it resumes only in the adult female at ovulation (Mira 1998). So a failure to segregate properly seems unlikely to account for degenerate ovaries.

Table 16.2: Studies on causes of infertility of hybrids

Species 1	Species 2	Sperm morphology	Sex of hybrids*	Reference
Reeves's Pheasant	"Japanese Pheasant"[1]	Sperm abnormal	Males and females: males sometimes fertile	Smith and Thomas 1913
Muscovy Duck	Mallard	No sperm; first meiotic division normal, second not	Males and females: males sometimes fertile	Mott et al. 1968
Rock Dove	Wonga Pigeon	Sperm were often large or twisted	No evidence for females	Smith 1912
Spotted Dove	African Collared-Dove	Sperm matured but were often abnormal	Males and females: males fertile	Shrigley 1940
Scaly-breasted Munia	Chestnut-breasted Munia	Sperm matured but were often abnormal	No evidence for females	Swan and Christidis 1987
Island Canary	European Goldfinch	No sperm; germ cells degenerated in testis	Males and females: males sometimes fertile	Swan 1985

*From Gray (1958) as well as original reference.
[1]Considered by Sibley and Monroe (1990) to be conspecific with the Common Pheasant.

In other examples, male infertility results from problems of sperm maturation and has little to do with chromosomal differences (Table 16.2). Thus, evidence that genetic differences are important seems fairly clear, and there is no convincing example in which structural differences in chromosomes drive infertility. Similar conclusions have been reached from much more detailed studies on *Drosophila* (Coyne and Orr 2004, pp. 259–265).

HALDANE'S RULE   In the simplest model of genetic incompatibilities, all F_1 individuals are the double heterozygote, *AaBb* (Figure 16.1). This means that, if the deleterious effects of an *a–B* interaction are to be expressed in hybrids, then both the *a* and the *B* alleles have to be expressed in the heterozygote. If the *a* allele is completely recessive to *A,* or if the *B* allele is completely recessive to *b,* the deleterious effects are masked. In fact, recessivity of deleterious effects appears to be common, at least in *Drosophila,* as can be shown by studies in the F_2 generation, where three kinds of genotypes arise. The first is the double heterozygote, like the F_1 (*AaBb*). The second is when an individual is heterozygous at one locus and homozygous at the other (*AaBB* or *aaBb*), and the third is the double homozygote (*aaBB*). In crosses between *Drosophila melanogaster* and *D. simulans,* Presgraves (2003) estimated the relative frequencies of each class causing hybrid inviability to be 0%, 12%, and 88%, respectively. Double recessive interactions causing hybrid lethality are more than seven times as common as recessive–dominant interactions, and no incompatibilities of the form *AaBb* are found in this cross. Note that, because recessive interactions are exposed in the F_2 generation, the F_2 typically has lower fitness than the F_1.

The strong role of recessivity in hybrid incompatibilities provides the most general explanation for Haldane's rule. When genes are recessive in their deleterious effects, as they commonly seem to be, the only time they can be expressed in the F_1 hybrid is if they happen to be on the sex chromosomes of the heterogametic sex (the female in birds). In particular, because most birds have a degenerate W chromosome with few active genes, many genes on the Z chromosome are expressed in the female, even when they are recessive. This is the "dominance theory" of Haldane's rule (Turelli and Orr 2000), and it probably accounts for some of the loss of fitness in females.

Other genetic effects may contribute to Haldane's rule in birds. First, genes on the female-restricted W chromosome from one species may interact poorly with genes from the other species (including the exposed Z). Equivalent Y-chromosome effects are common in *Drosophila* (Turelli and Orr 2000). Although most birds have a degenerate W chromosome, even in these species the W chromosome carries some active, W-specific genes. Mizuno et al. (2002) estimate there are about 20 such genes in the chicken. Models of meiotic drive described on p. 369 also predict Haldane's rule because meiotic drive is expected frequently be associated with the sex chromosomes (Frank 1991; Hurst and Pomiankowski 1991; Tao and Hartl 2003).

A final possible cause of Haldane's rule results from deleterious interactions between maternal contributions and the Z chromosome. In F_1 hybrid females, the egg cytoplasm and other forms of maternal care come from one species but the single Z chromosome comes from the other. Any maternal–Z interactions that are due to recessive effects on the Z chromosome will reduce fitness in hybrid females (Turelli and Orr 2000). In birds, evidence on all these points is lacking.

CHROMOSOMES

Chromosomal inversions may contribute to the development of genetic incompatibilities by reducing recombination between blocks of genes. It is also possible that rearrangements directly contribute to hybrid infertility, although, as noted, this has not been demonstrated. Here I describe chromosomal differences between species, their rate of accumulation, and evidence that chromosomal rearrangements reduce recombination in birds.

Bird karyotypes typically include both large and small chromosomes, termed the macrochromosomes and microchromosomes, respectively. The chicken has nine pairs of macrochromsomes, including the sex chromosomes, and 30 pairs of microchromosomes (Figure 16.5). In this species, the microchromosomes comprise about 25% of the DNA but encode 50% of the genes (Burt 2002). The Z chromosome is one of the larger macrochromosomes, and, according to the most recent sequence data available, it contains just under 3% of all genes (Hillier et al. 2004; www.ensembl.org). In most species, as in the chicken, the W chromosome is degenerate and carries few genes. In some groups, however,

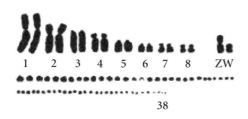
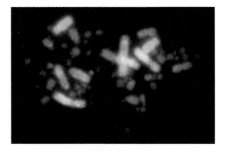

Figure 16.5 *Left*: The 39 pairs of chicken chromosomes (from Shibusawa et al. 2004b). *Right*: Dual color chromosome painting experiment on chicken chromosomes: Chromosome 2 is painted in red, chromosome 4 in green (courtesy of Lindsay Robertson and Darren K. Griffin, University of Kent, UK).

most notably the ratites (Ostrich, Emu etc.), the W is very similar to the Z and contains many active genes.

Species differ both in the number of chromosomes and in the order of the genes along the chromosomes. Chromosome number varies, but about two-thirds of 593 species fall between $N = 38$ and $N = 41$ pairs (Christidis 1990). Known extremes are $N = 20$ (the Eurasian Thick-knee, also known as the Stone-curlew) and $N = 66$ (the Common Kingfisher). The proportion of microchromosomes and macrochromosomes also differs among species; the Black-winged Kite ($N = 34$ pairs of chromosomes), for example, has only one microchromosome (Bed'Hom et al. 2003). Large-scale comparisons of gene order on chromosomes are now being made using whole chromosome "paints." A paint is a mixture of DNA that hybridizes to a single chromosome in the species from which it was made (Wienberg and Stanyon 1995; Guttenbach et al. 2003; Figure 16.5). Fluorescence in situ hybridization (FISH) of the paint to the chromosomes is used to detect homology across species. Paints have been developed for all the chromosomes of the chicken (Masabanda et al. 2004). A second method, termed comparative mapping, is to identify the locations of homologous DNA sequences on the chromosomes of different species.

The painting and mapping results have led to a general view of conservative evolution. For example, paints from each of chromosomes 1–9 of the chicken hybridize solely to the corresponding chromosomes in both the Japanese Quail and the Greylag Goose (Guttenbach et al. 2003), species that may be separated from the chicken by 30 million years (Slack et al. 2006) and more than 65 million years, respectively (Ericson et al. 2006; Slack et al. 2006). Chromosomes 1 and 4 of the chicken are split into two in all four passerine bird species studied so far, but the other macrochromosomes are conserved (Guttenbach et al. 2003; Derjusheva et al. 2004; Itoh and Arnold 2005). Passerines may be separated from the chicken by more than 80 million years (Slack et al. 2006; Ericson et al. 2006). The pattern of conservatism appears to be true also for the larger microchromosomes (Derjusheva et al. 2004).

Based on results from comparative mapping, Burt et al. (1999) identified just 72 large chromosomal rearrangements that separate humans from the chicken. Such rearrangements have occurred over perhaps 300 million years of independent evolution, and they imply a maximum rate of substitution of 0.16 per million years in the chicken lineage (Burt et al. 1999). Because of the slow rate of chromosomal evolution in the bird and human lineages and the high rate in the mouse lineage, we look more like a chicken than a mouse at this level of chromosomal organization (Burt et al. 1999; Ellegren 2005).

The appearance of a slow rate of chromosomal evolution in birds is misleading, for it masks variability not detected by the painting methods and comparative maps. Whole chromosome paints do not detect translocations smaller than about 10,000 bases, nor do they detect inversions within chromosomes. Comparative maps, which are based on a limited number of anchor points, grossly underestimate the number of inversions and small rearrangements (Eichler and Sankoff 2003). Other methods that preceded paints and comparative maps have identified other chromosomal differences. Most notably, various stains cause a characteristic banding pattern that can be compared across species (e.g., Figure 16.6). Giemsa stain has been especially effective in this regard. Rearrangements that are easiest to identify using these methods are pericentric inversions (inversions that include the centromere). Chromosomes that are identified as homologous based on their banding patterns, but which differ in centromere position, are assumed to differ because of a pericentric inversion. This assumption seems reasonable, and it has been confirmed in some cases using modern methods (e.g., Shibusawa et al. 2001, 2002, 2004a), but the centromere can change position without an inversion (Kasai et al. 2003). Paracentric inversions do not include the centromere. They can be detected only by comparing banding patterns, and it is quite probable that many have been missed.

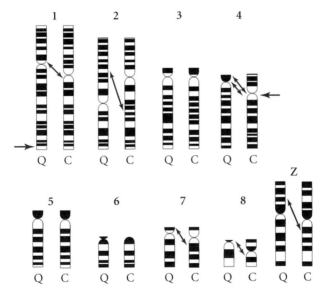

Figure 16.6 A comparison of the eight largest autosomes and the Z chromosome of the Japanese Quail (Q) and the chicken (C) (after Shibusawa et al. 2001). The banding patterns are those that appear when chromosomes are stained with Giemsa. Fluorescence in situ hybridization (FISH) of 45 chicken clones was used to identify inversions and translocations. Locations of an inferred seven inversions are shown (double-headed arrows). Inversions on chromosomes 7 and Z were not confirmed in this study and are based on the position of the G-bands and centromere. Inversions on chromosomes 1 and 2 were mapped more finely using an additional 13 chicken clones. At least two inversions are required to explain gene order and proximity to the centromere near the end of chromosome 4. The horizontal arrows indicate the locations of a sequence that is on chromosome 1 in the quail, but on chromosome 4 in the chicken.

Christidis (1983, 1986a, b, 1987) used classic staining techniques to compare different species of estrildid finches. He first demonstrated surprisingly large differences between two congeneric species, the Red-winged Pytilia and the Green-winged Pytilia (Christidis 1983). These species differ in chromosome number ($N = 28$ and $N = 38$, respectively), as well as at least ten inversions. Other patterns differ between the two species: in the Red-winged Pytilia, the microchromosomes and macrochromosomes fall into discrete size classes, whereas, in the Green-winged Pytilia, variation in chromosomal length is more continuous. In subsequent studies on 29 other estrildids, Christidis (1986a, b, 1987) identified an additional 55 pericentric inversions, three paracentric inversions, and one fission/fusion event.

Sorenson et al. (2004) obtained mitochondrial sequence data for 17 of the finch species studied by Christidis. I used these sequences to build a phylogeny for those species and thereby to estimate the rate of chromosomal evolution in the group (Figure 16.7). Following diagrams in Christidis (1986a, b; 1987), I placed chromosomal rearrangements on the tree, using the Zebra Finch as the karyotype against which changes were compared. Christidis assumed the Zebra Finch to carry the ancestral karotype, which is unlikely, but this should not alter the mapping substantially. Because of the differences between the tree Christidis used and the one presented here, some changes cannot be assigned to a specific branch and I show the approximate locations of these changes by a vertical bar. Christidis inferred that 57 inversions in total separate the 17 species on the tree. Sorenson and Payne (2001) place the root of the tree at ~12 million years (see Figure 13.8, p. 295). Given this calibration, the total branch length in the phylogeny sums to 155 million years, which yields an estimate of one inversion per 2.7 million years of evolution along a branch, or 1.35 million years of divergence between a pair of species. Excluding the two *Pytilia* species, the figure comes out to be one inversion for every 3.5 million years of evolution along a branch (1.75 million years of separation).

The chicken and the Japanese Quail both have $N = 39$ chromosomes, with nine pairs of macrochromosomes and 30 pairs of microchromosomes. Shibusawa et al. (2001) hybridized chicken fluorescent DNA clones both to chicken and to Japanese Quail chromosomes. Combined with classical banding techniques, they identified at least seven inversion differences between the species, as well as a small translocation between quail chromosome 1 and chicken chromosome 4 (Figure 16.6; Shibusawa et al. 2001). Given a date for the last common ancestor of 30 million years (Slack et al. 2006), this implies a rate of substitution of one inversion for 4.3 million years of separation. Differences between the Japanese Quail and the Blue-breasted Quail have also been analyzed at this level, using those same chicken clones (Shibusawa et al. 2001, 2004a). These species may be separated by about 12 million years (Randi 1996). They differ by at least four inversions, or one inversion for every 3 million years of separation.

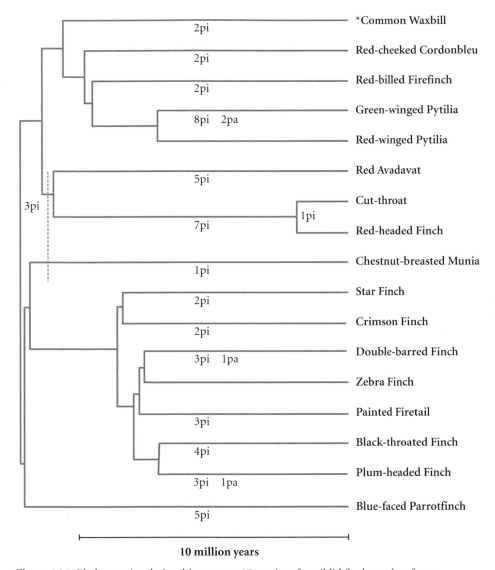

Figure 16.7 Phylogenetic relationships among 17 species of estrildid finches, taken from Sorenson et al. (2004). Using the topology in that paper, I estimated branch lengths based on 1049 bases of the coding regions of the ND2 and ND6 mitochondrial DNA genes, with the GTR-Γ + I model and parameters obtained from an analysis of 74 species in *MRBAYES*. The tree was then smoothed in *r8s* (see Appendix 2.2, p. 38). In the figure, some zero branch lengths at the base of the tree have been given small positive values, according to the topology of Sorenson et al. (2004). The calibration was taken from Sorenson et al. (2003), who placed the root at 12 million years (see Figure 13.8, p. 295). Chromosomal rearrangements identified by Christidis (1983, 1986a, b, 1987) with respect to the Zebra Finch are mapped on; pi = pericentric inversions, pa = paracentric inversions. *Chromosomes of the Common Waxbill were studied, but only the DNA sequence of the congeneric Fawn-breasted Waxbill is available. The tree differs from that suggested by Christidis, who constructed it using the chromosomes themselves, and ambiguous locations of chromosomal change are indicated by the dashed vertical bar. See text for further discussion.

Other differences probably remain to be discovered. For example, rearrangements within the microchromosomes have so far not been studied by any method, even though translocations between the microchromosomes and macrochromosomes are known (Shibusawa et al. 2002). Thus, a substitution rate of a major inversion on the order of roughly one for every 1.5–4 million years of separation is likely to be an underestimate. However, it should be noted that some lineages, most strikingly the Helmeted Guineafowl, in comparison with the chicken (Shibusawa et al. 2002), seem to separated by fewer inversions than expected given this substitution rate.

Chromosomal polymorphisms within species

In his review of 593 bird species, Christidis (1990, p. 79) identified 25 that contained polymorphisms in chromosome structure (i.e., two forms segregating within a single population). In total, there were 21 polymorphic pericentric inversions plus 10 polymorphisms for fusions, translocations, or deletions of chromosomal material. The frequency of polymorphism is likely to be grossly underestimated, because many studies were based on only one or a few individuals, and even then many rearrangements are likely to have been undetected. Recently discovered polymorphisms include a single Ostrich studied with modern techniques that was found to be heterozygous for a pericentric inversion on one of its chromosomes (Nanda et al. 2002) and the presence of a pericentric inversion within a captive colony of Zebra Finches (Itoh and Arnold 2005). The conclusion must be that chromosomal polymorphisms are common in birds.

A remarkable chromosomal polymorphism occurs in the White-throated Sparrow (Thorneycroft 1975). In this species, both the second chromosome and the third chromosome are polymorphic for a pericentric inversion. The polymorphism of the second chromosome is perfectly associated with the color of the crown stripe (white or tan) among breeding adults (Thorneycroft 1975), as well as with behavioral (Rising and Shields 1980; Tuttle 2003) and slight morphological (Tuttle 2003) differences. Individuals either have the genotype 2/2 (white crown stripes) or $2/2^m$ (tan crown stripes), where 2 is an acrocentric chromosome (with its centromere closer to one end) and 2^m is a metacentric chromosome (with its centromere close to the center). The frequency of these two types is approximately 50:50, and only one individual in 397 birds was found to be homozygous ($2^m/2^m$) (Thorneycroft 1975). Nearly all matings are disassortative, that is, between birds with tan crown stripes and birds with white crown stripes. Studies of meiosis in this species indicate that pairing is end-to-end in the heterozygote—only one chiasma forms, cf. the normal two in the homozygote (Figure 16.8). This likely leads to an absence of crossing-over between genes in the inversion, keeping the block of genes together.

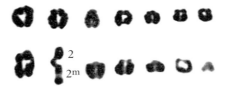

Figure 16.8 The seven largest chromosome pairs at meiosis in the White-throated Sparrow (from Thorneycroft 1975). *Above:* an individual homozygous for chromosome 2. *Below:* an individual heterozygous for chromosome 2 (i.e., 2/2^m). In the heterozygote, only one chiasma forms, and pairing is end to end.

The only other wild species in which more than 100 individuals have been karyotyped is the Dark-eyed Junco, which has polymorphisms for inversions on the second and fifth chromosomes (Shields 1973). The polymorphisms are shared across several subspecies as well as between two closely related species. Unlike the White-throated Sparrow, chromosome pairing is normal. Despite this, crossing-over in the heterozygotes is reduced (Shields 1976). For example, for chromosome 2, 54% of homozygotes had more than three chiasmata, whereas only 17% of the heterozygotes did ($N = 70$; Shields 1976).

The quite common presence of polymorphism suggests that chromosomal rearrangements can have little deleterious effect on fitness. Evidence from the White-throated Sparrow and the Dark-eyed Junco indicates they suppress crossing-over. It is worth noting that, even if crossing-over does occur within an inversion, it is expected to produce products that are deleterious, so only nonrecombinants get into the next generation (e.g., see Rieseberg 2001). Thus, inversions should suppress recombination, whether or not they suppress crossing-over.

Chromosomal differences between species may contribute to genetic differentiation, but why do these differences arise in the first place? The most obvious influence of a rearrangement is the suppression of recombination, and this may occasionally have beneficial effects on its bearer. Kirkpatrick and Barton (2006) show how migration between two populations in different environments favors reduced recombination between two alleles at different genetic loci, *A* and *B*, both of which are favored in one environment but not in the other. This therefore leads to selection favoring an increase in the frequency of any chromosomal inversion that includes both genes, because the inversion reduces recombination between them. In this way, ecological models of speciation become connected with nonecological models; for the inversion difference, once established between populations, will trap unconditionally favored alleles, enabling the build-up of incompatibilities (Navarro and Barton 2003).

CONCLUSIONS

In summarizing this chapter, I return to the main question, which is whether intrinsic loss of hybrid fitness contributes to speciation. Intrinsic loss of hybrid fitness can take a long time. A classic example is that of the Mallard and Northern Pintail, which may be separated by more than 4 million years, yet produce fully fertile hybrid offspring (Sharpe and Johnsgard 1966). These species continue to hybridize at low frequency in the wild (Panov 1989; McCarthy 2006). Given such examples, one may wonder whether intrinsic incompatibilities have any relevance to bird speciation. However, there are at least three reasons for believing that they do. First, loss of fertility and viability is gradual. Even a small loss of hybrid fitness could be greatly accentuated in natural conditions. For example, as noted in Chapter 15, in the Carrion Crow/Hooded Crow hybrid zone, some hybrid females lay eggs with thin shells, making the eggs susceptible to breakage. The ultimate causes of thin-shelled eggs may lie in intrinsic incompatibilities. Second, at the inferred end of adaptive radiations, when species are ecologically very similar and occupy abutting ranges, the process of speciation itself takes a long time, and genetic incompatibilities then contribute to reproductive isolation. A good example is that of the Western Meadowlark and the Eastern Meadowlark (Figure 2.6, p. 27), which have probably been separated for more than 2 million years but hybridize occasionally; the hybrids are largely infertile (Lanyon 1979a). A second example of low intrinsic fertility between hybridizing species is that of the Collared Flycatcher and the Pied Flycatcher (discussed in Chapter 14), with an estimated separation time of about 2 million years (Sætre et al. 2001). The F_1 females are sterile (Alatalo et al. 1990; Veen et al. 2001), and F_1 males have low fertility (Sætre et al. 2003). A final reason why intrinsic loss of hybrid fitness may contribute to reproductive isolation is that deleterious effects should show up in the F_2 generation and in backcross generations earlier than they do in the F_1, because it is only in the F_2 generation that recessive deleterious effects on the autosomes are exposed (Turelli and Orr 2000). F_2 fitness has rarely been studied.

It is likely that loss of hybrid fitness most often arises in multiple ways that together result in complete postmating isolation on a timescale shorter than that of the loss of fertility or viability. The Fischer's Lovebird and the Rosy-faced Lovebird are ~8% divergent in their mitochondrial DNA (Eberhard 1998), implying ~4 million years of separation using the usual clock. Female hybrids appear to be sterile (although male hybrids have motile sperm), and hybrids of both sexes develop several conditions that reduce survival, such as the tendency to get gout (Buckley 1969). This substantially lowers fitness, perhaps to zero in nature, but, in addition, the hybrids are well known for several behavioral problems associated with unusual nest-building habits (Buckley 1969). The Fischer's Lovebird carries strips of nest material in its beak, while the Rosy-faced Lovebird stuffs material in its rump. Hybrids do not have much of a clue about what to do, repeatedly stuffing, removing, and dropping nest material (illustrated in Buckley 1982). This must create some loss of fitness with respect to the parental species. The hybrids also show many other unusual behaviors—if females lay eggs, they cover them with nest material, and males often interfere with female attempts to carry such material.

Although hybrid fitness eventually declines to zero, the occasional successful backcross could lead to gene flow between species. Given that it may take millions of years to completely eliminate hybrid fertility, this seems to be a serious problem for speciation. However, even if some genes cross the species border, not all need do so (Wu 2001). Given that female hybrids are the less fit sex, and the mitochondrial DNA and the W chromosome are transmitted only through the female line, mitochondrial-DNA and W-chromosome gene flow should be cut off relatively early, even if autosomal and Z-linked traits can pass through the male line. Helbig et al. (2001) found a mitochondrial-DNA difference of 6% between hybridizing Common Chiffchaffs and Iberian Chiffchaffs, although nuclear markers were largely undifferentiated between the two species (see p.361). They suggested that gene flow was occurring only through male hybrids. Two studies of hybrid zones (Chapter 15) have found that genes on the Z chromosome fail to introgress. This may reflect a high concentration of infertility factors on the Z chromosome (Rice 1984; Gibson et al. 2002), plus a relatively low recombination rate, making it more difficult for neutral and weakly selected genes to become dissociated from the infertility factors (Servedio and Sætre 2003). A low recombination rate is expected because one-third of the time the Z chromosome is in the female, where it is alone, and cannot recombine with another Z chromosome.

Different alleles from the two species that result in low fitness when in a hybrid are unlikely to spread from one species to the other, as are different alleles that are favored for ecological reasons, when the two species occupy different environments. If the genes coding for such alternative alleles are involved in a chromosomal inversion, with one form of the inversion in one species and the other form in the other species, then, in the absence of recombination within the inversion, all the other genes in that inversion are stuck with it (Noor et al. 2001; Rieseberg 2001). New mutations that arise in one species will accumulate in the inversion because of their unbreakable association with the genes that are restricted to one species, even though they might be favored in the genetic background of other species, if there were any way for them to get into those species (Navarro and Barton 2003). As numbers of these mutations accumulate in the inversion, additional incompatibilities between the species can develop. Eventually, hybrid fitness gets reduced to zero, and no genes can get from one species into the other, wherever they arise in the genome.

While it is not known if chromosomal rearrangements are involved in the development of incompatibilities in birds, the analysis in this chapter implies that they arise on such a timescale that they could play a role. There is currently a tremendous surge of interest in the role of chromosomal rearrangements in speciation, driven by empirical findings (Noor et al. 2001; Rieseberg 2001; Lai et al. 2005), theory (Navarro and Barton 2003; Kirkpatrick and Barton 2006), and rapid advances in studies of the genome (Eichler and Sankoff 2003; Masabanda et al. 2004; Dawson et al. 2006). A first draft of the chicken genome is available (Hillier et al. 2004), and the Zebra Finch genome is on the way (see www.song-birdgenome.org). Soon, comparative maps of species other than the chicken will be available, and rapid progress towards identifying any possible role for chromosomal rearrangements in bird speciation is to be expected.

Intrinsic incompatibilities clearly take a long time to accumulate, and even low levels of hybrid fitness could result in the collapse of two species back into one, associated with selection for those gene combinations that work well together. Eventually, however, incompatibilities are complete. Once that happens, speciation becomes irreversible. Strong premating isolation is favored, and the only alternative to coexistence is extinction.

SUMMARY

Two species that are crossed may either not produce offspring or, if they do, the offspring have low viability or are infertile. This form of postmating isolation represents the last step in speciation, for once enforced, there is no possibility that hybridization will lead to a collapse of the two species back into one. Based on DNA-divergence dates, complete loss of hybrid fertility takes up to an average of 7 million years, and complete loss of viability takes up to an average of 11.5 million years. This is much longer than the timescales over which new species are formed. Indeed, speciation can be completed without any intrinsic loss of hybrid fitness, and some species pairs estimated to have separated more than 4 million years ago produce fully fertile hybrids (e.g., the Mallard and the Northern Pintail). Nevertheless, partial loss of fitness in the F_1 generation and further loss of fitness in the F_2 and backcross generations contribute to reproductive isolation between many incipient species. Evidence from finches, quail, and chickens suggests that detectable chromosomal rearrangements, notably inversions that include the centromere (pericentric inversions), become fixed in bird lineages on the order of one every 3–8 million years (i.e., every 1.5–4 million years of separation of divergent lines), but this is underestimated to an unknown extent because many rearrangements probably remain to be identified. Rearrangements restrict recombination and may prevent the spread of blocks of genes from one hybridizing species to the other. Given that the complete loss of hybrid fitness can take several times longer than the rate at which large rearrangements are fixed, such rearrangements could contribute to the accumulation of incompatibilities across hybridizing taxa.

CHAPTER SEVENTEEN

Conclusions

In this chapter, I briefly summarize the main conclusions from the book, considering the roles of selection, genetic drift, geographical isolation, and ecological differentiation, and the relationship between premating reproductive isolation and postmating reproductive isolation. Finally, I compare birds with other animal groups. Throughout this book I have defined species following field guides and the compilation of Sibley and Monroe (1990). These essentially use the biological species concept for sympatric taxa and roughly follow the criteria of Helbig et al. (2002) for related allopatric taxa—namely, taxa that are diagnostically different, and the differences approach the level seen among sympatric species.

Selection and genetic drift

The idea that random sampling of genes in small populations—genetic drift—has been important in speciation has been part of the thinking of many influential workers, when tackling the problem of how populations might diverge in (superficially) similar and constant environments (e.g., Mayr 1954; reviewed in Coyne et al. 1997). In a critical review, Coyne and Orr (2004, p. 410) came down in strong support of the importance of selection in the generation of population differences that contribute to reproductive isolation and argued for a negligible role for genetic drift. It is somewhat of an indictment of the state of research on speciation that this conclusion was disputed by Futuyma (2005), who stated "I cannot agree that this important debate [on the role of genetic drift] has been settled." However, material reviewed in this book supports Coyne and Orr's conclusion. I suggest that genetic drift is unlikely to play much of a role in the divergence of traits that contribute to reproductive isolation between taxa, largely because those traits are likely to have substantial effects on fitness within taxa. Instead, I propose that we should shift the emphasis towards determining the extent to which population divergence is driven by new mutations that are unconditionally favored by selection, and hence would spread throughout the species range if populations were connected by gene flow. This includes mutations that are favored under some forms of sexual selection, sexual conflict, and intragenomic conflict. In this case, divergence between populations may lead to reproductive isolation if different mutations arise and spread in different places and these mutations interfere with each other. This process of divergence can be contrasted with an alternative mode of population divergence resulting from different selection pressures in different environments; here gene flow between populations is less constraining on the rate at which they diverge. With the infusion of molecular genetics, it is beginning to be possible to investigate these alternative nonecological and ecological processes by studying the normal functions of genes that cause reproductive isolation (Presgraves et al. 2003; Orr et al. 2004), but we are some way from this in birds.

Geography and ecology

Except for the strong assortative mating that results from song learning and imprinting in the African parasitic finches (Sorenson et al. 2003) and from displacement in breeding times among the Band-rumped Storm-Petrels on the Azores (Friesen et al. 2007), sympatric speciation in birds is very rare (Chapter 2). Nearly all speciation events clearly involve spatial separation of populations, at least in the initial stages of divergence. Indeed, among animals in general, we have few compelling examples of sympatric speciation (Coyne and Orr 2004, their Chapter 4). This is not to say that barriers to dispersal must be absolute to allow speciation. Gene flow between populations may be reduced as a result of the low fitness of immigrants or their offspring (Chapter 4; Nosil et al. 2005). Nor need reproductive isolation be completed in geographically separated populations. Once a sufficiently high level of reproductive isolation has been achieved, populations can spread into sympatry. Such processes as ecological character displacement and reinforcement could then increase premating isolation.

Given the requirement of geographical separation, following Mayr (1947), I considered three steps in speciation (Chapter 2). First, populations must expand ranges and become established in new locations. Second, gene flow between populations must be restricted. Third, populations diverge in various traits that generate reproductive isolation. An important feature of this scheme is that any one of the three components—not just the acquisition of reproductive isolation—can limit the rate of speciation. In particular, successful range expansions may be more difficult later in an adaptive radiation rather than earlier, thus slowing the speciation rate (Chapter 6).

The three requirements for speciation—range expansion, barriers to gene flow, and subsequent divergence—may happen sequentially, with little temporal overlap. Following a range expansion, a vicariance event, such as the development of a mountain barrier, divides the species, limiting or cutting off gene flow between populations. After a million years or so, the populations diverge sufficiently that they are recognized as distinct species. In such cases, divergence may happen with relatively little ecological differentiation, as exemplified by many allospecies with largely nonoverlapping ranges. Similar patterns are seen in other groups. Pseudoscorpions in Central America are divided into genetically differentiated, geographically separated species that are morphologically very similar to one another (Zeh and Zeh 1994). *Anolis* lizards on each large Caribbean island consist of ecologically differentiated sympatric species, but, within at least one of these species on Hispaniola, genetically quite distinct taxa occur in different geographical locations; they are very similar morphologically and were not recognized as distinct until DNA sequence studies were conducted (Glor et al. 2003). Divergence in sexually selected traits and intrinsic loss of hybrid fitness probably contribute to the speciation process among these long-separated, ecologically similar taxa.

At the other extreme, strong divergent natural-selection pressures result in rapid divergence between populations in traits that lead to both premating and postmating isolation (ecological speciation), perhaps sometimes in the face of substantial gene flow. This rapid divergence may be especially likely early in an adaptive radiation, when a species finds itself in a depauperate environment (Rosenzweig 1995). Once sufficient reproductive isolation and ecological differentiation have developed, different populations can spread back into sympatry. The generation of young species such as Darwin's finches is related to strong, ecologically driven selection pressures. A similar example involves pairs of ecologically differentiated stickleback species occupying lakes in British Columbia, Canada, which were frozen solid only about 10,000 years ago (reviewed by McKinnon and Rundle 2002). These are young sympatric species, where strong ecological differentiation enables coexistence in sympatry, and the ecological differences create at least some reproductive isolation. Nevertheless, even in young species, differences that contribute to reproductive isolation may have accumulated for nonecological reasons. For example, songs in Darwin's finches, which are an important trait associated with premating isolation, have probably diverged partly as a result of the spread of different cultural mutations in geographically separated populations.

Under the biological species concept, acquisition of reproductive isolation between populations is the defining point in speciation. A recent surge of interest in ecological speciation highlights a difficulty with maintaining reproductive isolation among young, ecologically differentiated, biological species (Seehausen 2006). Such species are reproductively isolated primarily as a result of their ecological differences, meaning that both premating and postmating isolation are susceptible to breakdown when conditions change (Muller 1942, p. 83). In Darwin's finches, hybrids between the Common Cactus-Finch and the Medium Ground-Finch on Daphne Island have been surviving well and reproducing over the past ten years. It is only assortative mating based on song that prevents a more rapid fusion of the two species into one (Grant et al. 2004, 2005). In the sticklebacks, two species in one lake are collapsing into a hybrid swarm (Taylor et al. 2006; Gow et al. 2006). In this case, hybridization seems to be due to the recent introduction of a crayfish. The crayfish has caused ecological changes, including the destruction of vegetation and increased water turbidity. These changes may have increased stickleback hybridization rates and perhaps raised hybrid fitness relative to the parental forms.

Thus, many young biological species are at risk of losing their identity via hybridization as environments change. Most of these species would not be classified as species under the phylogenetic species concept, at least as determined by DNA sequences, because they have not been separated for long enough to become reciprocally monophyletic (see Chapter 1, Appendix 1.1). Instead, proponents of the phylogenetic species concept would probably refer to such species as "ecological races" and accord them a similar taxonomic level as geographical races. This has merit, given the susceptibility of both ecological races and geographical races to merging in response to altered environmental conditions, but it does lose focus on those factors that maintain the integrity of biological species when in sympatry. In addition, it is not always clear if rapid ecological speciation inevitably leads to species that are at risk of collapse through hybridization. In Chapter 5, I argued that differentiation along multiple ecological dimensions, especially if precipitated by a very unusual change in the environment, may make species less prone to collapse. In Chapter 6, I noted that short distances between speciation events recorded deep in phylogenetic trees may reflect a permanent role for ecological speciation as a driving force in the generation of reproductive isolation.

Despite these observations, little direct evidence supports a role for ecological speciation in the establishment of reproductive isolation between those long-divergent species that make up the majority of ecological communities. Nevertheless, premating isolation between most sympatric species is in part a result of ecological differences. For example, if species occur in different habitats they are less likely to encounter one another. Ecological differences also likely contribute to postmating isolation, because the environment in which a hybrid finds itself may result in lower fitness than the parental types. The presence of ecological contributions to reproductive isolation among coexisting species makes it plausible that such ecological factors contributed to the original development of reproductive isolation during the speciation process.

Muller (1942) suggested an additional role for ecological differences in speciation, in that the divergent selection pressures they generate should speed up the accumulation of genetic incompatibilities. He noted that environmental differences between populations cause adaptive divergence in various ecological traits, and these traits might lead to intrinsic incompatibilities as a side effect.

His specific example was that adaptation to different climatic regimes could lead to hybrid breakdown at whatever temperature the hybrids develop. Despite this statement from more than 60 years ago, the importance of ecology in driving the generation of intrinsic genetic incompatibilities remains largely unknown (Coyne and Orr 2004, their Chapter 8; Funk et al. 2006).

Premating and postmating isolation

Most sympatric species rarely hybridize, and assortment is strong. The reasons for this are fairly clear (Kirkpatrick and Ravigné 2002; Gavrilets 2003). If many individuals mate with members of other species and postmating isolation is high (i.e., hybrids have low fitness), the expected outcome is the extinction of one species (usually the rarer one, because a high proportion of individuals belonging to the rare species is involved in cross-species matings). If many individuals mate with members of the other species and postmating isolation is weak (i.e, hybrids survive and reproduce), the two species collapse into a hybrid swarm. On the other hand, if sufficient premating isolation has become established prior to the taxa coming together, and if substantial postmating isolation is also present, conspecific preferences should be strengthened (reinforcement), resulting in little cross-species mating.

It is theoretically possible that strong assortative mating generated in allopatry could produce species that coexist in sympatry, even in the complete absence of postmating isolation. Clearly, if assortative mating is so strong that females refuse to mate with males of other species, species integrity will be maintained without any postmating isolation. Because hybridization is rare among sympatric species, and hybrids produced in the laboratory are often fertile, the impression has been that the generation of premating isolation alone is indeed sufficient. I suggest, however, that postmating isolation is likely to be an essential part of species coexistence. This is because even a low frequency of hybridization should lead to the merging of two species back into one, if hybrids are perfectly fit. When hybrids are perfectly fit, selection opposing mating with heterospecifics is absent. Even when hybrids have reduced fitness, because of various search costs, mating with the first individual encountered, whether heterospecific or conspecific, may be selectively advantageous (Chapter 14). Thus, the reason hybridization is so rare among many coexisting species is perhaps because of reinforcement in the presence of strong postmating isolation, rather than the complete development of premating isolation between allopatric populations, prior to establishment of sympatry.

In the parasitic indigobirds, assortative mating is strong because of the way the species-recognition mechanism develops, and premating isolation should rapidly develop when a new host species is colonized (Chapter 13). This, then, seems to be an example of speciation in the absence of postmating isolation. However, subsequent adaptation to different hosts (in nestling mouth color) means that hybrids are likely to be at a disadvantage, and this postmating isolation could provide some selection to reinforce and maintain the premating isolation. Differentiation in nestling mouth color among the indigobirds is relatively small (Payne 2005b). The species are very similar in mitochondrial-DNA sequences, implying that they turn over rapidly (i.e., both speciation and extinction happen frequently) and/or they continue to hybridize successfully. The development of strong postzygotic isolation may happen only if different species remain for long periods in geographically separated locations, and this would then lead to selection favoring strong premating isolation if the species were to spread into sympatry.

Birds and other animal taxa

Both asexual and sexual animals share the property of being divided into discrete morphological groups separated by gaps (Coyne and Orr 2004, their Chapter 1). All animals consume resources and try to avoid being eaten, that is, they have an ecology. A good explanation for the presence of discrete morphological clusters (i.e., taxonomic species) is the presence of discrete ecological niches (Coyne and Orr 2004, pp. 49–54). A second property all animals share is DNA, which is subject to mutation. Among sexual animals, to which it is possible to apply the biological species concept, the inevitability of divergence due to the accumulation of new mutations in isolated groups means that the generation of reproductive isolation is also inevitable.

These are obvious commonalities across all animals. What are some of the differences between birds and other groups of sexual animals? I will describe four: the first difference is ecological, the second is behavioral, and the third and fourth are associated with genetics and development.

ECOLOGICAL DIFFERENCE Because of flight, birds can reach and colonize remote regions of the world more easily than many other groups of animals. This promotes speciation. For example, the moa-nalos on Hawaii filled a niche normally occupied elsewhere by tortoises or mammals (Chapter 6; Olson and James 1991). However, the fact that multiple species of birds adapted to different ecological niches can fly to isolated locations reduces the chances that any one species radiates to fill these niches. Indeed, adaptive radiations in birds seem to be less common than in some other groups. The *Anolis* lizards on each of the four main Caribbean islands have undergone separate adaptive radiations (Losos et al. 1998), and lizard radiations are found on several other oceanic archipelagos (e.g., Radtkey 1996). Snail radiations are even more widespread (Cowie 1996). By contrast, only the Galápagos Islands and the Hawaiian Islands provide examples of extensive bird radiations on oceanic archipelagoes.

BEHAVIORAL DIFFERENCE Learning, and more specifically, sexual imprinting, is a main cause of premating isolation in birds. This probably reflects an elaborate social system that favors individual recognition, as well as a brain that is large enough to allow the bird to actually carry out the task. The individual-recognition system becomes co-opted into a species-recognition system. Learning about conspecifics affects mating decisions in many other groups, including flies (Dukas 2004) and fish (Magurran and Ramnarine 2004), and its importance needs more study. However, in many taxa, species recognition clearly has a stronger genetic component than it does in birds. I argued that learning is one way in which the conflict between directional sexual selection and stabilizing species recognition could be resolved; new traits can invade a population and spread because they are attractive, but they can also become the focus of species recognition through imprinting. If learning is not important in other species, this mechanism is ruled out.

GENETIC AND DEVELOPMENTAL DIFFERENCES Several patterns in speciation seem likely to be explained by differences in genetics and development. The chromosomal complement in birds is unusual. One feature is that, unlike in mammals, flies, and many other taxa, the heterogametic sex is female. This is associated with the more rapid loss of female hybrid fitness (intrinsic infertility and inviability) than male hybrid fitness in species crosses (Haldane's rule). A second pattern in birds is that male hybrids are found between much more distant crosses among bird species than are either male or female hybrids in crosses among mammal species (Fitzpatrick 2004). An explanation for the relatively short time to viability loss in hybrids of both sexes in mammals is that one of the female's two X chromosomes is inactivated in each cell, which should make female F_1 hybrids as susceptible to recessive incompatibilities as male (XY) hybrids. In birds, genes on both the male's sex chromosomes (ZZ) are turned on in every cell, so male bird hybrids do not suffer from recessive incompatibilities in this way (Fitzpatrick 2004).

I started this book by noting that the history of the study of speciation is intricately tied to research on birds. The influential works of Mayr (1942, 1954, 1963) were dominated by bird examples. Major advances over the more than 40 years since Mayr's 1963 book have been adding the dimension of time to the study of patterns of speciation and the elucidation of various social, ecological, and genetic factors that contribute to reproductive isolation between divergent populations. If birds can be taken as representative animals, sexual and social selection are ubiquitous in leading to the divergence of many traits that form the basis of premating isolation, but whether sexual selection forms a rate-limiting step in speciation remains an open question. Multiple ecological causes affect the rate of speciation, including not only the rate of acquisition of reproductive isolation between separate populations (ecological speciation), but also limits on range expansions and limits on exchange of migrants between populations. It is important to recognize that the majority of sympatric species are very old and to consider the possibility that ecological speciation may have played a smaller role in their generation than it has in those young species that occupy young environments. We have not got very far with the genetics of speciation, and that is the area of speciation research that is likely to show the most rapid growth over the coming decades.

Glossary

Adaptive radiation The more or less simultaneous divergence of numerous lines from much the same adaptive type into different, also diverging, adaptive zones (Simpson 1953, p.233); the evolution of ecological and morphological diversity within a rapidly multiplying lineage (Schluter 2000, p. 10).

Adaptive peak A maximum in the adaptive surface.

Adaptive surface A plot of how the average fitness in the population changes as a function of the mean phenotype, assuming that the population variance is held constant. In theory, and in the absence of frequency-dependent selection, the mean phenotype should evolve to a position given by a maximum in the adaptive surface (not necessarily *the* maximum). See *adaptive peak*.

Allele An alternative form of a gene. Sometimes the word gene is used as shorthand for *allele*.

Allopatric speciation The origin of two species resulting from divergence of populations that are geographically separated from each other, with no movement of individuals between them.

Allopatry Geographical separation; individuals from different populations do not encounter each another.

Allospecies A group of entirely or essentially allopatric taxa that were once geographical races of a single species but that now have achieved species status (Amadon 1966). See *superspecies*.

Altricial species Species in which the young are blind and helpless upon hatching (this includes all passerine birds).

Associative learning Experiences reinforce one another and can be linked to enhance the learning process. As used here, the learning process that leads to generalization.

Assortative mating Nonrandom mating patterns, resulting in the generation of some premating isolation between populations. Assortment may result from active mate choice, or simply from being in a different habitat, or from breeding at a different time.

Autosomes All of the chromosomes in a karyotype except the sex chromosomes.

Backcross A cross of a hybrid to one of the parental species.

Badge of status A plumage patch whose degree of expression in an individual correlates with its social dominance.

Biogeography The study of the geographical distribution of animals and plants.

Biological species Groups of interbreeding populations reproductively isolated from other such groups.

Bounded hybrid superiority A hybrid zone maintained because hybrids have higher fitness, for ecological reasons, than the parental species in the center of the zone.

Centromere The constricted region of a chromosome to which spindle fibers attach during cell division.

Character displacement See *ecological character displacement* and *reproductive character displacement*.

Chiasma The physical site along the chromosome where crossing-over between homologous chromosomes takes place during meiosis.

Chromosomal inversion The removal of a chromosome segment, its rotation through 180°, and its reinsertion in the same location.

Chromosomal rearrangement Includes chromosomal inversions, chromosomal translocations (a piece of one chromosome detaches and becomes attached to another chromosome), chromosomal fission, and chromosomal fusion.

Chromosome paint A mixture of DNA that hybridizes to a single chromosome in the species from which it was made.

Clade A monophyletic group of organisms.

Cline A continuous change in the frequency of an allele, or in the mean of a character, over a geographic transect.

Coalescent A modelling framework in which two DNA sequence lineages converge in a common ancestral sequence, going backwards in time.

Comparative map A comparison of chromosomal rearrangements between a pair of species or individuals.

Competitive exclusion Competition for food or other resources is sufficiently intense that one species completely eliminates (or prevents the establishment of) another species.

Condition dependence As used here, the correlation between the expression of a trait and the overall condition, quality or health of an individual (Schluter and Price 1993).

Contingency The phenotype and genotype affect both the kinds of new visible mutations that arise and those that are likely to be favored.

Convergent evolution Similar features that evolved independently from ancestors that themselves appeared different from each other.

Cryptic species In birds, species that are so similar phenotypically that they have not generally been recognized as biological species based on plumage and morphology alone, but have instead been discovered using studies of DNA sequences and vocalizations.

Cultural drift Change in the frequency of a cultural variant that is independent of any special properties of the variant. Frequency may change because of differential copying (e.g., because the variant is associated with a high-status individual) or because of random mortality of individuals carrying the variant.

Cultural mutation A new song type or syllable type that arises in a population due either to copying errors or to spontaneous development.

Cultural speciation The development of premating isolation entirely through learning, with no contribution from genetic differentiation. This includes learning songs that are the subject of mate choice and learning to recognize those traits (e.g., through imprinting).

Depauperate A fauna that lacks many species that are found in similar climatic environments in other places or at different times.

Detectability As used here, the correlation between a female's preference for a male phenotype and the phenotype of male that she actually mates with (Schluter and Price 1993).

Dialect Song tradition shared by members of a local population of birds, with a boundary delineating it from other variant song traditions (Mundinger 1982).

Dispersal distance Distance between place of hatch and place of reproduction.

Distance metric As used in studies of DNA sequences, the distance between a pair of species can be summarized as the fraction of DNA bases that differ, or various metrics that take into account the possibility of multiple substitutions at one site and/or different rates of substitution at different sites.

Divergent selection Natural or sexual selection in different directions in different populations (e.g., large body size may be favored in one population and small body size may be favored in another).

DNA–DNA hybridization Formation of hybrid DNA molecules in the laboratory. The more similar the sequences, the stronger the binding and the higher the temperature at which they disassociate. The metric $T_{50}H$ is, roughly, the temperature at which 50% of the DNA from two species remains annealed (Sibley and Ahlquist 1990, p. 137).

Dobzhansky–Muller incompatibilities Interactions between alleles from different taxa that lower hybrid fitness, but that did not lower fitness when they were substituted in their respective taxa.

Dominant If one allele, A, is dominant to the other, a, the phenotype of the heterozygote Aa resembles that of the homozygote AA.

Ecological character displacement Evolutionary divergence between species, or incipient species, driven by competition for limited resources, including food and shelter. Also refers to the *pattern* of greater differences between a pair of taxa in ecologically related traits when they are found together in sympatry than expected based on differences in allopatry (see Goldberg and Lande 2006).

Ecological opportunity The presence of a diversity of underexploited, or inefficiently exploited, resources.

Ecological speciation The generation of reproductive isolation between populations as a result of ecologically based divergent selection pressures, including both natural selection and sexual selection.

Ecotone A transition zone between two distinct habitats that contains species from each area, and may also contain other species that do not occur in the two distinct habitats.

Endemic A taxon that occurs only in the particular region mentioned.

Endemism Fraction of all species (or other taxonomic group, if specified) that are endemic to the region.

F_1 hybrid The first generation of a cross between individuals drawn from different taxa.

F_2 hybrid Offspring from a cross among F_1 hybrids.

Family A taxonomic group of related animals forming a category above the genus and containing one or more genera.

Filial imprinting Learning process resulting in precocial chicks becoming attached to and following other organisms or objects to which they have been exposed and avoiding other organisms or objects.

Fitness As used here, the total fitness of an individual refers to the total number of offspring that individual produces. I also use the term to refer to a *component* of total fitness. For example, this may be the number of mates a male attracts, or survival over a certain period (so that an individual that survives has a fitness of 1, leaving one descendant (itself), and an individual that dies has a fitness of 0).

Founder effect Genetic drift that accompanies the colonization of a new location by a small number of individuals.

Frequency-dependent selection The fitness of a phenotype (i.e., the average fitness of all individuals of that phenotype) varies with the frequency of all individuals of that phenotype in the population.

Gene tree A tree depicting individual, population, or species relationships as defined by sequence variation at a particular gene.

Generalization If an organism's response is conditioned to one stimulus, the organism may also respond in a similar way to other stimuli learned by association with the first stimulus.

Genetic assimilation Refers to the process by which a phenotypic trait that is produced in response to some environmental stimulus becomes, as a result of selection, taken over by the genotype, so that is found even in the absence of the environmental influence that had at first been necessary to produce it (Waddington 1961).

Genetic drift Stochastic fluctuations in gene frequency produced by sampling.

Genomic conflict Divergent evolutionary interests of different parts of the genome (Hurst et al. 1996). Genomic conflict arises because different parts of the genome are transmitted in different ways (e.g., the W chromosome is transmitted only from mothers to daughters). This means that mutations that increase transmission frequency of one part of the genome (e.g., the W chromosome) can lower the fitness of another part of the genome.

Genotype The genetic makeup of an individual.

Guild A group of species having similar ecological-resource requirements or foraging strategies.

Haldane's rule If, in the F_1 offspring of two different animal races, one sex is absent, rare, or sterile, that sex is the heterogametic sex.

Haplotype A specific DNA sequence. For example, several individuals are said to share the same mitochondrial DNA haplotype if they have identical sequences.

Heterogametic sex The sex that carries two different sex chromosomes and therefore produces two kinds of gametes. In birds, the female is the heterogametic sex, carrying a Z chromosome and (usually a degenerate) W chromosome.

Heterozygote An individual that carries two different alleles at a particular genetic locus, here signified *Aa*.

Homozygote An individual that carries the same alleles at a particular genetic locus, here signified *AA* or *aa*, or, for a second gene, *BB* or *bb*.

Hybrid index A score assigned to an individual representing the degree to which the individual resembles one or the other parent species.

Hybrid swarm A variable population derived from the hybridization of two different taxa and consisting of multiple generations of crossing among the parentals and hybrids.

Hybrid zone Area in which genetically distinct populations meet, mate, and produce hybrids.

Hybridization Crossing of individuals from genetically distinct populations.

Individual fitness surface A plot of the average fitness of a phenotype against that phenotype (e.g., the average number of mates that males of a particular tail length receive against tail length).

Intrinsic incompatibility Hybrid inviability or infertility that is relatively independent of the social or ecological environment, generally arising from interactions between alleles from different species.

Introgression Movement of alleles from one species to another through hybrid intermediates.

Karyotype The chromosome complement of an organism.

Linked Refers to two genes nearby on the same chromosome, so that recombination between them is not free.

Macrochromosomes Typical large chromosomes in the karyotypes of birds.

Maternal effect Some aspect of the mother that affects the phenotype of her offspring, independent of the genes she has passed on to her offspring. (However, effects due to the mitochondrial genome, which is transmitted solely through the female line, are often classified as a maternal effect.)

Mean fitness The average fitness of all individuals in a population. Usually refers to a component of fitness.

Meiosis Cell division that results in the production of gametes, and a halving of chromosome number.

Meiotic drive Any mechanism that causes alleles to be recovered unequally in the gametes of a heterozygote.

Microchromosomes Relatively small chromosomes present in most bird karyotypes.

Microsatellite Tandem repetitions of short motifs of DNA (e.g., ATATATAT . . .) used as genetic markers because they commonly show length polymorphisms in populations.

Migration rate Fraction of all individuals in a population that are immigrants.

Mitochondrial DNA A circular ring of DNA found in mitochondria (16,755 base pairs long in the chicken, making up less than 0.001% of the total DNA sequence). It is passed only from the mother to the offspring.

Molecular clock The idea that the number of substitutions separating two homologous DNA sequences is proportional to the amount of time since they last shared an ancestor.

Monogamous A mating system in which one male mates with and stays with one female. This is now referred to as social monogamy, recognizing that individuals may mate outside the pairbond.

Monophyly Describes a group that contains all the descendants of their shared ancestor.

Monotypic Refers to a taxon containing only one taxon of the next lowest taxonomic category, e.g., a monotypic genus contains only one species, and a monotypic species contains only one subspecies.

Neural networks As used here, computer simulations designed to model the supposed structure and operation of the brain of an organism.

Niche The range of values of environmental factors that are necessary and sufficient to carry out an organism's life history. In this book, the concept is used more or less interchangeably with both the distribution of resources utilized by a species and the term *adaptive peak*.

Normal distribution A model of the frequency distribution of phenotypes in a population, when phenotypes are measured on a continuous scale (e.g., beak depth). The normal distribution is bell-shaped and symmetrical about the mean, and it is completely defined by the mean and the standard deviation (e.g., see Figure 5.3, p. 81). Empirically, many traits closely follow a normal distribution.

Note A continuous trace on a sonogram.

Null hypothesis The hypothesis that an observed difference (as between the means of two samples) is due to sampling alone and not due to a systematic cause.

Paracentric inversion A chromosomal inversion that does not include the centromere.

Parallel evolution The independent evolution of similar features from ancestors that themselves appeared similar to each other.

Parapatric speciation The evolution of reproductive isolation between two populations that were initially exchanging genes.

Parapatry Abutting geographical ranges.

Paraphyly Describes a group that contains those descendants of an ancestor that share certain features with the ancestor but that omits those descendants that have lost those features.

Parvorder A taxonomic category between family and order.

Passeriformes The largest order of birds, containing more than half of all living bird species.

Pericentric inversion A chromosomal inversion that includes the centromere.

Phenotype The appearance of an individual, including its morphology (size and shape), coloration, and behavior. Often, this is a quantitative measurement, e.g., beak depth.

Phenotypic plasticity The ability of an organism to alter its phenotype in response to the biotic or abiotic environment (Agrawal 2001).

Phylogenetic species The smallest diagnosable cluster of organisms within which there is a parental pattern of ancestry and descent. Now often considered to be interbreeding populations that are reciprocally monophyletic with respect to other related taxa, with monophyly determined using mitochondrial DNA.

Phylogeny Evolutionary relationships among taxonomic groups, generally species. In this book, the term is used interchangeably with the *estimate* of evolutionary relationships based on DNA sequences.

Phylogroup A reciprocally monophyletic geographic subdivision within a species. Currently, monophyly is usually defined with respect to mitochondrial-DNA sequences.

Polygamous A mating system in which an individual accepts more than one mate.

Polygenic trait A trait whose expression is influenced by more than one gene.

Polygynous A mating system in which a male mates with more than one female.

Polymorphism The presence of two or more alleles at a single locus segregating in a population.

Polytypic Refers to a taxon (e.g., species, family) with more than one subgroup at the next lowest taxonomic level.

Postmating isolation Reproductive isolation that occurs after mating; includes prezygotic isolation (failure of sperm to fertilize the egg) as well as postzygotic isolation (low viability, low fertility, or low mating success of hybrids).

Precocial species Species in which the chicks leave the nest soon after hatching. Includes many nonpasserines, such as the chicken. Compare with *altricial species.*

Premating isolation Reproductive isolation that occurs before mating; includes sexual behaviors (mate choice), occupancy of different habitats, and different breeding times.

Productivity The rate per unit area at which biomass consumable as food by other organisms is made by producers.

Promiscuous Refers to a mating behavior in which the male and female do not form lasting pair bonds.

Pure-birth model The pattern of lineage diversification through time, in which the probability of speciation remains constant and there is no extinction. Also known as the Yule model, it forms a null hypothesis against which to evaluate changing speciation rates through time.

Recessive If one allele, *a*, is recessive to the other, *A*, the phenotype of the heterozygote *Aa* resembles that of the homozygote *AA*, and is distinct from the phenotype of the homozygote, *aa*.

Recombination The process whereby a gamete carries a mixture of genes derived from both the father and the mother. Recombination is due to (1) independent assortment of chromosomes, so an individual's gamete may carry one chromosome from his or her father and another from his or her mother, and (2) crossing-over at meiosis so that one chromosome in the gamete carries parts from the father and parts from the mother. In molecular genetics, recombination is sometimes restricted to the second process, but that is not the way it is used in this book.

Regiolect Similar to dialects, but they are often more distinctive from one another, they define a group that is usually partially reproductively isolated from another group, and they extend over a larger geographical range.

Reinforcement Adaptive strengthening of premating isolation between taxa that have already developed some postmating isolation.

Reproductive character displacement As used in this book, the pattern of greater divergence in sexual traits between taxa when they are in sympatry rather than that predicted from trait measurements in allopatry (see Goldberg and Lande 2006). It is often attributed to reinforcement, but may arise for several other reasons (Noor 1999).

Restriction fragments Fragments of DNA produced by enzymes (restriction enzymes) that cut the DNA at specific sites as determined by base sequence.

RFLP Restriction fragment length polymorphism: variability in the length of DNA fragments, usually because of a mutation in the sequence that is recognized by the restriction enzyme.

Ring species Two reproductively isolated forms connected by a chain of interbreeding populations, wrapped around a geographical barrier.

Root The point of divergence between the basal two lineages in a phylogeny; the node that has offspring lineages, but no parents.

Runaway sexual selection Also known as the "Fisher process," this refers to a model in which female preferences for a male sexually selected trait evolve as a result of their correlation with the male trait. The altered preferences, in turn, affect selection on the male trait. This can lead to rapid, divergent evolution between populations.

Sex chromosomes The chromosomes carrying the sex-determining genes. In birds, they are signified as Z and W; ZW individuals are female, and ZZ individuals are male. The W chromosome is degenerate in most species.

Sexual conflict Divergent evolutionary interests of males and females.

Sexual imprinting The mating preference of an individual that develops based on learning the phenotypes of the individuals that rear it when young, during a short sensitive period. Sexual imprinting is often difficult to reverse.

Sexual selection Selection on traits resulting from differential mating success, including access to different numbers of mates, or to mates of different quality.

Sister species A pair of species that share a more recent common ancestor than either does with any other (extant) species.

Social recognition The process whereby animals become familiar with conspecifics and remember them (Mateo 2004).

Social selection Selection on traits resulting from the advantages that they bestow in social interactions, usually between members of the same species.

Song type Songs of a species that sound similar and look similar on a sonogram are classified as belonging to the same song type.

Sonogram A plot of sound frequency (hertz) against time (seconds). Also sometimes spelled sonagram, and identical with sound spectrogram.

Speciation clock The suggestion that intrinsic incompatibilities increase steadily with time. Renamed the intrinsic loss of fitness clock by Bolnick and Near (2005).

Speciation conflict Because of search costs, a rare species may be selected to sometimes mate with a more common species, but the more common species may be selected to avoid the rare species (Wilson and Hedrick 1982).

Standard deviation Square root of the variance. The standard deviation provides a measure of the variability of the population in units of the original measurements (e.g., centimeters, grams). It is one way to place different traits on a common scale of measurement. If the population is normally distributed, approximately 16% of all individuals lie more than one standard deviation above the mean, and a similar number lie more than one standard deviation below the mean.

Subsong A random, subdued warbling of longer duration than a bird's primary song.

Subspecies A population or group of populations that differ from other populations of the same species by one or more diagnostic features (also termed a geographical race).

Superspecies A collection of closely related allospecies. Often the criterion that the allospecies should be monophyletic is included. Demonstrating monophyly is difficult, because it depends on choice of gene (Appendix 1.1). However, if the allospecies are clearly not monophyletic some authors would not call the collection of allospecies a superspecies. Thus, based on mitochondrial DNA sequences, the Chestnut-bellied Monarch complex, which forms a collection of allospecies (Figures 1.1, 1.2), has the Island Monarch nested within it (Filardi and Smith 2005). Under the criterion of monophyly the Chestnut-bellied Monarch complex is not a superspecies. However, in this book I do consider examples such as these to be superspecies. First, monophyly has rarely been tested and often will be untestable (because gene sequences have not coalesced), and second, divergence among the allospecies that make up the superspecies is assumed to follow a similar process whether or not ecologically differentiated coexisting species have also been produced. (Note that extinction of the Island Monarch would result in classification of the Chestnut-bellied Monarch as a superspecies by all authorities.)

Syllable A collection of notes always sung together.

Syllable type Syllables that sound similar and that look similar on a sonogram are classified as belonging to the same syllable type.

Sympatric speciation Speciation that occurs within a single geographical area so that reproductive isolation arises between populations whose individuals have always had the opportunity to interbreed.

Sympatry Absence of geographical isolation, so that individuals of one species have a high chance of meeting individuals of another species.

Taxon Refers to any taxonomic category from subspecies to kingdom.

Taxon cycle Sequential phases of expansion and contraction of the ranges of species, associated with shifts in ecological distribution (Ricklefs and Bermingham 2002).

Tension zone A hybrid zone maintained by a balance between dispersal into the zone and low fitness of hybrids.

Trial-and-error learning Undirected behaviors with the repetition of those behaviors that are rewarded (West-Eberhard 2003, p. 39).

Tribe A taxonomic level between the genus and family.

Variance Average squared deviation from the mean; provides a measure of variability in the population. See *standard deviation*.

Vicariance The separation of a group of organisms by a geographic barrier, such as a mountain or a body of water, resulting in divergence.

Withdrawal of learning A burst of cultural mutation in song that is induced when adults are not available to tutor young birds.

Zoogeographic species A taxonomic category that includes both superspecies (which contain 2 or more allospecies) and species not divided into allospecies (Mayr and Diamond 2001).

References

Numbers in brackets after each reference indicate where it is cited in the text.

Abbott, I. 1978. The significance of morphological variation in the finch species on Gough, Inaccessible and Nightingale islands, South Atlantic Ocean. *Journal of Zoology* (London) 184:119–125. [85]

Adkisson, C. S. 1981. Geographic variation in vocalizations and evolution of North American pine grosbeaks. *Condor* 83:277–288. [190]

Adler, G. H. 1992. Endemism in birds of tropical Pacific islands. *Evolutionary Ecology* 6:296–306. [142, 143]

Adler, G. H. 1994. Avifaunal diversity and endemism on tropical Indian Ocean islands. *Journal of Biogeography* 21:85–95. [142, 143]

Agrawal, A. A. 2001. Phenotypic plasticity in the interactions and evolution of species. *Science* 294:321–326. [407]

Aitchison, J. 1991. *Language Change: Progress or Decay?* Cambridge University Press, Cambridge. [209–210, 296]

Alatalo, R. V., and L. Gustafsson. 1988. Genetic component of morphological differentiation in coal tits under competitive release. *Evolution* 42:200–203. [44, 137]

Alatalo, R. V., and J. Moreno. 1987. Body size, interspecific interactions, and use of foraging sites in tits (Paridae). *Ecology* 68:1773–1777. [128]

Alatalo, R. V., A. Carlson, A. Lundberg, and S. Ulfstrand. 1981. The conflict between male polygamy and female monogamy: The case of the pied flycatcher *Ficedula hypoleuca*. *American Naturalist* 117:738–753. [312]

Alatalo, R. V., L. Gustafsson, M. Linden, and A. Lundberg. 1985. Interspecific competition and niche shifts in tits and the goldcrest: An experiment. *Journal of Animal Ecology* 54:977–984. [128]

Alatalo, R. V., L. Gustafsson, and A. Lundberg. 1986a. Interspecific competition and niche changes in tits (*Parus* spp)—evaluation of non-experimental data. *American Naturalist* 127:819–834. [50]

Alatalo, R. V., A. Lundberg, and C. Glynn. 1986b. Female pied flycatchers choose territory quality and not male characteristics. *Nature* 323:152–153. [313]

Alatalo, R. V., D. Eriksson, L. Gustafsson, and A. Lundberg. 1990. Hybridization between pied and collared flycatchers—sexual selection and speciation theory. *Journal of Evolutionary Biology* 3:375–389. [307, 308, 316, 356, 390]

Alatalo, R. V., L. Gustafsson, and A. Lundberg. 1994. Male colouration and species recognition in sympatric flycatchers. *Proceedings of the Royal Society of London Series B—Biological Sciences* 256:113–118. [308–310, 314, 339, 356]

Albert, A. Y. K., and D. Schluter. 2004. Reproductive character displacement of male stickleback mate preference: Reinforcement or direct selection? *Evolution* 58:1099–1107. [314]

Aldrich, J. W. 1982. Rapid evolution in the house finch (*Carpodacus mexicanus*). *Journal of the Yamashina Institute for Ornithology* 14:179–186. [47]

Aldrich, J. W. 1984. Ecogeographical variation in size and proportions of song sparrows (*Melospiza melodia*). *Ornithological Monographs* 35:1–134. [33]

Aleixo, A. 2004. Historical diversification of a Terra-firme forest bird superspecies: A phylogeographic perspective on the role of different hypotheses of Amazonian diversification. *Evolution* 58:1303–1317. [27]

Alerstam, T., A. Hedenström, and S. Åkesson. 2003. Long-distance migration: Evolution and determinants. *Oikos* 103:247–260. [92]

Aliabadian, M., C. S. Roselaar, V. Nijman, R. Sluys, and M. Vences. 2005. Identifying contact zone hotspots of passerine birds in the Palearctic region. *Biology Letters* 1:21–23. [114, 324, 332]

Allen, E. S. 2002. Long-term hybridization and the maintenance of species identity in orioles (*Icterus*). Unpublished Ph.D. thesis, Indiana University, Bloomington. [365]

Allen, E. S., and K. E. Omland. 2003. Novel intron phylogeny supports plumage convergence in orioles (*Icterus*). *Auk* 120:961–970. [234]

Alström, P., and U. Olsson. 1999. The golden-spectacled warbler: A complex of sibling species, including a previously undescribed species. *Ibis* 141:545–568. [2, 315, 316]

Alström, P., and U. Olsson. 2000. Golden-spectacled warbler systematics. *Ibis* 142:495–500. [315, 316]

Alström, P., and R. Ranft. 2003. The use of sounds in avian systematics and the importance of bird sound archives. *Bulletin of the British Ornithologists' Club*, supplement 123A:114–135. [4, 271]

Amadon, D. 1966. The superspecies concept. *Systematic Zoology* 15:245–249. [3, 403]

Amadon, D. 1968. Further remarks on the superspecies concept. *Systematic Zoology* 17:345–346. [114]

Anderson, K. E., S. I. Rothstein, R. C. Fleischer, and A. L. O'Loghlen. 2005. Large-scale movement patterns between song dialects in brown–headed cowbirds *(Molothrus ater)*. *Auk* 122:803–818. [280]

Andersson, M. 1994. *Sexual Selection*. Princeton University Press, Princeton, NJ. [158, 160, 180]

Andrew, R. J. 1957. The aggressive and courtship behaviour of certain emberizines. *Behaviour* 10:255–308. [167]

Anon. 1998. *The American Standard of Perfection*. American Poultry Association, Inc., Mendon, MA. [222]

Arak, A., and M. Enquist. 1993. Hidden preferences and the evolution of signals. *Philosophical Transactions of the Royal Society Series B* 340:207–213. [171, 173]

Arak, A., and M. Enquist. 1995. Conflict, receiver bias and the evolution of signal form. *Philosophical Transactions of the Royal Society Series B* 349:339–344. [169, 180, 184]

Armstrong, E. A. 1947. *Bird Display and Behavior: An Introduction to the Study of Bird Psychology*. Oxford University Press, Oxford. [175, 240]

Armstrong, E. A. 1955. *The Wren*. Collins, London. [265]

Arnaiz-Villena, A., M. Alvarez-Tejado, V. Ruiz-del-Valle, C. Garcia-de-la-Torre, P. Varela, M. J. Recio, S. Ferre, and J. Martinez–Laso. 1998. Phylogeny and rapid Northern and Southern Hemisphere speciation of goldfinches during the Miocene and Pliocene epochs. *Cellular and Molecular Life Sciences* 54: 1031–1041. [103]

Arnold, M. L. 1997. *Natural Hybridization and Evolution*. Oxford University Press, Oxford. [323]

Arnqvist, R., and M. Kirkpatrick. 2005. The evolution of infidelity in socially monogamous passerines: The strength of direct and indirect selection on extrapair copulation behavior in females. *American Naturalist* 165:S26–S37. [249]

Arnqvist, G. A., and L. Rowe. 2005. *Sexual Conflict*. Princeton University Press, Princeton, NJ. [180, 239]

Aubin, T., and P. Jouventin. 2002. How to vocally identify kin in a crowd: The penguin model. *Advances in the Study of Behavior* 31:243–277. [284]

Avanzi, C. F., and R. D. Crawford. 1990. Mutations and major variants in muscovy ducks. Pp. 389–394 *in* R. D. Crawford, ed. *Poultry Breeding and Genetics*. Elsevier, Oxford, New York and Tokyo. [226]

Avise, J. C., and D. Walker. 1998. Pleistocene phylogeographic effects on avian populations and the speciation process. *Proceedings of the Royal Society of London Series B—Biological Sciences* 265:457–463. [31, 107]

Avise, J. C., and K. Wollenberg. 1997. Phylogenetics and the origin of species. *Proceedings of the National Academy of Sciences of the United States of America* 94:7748–7755. [37]

Badyaev, A. V. 1997. Altitudinal variation in sexual dimorphism: A new pattern and alternative hypotheses. *Behavioral Ecology* 8:675–690. [260]

Badyaev, A. V., and E. S. Leaf. 1997. Habitat associations of song characteristics in *Phylloscopus* and *Hippolais* warblers. *Auk* 114:40–46. [253, 254, 256]

Badyaev, A. V., and G. E. Hill. 2003. Avian sexual dichromatism in relation to phylogeny and ecology. *Annual Reviews of Ecology and Systematics* 34:27–49. [240, 258, 259, 263]

Badyaev, A. V., G. E. Hill, and B. V. Weckworth. 2002. Species divergence in sexually selected traits: Increase in song elaboration is related to decrease in plumage ornamentation in finches. *Evolution* 56:412–419. [255, 259, 261]

Baker, M. C. 1975. Song dialects and genetic differences in white-crowned sparrows *(Zonotrichia leucophrys)*. *Evolution* 29:226–241. [206]

Baker, A. J. 1980. Morphometric differentiation in New Zealand populations of the house sparrow *(Passer domesticus)*. *Evolution* 34:638–653. [46]

Baker, M. C. 1983. The behavioral response of female Nuttall's white-crowned sparrows to male song of natal and alien dialects. *Behavioral Ecology and Sociobiology* 12:309–315. [191, 196]

Baker, M. 1995. Environmental component of latitudinal clutch size variation in house sparrows *(Passer domesticus)*. *Auk* 112:249–252. [45]

Baker, M. C. 1996. Depauperate meme pool of vocal signals in an island population of singing honeyeaters. *Animal Behaviour* 51:853–858. [264, 267]

Baker, M. C. 2006. Differentiation of mating vocalizations in birds: Acoustic features in mainland and island populations and evidence of habitat-dependence in songs. *Ethology* 112:757-771.[267]

Baker, M. C., and A. E. M. Baker. 1990. Reproductive behavior of female buntings: Isolating mechanisms in a hybridizing pair of species. *Evolution* 44:332–338. [286, 288, 319]

Baker, R. J., and J. W. Bickham. 1986. Speciation by monobrachial centric fusions. *Proceedings of the National Academy of Sciences of the United States of America* 83:8245–8248. [380]

Baker, M. C., and J. T. Boylan. 1995. A catalogue of song syllables of indigo and lazuli buntings. *Condor* 97:1028–1040. [188, 192, 215–216, 365]

Baker, M. C., and J. T. Boylan. 1999. Singing behavior, mating associations and reproductive success in a population of hybridizing lazuli and indigo buntings. *Condor* 101:493–504. [216, 307, 319, 365]

Baker, A. J., and P. F. Jenkins. 1987. Founder effect and cultural evolution of songs in an isolated population of chaffinches, *Fringilla coelebs*, in the Chatham Islands. *Animal Behaviour* 35:1793–1803. [188, 204, 216–217, 265, 269]

Baker, M. C., and M. S. Johnson. 1998. Allozymic and morphometric comparisons among indigo and lazuli buntings and their hybrids. *Auk* 115:537–542. [365]

Baker, A. J., and A. Moeed. 1979. Evolution in the introduced New Zealand populations of the common myna, *Acridotheres tristis* (Aves: Sturnidae). *Canadian Journal of Zoology* 57:570–584. [46]

Baker, R. R., and G. A. Parker. 1979. The evolution of bird coloration. *Philosophical Transactions of the Royal Society*, B 287:63–130. [259, 260]

Baker, M. C., T. K. Bjerke, H. U. Lampe, and Y. O. Espmark. 1987. Sexual response of female yellowhammers to differences in regional song dialects and repertoire sizes. *Animal Behaviour* 35:395–401. [191, 212–213, 292]

Baker, M. C., M. S. A. Baker, and E. M. Baker. 2003. Rapid evolution of a novel song and an increase in repertoire size in an island population of an Australian songbird. *Ibis* 145:465–471. [269]

Baker, M. C., M. S. A. Baker, and L. M. Tilghman. 2006. Differing effects of isolation on evolution of bird songs: Examples from an island-mainland comparison of three species. *Biological Journal of the Linnean Society* 89:331–342. [264]

Balaban, E. 1988. Bird song syntax—learned intraspecific variation is meaningful. *Proceedings of the National Academy of Sciences of the United States of America* 85:3657–3660. [190, 191, 192, 205]

Baldwin, J. M. 1896. A new factor in evolution. *American Naturalist* 30:441–451; 536–553. [126]

Balph, M. H. 1977. Winter social behavior of dark eyed juncos—communication, social organization, and ecological implications. *Animal Behaviour* 25:859–884. [167, 283]

Baptista, L. F. 1975. Song dialects and demes in sedentary populations of the white-crowned sparrow. *University of California Publications in Zoology* 105:1–52. [193, 307]

Baptista, L. F. 1977. Geographic variation in song and dialects of the Puget Sound white-crowned sparrow. *Condor* 79:356–370. [193–194]

Baptista, L. F. 1981. Behavior genetics studies with birds. Pp. 217–249 *in* A. C. Risser, L. F. Baptista, S. R. Wylie, and N. B. Gale, eds. *Proceedings of the First International Birds in Captivity Symposium. 1978, Seattle, Washington.* International Foundation for the Conservation of Birds, Hollywood, California. North Hollywood, CA. [224]

Baptista, L. F. 1996. Nature and its nurturing in vocal development. Pp. 39–60 *in* D. E. Kroodsma and E. H. Miller, eds. *Ecology and Evolution of Acoustic Communication in Birds.* Cornell University Press, Ithaca, NY and London. [187]

Baptista, L. F., and M. Abs. 1983. Vocalizations. Pp. 309–325 *in* M. Abs, ed. *Physiology and Behaviour of the Pigeon.* Academic Press, London. [228]

Baptista, L. F., and R. B. Johnson. 1982. Song variation in insular and mainland California brown creepers (*Certhia familiaris*). *Journal für Ornithologie* 123:131–144. [267]

Baptista, L. F., and D. E. Kroodsma. 2001. Avian bioacoustics. Pp. 11–52 *in* J. del Hoyo, A. Elliott, and J. Sargatal, eds. *Handbook of Birds of the World.* Lynx Editions, Barcelona. [187]

Baptista, L. F., and P. W. Trail. 1992. The role of song in the evolution of passerine diversity. *Systematic Biology* 41:242–247. [192, 213]

Baptista, L. F., D. A. Bell, and P. W. Trail. 1993. Song learning and production in the white-crowned sparrow—parallels with sexual imprinting. *Netherlands Journal of Zoology* 43:17–33. [28]

Baptista, L. F., P. W. Trail, and H. M. Horblit. 1997. Order Columbiformes. Pp. 60–243 *in* J. del Hoyo, A. Elliott, and J. Sargatal, eds. *Handbook of Birds of the World.* Lynx Editions, Barcelona. [229, 230]

Barker, F. K., A. Cibois, P. Schikler, J. Feinstein, and J. Cracraft. 2004. Phylogeny and diversification of the largest avian radiation. *Proceedings of the National Academy of Sciences of the United States of America* 101:11040–11045. [38, 115, 116, 122]

Barlow, J. C. 1978. Another colony of the Guadeloupe house wren. *Wilson Bulletin* 90:635–637. [267]

Barraclough, T. G., P. H. Harvey, and S. Nee. 1995. Sexual selection and taxonomic diversity in passerine birds. *Proceedings of the Royal Society of London Series B—Biological Sciences* 259:211–215. [270, 271]

Barraclough, T. G., and A. P. Vogler. 2000. Detecting the geographical pattern of speciation from species-level phylogenies. *American Naturalist* 155:419–434. [34]

Barrowclough, G. F. 1980. Genetic and phenotypic differentiation in a wood warbler (Genus *Dendroica*) hybrid zone. *Auk* 97:655–668. [363]

Barry, K. L., and A. Göth. 2006. Call recognition in chicks of the Australian brush turkey (*Alectura lathami*). *Animal Cognition* 9:47–54. [280]

Barton, N. H. 1979. Gene flow past a cline. *Heredity* 43:333–339. [30, 208, 345]

Barton, N. H. 1992. On the spread of new gene combinations in the 3rd phase of Wright's shifting balance. *Evolution* 46:551–557. [58]

Barton, N. H. 2001. The role of hybridization in evolution. *Molecular Ecology* 10:551–568. [323, 346]

Barton, N. H., and G. M. Hewitt. 1981. Hybrid zones and speciation. Pp. 109–145 *in* W. R. Atchley and D. S. Woodruff, eds. *Evolution and Speciation: Essays in Honor of M. J. D. White.* Cambridge University Press, Cambridge. [208, 347]

Barton, N. H., and G. M. Hewitt. 1985. Analysis of hybrid zones. *Annual Review of Ecology and Systematics* 16:113–148. [67, 324, 328, 331, 336]

Barton, N. H., and G. M. Hewitt. 1989. Adaptation, speciation and hybrid zones. *Nature* 341:497–503. [207, 208, 328, 331, 332, 347]

Bates, J. M., and R. M. Zink. 1994. Evolution into the Andes—molecular evidence for species relationships in the Genus *Leptopogon. Auk* 111:507–515. [118]

Bateson, P. P. G. 1966. The characteristics and context of imprinting. *Biological Reviews* 41:177–220. [282]

Bazykin, A. D. 1969. Hypothetical mechanism of speciation. *Evolution* 23:685–687. [328]

Bearhop, S., W. Fiedler, R. W. Furness, S. C. Votier, S. Waldron, J. Newton, G. J. Bowen, P. Berthold, and K. Farnsworth. 2005. Assortative mating as a mechanism for rapid evolution of a migratory divide. *Science* 310:502–504. [94]

Becker, P. H. 1977. Geographische Variation des Gesanges von Winter-und Sommergoldhähnchen *(Regulus regulus, R. ignicapillus). Die Vogelwarte* 29:1–37. [190, 267]

Becker, P. H. 1978. Der Einfluss des Lernens auf einfache und komplexe Gesangsstrophen der Sumpfmeise *(Parus palustris). Journal für Ornithologie* 119:388–411. [201, 202]

Becker, P. H. 1982. The coding of species-specific characteristics of bird sounds. Pp. 213–252 *in* D. E. Kroodsma and E. H. Miller, eds. *Acoustic Communication in Birds.* Academic Press, New York and London. [198, 201]

Becker, J. 1995. Sympatrisches Vorkommen und Hybridisierung von Sprosser *Luscinia luscinia* und Nachtigall *L. megarhynchos* bei Frankfurt (Oder), Brandenburg. *Vogelwelt* 116:109–118. [357]

Becker, P. H., G. Thielcke, and K. Wüstenberg. 1980. Versuche zum angenommenen Kontrastverlust im Gesang der Blaumeise *(Parus caeruleus)* auf Teneriffa. *Journal für Ornithologie* 121:81–95. [267]

Bed'Hom, B., P. Coullin, Z. Guillier-Gencik, S. Moulin, A. Bernheim, and V. Volobouev. 2003. Characterization of the atypical karyotype of the black-winged kite *Elanus caeruleus* (Falconiformes: Accipitridae) by means of classical and molecular cytogenetic techniques. *Chromosome Research* 11:335–343. [384]

Beecher, M. D. 1982. Signature systems and kin recognition. *American Zoologist* 22:477–490. [284]

Beecher, M. D., and E. A. Brenowitz. 2005. Functional aspects of song learning in songbirds. *Trends in Ecology and Evolution* 20:143–149. [189, 201, 211, 281]

Beecher, M. D., S. E. Campbell, and P. K. Stoddard. 1994. Correlation of song learning and territory establishment strategies in the song sparrow. *Proceedings of the National Academy of Sciences of the United States of America* 91:1450–1454. [199, 200]

Beecher, M. D., P. K. Stoddard, S. E. Campbell, and C. L. Horning. 1996. Repertoire matching between neighboring song sparrows. *Animal Behaviour* 51:917–923. [202]

Beecher, M. D., S. E. Campbell, J. M. Burt, C. E. Hill, and J. C. Nordby. 2000. Song-type matching between neighbouring song sparrows. *Animal Behaviour* 59:21–27. [202]

Beehler, B. 1981. Ecological structuring of forest bird communities in New Guinea. *Monographiae Biologicae* 42:837–861. [101]

Bélichon, S., J. Clobert, and M. Massot. 1996. Are there differences in fitness components between philopatric and dispersing individuals? *Acta Œcologia* 17:503–517. [68]

Bell, D. A. 1996. Genetic differentiation, geographic variation and hybridization in gulls of the *Larus glaucescens-occidentalis* complex. *Condor* 98:527–546. [353]

Belliure, J., G. Sorci, A. P. Møller, and J. Clobert. 2000. Dispersal distances predict subspecies richness in birds. *Journal of Evolutionary Biology* 13:480–487. [33, 58, 59]

Beltman, J. B., and P. Haccou. 2005. Speciation through the learning of habitat features. *Theoretical Population Biology* 67:189–202. [299]

Benkman, C. W. 1993. Adaptation to single resources and the evolution of crossbill *(Loxia)* diversity. *Ecological Monographs* 63:305–325. [47, 77, 79]

Benkman, C. W. 1999. The selection mosaic and diversifying coevolution between crossbills and lodgepole pine. *American Naturalist* 153:S75–S91. [43, 47, 80]

Benkman, C. W. 2003. Divergent selection drives the adaptive radiation of crossbills. *Evolution* 57:1176–1181. [77, 79]

Benkman, C. W., W. C. Holimon, and J. W. Smith. 2001. The influence of a competitor on the geographic mosaic of coevolution between crossbills and lodgepole pine. *Evolution* 55:282–294. [69, 77, 79, 80]

Bennett, R. B. 1961. *Care and Breeding of Budgerigars, Canaries and Foreign Finches.* Arco, New York. [222, 227]

Bensch, S., and D. Hasselquist. 1992. Evidence for active choice in a polygynous warbler. *Animal Behaviour* 44:301–311. [288]

Bensch, S., D. Hasselquist, B. Nielsen, and B. Hansson. 1998. Higher fitness for philopatric than for immigrant males in a semi-isolated population of great reed warblers. *Evolution* 52:877–883. [68, 72]

Bensch, S., T. Andersson, and S. Akesson. 1999. Morphological and molecular variation across a migratory divide in willow warblers, *Phylloscopus trochilus. Evolution* 53:1925–1935. [325, 335, 336, 360]

Bensch, S., A. J. Helbig, M. Salomon, and I. Siebold. 2002. Amplified fragment length polymorphism analysis identifies hybrids between two subspecies of warblers. *Molecular Ecology* 11:473–481. [345, 361]

Bensch, S., D. E. Irwin, J. H. Irwin, L. Kvist, and S. Akesson. 2006. Conflicting patterns of mitochondrial and nuclear DNA diversity in *Phylloscopus* warblers. *Molecular Ecology* 15:161–171. [11, 12]

Bentley, G. E., J. C. Wingfield, M. L. Morton, and G. F. Ball. 2000. Stimulatory effects on the reproductive axis in female songbirds by conspecific and heterospecific male song. *Hormones and Behavior* 37:179–189. [163]

Bergmann, C. 1847. Ueber die Verhältnisse der Wärmeökonomie der Thiere zu ihrer Grösse. *Göttinger Studien* 3, pt 1:595–708. [43]

Bergmann, H. H. 1976. Konstitutionsbedingte Merkmale in Gesängen und Rufen europäischer Grasmücken (Gattung *Sylvia*). *Zeitschrift für Tierpsychologie* 42:315–329. [254]

Bertelli, S., and P. L. Tubaro. 2002. Body mass and habitat correlates of song structure in a primitive group of birds. *Biological Journal of the Linnean Society* 77:423–430. [254]

Berthold, P., and U. Querner. 1981. Genetic basis of migratory behavior in European warblers. *Science* 212:77–79. [94]

Berthold, P., and U. Querner. 1982. Genetic basis of moult, wing length and body weight in a migratory bird species, *Sylvia atricapilla. Experientia* 38:801–802. [44]

Berthold, P., A. J. Helbig, G. Mohr, and U. Querner. 1992. Rapid microevolution of migratory behavior in a wild bird species. *Nature* 360:668–670. [46, 93]

Bertram, B. 1970. The vocal behaviour of the Indian hill mynah, *Gracula religiosa. Animal Behaviour Monographs* 3:81–192. [190]

Binks, G. S. 1997. *The Challenge: Breeding Championship Budgerigars.* Osprey International, Tanglewood, Virginia Water, UK. [222, 224–226]

Birkhead, T. R. 1978. Behavioural adaptations to high density nesting in the Common guillemot *Uria aalge. Animal Behaviour* 26:321–331. [168]

Birkhead, T. R. 2003. *The Red Canary.* Weidenfeld and Nicholson, London. [135, 221]

Birkhead, T. R., and J.-P. Brillard. 2007. Reproductive isolation in birds: postcopulatory prezygotic barriers. *Trends in Ecology and Evolution.* In press. [371, 372]

Birkhead, T. R., and T. Pizzari. 2002. Postcopulatory sexual selection. *Nature Reviews Genetics* 3:262–273. [372]

Birkhead, T. R., S. Immler, E. J. Pellatt, and R. Freckleton. 2006. Unusual sperm morphology in the Eurasian Bullfinch *(Pyrrhula pyrrhula). Auk* 123:383–392. [372]

Birkhead, T. R., E. J. Pellat, R. Yeates, and E. Schut. 2007. Post-copulatory reproductive barriers in birds: An experimental test. *Reproduction.* In press. [372]

Bischof, H. J., and N. Clayton. 1991. Stabilization of sexual preferences by sexual experience in male zebra finches *Taeniopygia guttata castanotis. Behaviour* 118:144–155. [278]

Bischof, H. J., J. Böhner, and R. Sossinka. 1981. Influence of external stimuli on the quality of song of the zebra finch *(Taeniopygia guttata castanotis* Gould). *Zeitschrift für Tierpsychologie* 57:261–267. [164]

Bitterbaum, E., and L. F. Baptista. 1979. Geographical variation in songs of California house finches *(Carpodacus mexicanus). Auk* 96:462–474. [188, 193, 206, 267, 269]

Bleiweiss, R. 1997. Covariation of sexual dichromatism and plumage colours in lekking and nonlekking birds: A comparative analysis. *Evolutionary Ecology* 11:217–235. [160, 250]

Bleiweiss, R. 1998. Slow rate of molecular evolution in high-elevation hummingbirds. *Proceedings of the National Academy of Sciences of the United States of America* 95:612–616. [36]

Blem, C. R. 1981. Geographic variation in mid-winter body composition of starlings. *Condor* 83:370–376. [46]

Blondel, J., P. C. Dias, P. Perret, M. Maistre, and M. M. Lambrechts. 1999. Selection-based biodiversity at a small spatial scale in a low-dispersing insular bird. *Science* 285:1399–1402. [44, 68]

Boag, P. T., and P. R. Grant. 1981. Intense natural selection in a population of Darwin's finches (Geospizinae) in the Galápagos. *Science* 214:82–85. [47]

Boag, P. T., and P. R. Grant. 1984. Darwin's finches *(Geospiza)* on Isla Daphne Major, Galápagos—breeding and feeding ecology in a climatically variable environment. *Ecological Monographs* 54:463–489. [127]

Boecklen, W. J. 1997. Nestedness, biogeographic theory, and the design of nature reserves. *Oecologia* 112:123–142. [54, 56]

Böhning-Gaese, K., M. D. Schuda, and A. J. Helbig. 2003. Weak phylogenetic effects on ecological niches of *Sylvia* warblers. *Journal of Evolutionary Biology* 16:956–965. [102]

Bolhuis, J. J. 1991. Mechanisms of avian imprinting: A review. *Biological Reviews* 66:303–345. [281–283]

Bolhuis, J. J., and H. S. van Kampen. 1992. An evaluation of auditory learning in filial imprinting. *Behaviour* 122:195–230. [282, 284]

Bolnick, D. I., and T. J. Near. 2005. Tempo of hybrid inviability in centrarchid fishes (Teleostei: Centrarchidae). *Evolution* 59:1754 1767. [376, 408]

Borge, T., K. Lindroos, P. Nádvornik, A. C. Syvänen, and G. P. Sætre. 2005. Amount of introgression in fly-catcher hybrid zones reflects regional differences in pre and post-zygotic barriers to gene exchange. *Journal of Evolutionary Biology* 18:1416–1424. [345, 356]

Borgia, G. 1985. Bower destruction and sexual competition in the satin bowerbird *(Ptilonorhynchus violaceus)*. *Behavioral Ecology and Sociobiology* 18:91–100. [185]

Borgia, G., and M. A. Gore. 1986. Feather stealing in the satin bowerbird *(Ptilonorhynchus violaceus):* Male competition and the quality of display. *Animal Behaviour* 34:727–738. [185]

Bowman, R. I. 1979. Adaptive morphology of song dialects in Darwin's finches. *Journal für Ornithologie* 120:353–389. [254, 261, 262]

Bowman, R. I., and S. I. Billeb. 1968. Blood eating in a Galápagos finch. *Living Bird* 4:29–44. [127]

Bradbury, J. W. 1981. The evolution of leks. Pp. 138–169 *in* R. D. Alexander and D. W. Tinkle, eds. *Natural Selection and Social Behavior.* Chiron Press, New York and Concord. [248–249]

Bradbury, J. W., and S. L. Vehrencamp. 2000. Economic models of animal communication. *Animal Behaviour* 59:259–268. [179, 184]

Braun, M. D., G. B. Kitto, and M. J. Braun. 1984. Molecular population genetics of tufted and black-crested forms of *Parus bicolor. Auk* 101:170–173. [359]

Brewer, R. 1963. Ecological and reproductive relationships of black-capped and Carolina chickadees. *Auk* 80:9–47. [358]

Briskie, J. V., and R. Montgomerie. 1992. Sperm size and sperm competition in birds. *Proceedings of the Royal Society* B—Biological Sciences 247:89–95. [372]

Brockway, B. F. 1965. Stimulation of ovarian development and egg laying by male courtship vocalization in budgerigars *(Melopsittacus undulatus). Animal Behaviour* 13:575–578. [162]

Brodsky, L. M., C. D. Ankney, and D. G. Dennis. 1988. The influence of male dominance on social interactions in black ducks and mallards. *Animal Behaviour* 36:1371–1378. [313]

Bromham, L. 2003. Molecular clocks and explosive radiations. *Journal of Molecular Evolution* 57:S13–S20. [15]

Bronson, C. L., T. C. Grubb, and M. J. Braun. 2003a. A test of the endogenous and exogenous selection hypotheses for the maintenance of a narrow avian hybrid zone. *Evolution* 57:630–637. [335, 358]

Bronson, C. L., T. C. Grubb, G. D. Sattler, and M. J. Braun. 2003b. Mate preference: A possible causal mechanism for a moving hybrid zone. *Animal Behaviour* 65:489–500. [313, 340, 358]

Bronson, C. L., T. C. Grubb Jr., G. D. Sattler, and M. J. Braun. 2005. Reproductive success across the black-capped chickadee *(Poecile atricapillus)* and Carolina chickadee *(P. carolinensis)* hybrid zone in Ohio. *Auk* 122:759–772. [335, 345, 358]

Brooke, M. D. L. 1978. Sexual differences in the voice and individual vocal recognition in the Manx shearwater *(Puffinus puffinus). Animal Behaviour* 26:622–629. [284]

Brown, T. J., and P. Handford. 1996. Acoustic signal amplitude patterns: A computer simulation investigation of the acoustic adaptation hypothesis. *Condor* 98:608–613. [256]

Brumfield, R. T. 2005. Mitochondrial variation in Bolivian populations of the variable antshrike *(Thamnophilus caerulescens). Auk* 122:414–432. [354]

Brumfield, R. T., R. W. Jernigan, D. B. McDonald, and M. J. Braun. 2001. Evolutionary implications of divergent clines in an avian *(Manacus:* Aves) hybrid zone. *Evolution* 55:2070–2087. [338, 341, 346, 347, 353, 354]

Brush, A. H. 1990. Metabolism of carotenoid pigments in birds. *FASEB Journal* 4:2969–2977. [135]

Bubier, N. E., C. G. M. Paxton, P. Bowers, and D. C. Deeming. 1998. Courtship behaviour of ostriches *(Struthio camelus)* towards humans under farming conditions in Britain. *British Poultry Science* 39:477–481. [277]

Buckley, P. A. 1969. Disruption of species typical behavior patterns in F1 *Agapornis* parrots. *Zeitschrift für Tierpsychologie* 26:737–743. [83, 390, 391]

Buckley, P. A. 1982. Avian genetics. Pp. 21–110 *in* M. Petrak, ed. *Diseases of Cage and Aviary Birds,* 2nd ed. Lea and Febiger, Philadelphia. [315, 391]

Burbidge, M. L., R. M. Colbourne, H. A. Robertson, and A. J. Baker. 2003. Molecular and other biological evidence supports the recognition of at least three species of brown kiwi. *Conservation Genetics* 4:167–177. [113]

Burley, N. 1986a. Comparison of the band-colour preferences of two species of estrildid finches. *Animal Behaviour* 34:1732–1741. [173, 174]

Burley, N. 1986b. Sexual selection for aesthetic traits in species with biparental care. *American Naturalist* 127:415–445. [173, 174, 180]

Burley, N. 1988. The differential-allocation hypothesis: An experimental test. *American Naturalist* 132:611–628. [162, 164, 173, 174, 180]

Burley, N. T., and R. Symanski. 1998. "A taste for the beautiful": Latent aesthetic mate preferences for white crests in two species of Australian grassfinches. *American Naturalist* 152:792–802. [173, 174]

Burley, N. T., P. G. Parker, and K. Lundy. 1996. Sexual selection and extra-pair fertilization in a socially monogamous passerine, the zebra finch *(Taeniopygia guttata)*. *Behavioral Ecology* 7:218–226. [174]

Burns, K. J. 1997. Molecular systematics of tanagers (Thraupinae): Evolution and biogeography of a diverse radiation of neotropical birds. *Molecular Phylogenetics and Evolution* 8:334–348. [115, 116]

Burns, K. J. 1998. A phylogenetic perspective on the evolution of sexual dichromatism in tanagers (Thraupidae): The role of female versus male plumage. *Evolution* 52:1219–1224. [240]

Burns, K. J., and K. Naoki. 2004. Molecular phylogenetics and biogeography of Neotropical tanagers in the genus *Tangara*. *Molecular Phylogenetics and Evolution* 32:838–854. [102, 119]

Burns, K. J., S. J. Hackett, and N. K. Klein. 2002. Phylogenetic relationships and morphological diversity in Darwin's finches and their relatives. *Evolution* 56:1240–1252. [24]

Burt, D. W. 2002. Origin and evolution of avian microchromosomes. *Cytogenetic and Genome Research* 96:97–112. [383]

Burt, D. W., C. Bruley, I. C. Dunn, C. T. Jones, A. Ramage, A. S. Law, D. R. Morrice, I. R. Paton, J. Smith, D. Windsor, A. Sazanov, R. Fries, and D. Waddington. 1999. The dynamics of chromosome evolution in birds and mammals. *Nature* 402:411–413. [384]

Burton, T. C., and A. A. Martin. 1976. Analysis of hybridization between black-headed and white-backed magpies in south-eastern Australia. *Emu* 76:30–37. [355]

Butlin, R. K. 1987. Species, speciation, and reinforcement. American Naturalist 130:461–464. [303]

Calder, W. C. 1978. The kiwi. *Scientific American* 239 (July):132–143. [113]

Cardillo, M., C. D. L. Orme, and I. P. F. Owens. 2005. Testing for latitudinal bias in diversification rates: an example using New World birds. *Ecology* 86:2278–2287. [117]

Caro, S. P., M. M. Lambrechts, J. Balthazart, and P. Peret. 2007. Non-photoperiodic factors and timing of breeding in blue tits: Impact of environmental and social influences in semi-natural conditions. *Behavioural Processes*. In press. [45]

Case, T. J. 1978. A general explanation for insular body size trends in terrestrial vertebrates. *Ecology* 59:1–18. [48]

Case, T. J. 1990. Invasion resistance arises in strongly interacting species-rich model competition communities. *Proceedings of the National Academy of Sciences of the United States of America* 87:9610–9614. [152]

Case, T. J., and M. L. Taper. 2000. Interspecific competition, environmental gradients, gene flow, and the coevolution of species' borders. *American Naturalist* 155:583–605. [109, 111, 112, 332, 333]

Cassey, P., T. M. Blackburn, S. Sol, R. P. Duncan, and J. L. Lockwood. 2004. Global patterns of introduction effort and establishment success in birds. *Proceedings of the Royal Society of London Series B—Biological Sciences* 271:S405–S408. [131]

Catchpole, C. K. 1980. Sexual selection and the evolution of complex songs among European warblers of the genus *Acrocephalus*. *Behaviour* 74:149–166. [251]

Catchpole, C. K. 2000. Sexual selection and the evolution of song and brain structure in *Acrocephalus* warblers. *Advances in the Study of Behaviour* 29:45–97. [251]

Catchpole, C. K., and J. Komdeur. 1993. The song of the Seychelles warbler *Acrocephalus sechellensis*, an island endemic. *Ibis* 135:190–195. [264]

Catchpole, C. K., and P. J. B. Slater. 1995. *Bird Song: Biological Themes and Variations*. Cambridge University Press, Cambridge. [187, 189, 238, 252]

Chan, K. M. A., and S. A. Levin. 2005. Prezygotic isolation and porous genomes: Rapid introgression of maternally inherited DNA. *Evolution* 59:720–729. [335]

Chang, J. T. 1999. Recent common ancestors of all present-day individuals. *Advances in Applied Probability* 31:1002-1026. [11]

Chapin, J. P. 1948. Variation and hybridization among the paradise flycatchers of Africa. *Evolution* 2:111–126. [310, 347]

Chappuis, C. 1971. Un example de l'influence du milieu sur les émissions vocales des oiseaux: L'évolution des chantes en forêt équatoriale. *Terre Vie* 25:183–202. [256]

Cheke, A. S. 1969. Mechanism and consequences of hybridization in sparrows, *Passer*. *Nature* 222:179–180. [276]

Cheng, K. M., R. N. Shoffner, R. E. Phillips, and F. B. Lee. 1979. Mate preference in wild and domesticated (game-farm) mallards: II. Pairing success. *Animal Behaviour* 27:417–425. [277, 279]

Cherry, J. L., F. R. Adler, and K. P. Johnson. 2002. Islands, equilibria, and speciation. *Science* 296:975a. [151]

Chesser, R. T. 2000. Evolution in the high Andes: The phylogenetics of *Muscisaxicola* ground-tyrants. *Molecular Phylogenetics and Evolution* 15:369–380. [69, 118]

Chilton, G., and M. R. Lein. 1996. Long-term changes in songs and song dialect boundaries of Puget Sound white-crowned sparrows. *Condor* 98:567–580. [194, 206]

Chorowicz, J. 2005. The East African rift system. Journal of African Earth Sciences 43:379–410. [119]

Christidis, L. 1983. Extensive chromosomal repatterning in two congeneric species: *Pytilia melba*, L. and *Pytilia phoenicoptera* Swainson (Estrildidae; Aves). *Cytogenetics and Cell Genetics* 36:641–648. [386, 387]

Christidis, L. 1986a. Chromosomal evolution within the family Estrildidae (Aves). 1. The Poephilae. *Genetica* 71:81–97. [386, 387]

Christidis, L. 1986b. Chromosomal evolution within the family Estrildidae (Aves). 2. The Lonchurae. *Genetica* 71:99–113. [386, 387]

Christidis, L. 1987. Chromosomal evolution within the family Estrildidae (Aves). 3. The Estrildae (waxbill finches). *Genetica* 72:93–100. [386, 387]

Christidis, L. 1990. Aves. Pp. 116 *in* B. John, ed. *Animal Cytogenetics*. Gebrüder Bornträger, Berlin and Stuttgart. [380, 384, 388]

Christidis, L., R. Schodde, and P. R. Baverstock. 1988. Genetic and morphological differentiation and phylogeny in the Australo-papuan scrubwrens *(Sericornis,* Acanthizidae). *Auk* 105:616–629. [233]

Cicero, C., and N. K. Johnson. 1995. Speciation in sapsuckers *(Sphyrapicus)* 3. Mitochondrial DNA sequence divergence at the cytochrome *b* locus. *Auk* 112:547–563. [351]

Clarke, J. A., C. P. Tambussi, J. I. Noriega, G. M. Erickson, and R. A. Ketcham. 2005. Definitive fossil evidence for the extant avian radiation in the Cretaceous. *Nature* 433:305–308. [15, 16]

Clayton, N. S. 1990a. Assortative mating in zebra finch subspecies, *Taeniopygia guttata guttata* and *T. g. castanotis*. *Philosophical Transactions of the Royal Society of London,* B 330:351–370. [45, 273–275, 292]

Clayton, N. S. 1990b. Subspecies recognition and song learning in zebra finches. *Animal Behaviour* 40:1009–1017. [253, 273, 274, 276, 292, 318, 319]

Clayton, N. S. 1990c. The effects of cross-fostering on assortative mating between Zebra finch subspecies. *Animal Behaviour* 40:1102–1110. [273, 274, 292]

Clayton, N. S. 1990d. Mate choice and pair formation in Timor and Australian mainland Zebra finches. *Animal Behaviour* 39:474–480. [273–275]

Clayton, N. S. 1993. Song, sex and sensitive phases in the behavioural development of birds. *Trends in Ecology and Evolution* 4:82–84. [276, 317]

Clayton, N., D. Hodson, and R. A. Zann. 1991. Geographic variation in zebra finch subspecies. *Emu* 91:2–11. [274]

Clegg, S. M., and I. P. F. Owens. 2002. The 'island rule' in birds: Medium body size and its ecological explanation. *Proceedings of the Royal Society of London Series B—Biological Sciences* 269:1359–1365. [51]

Clegg, S. M., S. M. Degnan, C. Moritz, A. Estoup, J. Kikkawa, and I. P. F. Owens. 2002a. Microevolution in island forms: The roles of drift and directional selection in morphological divergence of a passerine bird. *Evolution* 56:2090–2099. [43, 46, 47]

Clegg, S. M., S. M. Degnan, J. Kikkawa, C. Moritz, A. Estoup, and I. P. F. Owens. 2002b. Genetic consequences of sequential founder events by an island-colonizing bird. *Proceedings of the National Academy of Sciences of the United States of America* 99:8127–8132. [125]

Cockburn, A. 2003. Cooperative breeding in oscine passerines: Does sociality inhibit speciation? *Proceedings of the Royal Society of London Series B–Biological Sciences* 270:2207–2214. [125]

Cody, M. L. 1975. Towards a theory of continental species diversities. Pp. 214–257 *in* M. L. Cody and J. Diamond, eds. *Ecology and Evolution of Communities*. Belknap Press of Harvard University Press, Cambridge, MA. [101]

Cody, M. L., and H. A. Mooney. 1978. Convergence versus non-convergence in Mediterranean climate ecosystems. *Annual Review of Ecology and Systematics* 9:265–321. [101]

Coleman, S. W., G. L. Patricelli, and G. Borgia. 2004. Variable female preferences drive complex male displays. *Nature* 428:742–745. [168]

Collias, N. E. 1997. On the origin and evolution of nest building by passerine birds. *Condor* 99:253-270. [139]

Collins, S. A. 1999. Is female preference for male repertoires due to sensory bias? *Proceedings of the Royal Society of London Series B—Biological Sciences* 266:2309–2314. [163, 205, 212]

Collins, S. 2004. Vocal fighting and flirting: The functions of birdsong. Pp. 39–79 *in* P. Marler and H. Slabbekoorn, eds. *Nature's Music: the Science of Bird Song*. Academic Press, London. [187]

Collins, S. A., and S. T. Luddem. 2002. Degree of male ornamentation affects female preference for conspecific versus heterospecific males. *Proceedings of the Royal Society of London Series B—Biological Sciences* 269:111–117. [273]

Conant, S. 1988. Geographic variation in the Laysan finch *(Telespyza cantans)*. *Evolutionary Ecology* 2:270–282. [47]

Confer, J. L., and J. L. Larkin. 1998. Behavioral interactions between golden-winged and blue-winged warblers. *Auk* 115:209–214. [363]

Confer, J. L., J. L. Larkin, and P. E. Allen. 2003. Effects of vegetation, interspecific competition, and brood parasitism on golden-winged warbler *(Vermivora chrysoptera)* nesting success. *Auk* 120:138–144. [363]

Conrads, K. 1966. Der Egge Dialekt des Buchfinken *(Fringilla coelebs)*—ein Beitrag zur geographischen Gesangsvariation. *Vogelwelt* 87:176–182. [206]

Cook, A. 1975. Changes in a carrion-hooded crow hybrid zone and possible importance of climate. *Bird Study* 22:165–168. [342, 355]

Cooke, F., and C. M. McNally. 1976. Mate selection and colour preferences in lesser snow geese. *Behaviour* 103:151–170. [277]

Cooke, F., D. T. Parkin, and R. F. Rockwell. 1988. Evidence of former allopatry of the two color phases of lesser snow geese *(Chen caerulescens cearulescens)*. *Auk* 105:467–479. [299]

Cooper, A., and D. Penny. 1997. Mass survival of birds across the Cretaceous-Tertiary boundary: Molecular evidence. *Science* 275:1109–1113. [378]

Cowie, R. H. 1996. Pacific island land snails: Relationships, origins and determinants of diversity. Pp. 347–372 *in* A. Keast and S. E. Miller, eds. *The Origin and Evolution of Pacific Island Biotas, New Guinea to Eastern Polynesia: Patterns and Processes.* SPB Academic Publishing, Amsterdam. [401]

Cox, G. W. 1985. The evolution of avian migration systems between temperate and tropical regions of the New World. *American Naturalist* 126:451–474. [92]

Coyne, J. A. 1992. Genetics and speciation. *Nature* 355:511–515. [3]

Coyne, J. A., and H. A. Orr. 1989. Patterns of speciation in *Drosophila*. *Evolution* 43:362–381. [367]

Coyne, J. A., and H. A. Orr. 1997. "Patterns of speciation in *Drosophila*" revisited. *Evolution* 51:295–303. [367]

Coyne, J. A., and H. A. Orr. 2004. *Speciation*. Sinauer, Sunderland, MA. [1, 2, 3, 8, 65–67, 72, 83, 300, 375, 382, 395, 396, 399, 400]

Coyne, J. A., and T. D. Price. 2000. Little evidence for sympatric speciation in island birds. *Evolution* 54:2166–2171. [19, 89]

Coyne, J. A., N. H. Barton, and M. Turelli. 1997. Perspective: A critique of Sewall Wright's shifting balance theory of evolution. *Evolution* 51:643–671. [49, 395]

Cracraft, J. 1983. Species concepts and speciation analysis. Pp. 159-187 *in* R. F. Johnston, ed. *Current Ornithology*. Plenum Press, New York. [6]

Cracraft, J. 1986. Origin and evolution of continental biotas—speciation and historical congruence within the Australian avifauna. *Evolution* 40:977–996. [27]

Cracraft, J. 1992. The species of birds of paradise (Paradisaeidae). Applying the phylogenetic species concept to a complex pattern of diversification. *Cladistics* 8:1–43. [6, 7]

Cracraft, J., and R. O. Prum. 1988. Patterns and processes of diversification-speciation and historical congruence in some neotropical birds. *Evolution* 42:603–620. [27]

Cramp, S. 1992. *The Birds of the Western Palearctic,* Vol. 6. Oxford University Press, Oxford and New York. [92]

Cramp, S., and C. M. Perrins. 1993. *The Birds of the Western Palearctic,* Vol. 7. Oxford University Press, Oxford. [167]

Cramp, S., and C. M. Perrins. 1994. *The Birds of the Western Palearctic,* Vol. 8. Oxford University Press, Oxford. [372]

Cramp, S., and K. E. L. Simmons. 1977. *The Birds of the Western Palearctic,* Vol. 1. Oxford University Press, Oxford. [166]

Crawford, R. D. 1990. Origin and history of poultry species. Pp. 1–41 *in* R. D. Crawford, ed. *Poultry Breeding and Genetics*. Elsevier, Oxford, New York and Tokyo. [221–223]

Crochet, P. A., J. J. Z. Chen, J. M. Pons, J. D. Lebreton, P. D. N. Hebert, and F. Bonhomme. 2003. Genetic differentiation at nuclear and mitochondrial loci among large white-headed gulls: Sex-biased interspecific gene flow? *Evolution* 57:2865–2878. [345]

Crook, J. H. 1964. The evolution of social organisation and visual communication in the weaver birds (Ploceinae). *Behaviour Supplement* 10:1–178. [243, 244]

Cuervo, J. J., and A. P. Møller. 1999. Ecology and evolution of extravagant feather ornaments. *Journal of Evolutionary Biology* 12:986–998. [249]

Cunningham, E. J. A., and A. F. Russell. 2000. Egg investment is influenced by male attractiveness in the mallard. *Nature* 404:74–77. [162]

Curry, R. L. 2005. Hybridization in chickadees: Much to learn from familiar birds. *Auk* 122:747–758. [340]

Dale, J., D. B. Lank, and H. K. Reeve. 2001. Signaling individual identity versus quality: A model and case studies with ruffs, queleas, and house finches. *American Naturalist* 158:75–86. [284]

Dantzker, M. S., G. B. Deane, and J. W. Bradbury. 1999. Directional acoustic radiation in the strut display of male sage grouse *Centrocercus urophasianus*. *Journal of Experimental Biology* 202:2893–2909. [160]

Darwin, C. 1859. *On the Origin of Species by Means of Natural Selection.* J. Murray, London. [References from the reprinted version in: *On the Origin of Species. A Facsimile of the 1st Edition, with an Introduction by Ernst Mayr,* 1964: Harvard University Press, Cambridge, MA.]. [48, 75, 112, 152]

Darwin, C. 1868. *The Variation of Animals and Plants under Domestication.* O. Judd and Company, New York. [220, 221, 228, 229]

Darwin, C. 1871. *The Descent of Man, and Selection in Relation to Sex,* Vol. 2. John Murray, London. [References from the reprint of 1981, Princeton University Press, Princeton NJ.] [157, 160, 168, 175, 220, 259]

Davies, N. B. 1981. Calling as an ownership convention on pied wagtail territories. *Animal Behaviour* 29: 529–534. [285]

Davies, N. B. 1991. Mating systems. Pp. 263–299 *in* J. R. Krebs and N. B. Davies, eds. *Behavioural Ecology,* 3rd ed. Blackwell Scientific, Oxford. [248–249]

Davies, N. B., J. R. Madden, and S. H. M. Butchart. 2004. Learning fine-tunes a specific response of nestlings to the parental alarm calls of their own species. *Proceedings of the Royal Society of London Series B—Biological Sciences* 271:2297–2304. [281]

Davis, J. M., and J. A. Stamps. 2004. The effect of natal experience on habitat preferences. *Trends in Ecology and Evolution* 19:411–416. [299]

Dawkins, M. S., and T. Guilford. 1995. An exaggerated preference for simple neural network models of signal evolution. *Proceedings of the Royal Society of London Series B—Biological Sciences* 261:357–360. [172]

Dawkins, R., and J. R. Krebs. 1978. Animal signals: Information or manipulation? Pp. 282–309 *in* J. R. Krebs and N. B. Davies, eds. *Behavioural Ecology.* Blackwell Scientific, Oxford. [180]

Dawson, D. A., T. Burke, B. Hansson, J. Pandahl, M. C. Hale, G. N. Hinten, and J. Slate. 2006. A predicted microsatellite map of the passerine genome based on chicken-passerine sequence similarity. *Molecular Ecology* 15:1299–1320. [226, 392]

Day, T. 2000. Sexual selection and the evolution of costly female preferences: Spatial effects. *Evolution* 54:715–730. [239]

de Kort, S. R., P. M. den Hartog, and C. ten Cate. 2002. Diverge or merge? The effect of sympatric occurrence on the territorial vocalizations of the vinaceous dove *Streptopelia vinacea* and the ring-necked dove *S. capicola. Journal of Avian Biology* 33:150–158. [192]

de Lope, F., and A. P. Møller. 1993. Female reproductive effort depends on the degree of ornamentation of their mates. *Evolution* 47:1152–1160. [162]

de Queiroz, K. 1998. The general lineage concept of species, species criteria, and the process of speciation. Pp. 57–75 *in* D. J. Howard and S. H. Berlocher, eds. *Endless Forms: Species and Speciation.* Oxford University Press, New York. [11]

del Hoyo, J., A. Elliott, and J. Sargatal. 1999. *Handbook of Birds of the World,* Vol. 5. Lynx Editions, Barcelona. [118]

del Hoyo, J., A. Elliott, and D. A. Christie. 2004. *Handbook of Birds of the World,* Vol. 9. Lynx Editions, Barcelona. [159]

Delport, W., A. C. Kemp, and J. W. H. Ferguson. 2004. Structure of an African red-billed hornbill *(Tockus erythrorhynchus rufirostris* and *T. e. damarensis)* hybrid zone as revealed by morphology, behavior, and breeding biology. *Auk* 121:565-586. [352]

Deregnaucourt, S., J. C. Guyomarc'h, and N. J. Aebischer. 2002. Hybridization between European quail *Coturnix coturnix* and Japanese quail *Coturnix japonica. Ardea* 90:15–21. [379]

Derjusheva, S., A. Kurganova, F. Habermann, and E. Gaginskaya. 2004. High chromosome conservation detected by comparative chromosome painting in chicken, pigeon and passerine birds. *Chromosome Research* 12:715–723. [384]

Diamond, J. M. 1969. Avifaunal equilibria and species turnover rates on the Channel Islands of California. *Proceedings of the National Academy of Sciences of the United States of America* 64:57–63. [152]

Diamond, J. M. 1970. Ecological consequences of island colonization by southwest Pacific birds, 1. Types of niche shifts. *Proceedings of the National Academy of Sciences of the United States of America* 67:529–536. [48, 50]

Diamond, J. M. 1971. Comparison of faunal equilibrium turnover rates on a tropical island and a temperate island. *Proceedings of the National Academy of Sciences of the United States of America* 68:2742–2745. [152]

Diamond, J. M. 1972. Biogeographic kinetics: Estimation of relaxation times for avifaunas of southwest Pacific islands. *Proceedings of the National Academy of Sciences of the United States of America* 69:3199–3203. [70, 152]

Diamond, J. M. 1973. Distributional ecology of New Guinea birds. *Science* 179:759–769. [109–111]

Diamond, J. M. 1977. Continental and insular speciation in Pacific land birds. *Systematic Zoology* 26:263–268. [33, 143]

Diamond, J. M. 1980. Species turnover in island bird communities. *Proceedings of the 17th International Ornithological Congress* 2:777–782. [141, 142, 144, 145]

Diamond, J. M. 1984. "Normal" extinctions of isolated populations. Pp. 191–246 *in* M. H. Nitecki, ed. *Extinctions.* University of Chicago Press, Chicago. [141, 142, 144, 145]

Diamond, J. M. 1986. Evolution of ecological segregation in the New Guinea montane avifauna. Pp. 98–125 *in* J. M. Diamond and T. J. Case, eds. *Community Ecology.* Harper and Row, New York. [87]

Diamond, J. 1988. Experimental study of bower decoration by the bowerbird *Amblyornis inornatus* using colored poker chips. *American Naturalist* 131:631–653. [184]

Diamond, J. M. 1991. A new species of rail from the Solomon Islands, South Pacific Ocean and convergent evolution of insular flightlessness. *Auk* 108:461–470. [70, 71]

Diamond, J., S. L. Pimm, M. E. Gilpin, and M. Lecroy. 1989. Rapid evolution of character displacement in Myzomelid honeyeaters. *American Naturalist* 134:675–708. [46]

Dieckmann, U., and M. Doebeli. 1999. On the origin of species by sympatric speciation. *Nature* 400:354–357. [19, 34, 83, 314]

Dimcheff, D. E., S. V. Drovetski, and D. P. Mindell. 2002. Phylogeny of Tetraoninae and other galliform birds using mitochondrial 12S and ND2 genes. *Molecular Phylogenetics and Evolution* 24:203–215. [378]

Dixon, K. L. 1955. An ecological analysis of the interbreeding of crested titmice. *University of California Publications in Zoology* 54:125–205. [359]

Dixon, K. L. 1990. Constancy of margins of the hybrid zone in titmice of the *Parus bicolor* complex in coastal Texas. *Auk* 107:184–187. [359]

Dobkin, D. S. 1979. Functional and evolutionary relationships of vocal copying phenomena in birds. *Zeitschrift für Tierpsychologie* 50:348–363. [307]

Dobzhansky, T. 1937. *Genetics and the Origin of Species*. Columbia University Press, New York. [1]

Dobzhansky, T. 1940. Speciation as a stage in evolutionary divergence. *American Naturalist* 74:312–321. [300]

Dobzhansky, T. 1951. *Genetics and the Origin of Species,* 3rd ed. Columbia University Press, New York. [1]

Doebeli, M. 1996. An explicit genetic model for ecological character displacement. *Ecology* 77:510–520. [81]

Doebeli, M., and U. Dieckmann. 2003. Speciation along environmental gradients. *Nature* 421:259–264. [67]

Dorst, J., and F. Vuilleumier. 1986. Convergences in bird communities at high altitudes in the tropics (especially the Andes and Africa) and at temperate latitudes (Tibet). Pp. 120–149 *in* F. Vuilleumier and M. Monasterio, eds. *High Altitude Tropical Biogeography*. Oxford University Press, Oxford. [101, 114]

Douglis, M. B. 1948. Social factors influencing the hierarchies of small flocks of the domestic hen: Interactions between members and part-time members of organized flocks. *Physiological Zoology* 21:147–182. [283, 284]

Dowling, T. E., and C. L. Secor. 1997. The role of hybridization and introgression in the diversification of animals. *Annual Review of Ecology and Systematics* 28:593–619. [323]

Dowsett-Lemaire, F. 1979. The imitative range of the song of the marsh warbler *Acrocephalus palustris*, with special reference to imitations of African birds. *Ibis* 121:453–468. [252]

Drăgănoiu, T. I., L. Nagle, and M. Kreutzer. 2002. Directional female preference for an exaggerated male trait in canary *(Serinus canaria)* song. *Proceedings of the Royal Society of London Series B—Biological Sciences* 269:2525–2531. [163]

Duckworth, R. A. 2007. Behavioral integration and the rapid range expansion of a passerine bird. Unpublished manuscript. [342, 344]

Dukas, R. 2004. Male fruit flies learn to avoid interspecific courtship. *Behavioral Ecology* 15:695–698. [401]

Duncker, H. 1931. Erblichkeitsverhältnisse bei Vögeln. Pp. 215–243. *Proceedings of the VIIth International Orthnithological Congress at Amsterdam* 1930. C. de Boer Jr., Den Helder, the Netherlands. [227]

Dunham, D. W. 1966. Agonistic behavior in captive rose-breasted grosbeaks, *Pheucticus ludovicianus* (L.). *Behaviour* 27:160–173. [177]

Dunn, P. O., L. A. Whittingham, and T. E. Pitcher. 2001. Mating systems, sperm competition, and the evolution of sexual dimorphism in birds. *Evolution* 55:161–175. [160, 249, 250, 260]

Eaton, M. D. 2005. Human vision fails to distinguish widespread sexual dichromatism among sexually "monochromatic" birds. *Proceedings of the National Academy of Sciences of the United States of America* 102:10942–10946. [258]

Eaton, M. D. 2006. A phylogenetic perspective on the evolution of chromatic ultraviolet plumage coloration in grackles and allies (Icteridae). *Auk* 123:211-234. [233]

Eberhard, J. R. 1998. Evolution of nest-building behavior in *Agapornis* parrots. *Auk* 115:455–464. [139, 377, 390]

Edgar, R. C. 2004. MUSCLE: Multiple sequence alignment with high accuracy and high throughput. *Nucleic Acids Research* 32:1792–1797. [40]

Edinger, B. B. 1985. Limited hybridization and behavioral differences among sympatric Baltimore and bullock's orioles. Unpublished M.S. thesis, University of Minnesota. [328, 365]

Edwards, S. V., and P. Beerli. 2000. Perspective: Gene divergence, population divergence, and the variance in coalescence time in phylogeographic studies. *Evolution* 54:1839–1854. [11, 37]

Edwards, S. V., S. B. Kingan, J. D. Calkins, C. N. Balakrishnan, W. B. Jennings, W. J. Swanson, and M. D. Sorenson. 2005. Speciation in birds: Genes, geography, and sexual selection. *Proceedings of the National Academy of Sciences of the United States of America* 102:6550–6557. [11]

Eichler, E. E., and D. Sankoff. 2003. Structural dynamics of eukaryotic chromosome evolution. *Science* 301:793–797. [385, 392]

Ellegren, H. 2005. The avian genome uncovered. *Trends in Ecology and Evolution* 20:180–186. [384]

Ellegren, H., and A. K. Fridolfsson. 1997. Male-driven evolution of DNA sequences in birds. *Nature Genetics* 17:182–184. [227]

Elliott, J. J., and R. S. Arbib Jr. 1953. Origin and status of the house finch in the eastern United States. *Auk* 70:31–37. [206]

Ellis, C. R. 1966. Agonistic behavior in the male starling. *Wilson Bulletin* 78:208–224. [177]

Emlen, S. T. 1971. Geographic variation in indigo bunting song. *Animal Behaviour* 19:407–408. [215]

Emlen, S. T., and L. W. Oring. 1977. Ecology, sexual selection, and the evolution of mating systems. *Science* 197:215–223. [248]

Emlen, S. T., J. D. Rising, and W. L. Thompson. 1975. A behavioral and morphological study of sympatry in the Indigo and Lazuli buntings of the Great Plains. *Wilson Bulletin* 87:145–179. [216, 307, 365]

Endler, J. A. 1977. *Geographic Variation, Speciation, and Clines.* Princeton University Press, Princeton, NJ. [63, 67, 324]

Endler, J. A. 1982. Problems in distinguishing historical from ecological factors in biogeography. *American Zoologist* 22:441–452. [1]

Endler, J. A. 1986. *Natural Selection in the Wild.* Princeton University Press, Princeton, NJ. [269]

Endler, J. A. 1992. Signals, signal conditions, and the direction of evolution. *American Naturalist* 139: S125–S153. [240, 259, 268]

Endler, J. A., and M. Théry. 1996. Interacting effects of lek placement, display behavior, ambient light, and color patterns in three neotropical forest-dwelling birds. *American Naturalist* 148:421–452. [249, 258, 259, 270]

Enquist, M., and A. Arak. 1993. Selection of exaggerated male traits by female aesthetic senses. *Nature* 361:446–448. [169, 170]

Enquist, M., and A. Arak. 1998. Neural representation and the evolution of signal form. Pp. 21–87 *in* R. Dukas, ed. *Cognitive Ecology: The Evolutionary Ecology of Information Processing and Decision Making.* University of Chicago Press, Chicago. [169, 172]

Entrikin, R. K., and L. C. Erway. 1972. A genetic investigation of roller and tumbler pigeons. *Journal of Heredity* 63:351–354. [228]

Erard, C. 1989. Bird community structure in two rainforests: Africa (Gabon) and South America (French Guiana)—a comparison. Pp. 89–122 *in* M. L. Harmelin-Vivien and F. Bourliere, eds. *Vertebrates in Complex Tropical Systems.* Springer-Verlag, New York. [98, 101]

Erickson, C. J., and D. S. Lehrman. 1964. Effects of castration of male ring doves upon ovarian activity of females. *Journal of Comparative and Physiological Psychology* 58:164–166. [162, 165]

Ericson, P. G. P., C. L. Anderson, T. Britton, A. Elzanowski, U. S. Johansson, M. Källersjö, J. I. Ohlson, T. J. Parsons, D. Zuccon, and G. Mayr. 2006. Diversification of Neoaves: Integration of molecular sequence data and fossils. *Biology Letters* 2: 543-547. [15, 364]

Evans, K. L., P. H. Warren, and K. J. Gaston. 2005. Species–energy relationships at the macroecological scale: A review of the mechanisms. *Biological Reviews* 80:1–25. [98]

Evans, M. R. 1998. Selection on swallow tail streamers. *Nature* 394:233–234. [161]

Fabricius, E. 1991. Interspecific mate choice following cross-fostering in a mixed colony of Grey lag geese *(Anser anser)* and Canada geese *(Branta canadensis)*. A study on development and persistence of species preferences. *Ethology* 88:287–296. [276, 279]

Fabricius, E., and A. M. Jansson. 1963. Laboratory observations on the reproductive behaviour of the pigeon *(Columba livia)* during the pre-incubation phase of the breeding cycle. Animal *Behaviour* 11:534–547. [165]

Faivre, B., J. Secondi, C. Ferry, L. Chastragnat, and F. Cezilly. 1999. Morphological variation and the recent evolution of wing length in the Icterine warbler: A case of unidirectional introgression? Journal of Avian Biology 30:152–158. [360]

Fear, K. K., and T. Price. 1998. The adaptive surface in ecology. Oikos 82:440–448. [76, 79, 134]

Feder, H. H., A. Storey, D. Goodwin, C. Reboulleau, and R. Silver. 1977. Testosterone and "5α-dihydrotestosterone" levels in peripheral plasma of male and female ring doves *(Streptopelia risoria)* during the reproductive cycle. *Biological Reproduction* 16:666–677. [164]

Feduccia, A. 1995. Explosive evolution in Tertiary birds and mammals. *Science* 267:637–638. [15. 138]

Felsenstein, J. 1985. Phylogenies and the comparative method. *American Naturalist* 125:1–15. [58, 59]

Ferns, P. N., and S. A. Hinsley. 2004. Immaculate tits: Head plumage pattern as an indicator of quality in birds. *Animal Behaviour* 67:261–272. [182]

Ficken, M. S. 1981. What is the song of the black-capped chickadee? *Condor* 83:384–386. [195]

Ficken, M. S., and R. W. Ficken. 1968. Courtship of blue-winged warblers, golden-winged warblers, and their hybrids. *Wilson Bulletin* 80:161–172. [319]

Fiedler, W. 2003. Recent changes in migratory behaviour of birds: A compilation of field observations and ringing data. Pp. 21–38 *in* P. Berthold, E. Gwinner and E. Sonnenscheim, eds. *Avian Migration.* Springer-Verlag, Berlin. [93]

Figuerola, J., and A. J. Green. 2000. The evolution of sexual dimorphism in relation to mating patterns, cavity nesting, insularity and sympatry in the Anseriformes. *Functional Ecology* 14:701–710. [240, 264]

Filardi, C. E., and R. G. Moyle. 2005. Single origin of a pan-Pacific bird group and upstream colonization of Australasia. *Nature* 438:216–219. [148, 151]

Filardi, C. E., and C. E. Smith. 2005. Molecular phylogenetics of monarch flycatchers (genus *Monarcha*) with emphasis on Solomon Island endemics. *Molecular Phylogenetics and Evolution* 37:776–788.[5, 6, 153, 409]

Fisher, J., and R. A. Hinde. 1949. The opening of milk bottles by birds. *British Birds* 42:347–357. [128]

Fisher, R. A. 1915. The evolution of sexual preference. *Eugenics Review* 7:184–192. [180]

Fisher, R. A. 1930. *The Genetical Theory of Natural Selection.* The Clarendon Press, Oxford. [63, 67]

Fisher, R. A. 1937. The wave of advance of advantageous genes. *Annals of Eugenics* 7:355–369. [207]

Fitze, P. S., and H. Richner. 2002. Differential effects of a parasite on ornamental structures based on melanins and carotenoids. *Behavioral Ecology* 13:401–407. [182]

Fitzpatrick, J. W. 1988. Why so many passerine birds—response. *Systematic Zoology* 37:71–76. [16, 213]

Fitzpatrick, S. 1994. Colorful migratory birds—evidence for a mechanism other than parasite resistance for the maintenance of good genes sexual selection. *Proceedings of the Royal Society of London Series B—Biological Sciences* 257:155–160. [260]

Fitzpatrick, S. 1998a. Birds' tails as signalling devices: Markings, shape, length, and feather quality. *American Naturalist* 151:157–173. [176]

Fitzpatrick, S. 1998b. Intraspecific variation in wing length and male plumage coloration with migratory behaviour in continental and island populations. *Journal of Avian Biology* 29:248–256. [240, 262, 263]

Fitzpatrick, B. M. 2004. Rates of evolution of hybrid inviability in birds and mammals. *Evolution* 58:1865–1870. [401]

Fjeldså, J. 1981. Comparative ecology of Peruvian grebes—a study of the mechanisms of evolution of ecological isolation. *Videnskabelige meddelelser fra Dansk Naturhistorisk Forening* 143:125–249. [46]

Fjeldså, J. 1983. Ecological character displacement and character release in grebes Podicipedidae. *Ibis* 125: 463–481. [19, 46, 48, 50, 52]

Fjeldså, J. 1994. Geographical patterns for relict and young species of birds in Africa and South America and implications for conservation priorities. *Biodiversity and Conservation* 3:207–226. [115–117]

Fjeldså, J., and J. C. Lovett. 1997. Geographical patterns of old and young species in African forest biota: The significance of specific montane areas as evolutionary centres. *Biodiversity and Conservation* 6:325–346. [119]

Fleischer, R. C., C. E. McIntosh, and C. L. Tarr. 1998. Evolution on a volcanic conveyor belt: Using phylogeographic reconstructions and K-Ar-based ages of the Hawaiian Islands to estimate molecular evolutionary rates. *Molecular Ecology* 7:533–545. [18, 36, 37, 38]

Fleming, T. H., R. Breitwisch, and G. H. Whitesides. 1987. Patterns of tropical vertebrate frugivore diversity. *Annual Review of Ecology and Systematics* 18:91–109. [115]

Ford, J. 1978. Hybridization between the white-vented and black-vented forms of the black-faced woodswallow. *Emu* 78:105–114. [356]

Ford, J. 1980. Hybridization between contiguous subspecies of the varied sittella in Queensland. *Emu* 80:1–12. [354]

Ford, J. 1982. Hybrid phenotypes in male figbirds *Sphecotheres viridis* in Queensland. *Emu* 82:126–130. [356]

Ford, J. 1986. Avian hybridization and allopatry in the region of the Einasleigh Uplands and Burdekin-Lynd Divide, Northeastern Queensland. *Emu* 86:87–110. [361]

Ford, J. 1987. Hybrid zones in Australian birds. *Emu* 87:158–178. [324]

Ford, J. R., and R. E. Johnstone. 1991. Hybridization between *Malurus lamberti rogersi* and *Malurus lamberti assimilis* in North-Western Australia. *Emu* 91:251–254. [354]

Forshaw, J. M. 1969. *Australian Parrots.* Lansdowne Press, Melbourne. [347]

Forstmeier, W., and I. Weiss. 2004. Adaptive plasticity in nest-site selection in response to changing predation risk. *Oikos* 104:487–499. [137]

Forstmeier, W., B. Kempenaers, A. Meyer, and B. Leisler. 2002. A novel song parameter correlates with extra-pair paternity and reflects male longevity. *Proceedings of the Royal Society of London Series B—Biological Sciences* 269:1479–1485.[160, 205]

Fowlie, M. K., and O. Kruger. 2003. The evolution of plumage polymorphism in birds of prey and owls: The apostatic selection hypothesis revisited. *Journal of Evolutionary Biology* 16:577–583. [315]

Frank, S. A. 1991. Divergence of meiotic drive-suppression systems as an explanation for sex-biased hybrid sterility and inviability. *Evolution* 45:262–267. [369, 370, 383]

Freeberg, T. M. 1996. Assortative mating in captive cowbirds is predicted by social experience. *Animal Behaviour* 52:1129–1142. [280]

Freeberg, T. M., A. P. King, and M. J. West. 1995. Social malleability in cowbirds *(Molothrus ater artemisiae)*—species and mate recognition in the first 2 years of life. *Journal of Comparative Psychology* 109:357–367. [280]

Freeberg, T. M., S. D. Duncan, T. L. Kast, and D. A. Enstrom. 1999. Cultural influences on female mate choice: An experimental test in cowbirds, *Molothrus ater. Animal Behaviour* 57:421–426. [280]

Freeberg, T. M., A. P. King, and M. J. West. 2001. Cultural transmission of vocal traditions in cowbirds *(Molothrus ater)* influences courtship patterns and mate preferences. *Journal of Comparative Psychology* 115:201–211. [280]

Freeland, J. R., and P. T. Boag. 1999a. The mitochondrial and nuclear genetic homogeneity of the phenotypically diverse Darwin's ground finches. *Evolution* 53:1553–1563. [22]

Freeland, J. R., and P. T. Boag. 1999b. Phylogenetics of Darwin's finches: Paraphyly in the tree-finches, and two divergent lineages in the Warbler Finch. *Auk* 116:577–588. [24]

Friesen, V. L., A. J. Baker, and J. F. Piatt. 1996. Phylogenetic relationships within the Alcidae (Charadriiformes: Aves) inferred from total molecular evidence. *Molecular Biology and Evolution* 13:359–367. [102]

Friesen, V. L., A. L. Smith, R. W. Furness, and M. Bolton. 2007. Sympatric speciation by allochrony in the band-rumped storm petrel. Unpublished manuscript. [21, 396]

Fritts, T. H., and D. Leasman-Tanner. 2001. The Brown Tree Snake on Guam: How the arrival of one invasive species damaged the ecology, commerce, electrical systems, and human health on Guam: A comprehensive information source. Available online: http://www.fort.usgs.gov/resources/education/bts/bts_home.asp [137]

Fritts, T. H., and G. H. Rodda. 1998. The role of introduced species in the degradation of island ecosystems: A case history of Guam. *Annual Review of Ecology and Systematics* 29:113–140. [137]

Fumihito, A., T. Mikaye, M. Takada, R. Shingu, T. Endo, T. Gojbori, N. Kondo, and O. Susumu. 1996. Monophyletic origin and unique dispersal patterns of domestic fowls. *Proceedings of the National Academy of Sciences of the United States of America* 93:6792–6795. [221]

Funk, D. J., P. Nosil, and W. J. Etges. 2006. Ecological divergence exhibits consistently positive associations with reproductive isolation across disparate taxa. *Proceedings of the National Academy of Sciences of the United States of America* 103:3209–3213. [369, 399]

Futuyma, D. J. 2005. Progress on the origin of species. *PLOS Biology* 3:0197–0199. [395]

Galbraith, I. C. J. 1956. Variation, relationships and evolution in the *Pachycephala pectoralis* superspecies (Aves, Muscicapidae). *Bulletin of the British Museum of Natural History (Zoology)* 4:133–222. [314, 347]

Garant, D., L. E. B. Kruuk, T. A. Wilkin, R. H. McCleery, and B. C. Sheldon. 2005. Evolution driven by differential dispersal within a wild bird population. *Nature* 433:60–65. [43]

García-Moreno, J. 2004. Is there a universal mtDNA clock for birds? *Journal of Avian Biology* 35:465–468. [18, 36, 38]

García-Moreno, J., P. Arctander, and J. Fjeldså. 1999. Strong diversification at the treeline among *Metallura* hummingbirds. *Auk* 116:702–711. [118]

Garvin, M. C., and J. V. Remsen. 1997. An alternative hypothesis for heavier parasite loads of brightly colored birds: Exposure at the nest. *Auk* 114:179–191. [259, 260]

Gaston, K. J., and T. M. Blackburn. 1996. The tropics as a museum of biological diversity: Analysis of the New World avifauna. *Proceedings of the Royal Society of London Series B—Biological Sciences* 263:63–68. [100]

Gavrilets, S. 2003. Perspective: Models of speciation: What have we learned in 40 years? *Evolution* 57:2197–2215. [67, 399]

Gavrilets, S. 2004. *Fitness Landscapes and the Origin of Species.* Princeton University Press, Princeton, NJ. [20, 29, 30, 34, 83, 94]

Gavrilets, S. 2005. "Adaptive speciation"—it is not that easy: A reply to Doebeli et al. *Evolution* 59:696–699. [20, 34]

Gavrilets, S., and A. Vose. 2005. Dynamic patterns of adaptive radiation. *Proceedings of the National Academy of Sciences of the United States of America* 102:18040–18045. [120]

Gay, L., D. A. Bell, and P. A. Crochet. 2005. Additional data on mitochondrial DNA of North American large gull taxa. *Auk* 122:684–688. [353]

Gee, J. M. 2003. How a hybrid zone is maintained: Behavioral mechanisms of interbreeding between California and Gambel's quail *(Callipepla californica and C. gambelii). Evolution* 57:2407–2415. [313, 318, 351]

Gee, J. M. 2004. Gene flow across a climatic barrier between hybridizing avian species, California and Gambel's quail *(Callipepla californica* and *C. gambelii). Evolution* 58:1108–1121. [351]

Gibbs, H. L. 1990. Cultural evolution of male song types in Darwin's medium ground finches, *Geospiza fortis. Animal Behaviour* 39:253–263. [316]

Gibbs, H. L., and P. R. Grant. 1987. Oscillating selection on Darwin's finches. *Nature* 327:511–513. [90]

Gibson, R. M., J. W. Bradbury, and S. L. Vehrencamp. 1991. Mate choice in lekking sage grouse revisited: The roles of vocal display, female site fidelity, and copying. *Behavioral Ecology* 2:165–180. [160]

Gibson, J. R., A. K. Chippindale, and W. R. Rice. 2002. The X chromosome is a hot spot for sexually antagonistic fitness variation. *Proceedings of the Royal Society of London Series B—Biological Sciences* 269:499–505. [391]

Gil, D. 1997. Increased response of the short-toed tree-creeper *Certhia brachydactyla* in sympatry to the playback of the song of the common treecreeper *C. familiaris. Ethology* 103:632–641. [286]

Gil, D., G. Leboucher, A. Lacroix, R. Cue, and M. Kreutzer. 2004. Female canaries produce eggs with greater amounts of testosterone when exposed to preferred male song. *Hormones and Behavior* 45:64–70. [163]

Gilbert, L., K. A. Williamson, N. Hazon, and J. A. Graves. 2006. Maternal effects due to male attractiveness affect offspring development in the zebra finch. *Proceedings of the Royal Society* B—Biological Sciences 273:1765-1771. [163]

Gill, F. B. 1967. Birds of Rodriguez Island (Indian Ocean). *Ibis* 109:383–390. [142]

Gill, F. B. 1970. Hybridization in Norfolk Island white-eyes. *Condor* 72:481–482. [311]

Gill, F. B. 1980. Historical aspects of hybridization between blue-winged and golden-winged warblers. *Auk* 97:1–18. [363]

Gill, F. B. 1997. Local cytonuclear extinction of the golden-winged warbler. *Evolution* 51:519–525. [343]

Gill, F. B. 2004. Blue-winged warblers *(Vermivora pinus)* versus golden-winged warblers *(V. chrysoptera). Auk* 121:1014–1018. [335, 363]

Gill, F. B., and B. G. Murray. 1972. Discrimination behavior and hybridization of blue winged and golden winged warblers. *Evolution* 26:282–293. [288]

Gill, F. B., A. M. Mostrom, and A. L. Mack. 1993. Speciation in North American Chickadees.1. Patterns of mt DNA genetic divergence. *Evolution* 47:195–212. [43]

Gill, F. B., R. A. Canterbury, and J. L. Confer. 2001. Blue-winged warbler *(Vermivora pinus). In* A. Poole and F. B. Gill, eds. *The Birds of North America,* no. 584. Birds of North America, Inc., Philadelphia, Pa. [363]

Gill, F. B., B. Slikas, and F. H. Sheldon. 2005. Phylogeny of titmice (Paridae): II. Species relationships based on sequences of the mitochondrial cytochrome-b gene. *Auk* 122:121–143. [26, 31]

Gillespie, R. 2004. Community assembly through adaptive radiation in Hawaiian spiders. *Science* 303: 356–359. [120]

Gilliard, E. T. 1969. *Birds of Paradise and Bower Birds.* Natural History Press, Garden City, NY. [184]

Glor, R. E., J. J. Kolbe, R. Powell, A. Larson, and J. B. Losos. 2003. Phylogenetic analysis of ecological and morphological diversification in Hispaniolan trunk-ground anoles *(Anolis cybotes* group). *Evolution* 57: 2383–2397. [397]

Godard, R. 1991. Long-term memory of individual neighbors in a migratory songbird. *Nature* 350:228–229. [284]

Goldberg, E. E., and R. Lande. 2006. Ecological and reproductive character displacement on an environmental gradient. *Evolution* 60:1344–1357. [114, 303, 333, 408]

Gomez, D., and M. Théry. 2004. Influence of ambient light on the evolution of colour signals: Comparative analysis of a Neotropical rainforest bird community. *Ecology Letters* 7:279–284. [258–260]

Gontard-Danek, M. C., and A. P. Møller. 1999. The strength of sexual selection: A meta-analysis of bird studies. *Behavioral Ecology* 10:476–486. [160]

Good, T. P., J. C. Ellis, C. A. Annett, and R. Pierotti. 2000. Bounded hybrid superiority in an avian hybrid zone: Effects of mate, diet, and habitat choice. *Evolution* 54:1774–1783. [84, 86, 317, 338, 353]

Goodwin, D. 1983. *Pigeons and Doves of the World,* 3rd ed. Cornell University Press, Ithaca, NY. [175]

Göth, A., and C. S. Evans. 2004. Social responses without early experience: Australian brush-turkey chicks use specific visual cues to aggregate with conspecifics. *Journal of Experimental Biology* 207:2199–2208. [280]

Gottlieb, G. 1978. Development of species identification in ducklings: IV. Change in species-specific perception caused by auditory deprivation. *Journal of Comparative and Physiological Psychology* 92:375–387. [282]

Gow, J. L., C. L. Peichel, and E. B. Taylor. 2006. Contrasting hybridization rates between sympatric three-spined sticklebacks highlight the fragility of reproductive barriers between evolutionarily young species. *Molecular Ecology* 15:739–752. [398]

Grafen, A. 1990. Do animals really recognize kin? *Animal Behaviour* 39:42–54. [283]

Grant, P. R. 1965a. The adaptive significance of some size trends in island birds. *Evolution* 19:355–367. [48]

Grant, P. R. 1965b. Plumage and the evolution of birds on islands. *Systematic Zoology* 14:47–52. [262]

Grant, P. R. 1966. Ecological compatability of bird species on islands. *American Naturalist* 100:451–462. [89]

Grant, P. R. 1968. Bill size, body size and ecological adaptations of bird species to competitive situations on islands. *Systematic Zoology* 17:319–333. [50, 51]

Grant, P. R. 1998. Speciation. Pp. 83–101 *in* P. R. Grant, ed. *Evolution on Islands.* Oxford University Press, Oxford. [90]

Grant, P. R. 1999. *Ecology and Evolution of Darwin's Finches,* 2nd. ed. Princeton University Press, Princeton, NJ. [22, 80]

Grant, P. R. 2001. Reconstructing the evolution of birds on islands: 100 years of research. *Oikos* 92:385–403. [57]

Grant, P. R., and B. R. Grant. 1992. Hybridization of bird species. *Science* 256:193–197. [22, 78, 300, 317, 373]

Grant, B. R., and P. R. Grant. 1993. Evolution of Darwin's finches caused by a rare climatic event. *Proceedings of the Royal Society of London Series B—Biological Sciences* 251:111–117. [69, 84, 91, 373]

Grant, P. R., and B. R. Grant. 1994. Phenotypic and genetic effects of hybridization in Darwin's finches. *Evolution* 48:297–316. [78, 84, 323, 324]

Grant, P. R., and B. R. Grant. 1995. Predicting microevolutionary responses to directional selection on heritable variation. *Evolution* 49:241–251. [47, 91]

Grant, B. R., and P. R. Grant. 1996a. High survival of Darwin's finch hybrids: Effects of beak morphology and diets. *Ecology* 77:500–509. [69, 84]

Grant, P. R., and B. R. Grant. 1996b. Speciation and hybridization in island birds. *Philosophical Transactions of the Royal Society of London Series B—Biological Sciences* 351:765–772. [323]

Grant, B. R., and P. R. Grant. 1996c. Cultural evolution of song and its role in the evolution of Darwin's finches. *Evolution* 50:2471–2487. [34, 189, 198, 316]

Grant, P. R., and B. R. Grant. 1997a. Hybridization, sexual imprinting, and mate choice. *American Naturalist* 149:1–28. [83, 306, 308, 310, 373]

Grant, P. R., and B. R. Grant. 1997b. Genetics and the origin of bird species. *Proceedings of the National Academy of Sciences of the United States of America* 94:7768–7775. [83, 314]

Grant, B. R., and P. R. Grant. 1998. Hybridization and speciation in Darwin's finches. Pp. 404–422 *in* D. Howard and S. Berlocher, eds. *Endless Forms: Species and Speciation*. Oxford University Press, Oxford. [78, 90, 91, 317]

Grant, B. R., and P. R. Grant. 2002a. Lack of premating isolation at the base of a phylogenetic tree. *American Naturalist* 160:1–19. [24]

Grant, P. R., and B. R. Grant. 2002b. Unpredictable evolution in a 30-year study of Darwin's finches. *Science* 296:707–711. [47, 90, 324]

Grant, B. R., and P. R. Grant. 2002c. Simulating secondary contact in allopatric speciation: An empirical test of premating isolation. *Biological Journal of the Linnean Society* 76:545–556. [191]

Grant, P. R., and B. R. Grant. 2006. Evolution of character displacement in Darwin's finches. *Science* 313: 224–226. [47]

Grant, P. R., B. R. Grant, and K. Petren. 2000. The allopatric phase of speciation: The sharp-beaked ground finch *(Geospiza difficilis)* on the Galápagos Islands. *Biological Journal of the Linnean Society* 69: 287–317. [147, 261]

Grant, P. R., B. R. Grant, J. A. Markert, L. F. Keller, and K. Petren. 2004. Convergent evolution of Darwin's finches caused by introgressive hybridization and selection. *Evolution* 58:1588–1599. [57, 324, 398]

Grant, P. R., B. R. Grant, and K. Petren. 2005. Hybridization in the recent past. *American Naturalist* 166:56–67. [22, 57, 78, 323, 324, 373, 398]

Graur, D., and W. Martin. 2004. Reading the entrails of chickens: Molecular timescales of evolution and the illusion of precision. *Trends in Genetics* 20:80–86. [16, 18, 36]

Graves, G. R. 1985. Elevational correlates of speciation and intraspecific geographic variation in plumage in Andean forest birds. *Auk* 102:556–579. [33, 99]

Graves, G. R. 1992. The endemic landbirds of Henderson Island, southeastern Polynesia: Notes on natural history and conservation. *Wilson Bulletin* 104:32–43. [264]

Gray, A. P. 1958. *Bird Hybrids, a Check-list with Bibliography*. Commonwealth Agricultural Bureaux, Farnham Royal, UK. [367, 373, 376, 381]

Greenberg, R. S. 1990. Ecological plasticity, neophobia, and resource use in birds. *Studies in Avian Biology* 13:431–437. [131]

Greenberg, R., and C. Mettke-Hofmann. 2001. Ecological aspects of neophobia and neophilia in birds. Pp. 119–178 *in* V. Nolan Jr. and C. F. Thompson, eds. *Current Ornithology*. Kluwer Academic, New York. [132]

Greene, E., B. Lyon, V. Muehter, L. Ratcliffe, S. Oliver, and P. Boag. 2000. Disruptive sexual selection for plumage coloration in a passerine bird. *Nature* 407:1000–1003. [160, 308]

Greenslade, P. J. M. 1968. Island patterns in the Solomon Islands bird fauna. *Evolution* 22:751–761. [147]

Gregory-Wodzicki, K. M. 2000. Uplift history of the central and northern Andes: A review. *GSA Bulletin* 112:1091–1105. [119]

Griffith, S. C. 2000. High fidelity on islands: A comparative study of extrapair paternity in passerine birds. *Behavioral Ecology* 11:265–273. [263, 265]

Griffith, S. C., I. P. F. Owens, and K. A. Thuman. 2002. Extra-pair paternity in birds: A review of interspecific variation and adaptive function. *Molecular Ecology* 11:2195–2212. [160, 249]

Gross, S., and T. Price. 2000. Determinants of the northern and southern range limits of a warbler. *Journal of Biogeography* 27:869–878. [342]

Groth, J. G. 1993a. Evolutionary differentiation in morphology, vocalizations, and allozymes among nomadic sibling species in the North American red crossbill *(Loxia curvirostra)* complex. *University of California Publications in Zoology* 127:1–143. [77]

Groth, J. G. 1993b. Call matching and positive assortative mating in red crossbills. *Auk* 110:398–401. [77, 78]

Grubb, T. C., and V. V. Pravosudov. 1994. Tufted titmouse *(Parus bicolor)*. *In* A. Poole and F. B. Gill, eds. *The Birds of North America*, no. 86. Academy of Natural Sciences and The American Ornithologist's Union, Philadelphia and Washington, DC. [359]

Guhl, A. M., and I. L. Ortman. 1953. Visual patterns in the recognition of individuals among chickens. *Condor* 55:287–298. [173, 284]

Gulick, J. T. 1890a. Unstable adjustments as affected by isolation. *Nature* 42:28–29. [29]

Gulick, J. T. 1890b. Indiscriminate separation under the same environment, a cause of divergence. *Nature* 42:369–370. [29]

Gulick, J. T. 1905. *Evolution, Racial and Habitudinal.* Carnegie Institution of Washington, Washington, DC. [29]

Gustafsson, L. 1988. Foraging behaviour of individual coal tits, *Parus ater,* in relation to their age, sex and morphology. *Animal Behaviour* 36:696–704. [137]

Guttenbach, M., I. Nanda, W. Feichtinger, J. S. Masabanda, D. K. Griffin, and M. Schmid. 2003. Comparative chromosome painting of chicken autosomal paints 1–9 in nine different bird species. *Cytogenetic and Genome Research* 103:173–184. [384]

Güttinger, H. R. 1985. Consequences of domestication on the song structures in the canary. *Behaviour* 94:254–278. [224]

Gwinner, E., and V. Neusser. 1985. Postjuvenile moult of European and African stonechats *(Saxicola torquata torquata* and *S. t. axillaris)* and their F1 hybrids. *Journal für Ornithologie* 126:219–220. [44]

Haas, F., and A. Brodin. 2005. The crow *Corvus corone* hybrid zone in southern Denmark and northern Germany. *Ibis* 147:649–656. [342, 355]

Haavie, J., T. Borge, S. Bures, L. Z. Garamszegi, H. M. Lampe, J. Moreno, A. Qvarnström, J. Torok, and G. P. Saetre. 2004. Flycatcher song in allopatry and sympatry—convergence, divergence and reinforcement. *Journal of Evolutionary Biology* 17:227–237. [307, 308]

Haffer, J. 1969. Speciation in Amazonian forest birds. *Science* 165:131–137. [16, 27]

Haffer, J. 1977. Secondary contact zones of birds in northern Iran. *Bonner Zoologische Monographien* 10:1–64. [27, 357]

Haffer, J. 1992. Parapatric species of birds. *Bulletin of the British Ornithologist's Club* 112:250–264. [324, 332]

Haffer, J. 1997. Contact zones between birds of southern Amazonia. *Ornithological Monographs* 48:281–305. [16, 346]

Haffer, J. 2002. A rare hybrid manakin (Aves, Pipridae) and the origin of vertebrate species in Amazonia. *Rudolstädter naturhistorische Schriften* Supplement 4:47–73. [27, 346]

Haldane, J. B. S. 1924. A mathematical theory of natural and artificial selection, part I. *Transactions of the Cambridge Philosophical Society* 23:19–41. [227]

Haldane, J. B. S. 1930. A mathematical theory of natural and artificial selection (part VI, isolation). *Proceedings of the Cambridge Philosophical Society* 26:220–230. [58]

Haldane, J. B. S. 1954. Introducing Douglas Spalding. *British Journal of Animal Behaviour* 2:1. [133]

Hall, B. P., and R. E. Moreau. 1970. *An Atlas of Speciation in African Passerine Birds.* British Museum (National History), London. [26]

Hall, B. P., R. E. Moreau, and I. C. J. Galbraith. 1966. Polymorphism and parallelism in the African Bushshrikes of the genus *Malaconotus* (including *Chlorophoneus*). *Ibis* 108:161–181. [315]

Halpin, Z. T. 1991. Kin recognition cues of vertebrates. Pp. 220–259 *in* P. G. Hepper, ed. *Kin Recognition.* Cambridge University Press, Cambridge. [282]

Hamao, S., and K. Ueda. 2000. Simplified song in an island population of the bush warbler *Cettia diphone*. *Journal of Ethology* 18:53–57. [264, 267]

Hamilton, W. D., and M. Zuk. 1982. Heritable true fitness and bright birds: A role for parasites? *Science* 218:384–387. [180, 259]

Hancock, A. 2003. The Radio Ham. ("Hancock's Half Hour: 'The Blood Donor,' 'The Radio Ham,' and Two Other TV Episodes."). BBC Audiobooks, London. [125]

Handford, P. 1988. Trill rate dialects in the rufous-collared sparrow, *Zonotricha capensis,* in northwestern Argentina. *Canadian Journal of Zoology* 66:2658–2670. [256–258]

Handford, P. 2005. Latin accents: Song dialects of a South American sparrow. *Birding* September/October: 510–519. [256–258]

Handford, P., and S. C. Lougheed. 1991. Variation in duration and frequency characters in the song of the rufous-collared sparrow *Zonotrichia capensis,* with respect to habitat, trill dialects and body size. *Condor* 93:644–658. [258]

Hansell, J. 1998. *The Pigeon in History.* Millstream, Bath. [222]

Hansson, B., S. Bensch, and D. Hasselquist. 2004. Lifetime fitness of short- and long-distance dispersing great reed warblers. *Evolution* 58:2546–2557. [68, 69]

Harper, D. G. C. 1999. Feather mites, pectoral muscle condition, wing length and plumage coloration of passerines. *Animal Behaviour* 58:553–562. [180]

Harris, M. 1970. Abnormal migration and hybridization of *Larus argentatus* and *L. fuscus* after interspecies fostering experiments. *Ibis* 112:488–498. [95, 276]

Harris, M. A., and R. E. Lemon. 1974. Songs of song sparrows: Reactions of males to songs of different localities. *Condor* 76:33–44. [190]

Harris, M. P., C. Morley, and G. H. Green. 1978. Hybridization of herring and lesser black-backed gulls in Britain. *Bird Study* 25:161–166. [94]

Harrison, C. J. O. 1963a. The post-ocular green stripe as a plumage character of the Anatinae. *Bulletin of the British Ornithologist's Club* 83:15–20. [231]

Harrison, C. J. O. 1963b. The incidence and origin of spotted patterns in the Estrildidae. *Ibis* 105:145–155. [231]

Harrison, R. G. 1998. Linking evolutionary pattern and process: The relevance of species concepts for the study of speciation. Pp. 19–31 *in* D. J. Howard and S. H. Berlocher, eds. *Endless Forms: Species and Speciation.* Oxford University Press, New York. [8]

Harrison, J. M., and J. G. Harrison. 1963a. Comments on a hybrid red shoveler × northern shoveler. *Bulletin of the British Ornithologist's Club* 83:21–25. [231]

Harrison, J. M., and J. G. Harrison. 1963b. A gadwall with a white neck ring and a review of plumage variants in wildfowl. *Bulletin of the British Ornithologist's Club* 83:101–108. [231]

Harrison, J. M., and J. G. Harrison. 1971. A back-cross hybrid involving cinnamon teal and northern shoveler, and remarks on hybrid characters in the group of "blue-winged ducks." *Bulletin of the British Ornithologist's Club* 91:99–103. [231]

Harrison, G. L., P. A. McLenachan, M. J. Phillips, K. E. Slack, A. Cooper, and D. Penny. 2004. Four new avian mitochondrial genomes help get to basic evolutionary questions in the late Cretaceous. *Molecular Biology and Evolution* 21:974–983. [16]

Harvey, P. H., and M. D. Pagel. 1991. *The Comparative Method in Evolutionary Biology.* Oxford University Press, Oxford. [58]

Harvey, P. H., and A. Rambaut. 2000. Comparative analyses for adaptive radiations. *Philosophical Transactions of the Royal Society of London Series B—Biological Sciences* 355:1599–1605. [59]

Hasegawa, M., H. Kishino, and T. Yano. 1985. Detection of the human-ape splitting by a molecular clock of mitochondrial DNA. *Journal of Molecular Evolution* 22:160–174. [39]

Hasselquist, D., S. Bensch, and T. von Schantz. 1996. Correlation between male song repertoire, extra-pair paternity and offspring survival in the great reed warbler. *Nature* 381:229–232. [160]

Hasson, O. 1991. Sexual displays as amplifiers: Practical examples with an emphasis on feather decorations. *Behavioral Ecology* 2:189–197. [173]

Hauber, M. E., S. A. Russo, and P. W. Sherman. 2001. A password for species recognition in a brood-parasitic bird. *Proceedings of the Royal Society of London Series B—Biological Sciences* 268:1041–1048. [279, 280]

Hawkins, B. A., E. E. Porter, and J. A. F. Diniz. 2003a. Productivity and history as predictors of the latitudinal diversity gradient of terrestrial birds. *Ecology* 84:1608–1623. [98–100, 115]

Hawkins, B. A., R. Field, H. V. Cornell, D. J. Currie, J. F. Guegan, D. M. Kaufman, J. T. Kerr, G. G. Mittelbach, T. Oberdorff, E. M. O'Brien, E. E. Porter, and J. R. G. Turner. 2003b. Energy, water, and broad-scale geographic patterns of species richness. *Ecology* 84:3105–3117. [98]

Hawkins, B. A., J. A. F. Diniz-Filho, C. A. Jaramillo, and S. A. Soeller. 2006. Post-Eocene climate change and the latitudinal diversity gradient of New World birds. *Journal of Biogeography* 33:770–780. [99, 100]

Hawkins, B. A., J. A. F. Diniz-Filho, C. A. Jaramillo, and S. A. Soeller. 2007. Climate, niche conservatism, and the global bird diversity gradient. *American Naturalist.* In press. [97–100]

Hay, W. W., E. Soeding, R. M. DeConto, and C. N. Wold. 2002. The Late Cenozoic uplift—climate change paradox. *International Journal of Earth Sciences* 91:746–774. [119]

Hebert, P. D. N., M. Y. Stoeckle, T. S. Zemlak, and C. M. Francis. 2004. Identification of birds through DNA barcodes. *PLOS Biology* 2:1657–1663. [12]

Hedges, S. B., P. H. Parker, C. G. Sibley, and S. Kumar. 1996. Continental breakup and the ordinal diversification of birds and mammals. *Nature* 381:226–229. [15]

Hedrick, P. W. 1983. *Genetics of Populations.* Jones and Bartlett, London. [57, 58]

Heindl, M., and H. Winkler. 2003. Vertical lek placement of forest-dwelling manakin species (Aves, Pipridae) is associated with vertical gradients of ambient light. *Biological Journal of the Linnean Society* 80:647–658. [259]

Helb, H. W., F. Dowsett-Lemaire, H. H. Bergmann, and K. Conrads. 1985. Mixed singing in European songbirds—a review. *Zeitschrift für Tierpsychologie* 69:27–41. [307]

Helbig, A. J. 1991. Inheritance of migratory direction in a bird species: A cross-breeding experiment with SE- and SW-migrating blackcaps *(Sylvia atricapilla). Behavioral Ecology and Sociobiology* 28:9–12. [45, 92]

Helbig, A. J., and I. Seibold. 1999. Molecular phylogeny of Palearctic-African *Acrocephalus* and *Hippolais* warblers (Aves: Sylviidae). *Molecular Phylogenetics and Evolution* 11:246–260. [102, 360]

Helbig, A. J., P. Berthold, G. Mohr, and U. Querner. 1994. Inheritance of a novel migratory direction in central European blackcaps. *Naturwissenschaften* 81:184–186. [46, 93]

Helbig, A. J., J. Martens, I. Seibold, F. Henning, B. Schottler, and M. Wink. 1996. Phylogeny and species limits in the Palearctic chiffchaff *Phylloscopus collybita* complex: Mitochondrial genetic differentiation and bioacoustic evidence. *Ibis* 138:650–666. [197, 261]

Helbig, A. J., M. Salomon, and S. Bensch. 2001. Male-biased gene flow across an avian hybrid zone: Evidence from mitochondrial and microsatellite DNA. *Journal of Evolutionary Biology* 14:277–287. [307, 345, 361,391]

Helbig, A. J., A. G. Knox, D. T. Parkin, G. Sangster, and M. Collinson. 2002. Guidelines for assigning species rank. *Ibis* 144:518–525. [3, 4, 5, 12, 395]

Helm-Bychowski, K. M., and A. C. Wilson. 1986. Rates of nuclear DNA evolution in pheasant-like birds: Evidence from restriction maps. *Proceedings of the National Academy of Sciences of the United States of America* 83:688–692. [38]

Hendry, A. P., and T. Day. 2005. Population structure attributable to reproductive time: Isolation by time and adaptation by time. *Molecular Ecology* 14:901–916. [21]

Hendry, A. P., T. Day, and E. B. Taylor. 2001. Population mixing and the adaptive divergence of quantitative traits in discrete populations: A theoretical framework for empirical tests. *Evolution* 55:459–466. [58]

Herzog, S. K., M. Kessler, and K. Bach. 2005. The elevational gradient in Andean bird species richness at the local scale: A foothill peak and a high-elevation plateau. *Ecography* 28:209–222. [99]

Hespenheide, H. A. 1975. Prey characteristics and predator niche width. Pp. 158–180 *in* M. L. Cody and J. M. Diamond, eds. *Ecology and Evolution of Communities.* Belknap Press of Harvard University Press. Cambridge, MA. [13]

Hewitt, G. 2000. The genetic legacy of the Quaternary ice ages. *Nature* 405:907–913. [345]

Hickerson, M., C. Meyer, and C. Moritz. 2006. DNA barcoding will often fail to discover new animal species over broad parameter space. *Systematic Biology* 55:729-739. [12]

Hill, G. E. 1993. Geographic variation in the carotenoid plumage pigmentation of male house finches *(Carpodacus mexicanus). Biological Journal of the Linnean Society* 49:63–86. [47, 135, 136]

Hill, G. E. 1994. Geographic variation in male ornamentation and female mate preference in the house finch: A comparative test of models of sexual selection. *Behavioral Ecology* 5:64–73. [135, 136, 160]

Hill, G. E. 2002. *A Red Bird in a Brown Bag: The Function and Evolution of Colorful Plumage in the House Finch.* Oxford University Press, Oxford. [206]

Hillier, L. W., W. Miller, E. Birney, and others 2004. Sequence and comparative analysis of the chicken genome provide unique perspectives on vertebrate evolution. *Nature* 432:695–716. [225, 393, 392]

Hinde, R. A. 1952. The social behaviour of the great tit *(Parus major)* and some other related species. *Behaviour Supplement* 2:1–201. [167, 175, 176, 264]

Hinde, R. A. 1958. Alternative motor patterns in chaffinch song. *Animal Behaviour* 6:211–218. [216]

Hinde, R. A. 1961. The establishment of the parent-offspring relation in birds, with some mammalian analogies. Pp. 175–193 *in* W. H. Thorpe and O. Zangwill, eds. *Current Problems in Animal Behaviour.* Cambridge University Press, Cambridge. [282–284]

Hinde, R. A. 1962. Some aspects of the imprinting problem. *Symposia of the Zoological Society of London* 8:129–138. [281]

Hinde, R. A., and J. Fisher. 1951. Further observations on the opening of milk bottles by birds. *British Birds* 44:393–396. [128]

Hoekstra, H. E., and T. Price. 2004. Parallel evolution is in the genes. *Science* 303:1779–1780. [234]

Höglund, J., R. V. Alatalo, R. M. Gibson, and A. Lundberg. 1995. Mate-choice copying in black grouse. *Animal Behaviour* 49:1627–1633. [279]

Holland, B., and W. R. Rice. 1998. Perspective: Chase away sexual selection: Antagonistic seduction versus resistance. *Evolution* 52:1–7. [180]

Honey, R. C., P. Bateson, and G. Horn. 1994. The role of stimulus comparison in perceptual learning: An investigation with the domestic chick. *Quarterly Journal of Experimental Psychology* 47B:83–103. [283]

Howard, D. J. 1993. Reinforcement: Origin, dynamics, and fate of an evolutionary hypothesis. Pp. 46–69 *in* R. G. Harrison, ed. *Hybrid Zones and the Evolutionary Process.* Oxford University Press, Oxford. [300, 304, 313]

Howard, R., and A. Moore. 1991. *A Complete Checklist of Birds of the World.* Academic Press, London. [59]

Hubbard, J. P. 1969. Relationships and evolution of the *Dendroica coronata* complex. *Auk* 86:393–432. [363]

Hudson, R. R., and J. A. Coyne. 2002. Mathematical consequences of the genealogical species concept. *Evolution* 56:1557–1565. [11, 12, 39]

Hudson, R. R., and M. Turelli. 2003. Stochasticity overrules the "three-times rule": Genetic drift, genetic draft, and coalescence times for nuclear loci versus mitochondrial DNA. *Evolution* 57:182–190. [11, 12, 208]

Huelsenbeck, J. P., and F. Ronquist. 2001. MRBAYES: Bayesian inference of phylogenetic trees. *Bioinformatics* 17:754–755. [40]

Huey, R. B., P. E. Hertz, and B. Sinervo. 2003. Behavioral drive versus behavioral inertia in evolution: A null model approach. *American Naturalist* 161:357–366. [134]

Hughes, C., and R. Eastwood. 2006. Island radiation on a continental scale: Exceptional rates of plant diversification after uplift of the Andes. *Proceedings of the National Academy of Sciences (USA)* 103: 10334–10339. [118]

Hughes, J. M., A. M. Baker, G. De Zylva, and P. B. Mather. 2001. A phylogeographic analysis of southern and eastern populations of the Australian magpie: Evidence for selection in maintenance of the distribution of two plumage morphs. *Biological Journal of the Linnean Society* 74:25–34. [355]

Hughes, J. M., C. L. Lange, P. B. Mather, and A. Robinson. 2002. A comparison of fitness components among different plumage morphs of the Australian magpie, *Gymnorhina tibicen*. *Emu* 102:331–338. [355]

Hughes, J. M., P. B. Mather, A. Toon, J. Ma, I. Rowley, and E. Russell. 2003. High levels of extra-group paternity in a population of Australian magpies *Gymnorhina tibicen*: Evidence from microsatellite analysis. *Molecular Ecology* 12:3441–3450. [355]

Hunt, J. S., E. Bermingham, and R. E. Ricklefs. 2001. Molecular systematics and biogeography of Antillean thrashers, tremblers, and mockingbirds (Aves: Mimidae). *Auk* 118:35–55. [24, 31]

Hunter, M. L., Jr., and J. R. Krebs. 1979. Geographical variation in the song of the great tit *(Parus major)* in relation to ecological factors. *Journal of Animal Ecology* 48:759–785. [253]

Huntington, C. E. 1952. Hybridization in the purple grackle *Quiscalus quiscula*. *Systematic Zoology* 1:149–170. [366]

Hurd, P. L., and M. Enquist. 2001. Threat display in birds. *Canadian Journal of Zoology* 79:931–942. [167, 170]

Hurd, P. L., C. A. Wachtmeister, and M. Enquist. 1995. Darwin's principle of antithesis revisited: A role for perceptual biases in the evolution of intraspecific signals. *Proceedings of the Royal Society of London Series B—Biological Sciences* 259:201–205. [170–172]

Hurst, L. D., and A. Pomiankowski. 1991. Causes of sex ratio bias may account for unisexual sterility in hybrids—a new explanation of Haldane's rule and related phenomena. *Genetics* 128:841–858. [369, 370, 381]

Hurst, L. D., A. Atlan, and B. O. Bengtsson. 1996. Genetic conflicts. *Quarterly Review of Biology* 71:317–364. [30, 369, 370, 405]

Huxley, J. S. 1914. The courtship habits of the great crested grebe *(Podiceps cristatus)*. *Proceedings of the Zoological Society of London* 35:491–562. [161, 166, 240]

Huxley, J. S. 1923. Courtship activities in the red-throated diver; together with a discussion of the evolution of courtship in birds. *Journal of the Linnean Society (Zoology)* 35:253–292. [161, 166, 168, 176]

Huxley, J. S. 1938. Darwin's theory of sexual selection and the data subsumed by it, in the light of recent research. *American Naturalist* 72:416–433. [158, 170, 252]

Huxley, J. 1942. *Evolution, the Modern Synthesis*. Allen and Unwin, London. [1, 17, 19]

Immelmann, K. 1959. Experimentelle Untersuchungen über die biologische Bedeutung artspecifischer Merkmale beim Zebrafinken, *Taeniopygia castonitis*. *Zoologische Jahrbücher Abteilung für Systematik Okologie und Geographie der Tiere* 86:437–592. [284]

Immelmann, K. 1972. Sexual and other long-term aspects of imprinting in birds and other species. *Advances in the Study of Behaviour* 4:147–174. [317]

Ingolfsson, A. 1970. Hybridization of glaucous gulls *Larus hyperboreus* and herring gulls *L. argentatus* in Iceland. *Ibis* 112:340–362. [317]

Irwin, R. E. 1994. The evolution of plumage dichromatism in the New World Blackbirds—social selection on female brightness? *American Naturalist* 144:890–907. [240, 264]

Irwin, D. E. 2000. Song variation in an avian ring species. *Evolution* 54:998–1010. [188, 236, 237]

Irwin, D. E. 2002. Phylogeographic breaks without barriers to gene flow. *Evolution* 56:2383–2394. [12]

Irwin, D. E., and J. H. Irwin. 2004. Siberian migratory divides: The role of seasonal migration in speciation. Pp. 27–40 *in* R. Greenberg and P. Marra, eds. *Birds of Two Worlds*. Johns Hopkins University Press, Baltimore, MD. [29, 92, 93]

Irwin, D. E., and T. Price. 1999. Sexual imprinting, learning, and speciation. *Heredity* 82:347–354. [198, 285, 286, 288]

Irwin, D. E., P. Alström, U. Olsson, and Z. M. Benowitz-Fredericks. 2001a. Cryptic species in the genus *Phylloscopus*. *Ibis* 143:233–247. [2, 4, 196, 271]

Irwin, D. E., S. Bensch, and T. D. Price. 2001b. Speciation in a ring. *Nature* 409:333–337. [66, 236, 237]

Irwin, D. E., J. H. Irwin, and T. D. Price. 2001c. Ring species as bridges between microevolution and speciation. *Genetica* 112:223–243. [236]

Irwin, D. E., S. Bensch, J. H. Irwin, and T. D. Price. 2005. Speciation by distance in a ring species. *Science* 307:414–416. [65, 66, 236, 237]

Isler, M. L., P. R. Isler, and B. M. Whitney. 1998. Use of vocalizations to establish species limits in antbirds (Passeriformes: Thamnophilidae). *Auk* 115:577–590. [4, 271]

Isler, M. L., P. R. Isler, B. M. Whitney, and B. Walker. 2001. Species limits in antbirds: The *Thamnophilus punctatus* complex continued. *Condor* 103:278–286. [2, 192]

Isler, M. L., P. R. Isler, and R. T. Brumfield. 2005. Clinal variation in vocalizations of an antbird (Thamnophilidae) and implications for defining species limits. *Auk* 122:433–444. [192, 347, 354]

Itoh, Y., and A. P. Arnold. 2005. Chromosomal polymorphism and comparative painting analysis in the zebra finch. *Chromosome Research* 13:47–56. [384, 388]

Jackson, P. S., and P. P. G. Bateson. 1974. Imprinting and exploration of slight novelty in chicks. *Nature* 251:609–610. [282]

James, F. C. 1970. Geographic size variation in birds and its relationship to climate. *Ecology* 51:365–390. [43]

James, F. C. 1983. Environmental component of morphological differentiation in birds. *Science* 221:184–186. [42, 45]

James, F. C., and C. Nesmith. 1986. Nongenetic effects in geographic differences among nestling populations of red-winged blackbirds. Pp. 1424–1433 *in* H. Ouellet, ed. *Acta XIX International Ornithological Congress*. University of Ottawa Press, Ottawa. [42, 45]

Järvi, T., Ø. Walso and M. Bakken. 1987. Status signalling by *Parus major:* An experiment in deception. *Ethology* 76:334–342. [181]

Jenkins, P. F., and A. J. Baker. 1984. Mechanisms of song differentiation in introduced populations of chaffinches *Fringilla coelebs* in New Zealand. *Ibis* 126:510–524. [217]

Jennings, W. B., and S. V. Edwards. 2005. Speciational history of Australian grass finches *(Poephila)* inferred from thirty gene trees. *Evolution* 59:2033–2047. [11, 12]

Jetz, W., and C. Rahbek. 2002. Geographic range size and determinants of avian species richness. *Science* 297:1548–1551. [97]

Jiggins, C. D., and J. Mallet. 2000. Bimodal hybrid zones and speciation. *Trends in Ecology and Evolution* 15:250–255. [333]

Johansson, U. S., P. Alström, U. Olsson, P. G. P. Ericson, P. Sundberg, and T. D. Price. 2006. Build-up of the Himalayan avifauna through immigration: a biogeographical analysis of the *Phylloscopus* and *Seicercus* warblers. *Evolution* 61:324–333. [89]

Johnsgard, P. A. 1971. Experimental hybridization of the New World quail (Odontophorinae). *Auk* 88:264–275. [231, 318]

Johnson, N. K. 1966. Morphologic stability versus adaptive variation in the Hammond's flycatcher. *Auk* 83:179–200. [41]

Johnson, N. K. 1975. Controls of number of bird species on montane islands in the Great Basin. *Evolution* 29:545–567. [56]

Johnson, S. G. 1991. Effects of predation, parasites, and phylogeny on the evolution of bright coloration in North American male passerines. *Evolutionary Ecology* 5:52–62. [260]

Johnson, N. K., and C. Cicero. 2002. The role of ecologic diversification in sibling speciation of *Empidonax* flycatchers (Tyrannidae): Multigene evidence from mtDNA. *Molecular Ecology* 11:2065–2081. [102]

Johnson, N. K., and C. Cicero. 2004. New mitochondrial DNA data affirm the importance of Pleistocene speciation in North American birds. *Evolution* 58:1122–1130. [32, 33]

Johnson, N. K., and C. B. Johnson. 1985. Speciation in sapsuckers *(Sphyrapicus)*. 2. Sympatry, hybridization, and mate preference in *Sphyrapicus ruber daggetti* and *Sphyrapicus nuchalis. Auk* 102:1–15. [351]

Johnson, M. H., and G. Horn. 1988. Development of filial preferences in dark-reared chicks. *Animal Behaviour* 36:675–683. [282]

Johnson, K. P., and S. M. Lanyon. 2000. Evolutionary changes in color patches of blackbirds are associated with marsh nesting. *Behavioral Ecology* 11:515–519. [233, 251, 260]

Johnson, K. P., and M. D. Sorenson. 1999. Phylogeny and biogeography of dabbling ducks (genus: *Anas*): A comparison of molecular and morphological evidence. *Auk* 116:792–805. [102, 377]

Johnson, K. P., S. De Kort, K. Dinwoodey, A. C. Mateman, C. ten Cate, C. M. Lessells, and D. H. Clayton. 2001. A molecular phylogeny of the dove genera *Streptopella* and *Columba. Auk* 118:874–887. [377]

Johnston, R. F. 1969. Taxonomy of house sparrows and their allies in Mediterranean basin. *Condor* 71:129–139. [347, 361]

Johnston, R., and M. Janiga. 1995. *Feral pigeons.* Oxford University Press, New York. [222]

Johnston, R. F., and R. K. Selander. 1971. Evolution in the house sparrow. II. Adaptive differentiation in North American populations. *Evolution* 25:1–28. [46]

Johnstone, R. A. 1995. Sexual selection, honest advertisement and the handicap principle: Reviewing the evidence. *Biological Reviews* 70:1–65. [180, 181]

Jones, I. L., and F. M. Hunter. 1998. Heterospecific mating preferences for a feather ornament in least auklets. *Behavioral Ecology* 9:187–192. [273]

Joseph, L., T. Wilke, E. Bermingham, D. Alpers, and R. Ricklefs. 2004. Towards a phylogenetic framework for the evolution of shakes, rattles, and rolls in *Myiarchus* tyrant-flycatchers (Aves: Passeriformes: Tyrannidae). Molecular Phylogenetics and Evolution 31:139–152. [103]

Junco, F. 1988. Filial imprinting in an altricial bird: The blackbird *(Turdus merula)*. *Behaviour* 106:25–42. [282]

Kallioinen, R. U. O., J. M. Hughes, and P. B. Mather. 1995. Significance of back colour in territorial interactions in the Australian magpie. *Australian Journal of Zoology* 43:665–673. [355]

Kaltenhäuser, D. 1971. Über Evolutionsvorgänge in der Schwimmentenbalz. Zeitschrift für Tierpsychologie 29:481–540. [229]

Kasai, F., C. Garcia, M. V. Arruga, and M. A. Ferguson-Smith. 2003. Chromosome homology between chicken *(Gallus gallus domesticus)* and the red-legged partridge *(Alectoris rufa)*; evidence of the occurrence of a neocentromere during evolution. *Cytogenetic and Genome Research* 102:326–330. [385]

Kattan, G. H., P. Franco, V. Rojas, and G. Morales. 2004. Biological diversification in a complex region: A spatial analysis of faunistic diversity and biogeography of the Andes of Colombia. *Journal of Biogeography* 31:1829–1839. [115, 119]

Keast, A. 1961. Bird speciation on the Australian continent. *Bulletin of the Museum of Comparative Zoology at Harvard College* 123:305–495. [25]

Keast, A. 1968. Competitive interactions and the evolution of ecological niches as illustrated by the Australian honeycreeper genus *Melithreptus* (Meliphagidae). *Evolution* 22:762–784. [50]

Keast, J. A. 1974. Avian speciation in Africa and Australia: Some comparisons. *Emu* 74:261–269. [26]

Kegl, J., A. Senghas, and M. Coppola. 1999. Creation through contact: Sign language emergence and sign language change in Nicaragua. Pp. 179–237 *in* M. deGraff, ed. *Language Creation and Language Change.* MIT Press, Cambridge, MA. [209]

Keigwin, L. D. 1978. Pliocene closing of Isthmus of Panama, based on biostratigraphic evidence from nearby Pacific Ocean and Caribbean Sea cores. *Geology* 6:630–634. [115]

Keller, L. F., K. J. Jeffery, P. Arcese, M. A. Beaumont, W. M. Hochachka, J. N. M. Smith, and M. W. Bruford. 2001. Immigration and the ephemerality of a natural population bottleneck: Evidence from molecular markers. *Proceedings of the Royal Society of London Series B—Biological Sciences* 268:1387–1394. [69]

Kelly, J. K., and M. A. F. Noor. 1996. Speciation by reinforcement: A model derived from studies of *Drosophila. Genetics* 143:1485–1497. [301]

Kelly, C., and T. D. Price. 2004. Comparative methods based on species mean values. *Mathematical Biosciences* 187:135–154. [59]

Kempanaers, B., G. R. Verheyen, and A. A. Dondt. 1997. Extrapair paternity in the blue tit *(Parus caeruleus)*: Female choice, male characteristics, and offspring quality. *Behavioral Ecology* 8:481–492. [160]

Ketterson, E. D., V. Nolan Jr., J. M. Casto, C. A. Buerkle, E. Clotfelter, J. L. Grindstaff, K. L. Jones, J. L. Lipar, F. M. A. McNab, D. L. Neudorf, I. Parker-Renga, S. J. Schoech, and E. Snajdr. 2001. Testosterone, phenotype and fitness: A research program in evolutionary behavioral endocrinology. Pp. 19–40 *in* A. Dawson and C. M. Chaturvedi, eds. *Avian Endocrinology.* Narosa Publishing House, New Delhi. [163]

Kimball, R. T., E. L. Braun, P. W. Zwartjes, T. M. Crowe, and J. D. Ligon. 1999. A molecular phylogeny of the pheasants and partridges suggests that these lineages are not monophyletic. *Molecular Phylogenetics and Evolution* 11:38–54. [377]

King, M. 1993. *Species Evolution: The Role of Chromosome Change.* Cambridge University Press, Cambridge and New York. [380]

King, A. P., and M. J. West. 1977. Species identification in the North American cowbird: Appropriate responses to abnormal song. *Science* 195:1002–1004. [279]

King, A. P., M. J. West, and D. H. Eastzer. 1980. Song structure and song development as potential contributors to reproductive isolation in cowbirds *(Molothrus ater)*. *Journal of Comparative and Physiological Psychology* 94:1028–1039. [191, 207, 280]

Kirkpatrick, M. 1982. Sexual selection and the evolution of female choice. *Evolution* 36:1–12. [239]

Kirkpatrick, M., and N. H. Barton. 2006. Chromosome inversions, local adaptation, and speciation. *Genetics* 173:419–434. [389, 392]

Kirkpatrick, M., and D. W. Hall. 2004. Male-biased mutation, sex linkage, and the rate of adaptive evolution. *Evolution* 58:437–440. [227]

Kirkpatrick, M., and V. Ravigné. 2002. Speciation by natural and sexual selection: Models and experiments. *American Naturalist* 159:S22–S35. [399]

Kirkpatrick, M., and M. Servedio. 1999. The reinforcement of mating preferences on an island. *Genetics* 151:865–884. [304]

Kirkpatrick, M., T. Price, and S. J. Arnold. 1990. The Darwin–Fisher theory of sexual selection in monogamous birds. *Evolution* 44:180–193. [160, 245]

Klein, N. K., and R. B. Payne. 1998. Evolutionary associations of brood parasitic finches *(Vidua)* and their host species: Analyses of mitochondrial DNA restriction sites. *Evolution* 52:566–582. [294]

Klicka, J., and R. M. Zink. 1997. The importance of recent ice ages in speciation: A failed paradigm. *Science* 277:1666–1669. [18, 25, 215, 364]

Klicka, J., and R. M. Zink. 1999. Pleistocene effects on North American songbird evolution. *Proceedings of the Royal Society of London Series B—Biological Sciences* 266:695–700. [37]

Kluijver, H. N. 1951. The population ecology of the great tit, *Parus m. major* L. *Ardea* 39:1–135. [167, 176]

Kochmer, J. P., and R. H. Wagner. 1988. Why are there so many kinds of passerine birds? Because they are small. A reply to Raikow. *Systematic Zoology* 37:68–69. [16]

Kondrashov, A. S. 2003. Accumulation of Dobzhansky–Muller incompatibilities within a spatially structured population. *Evolution* 57:151–153. [30]

Kondrashov, A. S., L. Y. Yampolsky, and S. A. Shabalina. 1998. On the sympatric origin of species by means of natural selection. Pp. 90–98 *in* D. Howard and S. Berlocher, eds. *Endless Forms: Species and Speciation.* Oxford University Press, Oxford. [31]

König, C., F. Weick, and J. H. Becking. 1999. *Owls: A Guide to the Owls of the World.* Yale University Press, New Haven, CT. [2]

Kovach, J. 1980. Mendelian units of inheritance control color preferences in quail chicks *(Coturnix coturnix japonica). Science* 207:549–551. [289]

Kovach, J. 1990. Nonspecific imprintability of quail to colors: response to artificial selection. *Behavior genetics* 20:91–96. [289, 290]

Kovach, J. 1993. Constitution–environment interactions modeled by artificially selected colour preferences and imprinting in quail. *Netherlands Journal of Zoology* 43:46–67. [289, 290]

Krabbe, N., and T. S. Schulenberg. 1997. Species limits and natural history of *Scytalopus* tapaculos (Rhinocryptidae), with descriptions of the Ecuadorean taxa, including three new species. *Ornithological Monographs* 48:47–88. [2]

Krajewski, C., and D. G. King. 1996. Molecular divergence and phylogeny—rates and patterns of cytochrome *b* evolution in cranes. *Molecular Biology and Evolution* 13:21–30. [18, 102]

Krakauer, D. C., and R. A. Johnstone. 1995. The evolution of exploitation and honesty in animal communication—a model using artificial neural networks. *Philosophical Transactions of the Royal Society of London Series B* 348:355–361. [182–184]

Krebs, J. R., R. Ashcroft, and K. van Orsdol. 1981. Song matching in the great tit *Parus major* L. *Animal Behaviour* 29:918–923. [202]

Kreutzer, M. 1974. Réponses comportementales des males troglodytes (Passeriformes) a des chants spécifiques de dialectes différents. *Revue du Comportement Animal* 6:287–295. [190]

Kroodsma, R. L. 1975. Hybridization in buntings *(Passerina)* in North Dakota and eastern Montana. *Auk* 92:66–80. [365]

Kroodsma, D. E. 1976. Reproductive development in a female songbird: Differential stimulation by quality of male song. *Science* 192:574–575. [163, 164]

Kroodsma, D. E. 1985. Geographic variation in songs of the Bewick's wren: A search for correlations with avifaunal complexity. *Behavioral Ecology and Sociobiology* 16:143–150. [267]

Kroodsma, D. 2004. The diversity and plasticity of birdsong. Pp. 108–131 *in* P. Marler and H. Slabbekoorn, eds. *Nature's Music: The Science of Bird Song.* Academic Press, London. [187]

Kroodsma, D. E., and R. A. Canady. 1985. Differences in repertoire size, singing behavior, and associated neuroanatomy among marsh wren populations have a genetic basis. *Auk* 102:439–446. [211]

Kroodsma, D. E., D. J. Albano, P. W. Houlihan, and J. A. Wells. 1995. Song development by black-capped chickadees *(Parus atricapillus)* and Carolina chickadees *(P. carolinensis). Auk* 112:29–43. [207]

Kroodsma, D. E., P. W. Houlihan, P. A. Fallon, and J. A. Wells. 1997. Song development by gray catbirds. *Animal Behaviour* 54:457–464. [201]

Kroodsma, D. E., B. E. Byers, S. L. Halkin, C. Hill, D. Minis, J. R. Bolsinger, J. A. Dawson, E. Donelan, J. Farrington, F. B. Gill, P. Houlihan, D. Innes, G. Keller, L. Macaulay, C. A. Marantz, J. Ortiz, P. K. Stoddard, and K. Wilda. 1999. Geographic variation in black-capped chickadee songs and singing behavior. *Auk* 116:387–402. [196, 207, 267]

Kruijt, J. P., and G. B. Meeuwissen. 1993. Consolidation and modification of sexual preferences in adult male Zebra finches. *Netherlands Journal of Zoology* 43:68–79. [277]

Kruijt, J. P., I. Bossema, and G. J. Lammers. 1982. Effects of early experience and male activity on mate choice in mallard females *(Anas platyrhynchos). Behaviour* 80:32–43. [279]

Kruijt, J. P., C. J. ten Cate, and G. B. Meeuwissen. 1983. The influence of siblings on the development of sexual preferences in male zebra finches. *Developmental Psychobiology* 16:233–239. [277]

Kruuk, L. E. B., S. J. E. Baird, K. S. Gale, and N. H. Barton. 1999. A comparison of multilocus clines maintained by environmental adaptation or by selection against hybrids. *Genetics* 153:1959–1971. [331, 332, 347]

Kryukov, A. P., and H. Suzuki. 2000. Phylogeography of carrion, hooded, and jungle crows (Aves, Corvidae) inferred from partial sequencing of the mitochondrial DNA cytochrome *b* gene. *Russian Journal of Genetics* 36:922–929. [355]

Kuhl, P. K. 2003. Human speech and birdsong: Communication and the social brain. *Proceedings of the National Academy of Sciences of the United States of America* 100:9645–9646. [209]

Kusmierski, R., G. Borgia, A. Uy, and R. H. Crozier. 1997. Labile evolution of display traits in bowerbirds indicates reduced effects of phylogenetic constraint. *Proceedings of the Royal Society of London Series B— Biological Sciences* 264:307–313. [185]

Lack, D. 1943. *The Life of the Robin.* Witherby, London. [158, 187]

Lack, D. 1944. Ecological aspects of species formation in passerine birds. *Ibis* 86:260–286. [89]

Lack, D. 1947. *Darwin's Finches.* Cambridge University Press, Cambridge. [1]

Lack, D. 1949. The significance of ecological isolation. Pp. 299–308 *in* G. L. Jepsen, E. Mayr and G. G. Simpson, eds. *Genetics, Paleontology, and Evolution.* Princeton University Press, Princeton, NJ. [152]

Lack, D. 1968. *Ecological Adaptations for Breeding in Birds.* Methuen, London. [243, 244, 264]

Lack, D. 1970. The endemic ducks of remote islands. *Wildfowl* 21:5–10. [263]

Lack, D. 1976. *Island Biology, Illustrated by the Land Birds of Jamaica.* University of California Press, Berkeley. [1, 16, 60, 152]

Lai, Z., T. Nakazato, M. Salmaso, J. M. Burke, S. X. Tang, S. J. Knapp, and L. H. Rieseberg. 2005. Extensive chromosomal repatterning and the evolution of sterility barriers in hybrid sunflower species. *Genetics* 171:291–303. [380, 391]

Laland, K. 1994. On the evolutionary significance of imprinting. *Evolution* 48:477–489. [273, 292]

Lambrechts, M. M., and P. C. Dias. 1993. Differences in the onset of laying between island and mainland Mediterranean blue tits *Parus caeruleus*: Phenotypic plasticity or genetic differences? *Ibis* 135: 451–455. [44]

Lampe, H. M., and T. Slagsvold. 1998. Female pied flycatchers respond differently to songs of mates, neighbours and strangers. *Behaviour* 135:269–285. [284]

Lancaster, F. M. 1963. The inheritance of plumage colour in the common duck *(Anas platyrhynchos* Linné). *Bibliographia Genetica* 19:317–404. [226, 227]

Lande, R. 1976. Natural selection and random genetic drift in phenotypic evolution. *Evolution* 30:314–334. [76]

Lande, R. 1980. Genetic variation and phenotypic evolution during allopatric speciation. *American Naturalist* 116:463–479. [49]

Lande, R. 1981a. Models of speciation by sexual selection on polygenic traits. *Proceedings of the National Academy of Sciences of the United States of America* 78:3721–3725. [29, 239, 268, 270]

Lande, R. 1981b. The minimum number of genes contributing to quantitative variation between and within populations. *Genetics* 99:541–553. [228]

Lande, R. 1996. Statistics and partitioning of species diversity, and similarity among multiple communities. *Oikos* 76:5–13. [54, 55, 56]

Lander, P., and B. Partridge. 1998. *Greenfinches.* Kingdom Books, Waterlooville, UK. [222, 225]

Lang, A. L., and J. C. Barlow. 1987. Syllable sharing among North American populations of the Eurasian tree sparrow. *Condor* 89:746–751. [204[

Lang, A. L., and J. C. Barlow. 1997. Cultural evolution in the Eurasian tree sparrow: Divergence between introduced and ancestral populations. *Condor* 99:413–423. [204, 269]

Lanyon, W. E. 1956. Ecological aspects of the sympatric distribution of meadowlarks in the north-central states. *Ecology* 37:98–108. [29, 114[

Lanyon, W. E. 1966. Hybridization in meadowlarks. *Bulletin of the American Museum of Natural History* 134:1–26. [25]

Lanyon, W. E. 1978. Revision of the *Myiarchus* flycatchers of South America. *Bulletin of the American Museum of Natural History* 161:429–627. [192, 253]

Lanyon, W. E. 1979a. Hybrid sterility in meadowlarks. *Nature* 279:557–558. [25, 390]

Lanyon, W. E. 1979b. Development of song in the wood thrush *(Hylocichla mustelina)*, with notes on a technique for hand-rearing passerines from the egg. *American Museum Novitates* 2666:1–27. [201]

Lanyon, S. M., and K. E. Omland. 1999. A molecular phylogeny of the blackbirds (Icteridae): Five lineages revealed by cytochrome-*b* sequence data. *Auk* 116:629–639. [25]

Laurie, C. C. 1997. The weaker sex is heterogametic: 75 years of Haldane's rule. *Genetics* 147:937–951. [375]

Leader, N., J. Wright, and Y. Yom-Tov. 2002. Dialect discrimination by male orange-tufted sunbirds *(Nectarinia osea)*: Reactions to own vs. neighbor dialects. *Ethology* 108:367–376. [190]

Leafloor, J. O., C. D. Ankney, and D. H. Rusch. 1998. Environmental effects on body size of Canada geese. *Auk* 115:26–33. [42, 44]

Leboucher, G., V. Depraz, M. Kreutzer, and L. Nagle. 1998. Male song stimulation of female reproduction in canaries: Features relevant to sexual displays are not relevant to nest-building or egg-laying. *Ethology* 104:613–624. [164]

Lefebvre, L. 1986. Cultural diffusion of a novel food finding behaviour in urban pigeons: An experimental field test. *Ethology* 71:295–304. [128]

Lefebvre, L. 1995. The opening of milk bottles by birds: Evidence for accelerating learning rates, but against the wave-of-advance model of cultural transmission. *Behavioural Processes* 34:43–54. [128]

Lefebvre, L. 2000. Feeding innovations and their cultural transmission in bird populations. Pp. 311–328 *in* C. Heyes and L. Huber, eds. *The Evolution of Cognition.* MIT press, Cambridge, MA. [128, 129]

Lefebvre, L., and L. A. Giraldeau. 1994. Cultural transmission in pigeons is affected by the number of tutors and bystanders present. *Animal Behaviour* 47:331–337. [128]

Lefebvre, L., P. Whittle, E. Lascaris, and A. Finkelstein. 1997. Feeding innovations and forebrain size in birds. *Animal Behaviour* 53:549–560. [127, 129]

Lefebvre, L., A. Gaxiola, S. Dawson, S. Timmermans, L. Rosza, and P. Kabai. 1998. Feeding innovations and forebrain size in Australasian birds. *Behaviour* 135:1077–1097. [127, 129]

Lefebvre, L., N. Juretic, N. Nicolakakis, and S. Timmermans. 2001. Is the link between forebrain size and feeding innovations caused by confounding variables? A study of Australian and North American birds. *Animal Cognition* 4:91–97. [127, 129]

Lefebvre, L., N. Nicolakakis, and D. Boire. 2002. Tools and brains in birds. *Behaviour* 139:939–973. [127, 135]

Lefebvre, L., S. M. Reader, and D. Sol. 2004. Brains, innovations and evolution in birds and primates. *Brain, Behavior and Evolution* 63:233–246. [127, 129, 132]

Lehrman, D. S. 1959. Hormonal responses to external stimuli in birds. *Ibis* 101:478–496. [165]

Lehrman, D. S., P. N. Brody, and R. P. Wortis. 1961. The presence of the mate and of nesting material as stimuli for the development of incubation behavior and for gonadotropin secretion in the ring dove *(Streptopelia risoria). Endocrinology* 68:507–516. [165]

Lemon, R. E. 1967. Response of cardinals to songs of different dialects. *Animal Behaviour* 15:538–545. [191]

Levi, W. M. 1963. *The Pigeon.* Bryan, Columbia, SC. [220, 226]

Levi, W. M. 1965. *Encyclopedia of Pigeon Breeds.* TFH Publications, Jersey City, NJ. [222, 226]

Li, C. C. 1976. *First Course in Population Genetics.* Boxwood Press, Pacific Grove, CA. [57]

Liebers, D., P. de Kniff, and A. J. Helbig. 2004. The herring gull complex is not a ring species. *Proceedings of the Royal Society of London Series B—Biological Sciences* 271:893–901. [66, 236]

Lijtmaer, D. A., B. Mahler, and P. L. Tubaro. 2003. Hybridization and postzygotic isolation patterns in pigeons and doves. *Evolution* 57:1411–1418. [376, 378, 379]

Lille, R. 1988. Art- und Michgesang von Nachtigall und Sprosser *(Luscinia megarhynchos, L. luscinia). Journal für Ornithologie* 129:133–159. [307]

Liou, L. W., and T. D. Price. 1994. Speciation by reinforcement of premating isolation. *Evolution* 48:1451–1459. [301–303]

Liu, Y. P., G. S. Wu, Y. G. Yao, Y. W. Miao, G. Luikart, M. Baig, A. Beja-Pereira, Z. L. Ding, M. G. Palanichamy, and Y. P. Zhang. 2006. Multiple maternal origins of chickens: Out of the Asian jungles. *Molecular Phylogenetics and Evolution* 38:12–19. [221]

Livezey, B. C. 2003. Evolution of flightlessness in rails (Gruiformes: Rallidae): Phylogenetic, ecomorphological, and ontogenetic perspectives. *Ornithological Monographs* 53:1–654. [70]

Lloyd, P., A. Craig, P. E. Hulley, M. F. Essop, P. Bloomer, and T. M. Crowe. 1997. Ecology and genetics of hybrid zones in the southern African *Pycnonotus* bulbul species complex. *Ostrich* 68:90–96. [359]

Lockley, A. K. 1992. The position of the hybrid zone between the house sparrow *Passer domesticus* and the Italian sparrow *Passer d. italiae* in the Alpes Maritimes. *Journal für Ornithologie* 133:77–82. [361]

Lockley, A. K. 1996. Changes in the position of the hybrid zone between the house sparrow *Passer domesticus* and the Italian sparrow *P. d. italiae* in the Alpes Maritimes. *Journal für Ornithologie* 137:243–248. [361]

Logan, C. A., L. E. Hyatt, and L. Gregorcyk. 1990. Song playback initiates nest building during clutch overlap in mockingbirds, *Mimus polyglottos. Animal Behaviour* 39:943–953. [162]

Lorenz, K. 1937. The companion in the bird's world. *Auk* 54:245–273. [276, 277, 317]

Lorenz, K. 1971. *Studies in Animal and Human Behaviour,* Vol. 2. Methuen, London. [176]

Losos, J. B., T. R. Jackman, A. Larson, K. de Queiroz, and L. Rodriguez-Schettino. 1998. Contingency and determinism in replicated adaptive radiations of island lizards. *Science* 279:2115–2118. [400]

Lougheed, S. C., P. Handford, and A. J. Baker. 1993. Mitochondrial DNA hyperdiversity and vocal dialects in a subspecies transition of the rufous-collared sparrow. *Condor* 95:889–895. [257, 258]

Lovette, I. J. 2004. Mitochondrial dating and mixed-support for the "2% rule" in birds. *Auk* 121:1–6. [18, 36, 38]

Lovette, I. J., and E. Bermingham. 1999. Explosive speciation in the New World *Dendroica* warblers. *Proceedings of the Royal Society of London Series B—Biological Sciences* 266:1629–1636. [16, 102, 107, 121, 271]

Lovette, I. J., E. Bermingham, G. Seutin, and R. E. Ricklefs. 1998. Evolutionary differentiation in three endemic West Indian Warblers. *Auk* 115:890–903. [31, 151]

Lovette, I. J., E. Bermingham, and R. E. Ricklefs. 2002. Clade-specific morphological diversification and adaptive radiation in Hawaiian songbirds. *Proceedings of the Royal Society of London Series B—Biological Sciences* 269:37–42. [86]

Lowther, P. E. 1993. Brown-headed cowbird *(Molothrus ater) in* A. Poole and F. B. Gill, eds. *The Birds of North America,* No. 47. The Academy of Natural Sciences, Philadelphia. [279]

Lynch, A. 1996. The population memetics of birdsong. Pp. 181–197 *in* D. E. Kroodsma and E. H. Miller, eds. *Ecology and Evolution of Acoustic Communication in Birds.* Cornell University Press, Ithaca, NY and London. [203, 206]

Lynch, A., and A. J. Baker. 1991. Increased vocal discrimination by learning in sympatry. *Behaviour* 116: 109–126. [207, 217]

Lynch, A., and A. J. Baker. 1993. A population memetics approach to cultural evolution in chaffinch song: Meme diversity within populations. *American Naturalist* 141:597–620. [204, 216, 267]

Lynch, A., and A. J. Baker. 1994. A population memetics approach to cultural evolution in chaffinch song: Differentiation among populations. *Evolution* 48:351–359. [217, 261, 264]

Lynch, A., G. M. Plunkett, A. J. Baker, and P. F. Jenkins. 1989. A model of cultural evolution of chaffinch song derived with the meme concept. *American Naturalist* 133:634–653. [216–217]

Ma, W., and D. Lambert. 1997. Minisatellite DNA markers reveal hybridisation between the endangered black robin and tomtit. *Electrophoresis* 18:1682–1687. [276]

MacArthur, R. H. 1958. Population ecology of some warblers of northeastern coniferous forests. *Ecology* 39:599–619. [13, 271]

MacArthur, R. H. 1969. Patterns of communities in the tropics. *Biological Journal of the Linnean Society* 1: 19–30. [16, 98]

MacArthur, R. H. 1972. *Geographical Ecology.* Harper and Row, New York. [31, 152]

MacArthur, R. H., and E. R. Pianka. 1966. On optimal use of a patchy environment. *American Naturalist* 100:603–609. [305]

MacArthur, R. H., and E. O. Wilson. 1967. *The Theory of Island Biogeography.* Princeton University Press, Princeton, NJ. [126, 144]

MacDougall-Shackleton, E. A., and S. A. MacDougall-Shackleton. 2001. Cultural and genetic evolution in mountain white-crowned sparrows: Song dialects are associated with population structure. *Evolution* 55:2568–2575. [209]

MacDougall-Shackleton, S. A., E. A. MacDougall-Shackleton, and T. P. Hahn. 2001. Physiological and behavioural responses of female mountain white-crowned sparrows to natal- and foreign-dialect songs. *Canadian Journal of Zoology* 79:325–333. [191, 198]

Madden, J. 2001. Sex, bowers and brains. *Proceedings of the Royal Society of London Series B—Biological Sciences* 268:833–838. [138]

Madden, J. R. 2002. Bower decorations attract females but provoke other male spotted bowerbirds: Bower owners resolve this trade-off. *Proceedings of the Royal Society of London Series B—Biological Sciences* 269:1347–1351. [185]

Madden, J. R., and K. Tanner. 2003. Preferences for coloured bower decorations can be explained in a nonsexual context. *Animal Behaviour* 65:1077–1083. [185]

Magurran, A. E., and I. W. Ramnarine. 2004. Learned mate recognition and reproductive isolation in guppies. *Animal Behaviour* 67:1077–1082. [401]

Mallet, J. 2005. Hybridization as an invasion of the genome. *Trends in Ecology and Evolution* 20:229–237. [346]

Marantz, C. A. 1997. Geographic variation of plumage patterns in the woodcreeper genus *Dendrocolaptes* (Dendrocolaptidae). *Ornithological Monographs* 48:399–429. [347]

Marchetti, K. 1993. Dark habitats and bright birds illustrate the role of the environment in species divergence. *Nature* 362:149–152. [176, 178, 260]

Margoliash, D., C. Staicer, and S. A. Inoue. 1994. The process of syllable acquisition in adult indigo buntings *(Passerina cyanea). Behaviour* 131:39–64. [199]

Marks, B. D., S. J. Hackett, and A. P. Capparella. 2002. Historical relationships among Neotropical lowland forest areas of endemism as determined by mitochondrial DNA sequence variation within the wedge-billed woodcreeper (Aves: Dendrocolaptidae: *Glyphorynchus spirurus). Molecular Phylogenetics and Evolution* 24:153–167. [31]

Marler, P. 1956. The voice of the chaffinch and its functions as a language. *Ibis* 98:231–261. [216]

Marler, P. 1970. Birdsong and speech development: Could there be parallels? *American Scientist* 58:669–673. [209]

Marler, P. 2004. Bird calls: A cornucopia for communication. Pp. 132–176 *in* P. Marler and H. Slabbekoorn, eds. *Nature's Music: The Science of Bird Song.* Academic Press, London. [282]

Marler, P., and S. Peters. 1977. Selective vocal learning in a sparrow. *Science* 198:519–521. [211]

Marler, P., and S. Peters. 1981. Sparrows learn adult song and more from memory. *Science* 213:780–782. [199]

Marler, P., and S. Peters. 1982. Subsong and plastic song: Their role in the vocal learning process. Pp. 25–50 *in* D. E. Kroodsma and E. H. Miller, eds. *Acoustic Communication in Birds*. Academic Press, New York and London. [199]

Marler, P., and S. Peters. 1988. The role of song phonology and syntax in vocal learning preferences in the song sparrow, *Melospiza melodia*. *Ethology* 77:125–149. [211]

Marler, P., and R. Pickert. 1984. Species-universal microstructure in the learned song of the swamp sparrow *(Melospiza georgiana)*. *Animal Behaviour* 32:673–689. [192]

Marler, P., and M. Tamura. 1962. Song "dialects" in three populations of white-crowned sparrows. *Condor* 64:368–377. [193]

Marler, P., M. Kreith, and M. Tamura. 1962. Song development in hand-raised Oregon juncos. *Auk* 79:12–30. [200, 267]

Marler, P., P. Mundinger, M. S. Waser, and A. Lutjen. 1972. Effects of acoustical stimulation and deprivation on song development in red-winged blackbirds *(Agelaius phoeniceus)*. *Animal Behaviour* 20:586–606. [199]

Marr, A. B., L. F. Keller, and P. Arcese. 2002. Heterosis and outbreeding depression in descendants of natural immigrants to an inbred population of song sparrows *(Melospiza melodia)*. *Evolution* 56:131–142. [68]

Marshall, H. D., and A. J. Baker. 1998. Rates and patterns of mitochondrial DNA sequence evolution in fringilline finches *(Fringilla* spp.) and the greenfinch *(Carduelis chloris)*. *Molecular Biology and Evolution* 15:638–646. [217]

Marten, K., D. Quine, and P. Marler. 1977. Sound transmission and its significance for animal vocalization. II. Tropical forest habitats. *Behavioral Ecology and Sociobiology* 2:291–302. [253]

Martens, J. 1975. Akustische Differenzierung verwandtschaftlicher Beziehungen in der *Parus (Periparus)*-Gruppe nach Untersuchungen im Nepal-Himalaya. *Journal für Ornithologie* 116:369–433. [194–195]

Martens, J. 1996. Vocalizations and speciation of Palearctic birds. Pp. 221–240 *in* D. E. Kroodsma and E. H. Miller, eds. *Ecology and Evolution of Acoustic Communication in Birds*. Cornell University Press, Ithaca, NY and London. [194, 196, 197, 207, 208, 261]

Martens, J., and G. Geduldig. 1990. Acoustic adaptations of birds living close to Himalayan torrents. Pp. 123–131 *in* R. van Elzen, K. L. Schuchmann and K. Schmidt-Koenig, eds. *Current Topics in Avian Biology* (Proceedings of the International 100. DO-G meeting, Bonn). Verlag DO-G, Garmisch-Partenkirchen. [255]

Martens, J., S. Eck, M. Päckert, and Y. H. Sun. 1999. The golden spectacled warbler *Seicercus burkii*—a species swarm (Aves: Passeriformes: Sylviidae). *Zoologische Abhandlungen Museum für Tierkunde Dresden* 50:281–327. [315]

Martens, J., D. T. Tietze, and Y.-H. Sun. 2006. Molecular phylogeny of *Parus (Periparus)*, a Eurasian radiation of tits (Aves: Passeriformes: Paridae). *Zoologische Abhandlungen Museum für Tierkunde Dresden* 55: 103–120. [194–195, 235]

Martin, T. E. 1995. Avian life-history evolution in relation to nest sites, nest predation, and food. *Ecological Monographs* 65:101–127. [127, 137]

Martin, T. E. 1996. Fitness costs of resource overlap among coexisting species. *Nature* 380:338–340. [75]

Martin, T. E., and A. V. Badyaev. 1996. Sexual dichromatism in birds: Importance of nest predation and nest location for females versus males. *Evolution* 50:2454–2460. [259, 260]

Martineau, F., and R. Mougeon. 2003. A sociolinguistic study of the origins of *ne* deletion in European and Quebec French. *Language* 79:118–152. [210]

Marzluff, J. M. 2001. Avian ecology and conservation in an urbanizing world. Pp. 19–47 *in* J. M. Marzluff, R. Bowman and R. Donnelly, eds. *Avian Ecology and Conservation in an Urbanizing World*. Kluwer Academic, Boston. [100]

Masabanda, J. S., D. W. Burt, P. C. M. O'Brien, A. Vignal, V. Fillon, P. S. Walsh, H. Cox, H. G. Tempest, J. Smith, F. Habermann, M. Schmid, Y. Matsuda, M. A. Ferguson-Smith, R. Crooijmans, M. A. M. Groenen, and D. K. Griffin. 2004. Molecular cytogenetic definition of the chicken genome: The first complete avian karyotype. *Genetics* 166:1367–1373. [392[

Mateo, J. M. 2004. Recognition systems and biological organization: The perception component of social recognition. *Annales Zoologici Fennici* 41:729–745. [273, 408]

Matyjasiak, P. 2005. Birds associate species-specific visual and acoustic cues: Recognition of heterospecific rivals in male blackcaps. *Behavioral Ecology* 16:467–471. [286]

Maynard-Smith, J., and D. Harper. 1988. The evolution of aggression: Can selection generate variability? *Philosophical Transactions of the Royal Society, Series B Biological Sciences* 319:557–570. [181–182]

Maynard-Smith, J., and D. Harper. 2003. *Animal Signals*. Oxford University Press, Oxford. [182]

Mayr, E. 1942. *Systematics and the Origin of Species from the Viewpoint of a Zoologist*. Columbia University Press, New York. [1, 5, 8, 9, 14, 49, 65, 89, 236, 401]

Mayr, E. 1945. Birds of paradise. *Natural History* 54:264–276. [300]

Mayr, E. 1947. Ecological factors in speciation. *Evolution* 1:263–288. [1, 14, 19, 29, 109, 120, 396]

Mayr, E. 1954. Change of genetic environment and evolution. Pp. 157–180 *in* J. Huxley, A. C. Hardy and E. B. Ford, eds. *Evolution as a Process*. Allen and Unwin, London. [49, 395, 401]

Mayr, E. 1963. *Animal Species and Evolution*. Belknap Press of Harvard University Press, Cambridge, MA. [1, 120, 127, 259, 299, 300, 401]

Mayr, E. 1965a. Avifauna: Turnover on islands. *Science* 150:1587–1588. [16, 141]

Mayr, E. 1965b. The nature of colonizations in birds. Pp. 29–47 *in* H. G. Baker and G. L. Stebbins, eds. *The Genetics of Colonizing Species*. Academic Press, New York and London. [132]

Mayr, E. 1969. Bird speciation in the tropics. *Biological Journal of the Linnean Society* 1:1–17. [143]

Mayr, E. 1982. *The Growth of Biological Thought: Diversity, Evolution, and Inheritance*. Belknap Press of Harvard University Press, Cambridge, MA. [3, 5]

Mayr, G. 2003. A new Eocene swift-like bird with a peculiar feathering. *Ibis* 145:382–391. [15]

Mayr, G. 2005. The Paleogene fossil record of birds in Europe. *Biological Reviews* 80:515–542. [138, 378]

Mayr, E., and J. M. Diamond. 2001. *The Birds of Northern Melanesia: Speciation, Ecology, and Biogeography*. Oxford University Press, New York. [1, 4, 5, 24, 26, 60–63, 70, 143, 144, 147, 152, 261, 311, 409]

Mayr, E., and E. T. Gilliard. 1952. Altitudinal hybridization in New Guinea honeyeaters. *Condor* 54:325–337. [315]

Mayr, E., and L. L. Short. 1970. *Species Taxa of North American Birds: A Contribution to Comparative Systematics*. Publications of the Nuttall Ornithological Club, No. 9, Cambridge, MA. [26, 41, 324, 332]

Mayr, E., and C. Vaurie. 1948. Evolution in the family Dicruridae (birds). *Evolution* 2:238–265. [51, 261]

McCall, R. A., S. Nee, and P. H. Harvey. 1998. The role of wing length in the evolution of avian flightlessness. *Evolutionary Ecology* 12:569–580. [70]

McCarthy, E. M. 2006. *Handbook of Avian Hybrids of the World*. Oxford University Press, Oxford. [300, 367, 372, 373, 378, 389]

McDonald, D. B., R. P. Clay, R. T. Brumfield, and M. J. Braun. 2001. Sexual selection on plumage and behavior in an avian hybrid zone: Experimental tests of male–male interactions. *Evolution* 55:1443–1451. [340–342, 346]

McGraw, K. J. 2005. The antioxidant function of many animal pigments: Are there consistent health benefits of sexually selected colourants? *Animal Behaviour* 69:757–764. [182]

McGraw, K. J., and G. E. Hill. 2000. Differential effects of endoparasitism on the expression of carotenoid- and melanin-based ornamental coloration. *Proceedings of the Royal Society of London Series B—Biological Sciences* 267:1525–1531. [180]

McGregor, P. K. 1983. The response of corn buntings to playback of dialects. *Zeitschrift für Tierpsychologie* 62:256–260. [190]

McKinnon, J. S., and H. D. Rundle. 2002. Speciation in nature: The threespine stickleback model systems. *Trends in Ecology and Evolution* 17:480–488. [397]

McKitrick, M. C., and R. M. Zink. 1988. Species concepts in ornithology. *Condor* 90:1–14. [5, 6, 8]

McNab, B. K. 1971. On the ecological significance of Bergmann's rule. *Ecology* 52:845–854. [48]

McNaught, M. K., and I. P. F. Owens. 2002. Interspecific variation in plumage colour among birds: Species recognition or light environment? *Journal of Evolutionary Biology* 15:505–514. [258, 260]

Meise, W. 1928. Die Verbreitung der Aaskrähe (Formenkreise *Corvus corone* L.). *Journal für Ornithologie* 76:1–203. [354, 355]

Meretsky, V., N. F. R. Snyder, S. R. Beissinger, D. A. Clendenen, and J. W. Wiley. 2000. Demography of the California condor: Implications for reestablishment. *Conservation Biology* 14:957–967. [277]

Meretsky, V., N. F. R. Snyder, S. R. Beissinger, D. A. Clendenen, and J. W. Wiley. 2001. Quantity versus quality in California condor reintroduction: Reply. *Conservation Biology* 15:1449–1451. [277]

Merilä, J. 1997. Quantitative trait and allozyme divergence in the greenfinch (*Carduelis chloris*, Aves: Fringillidae). *Biological Journal of the Linnean Society* 61:243–266. [43, 47]

Merilä, J., and B. C. Sheldon. 2001. Avian quantitative genetics. Pp. 179–255 *in* V. Nolan and C. F. Thompson, eds. *Current Ornithology*. Kluwer, New York. [44]

Merilä, J., M. Björklund, and A. J. Baker. 1997. Historical demography and present day population structure of the greenfinch, *Carduelis chloris*—an analysis of mt DNA control-region sequences. *Evolution* 51: 946–956. [47]

Merlen, G., and G. Davis-Merlen. 2000. Whish: More than a tool-using finch. *Noticias de Galápagos* 61:2–9. [134]

Mettke-Hofmann, C., H. Winkler, and B. Leisler. 2002. The significance of ecological factors for exploration and neophobia in parrots. *Ethology* 108:249–272. [131–133]

Milá, B., J. E. McCormack, G. Castañeda, R. K. Wayne, and T. B. Smith. 2007a. Rapid postglacial diversification in a songbird lineage. unpublished manuscript. [33]

Milá, B., T. B. Smith, and R. K. Warne. 2007b. Speciation and rapid phenotypic differentiation in the yellow-rumped Warbler *Dendroica coronata* complex. *Molecular Ecology* 16:159–173. [363]

Miller, A. H. 1941. Speciation in the avian genus *Junco*. Pp. 173–434. *University of California Publications in Zoology* 44:173–434. [33, 347]

Miller, E. H. 1996. Acoustic differentiation and speciation in shorebirds. Pp. 241–257 *in* D. E. Kroodsma and E. H. Miller, eds. *Ecology and Evolution of Acoustic Communication in Birds*. Cornell University Press, Ithaca, NY and London. [192]

Miller, P. M., S. Gavrilets, and W. R. Rice. 2006. Sexual conflict via maternal-effect genes in ZW species. *Science* 312:73. [369]

Millington, S. J., and T. D. Price. 1985. Song inheritance and mating patterns in Darwin's finches. *Auk* 102:342–346. [315]

Mira, A. 1998. Why is meiosis arrested? *Journal of Theoretical Biology* 194:275–287. [380]

Mirsky, E. N. 1976. Song divergence in hummingbird and junco populations on Guadalupe Island. *Condor* 78:230–235. [200, 267]

Mizuno, S., R. Kunita, O. Nakabayashi, Y. Kuroda, N. Arai, M. Harata, A. Ogawa, Y. Itoh, M. Teranishi, and T. Hori. 2002. Z and W chromosomes of chickens: Studies on their gene functions in sex determination and sex differentiation. *Cytogenetic and Genome Research* 99:236–244. [382]

Mobbs, A. J. 1992. The status of Australian finches in the United Kingdom. *Avicultural Magazine* 98:130–135. [227]

Møller, A. P. 1988. Female choice selects for male sexual tail ornaments in the monogamous swallow. *Nature* 332:640–642. [160–161]

Møller, A. P. 1990. Effects of a hematophagous mite on the barn swallow *(Hirundo rustica)*—a test of the Hamilton and Zuk Hypothesis. *Evolution* 44:771–784. [180]

Møller, A. P. 1992. Sexual selection in the monogamous barn swallow *(Hirundo rustica)*. 2. Mechanisms of sexual selection. *Journal of Evolutionary Biology* 5:603–624. [160–161]

Møller, A. P. 2000. Male parental care, female reproductive success, and extrapair paternity. *Behavioral Ecology* 11:161–168. [249]

Møller, A. P., and T. R. Birkhead. 1994. The evolution of plumage brightness in birds is related to extrapair paternity. *Evolution* 48:1089–1100. [249]

Møller, A. P., and J. J. Cuervo. 1998. Speciation and feather ornamentation in birds. *Evolution* 52:859–869. [33, 270]

Møller, A. P., and J. J. Cuervo. 2000. The evolution of paternity and paternal care in birds. *Behavioral Ecology* 11:472–485. [249]

Møller, A. P., and R. Thornhill. 1998. Male parental care, differential parental investment by females and sexual selection. *Animal Behaviour* 55:1507–1515. [181]

Molles, L. E., and S. L. Vehrencamp. 2001. Songbird cheaters pay a retaliation cost: Evidence for auditory conventional signals. *Proceedings of the Royal Society of London Series B—Biological Sciences* 268:2013–2019. [202, 203]

Monteiro, L. R., and R. W. Furness. 1998. Speciation through temporal segregation of Madeiran storm petrel *(Oceanodroma castro)* populations in the Azores? *Philosophical Transactions of the Royal Society of London Series B—Biological Sciences* 353:945–953. [21]

Moore, W. S. 1977. Evaluation of narrow hybrid zones in vertebrates. *Quarterly Review of Biology* 52:263–277. [332]

Moore, M. C. 1983. Effect of female sexual displays on the endocrine physiology and behaviour of male White-crowned sparrows, *Zonotrichia leucophrys*. *Journal of Zoology, London* 199:137–148. [164]

Moore, W. S. 1987. Random mating in the northern flicker hybrid zone—implications for the evolution of bright and contrasting plumage patterns in birds. *Evolution* 41:539–546. [317, 320, 337, 347, 352]

Moore, W. S. 1995. Inferring phylogenies from mtDNA variation: Mitochondrial-gene trees versus nuclear-gene trees. *Evolution* 49:718–726. [37, 39]

Moore, W. S., and D. B. Buchanan. 1985. Stability of the northern flicker hybrid zone in historical times—implications for adaptive speciation theory. *Evolution* 39:135–151. [352]

Moore, W. S., and W. Koenig. 1986. Comparative reproductive success of yellow-shafted, red-shafted and hybrid flickers across a hybrid zone. *Auk* 103:42–51. [319]

Moore, W. S., and J. T. Price. 1993. Nature of selection in the northern flicker hybrid zone and its implications for speciation theory. Pp. 196–225 *in* R. G. Harrison, ed. *Hybrid Zones and the Evolutionary Process*. Oxford University Press, Oxford. [329, 332, 337, 339, 347, 350, 352]

Moore, W. S., J. H. Graham, and J. T. Price. 1991. Mitochondrial DNA variation in the northern flicker *(Colaptes auratus*, Aves). *Molecular Biology and Evolution* 8:327–344. [352]

Moore, W. S., A. C. Weibel, and A. Agius. 2006. Mitochondrial DNA phylogeny of the woodpecker genus *Veniliornis* (Picidae, Picinae) and related genera implies convergent evolution of plumage patterns. *Biological Journal of the Linnean Society* 87:611–624. [233]

Morejohn, G. V. 1968. Breakdown of isolation mechanisms in two species of captive junglefowl. *Evolution* 22:576–582. [379]

Morgan, T. H. 1918. Inheritance of number of feathers of the fantail pigeon. *American Naturalist* 52:5–27. [227]

Moritz, C., J. L. Patton, C. J. Schneider, and T. B. Smith. 2000. Diversification of rainforest faunas: An integrated molecular approach. *Annual Review of Ecology and Systematics* 31:533–563. [27]

Morris, R. L., and C. J. Erickson. 1971. Pair bond maintenance in the ring dove *(Streptopelia risoria)*. *Animal Behaviour* 19:398–406. [284]

Morrow, E. H., T. E. Pitcher, and G. Arnqvist. 2003. No evidence that sexual selection is an 'engine of speciation' in birds. *Ecology Letters* 6:228–234. [270]

Morton, E. S. 1975. Ecological sources of selection on avian sounds. *American Naturalist* 109:17–34. [255, 256]

Morton, E. S. 1987. The effects of distance and isolation on song-type sharing in the Carolina wren. *Wilson Bulletin* 99:601–610. [202]

Mota, P. G., and V. Depraz. 2004. A test of the effect of male song on female nesting behaviour in the serin *(Serinus serinus)*: A field playback experiment. *Ethology* 110:841–850. [163]

Mott, C. L., L. H. Lockhart, and R. Rigden. 1968. Chromosomes of the sterile hybrid duck. *Cytogenetics* 7:403–412. [376, 380, 381]

Muller, H. 1942. Isolating mechanisms, evolution and temperature. *Biological Symposia* 6:71–125. [90, 369, 397, 398]

Mundinger, P. 1975. Song dialects and colonization in the house finch, *Carpodacus mexicanus,* on the east coast. *Condor* 77:407–422. [202, 269]

Mundinger, P. 1982. Microgeographic and macrogeographic variation in the acquired vocalizations of birds. Pp.147–208 *in* D. E. Kroodsma and E. H. Miller, eds. *Acoustic Communication in Birds.* Academic Press, New York and London. [192, 198, 404]

Mundy, N. I. 2005. A window on the genetics of evolution: MC1R and plumage colouration in birds. *Proceedings of the Royal Society of London Series B—Biological Sciences* 272:1633–1640. [232]

Mundy, N. I., N. S. Badcock, T. Hart, K. Scribner, K. Janssen, and N. J. Nadeau. 2004. Conserved genetic basis of a quantitative plumage trait involved in mate choice. *Science* 303:1870–1873. [229]

Murphy, R. C. 1938. The need of insular exploration as illustrated by birds. *Science* 88:533–539. [51]

Murton, R. K., and N. J. Westwood. 1977. *Avian Breeding Cycles.* Clarendon Press, Oxford. [164, 165]

Murton, R. K., R. Thearle, and B. Lofts. 1969. The endocrine basis of breeding behaviour in the feral pigeon *(Columba livia)*: Effects of exogenous hormones on the pre-incubation behaviour of intact males. *Animal Behaviour* 17:286–306. [162]

Nadeau, N. J., F. Minvielle, and N. I. Mundy. 2006. Association of a Glu92Lys substitution in MC1R with extended brown in Japanese quail *(Coturnix japonica)*. *Animal Genetics* 37:287-289. [229]

Nanda, I., D. Schrama, W. Feichtinger, T. Haaf, M. Schartl, and M. Schmid. 2002. Distribution of telomeric (TTAGGG)(n) sequences in avian chromosomes. *Chromosoma* 111:215–227. [388]

Naugler, C. T., and P. C. Smith. 1991. Song similarity in an isolated population of fox sparrows *(Passerella iliaca)*. *Condor* 93:1001–1003. [267]

Navarro, A., and N. H. Barton. 2003. Accumulating postzygotic isolation genes in parapatry: A new twist on chromosomal speciation. *Evolution* 57:447–459. [345, 371, 389, 391]

Nee, S., A. O. Mooers, and P. H. Harvey. 1992. Tempo and mode of evolution revealed from molecular phylogenies. *Proceedings of the National Academy of Sciences of the United States of America* 89:8322–8326. [213]

Nelson, D. A. 1988. Feature weighting in species song recognition by the field sparrow *(Spizella pusilla)*. *Behaviour* 106:158–182. [287]

Nelson, D. A. 1989. The importance of invariant and distinctive features in species recognition of bird sounds. *Condor* 91:120–130. [287]

Nelson, D. A. 1997. Social interaction and sensitive phases for song learning: A critical review. Pp. 7–22 *in* C. T. Snowdon and M. Hausberger, eds. *Social Influences on Vocal Development.* Cambridge University Press, New York. [199]

Nelson, D. A. 1998. Geographic variation in song of Gambel's white crowned sparrow. *Behaviour* 135:321–342. [189, 194, 196]

Nelson, D. A. 2000a. A preference for own-subspecies' song guides vocal learning in a song bird. *Proceedings of the National Academy of Sciences of the United States of America* 97:13348–13353. [211, 281]

Nelson, D. A. 2000b. Song overproduction, selective attrition and song dialects in the white-crowned sparrow. *Animal Behaviour* 60:887–898. [307]

Nelson, D. A., and P. Marler. 1994. Selection-based learning in bird song development. *Proceedings of the National Academy of Sciences of the United States of America* 91:10498–10501. [199]

Nelson, D. A., and J. A. Soha. 2004. Perception of geographical variation in song by male Puget Sound white–crowned sparrows, *Zonotrichia leucophrys pugetensis*. *Animal Behaviour* 68:395–405. [191]

Nelson, D. A., K. I. Hallberg, and J. A. Soha. 2004. Cultural evolution of Puget Sound white-crowned sparrow song dialects. *Ethology* 110:879–908. [192, 194]

Newton, I. 1998. *Population Limitation in Birds.* Academic Press, San Diego, CA. [208]

Newton, I. 2003. *The Speciation and Biogeography of Birds.* Academic Press, San Diego, CA. [27, 324, 345]

Nicholls, J. A., and A. W. Goldizen. 2006. Habitat type and density influence vocal design in Satin Bowerbirds. *Journal of Animal Ecology* 75:549–558. [254]

Nicolai, J. 1974. Mimicry in parasitic birds. *Scientific American* 231 (October):91–98. [294, 295]

Nicolai, J. 1976. Evolutive Neuerungen in der Balz von Haustaubenrassen *(Columba livia* var *domestica)* als Ergebnis menschlicher Zuchtwahl. *Zeitscrift für Tierpsychologie* 40:225–243. [228]

Nicolakakis, N., and L. Lefebvre. 2000. Forebrain size and innovation rate in European birds: Feeding, nesting and confounding variables. *Behaviour* 137:1415–1429. [129, 138]

Nicolakakis, N., D. Sol, and L. Lefebvre. 2003. Behavioural flexibility predicts species richness in birds, but not extinction risk. *Animal Behaviour* 65:445–452. [130]

Nobel, G. K. 1936. Courtship and sexual selection in the flicker *Colaptes auratus luteus. Auk* 53:269–282. [320]

Nolan, V., Jr., E. D. Ketterson, D. A. Cristol, C. M. Rogers, E. D. Clotfelter, R. C. Titus, S. J. Schoech, and E. Snajdr. 2002. Dark-eyed junco *(Junco hyemalis). In* A. Poole and F. Gill, eds. *The Birds of North America,* No. 716. The Academy of Natural Sciences, Philadelphia and the American Ornithologists' Union, Washington, DC. [33, 200]

Noor, M. A. F. 1999. Reinforcement and other consequences of sympatry. *Heredity* 83:503–508. [314, 408]

Noor, M. A. F., K. L. Grams, L. A. Bertucci, and J. Reiland. 2001. Chromosomal inversions and the reproductive isolation of species. *Proceedings of the National Academy of Sciences of the United States of America* 98:12084–12088. [371, 391, 392]

Nordby, J. C., S. E. Campbell, and M. D. Beecher. 1999. Ecological correlates of song learning in song sparrows. *Behavioral Ecology* 10:287–297. [199]

Nordby, J. C., S. E. Campbell, J. M. Burt, and M. D. Beecher. 2000. Social influences during song development in the song sparrow: A laboratory experiment simulating field conditions. *Animal Behaviour* 59: 1187–1197. [199, 207]

Nordby, J. C., S. E. Campbell, and M. D. Beecher. 2002. Adult song sparrows do not alter their song repertoires. *Ethology* 108:39–50. [209]

Nores, M. 1999. An alternative hypothesis for the origin of Amazonian bird diversity. *Journal of Biogeography* 26:475–485. [346]

Norris, K. J. 1990. Female choice and the quality of parental care in the great tit *Parus major. Behavioral Ecology and Sociobiology* 27:275–281. [182]

Nosil, P., T. Vines, and D. J. Funk. 2005. Perspective: Reproductive isolation caused by natural selection against migrants between divergent environments. *Evolution* 59:705–719. [73, 396]

Nottebohm, F. 1969. The song of the chingolo, *Zonotricha capensis,* in Argentina: Description and evaluation of a system of dialects. *Condor* 71:299–315. [187]

Nottebohm, F. 1975. Continental patterns of song variability in *Zonotrichia capensis:* Some possible ecological correlates. *American Naturalist* 109:605–624. [256–258]

Nowicki, S., and D. A. Nelson. 1990. Defining natural categories in acoustic signals—comparison of 3 methods applied to chick-a-dee call notes. *Ethology* 86:89–101. [188]

Nuechterlein, G. L., and D. Buitron. 1998. Interspecific mate choice by late-courting male western grebes. *Behavioral Ecology* 9:313–321. [310, 319]

O'Connell, M. E., C. Reboulleau, H. H. Feder, and R. Silver. 1981a. Social interactions and androgen levels in birds. 1. Female characteristics associated with increased plasma androgen levels in the male ring dove *(Streptopelia risoria). General and Comparative Endocrinology* 44:454–463. [164]

O'Connell, M. E., R. Silver, H. H. Feder, and C. Reboulleau. 1981b. Social interactions and androgen levels in birds. 2. Social factors associated with a decline in plasma androgen levels in male ring doves *(Streptopelia risoria). General and Comparative Endocrinology* 44:464–469. [164]

O'Loghlen, A. L., and M. D. Beecher. 1999. Mate, neighbour and stranger songs: A female song sparrow perspective. *Animal Behaviour* 58:13–20. [202, 203]

O'Loghlen, A. L., and S. I. Rothstein. 1995. Culturally correct song dialects are correlated with male age and female song preferences in wild populations of brown headed cowbirds. *Behavioral Ecology and Sociobiology* 36:251–259. [191, 202, 280]

Ödeen, A., and M. Björklund. 2003. Dynamics in the evolution of sexual traits: Losses and gains, radiation and convergence in yellow wagtails *(Motacilla flava). Molecular Ecology* 12:2113–2130. [235]

Oetting, S., E. Prove, and H. J. Bischof. 1995. Sexual imprinting as a two-stage process—mechanisms of information storage and stabilization. *Animal Behaviour* 50:393–403. [277, 278]

Okanoya, K. 2004. Song syntax in Bengalese finches: Proximate and ultimate analyses. *Advances in the Study of Behavior* 34:297–346. [163, 205, 224]

Olson, S. L. 1973. *Evolution of the Rails of the South Atlantic Islands (Aves: Rallidae)*. Smithsonian Institution Press, Washington, DC. [70]

Olson, S. L. 2001. Why so many kinds of passerine birds? *Bioscience* 51:268–269. [139]

Olson, S. L., and H. F. James. 1991. Descriptions of thirty-two new species of birds from the Hawaiian islands. I. Non–passeriformes. *Ornithological Monographs* 45:1–88. [113, 400]

Olson, V. A., and I. P. F. Owens. 2005. Interspecific variation in the use of carotenoid-based coloration in birds: Diet, life history and phylogeny. *Journal of Evolutionary Biology* 18:1534–1546. [259, 260]

Olsson, U., P. Alström, and P. Sundberg. 2004. Non-monophyly of the avian genus *Seicercus* (Aves: Sylviidae) revealed by mitochondrial DNA. *Zoologica Scripta* 33:501–510. [102, 316]

Olsson, U., P. Alström, P. G. P. Ericson, and P. Sundberg. 2005. Non-monophyletic taxa and cryptic species— evidence from a molecular phylogeny of leaf-warblers (*Phylloscopus*, Aves). *Molecular Phylogenetics and Evolution* 36:261–276. [102]

Omland, K. E., and S. M. Lanyon. 2000. Reconstructing plumage evolution in orioles *(Icterus):* Repeated convergence and reversal in patterns. *Evolution* 54:2119–2133. [233, 234, 235, 240]

Omland, K. E., S. M. Lanyon, and S. J. Fritz. 1999. A molecular phylogeny of the new world orioles *(Icterus):* The importance of dense taxon sampling. *Molecular Phylogenetics and Evolution* 12:224–239. [102]

Orenstein, R. I. 1973. Colourful plumage in tropical birds. *Avicultural Magazine* 79:119–122. [259]

Orians, G. H., and G. M. Christman. 1968. A comparative study of the behavior of red-winged, tricolored and yellow-headed blackbirds. *University of California Publications in Zoology* 84:1–81. [175, 177]

Orme, C. D. L., R. G. Davies, M. Burgess, F. Eigenbrod, N. Pickup, V. A. Olson, A. J. Webster, T. S. Ding, P. C. Rasmussen, R. S. Ridgely, A. J. Stattersfield, P. M. Bennett, T. M. Blackburn, K. J. Gaston, and I. P. F. Owens. 2005. Global hotspots of species richness are not congruent with endemism or threat. *Nature* 436:1016–1019. [97, 99]

Orr, H. A., J. P. Masly, and D. C. Presgraves. 2004. Speciation genes. *Current Opinion in Genetics and Development* 14:675–679. [396]

Orr, H. A., and M. Turelli. 2001. The evolution of postzygotic isolation: Accumulating Dobzhansky–Muller incompatibilities. *Evolution* 55:1085–1094. [370]

Owens, I. P. F., and I. R. Hartley. 1998. Sexual dimorphism in birds: Why are there so many kinds of dimorphism? *Proceedings of the Royal Society of London Series B—Biological Sciences* 265:397–407. [249]

Owens, I. P. F., P. M. Bennett, and P. H. Harvey. 1999. Species richness among birds: Body size, life history, sexual selection or ecology? *Proceedings of the Royal Society of London Series B—Biological Sciences* 266:933–939. [33, 131, 270]

Päckert, M., and J. Martens. 2004. Song dialects on the Atlantic islands: goldcrests of the Azores *(Regulus regulus azoricus, R. r. sanctae-mariae, R. r. inermis)*. *Journal für Ornithologie* 145:23–30. [261, 264, 267]

Päckert, M., J. Martens, and T. Hofmeister. 2001. Vocalisations of firecrests from the islands of Madeira and Mallorca *(Regulus ignicapillus madeirensis, R. i. balearicus)*. *Journal für Ornithologie* 142:16–29. [267]

Päckert, M., J. Martens, J. Kosuch, A. A. Nazarenko, and M. Veith. 2003. Phylogenetic signal in the song of crests and kinglets (Aves: *Regulus*). *Evolution* 57:616–629. [235, 240]

Päckert, M., J. Martens, Y. H. Sun, and M. Veith. 2004. The radiation of the *Seicercus burkii* complex and its congeners (Aves: Sylviidae): Molecular genetics and bioacoustics. *Organisms, Diversity and Evolution* 4:341–364. [315]

Päckert, M., J. Martens, S. Eck, A. A. Nazarenko, O. P. Valchuk, B. Petri, and M. Veith. 2005. The great tit *(Parus major)*—a misclassified ring species. *Biological Journal of the Linnean Society* 86:153–174. [66]

Padian, K., and L. M. Chiappe. 1998. The origin and early evolution of birds. *Biological Reviews of the Cambridge Philosophical Society* 73:1–42. [16]

Page, R. D. M. 1996. TreeView: An application to display phylogenetic trees on personal computers. *Computer Applications in the Biosciences* 12:357–358. [40]

Pagel, M. 2000. The history, rate and pattern of world linguistic evolution. Pp. 391–416 *in* C. Knight, M. Studdert-Kennedy and J. R. Hurford, eds. *The Evolutionary Emergence of Language*. Cambridge University Press, Cambridge. [210, 270]

Panov, E. N. 1989. *Natural Hybridization and Ethological Isolation in Birds*. Nauka, Moscow [in Russian]. [300, 380]

Panov, E. N. 1992. Emergence of hybridogenous polymorphism in the *Oenanthe picata* complex. *Bulletin of the British Ornithologist's Club Centenary Supplement* 112A:237–249. [323, 346, 347]

Panov, E. N., and V. V. Ivanitzky. 1975. Evolutionary and taxonomic relations between *Oenanthe hispanica* and *O. pleschanka*. *Zoologichesky Zhurnal* 54:1860–1873. [357]

Panov, E. N., A. S. Roubtsov, and D. G. Monzikov. 2003. Hybridization between yellowhammer and pine bunting in Russia. *Dutch Birding* 25:17–31. [347, 362]

Paradis, E., S. R. Baillie, W. J. Sutherland, and R. D. Gregory. 1998. Patterns of natal and breeding dispersal in birds. *Journal of Animal Ecology* 67:518–536. [59, 208, 336]

Parchman, T. I., C. W. Benkman, and S. C. Britch. 2006. Patterns of genetic variation in the adaptive radiation of New World crossbills (Aves: *Loxia*). *Molecular Ecology* 15:1873–1887. [78]

Parker, T. H. 2003. Genetic benefits of mate choice separated from differential maternal investment in red junglefowl *(Gallus gallus)*. *Evolution* 57:2157–2165. [162]

Parsons, T. J., S. L. Olson, and M. J. Braun. 1993. Unidirectional spread of secondary sexual plumage traits across an avian hybrid zone. *Science* 260:1643–1646. [338, 341, 353]

Pärt, T. 1991. Philopatry pays—a comparison between collared flycatcher sisters. *American Naturalist* 138:790–796. [68]

Pärt, T. 1994. Male philopatry confers a mating advantage in the migratory collared flycatcher, *Ficedula albicollis*. *Animal Behaviour* 48:401–409. [68]

Partecke, J., and E. Gwinner. 2007. Increased sedentariness of European blackbirds following urbanization: A consequence of local adaptation? *Ecology* 88:882–890. [44]

Patten, M. A., and P. Unitt. 2002. Diagnosability versus mean differences of sage sparrow subspecies. *Auk* 119:26–35. [3]

Patten, M. A., J. T. Rotenberry, and M. Zuk. 2004. Habitat selection, acoustic adaptation, and the evolution of reproductive isolation. *Evolution* 58:2144–2155. [191, 288]

Pautasso, M., and K. J. Gaston. 2005. Resources and global avian assemblage structure in forests. *Ecology Letters* 8:282–289. [101]

Pavlova, A., R. M. Zink, S. V. Drovetski, Y. Red'kin, and S. Rohwer. 2003. Phylogeographic patterns in *Motacilla flava* and *Motacilla citreola:* Species limits and population history. *Auk* 120:744–758. [12, 235]

Payne, R. B. 1973a. Behavior, mimetic songs and song dialects, and relationships of the parasitic indigobirds *(Vidua)* of Africa. *Ornithological Monographs* 11. [293]

Payne, R. B. 1973b. Vocal mimicry of the paradise whydahs *(Vidua)* and response of female whydahs to the songs of their hosts *(Pytilia)* and their mimics. *Animal Behaviour* 21:762–771. [293]

Payne, R. B. 1979. Song structure, behaviour, and sequence of song types in a population of village indigobirds, *Vidua chalybeata*. *Animal Behaviour* 27:997–1013. [293, 294]

Payne, R. B. 1981. Song learning and social interaction in indigo buntings. *Animal Behaviour* 29:688–697. [202, 215]

Payne, R. B. 1983. The social context of song mimicry: Song matching dialects in indigo buntings *(Passerina cyanea)*. *Animal Behaviour* 31:788–805. [215]

Payne, R. B. 1996. Song traditions in indigo buntings: Origin, improvisation, dispersal and extinction in cultural evolution. Pp. 198–220 *in* D. E. Kroodsma and E. H. Miller, eds. *Ecology and Evolution of Acoustic Communication in Birds*. Cornell University Press, Ithaca, NY and London. [188, 202, 215]

Payne, R. B. 2005a. *The Cuckoos*. Oxford University Press, Oxford. [138]

Payne, R. B. 2005b. Nestling mouth markings and colors of Old World Finches (Estrildidae): Mimicry and coevolution of nesting finches and their Vidua brood parasites. Miscellaneous Publications No. 194. Museum of Zoology, Ann Arbor, MI. [295, 296, 400]

Payne, R. B., and L. L. Payne. 1994. Song mimicry and species associations of west African indigobirds *Vidua* with quail-finch *Ortygospiza atricollis,* goldbreast *Amandava subflava* and brown twinspot *Clytospiza monteiri*. *Ibis* 136:291–304. [294]

Payne, R. B., and L. L. Payne. 1997. Field observations, experimental design, and the time and place of learned bird songs. Pp. 57–84 *in* C. T. Snowdon and M. Hausberger, eds. *Social Influences on Vocal Development*. Cambridge University Press, Cambridge. [200, 204, 215]

Payne, R. B., L. L. Payne, and S. M. Doehlert. 1988. Biological and cultural success of song memes in Indigo buntings. *Ecology* 69:104–117. [202, 215]

Payne, R. B., L. L. Payne, and J. L. Woods. 1998. Song learning in brood-parasitic indigobirds *Vidua chalybeata:* Song mimicry of the host species. *Animal Behaviour* 55:1537–1553. [293, 294]

Payne, R. B., L. L. Payne, J. L. Woods, and M. D. Sorenson. 2000. Imprinting and the origin of parasite-host associations in brood-parasitic indigobirds, *Vidua chalybeata*. *Animal Behaviour* 59:69–81. [20, 293, 294, 306]

Payne, R. B., J. L. Woods, and L. L. Payne. 2001. Parental care in estrildid finches: Experimental tests of a model of *Vidua* brood parasitism. *Animal Behaviour* 62:473–483. [294, 295]

Payne, R. B., K. Hustler, R. Stjernstedt, K. M. Sefc, and M. D. Sorenson. 2002. Behavioural and genetic evidence of a recent population switch to a novel host species in brood-parasitic indigobirds *Vidua chalybeata*. *Ibis* 144:373–383. [294]

Pearson, S. F. 2000. Behavioral asymmetries in a moving hybrid zone. *Behavioral Ecology* 11:84–92. [320, 340, 364]

Pearson, S. F., and S. Rohwer. 2000. Asymmetries in male aggression across an avian hybrid zone. *Behavioral Ecology* 11:93–101. [340, 341]

Pereira, S. J., and A. J. Baker. 2005. Multiple gene evidence for parallel evolution and retention of ancestral states in the shanks (Charadriiformes: Scolopacidae). *Condor* 107:514–526. [233]

Peterson, A. T. 1996. Geographical variation in sexual dichromatism in birds. *Bulletin of the British Ornithologist's Club* 116:156–172. [240, 262, 263]

Peterson, A. T., J. Soberon, and V. Sanchez-Cordero. 1999. Conservatism of ecological niches in evolutionary time. *Science* 285:1265–1267. [114]

Petracci, P. F., and K. Delhey. 2004. Nesting attempts of the cliff swallow *Petrochelidon pyrrhonota* in Buenos Aires Province, Argentina. *Ibis* 146:522–525. [69]

Petren, K., B. R. Grant, and P. R. Grant. 1999. A phylogeny of Darwin's finches based on microsatellite DNA length variation. *Proceedings of the Royal Society of London Series B—Biological Sciences* 266:321–329. [24]

Petren, K., P. R. Grant, B. R. Grant, and L. F. Keller. 2005. Comparative landscape genetics and the adaptive radiation of Darwin's finches: The role of peripheral isolation. *Molecular Ecology* 14:2943–2957. [22, 23, 24]

Petrie, M., and A. Williams. 1993. Peahens lay more eggs for peacocks with larger trains. *Proceedings of the Royal Society of London Series B—Biological Sciences* 251:127–131. [162]

Petrinovich, L., and T. L. Patterson. 1981. The responses of white crowned sparrows to songs of different dialects and subspecies. *Zeitschrift für Tierpsychologie* 57:1–14. [190]

Pfennig, K. S. 1998. The evolution of mate choice and the potential for conflict between species and mate-quality recognition. *Proceedings of the Royal Society of London Series B—Biological Sciences* 265:1743–1748. [273, 292]

Phillimore, A. B., R. P. Freckleton, C. D. L. Orme, and I. P. F. Owens. 2006. Ecology predicts large-scale patterns of diversification in birds. *American Naturalist* 168:220–229. [95, 125, 131, 270]

Phillips, J. C. 1915. Experimental studies of hybridization among ducks and pheasants. *Journal of Experimental Zoölogy* 18:69–112. [229]

Pianka, E. R. 1966. Latitudinal gradients in species diversity. *American Naturalist* 100:33–46. [99]

Pierotti, R., and C. A. Annett. 1993. Hybridization and male parental investment in birds. *Condor* 95:670–679. [304]

Platt, M. E., and M. S. Ficken. 1987. Organization of singing in house wrens. *Journal of Field Ornithology* 58:190–197. [267]

Plenge, M., E. Curio, and K. Witte. 2000. Sexual imprinting supports the evolution of novel male traits by transference of a preference for the colour red. *Behaviour* 137:741–758. [291, 292]

Podos, J. 2001. Correlated evolution of morphology and vocal signal structure in Darwin's finches. *Nature* 409:185–188. [253]

Podos, J., S. K. Huber, and B. Taft. 2004. Bird song: the interface of evolution and mechanism. *Annual Review of Ecology Evolution and Systematics* 35:55–87. [211, 214, 253]

Polechova, J., and N. H. Barton. 2005. Speciation through competition: A critical review. *Evolution* 59:1194–1210. [67]

Pomiankowski, A., Y. Iwasa, and S. Nee. 1991. The evolution of costly mate preferences. 1. Fisher and biased mutation. *Evolution* 45:1422–1430. [239]

Postma, E., and A. J. van Noordwijk. 2005. Gene flow maintains a large genetic difference in clutch size at a small spatial scale. *Nature* 433:65–68. [58, 68]

Prager, E. M., and A. C. Wilson. 1975. Slow evolutionary loss of potential for interspecific hybridization in birds—manifestation of slow regulatory evolution. *Proceedings of the National Academy of Sciences of the United States of America* 72:200–204. [377]

Pratt, H. D. 2005. *The Hawaiian Honeycreepers.* Oxford University Press, Oxford. [22]

Presgraves, D. C. 2003. A fine-scale genetic analysis of hybrid incompatibilities in *Drosophila. Genetics* 163:955–972. [370, 382]

Presgraves, D. C., L. Balagopalan, S. M. Abmayr, and H. A. Orr. 2003. Adaptive evolution drives divergence of a hybrid inviability gene between two species of *Drosophila. Nature* 423:715–719. [396]

Price, T. 1991. Morphology and ecology of breeding warblers along an altitudinal gradient in Kashmir, India. *Journal of Animal Ecology* 60:643–664. [87, 88]

Price, T. 1998. Sexual selection and natural selection in bird speciation. *Philosophical Transactions of the Royal Society of London, B* 353:251–260. [270, 271, 273]

Price, T. D. 2002. Domesticated birds as a model for the genetics of speciation by sexual selection. *Genetica* 116:311–327. [222, 223, 228, 229, 317]

Price, T. D. 2006. Phenotypic plasticity, sexual selection, and the evolution of colour patterns. *Journal of Experimental Biology* 209:2368–2376. [136, 238]

Price, T., and G. L. Birch. 1996. Repeated evolution of sexual color dimorphism in passerine birds. *Auk* 113:842–848. [240, 263]

Price, T. D., and M. M. Bouvier. 2002. The evolution of F1 postzygotic incompatibilities in birds. *Evolution* 56:2083–2089. [373, 374, 376, 377]

Price, J. J., and S. M. Lanyon. 2002. A robust phylogeny of the oropendolas: polyphyly revealed by mitochondrial sequence data. *Auk* 119:335–348. [102]

Price, J. J., and S. M. Lanyon. 2004. Patterns of song evolution and sexual selection in the oropendolas and caciques. *Behavioral Ecology* 15:485–497. [102, 235]

Price, T., and M. Pavelka. 1996. Evolution of a colour pattern: History, development and selection. *Journal of Evolutionary Biology* 9:451–470. [178, 233]

Price, T. D., P. R. Grant, and P. T. Boag. 1984. Genetic changes during the morphological differentiation of Darwin's finches. Pp. 49–66 *in* K. Wöhrmann and V. Loeschke, eds. *Population Biology and Evolution.* Springer-Verlag, Berlin. [90]

Price, J., S. Droege, and A. Price. 1995. *The Summer Atlas of North American Birds.* Princeton University Press, Princeton, NJ. Now available online at http://www.mbr–pwrc.usgs.gov/bbs/bbs.html [27, 114]

Price, T., H. L. Gibbs, L. de Sousa, and A. D. Richman. 1998. Different timing of the adaptive radiations of North American and Asian warblers. *Proceedings of the Royal Society of London Series B—Biological Sciences* 265:1969–1975. [16]

Price, T., I. J. Lovette, E. Bermingham, H. L. Gibbs, and A. D. Richman. 2000. The imprint of history on communities of North American and Asian warblers. *American Naturalist* 156:354–367. [88, 120, 271]

Price, T., A. Qvarnström, and D. E. Irwin. 2003. The role of phenotypic plasticity in driving genetic evolution. *Proceedings of the Royal Society of London Series B—Biological Sciences* 270:1433–1440. [133, 134]

Price, J. J., N. R. Friedman, and K. E. Omland. 2007. Song and plumage evolution in the New World orioles *(Icterus)* show similar lability and convergence in patterns. *Evolution* In press. [235, 240]

Prigogine, A. 1980. Étude de quelques contacts secondaires au Zaire oriental. *Le Gerfaut* 70:305–384. [332]

Prigogine, A. 1987. Hybridization between the megasubspecies *cailliautii* and *permista* of the green-backed woodpecker, *Campethera cailliautii. Le Gerfaut* 77:187–204. [352]

Pröve, E. 1978. Quantitative Untersuchungen zu Wechselbeziehungen zwischen Balzaktivität und Testosterontitern bei männlichen Zebrafinken (*Taeniopygia guttata castanotis* Gould). *Zeitschrift für Tierpsychologie* 48:47–67. [164]

Pruett-Jones, S. G., M. A. Pruett-Jones, and H. I. Jones. 1991. Parasites and sexual selection in a New Guinea avifauna. Pp. 213–245 *in* D. Power, ed. *Current Ornithology,* Vol. 8. Plenum Press, New York. [259, 260]

Pryke, S. R., and S. Andersson. 2002. A generalized female bias for long tails in a short-tailed widowbird. *Proceedings of the Royal Society of London Series B—Biological Sciences* 269:2141–2146. [160]

Pybus, O. G., and P. H. Harvey. 2000. Testing macro-evolutionary models using incomplete molecular phylogenies. *Proceedings of the Royal Society of London Series B—Biological Sciences* 267:2267–2272. [104, 106, 107]

Questiau, S., L. Gielly, M. Clouet, and P. Taberlet. 1999. Phylogeographical evidence of gene flow among Common Crossbill *(Loxia curvirostra,* Aves, Fringillidae) populations at the continental level. *Heredity* 83:196–205. [78]

Qvarnström, A., V. Blomgren, C. Wiley, and N. Svedin. 2004. Female collared flycatchers learn to prefer males with an artificial novel ornament. *Behavioral Ecology* 15:543–548. [292]

Qvarnström, A., J. E. Brommer, and L. Gustafsson. 2006a. Testing the genetics underlying the co-evolution of mate choice and ornament in the wild. *Nature* 441:84–86. [239]

Qvarnström, A., J. Haavie, S. A. Sæther, D. Eriksson, and T. Pärt. 2006b. Song similarity predicts hybridization in flycatchers. *Journal of Evolutionary Biology* 19:1202-1209. [307, 308]

Radtkey, R. R. 1996. Adaptive radiation of day-geckos *(Phelsuma)* in the Seychelles Archipelago: A phylogenetic analysis. *Evolution* 50:604–623. [400]

Rahbek, C. 1997. The relationship among area, elevation, and regional species richness in neotropical birds. *American Naturalist* 149:875–902. [101]

Rahbek, C., and G. R. Graves. 2001. Multiscale assessment of patterns of avian species richness. *Proceedings of the National Academy of Sciences of the United States of America* 98:4534–4539. [97, 99]

Rambaut, A. 1996. Se-Al: Sequence Alignment Editor. Available at http://evolve.zoo.ox.ac.uk/ [40]

Randi, E. 1996. A mitochondrial cytochrome *b* phylogeny of the *Alectoris* partridges. Molecular Phylogenetics and Evolution 6:214–227. [351, 386]

Randi, E., and A. Bernard-Laurent. 1999. Population genetics of a hybrid zone between the red-legged partridge and rock partridge. *Auk* 116:324–337. [350]

Randler, C. 2002. Avian hybridization, mixed pairing and female choice. *Animal Behaviour* 63:103–119. [310, 312]

Rando, J. C., M. López, and B. Seguí. 1999. A new species of extinct flightless Passerine (Emberizidae: *Emberiza)* from the Canary Islands. *Condor* 101:1–13. [71]

Rasmussen, P. C., and J. C. Anderton. 2005. *Birds of South Asia: The Ripley Guide.* Lynx editions, Barcelona, Spain. [4]

Rasner, C. A., P. Yeh, L. S. Eggert, K. E. Hunt, D. S. Woodruff, and T. D. Price. 2004. Genetic and morphological evolution following a founder event in the dark-eyed junco, *Junco hyemalis thurberi*. *Molecular Ecology* 13:671–681. [45, 47]

Ratcliffe, L. M., and P. R. Grant. 1983. Species recognition in Darwin's finches *(Geospiza,* Gould). I. Discrimination by morphological cues. *Animal Behaviour* 31:1139–1153. [24, 34, 83, 285]

Read, A. F. 1987. Comparative evidence supports the Hamilton and Zuk hypothesis on parasites and sexual selection. *Nature* 328:68–70. [259]

Read, A. F., and D. M. Weary. 1992. The evolution of bird song: Comparative analyses. *Philosophical Transactions of the Royal Society, B* 338:165–187. [249, 254]

Real, L. 1990. Search theory and mate choice. 1. Models of single-sex discrimination. *American Naturalist* 136:376–405. [304]

Remsen, J. V. 1984. High incidence of leapfrog pattern of geographic variation in Andean birds—implications for the speciation process. *Science* 224:171–173. [235]

Remsen, J. V. 2005. Pattern, process, and rigor meet classification. *Auk* 122:403–413. [3, 4, 7, 271]

Remsen, J. V., and W. S. Graves. 1995a. Distribution patterns of Buarremon brush-finches (Emberizinae) and interspecific competition in Andean birds. *Auk* 112:225–236. [50, 51]

Remsen, J. V., and W. S. Graves. 1995b. Distribution patterns and zoogeography of *Atlapetes* brush finches (Emberizinae) of the Andes. *Auk* 112:210–224. [235]

Rhymer, J. M. 1992. An experimental study of geographic variation in avian growth and development. *Journal of Evolutionary Biology* 5:289–306. [42, 44]

Rhymer, J. M., and D. Simberloff. 1996. Extinction by hybridization and introgression. *Annual Review of Ecology and Systematics* 27:83–109. [299, 335]

Rice, W. R. 1984. Sex chromosomes and the evolution of sexual dimorphism. *Evolution* 38:735–742. [227, 391]

Rice, W. R. 1998. Intergenomic conflict, interlocus antagonistic evolution, and the evolution of reproductive isolation. Pp 261–270 *in* D. J. Howard and S. H. Berlocher, eds. *Endless Forms: Species and Speciation.* Oxford University Press, New York. [369, 370, 379]

Rice, W. R., and A. K. Chippindale. 2002. The evolution of hybrid infertility: Perpetual coevolution between gender-specific and sexually antagonistic genes. *Genetica* 116:179–188. [376, 379]

Rice, W. R., and E. E. Hostert. 1993. Laboratory experiments on speciation: What have we learned in 40 years? *Evolution* 47:1637–1653. [83, 120, 301]

Richards, D. G. 1979. Recognition of neighbors by associative learning in rufous sided towhees. *Auk* 96: 688–693. [286]

Richman, A. D. 1996. Ecological diversification and community structure in the Old World leaf warblers (genus *Phylloscopus*): A phylogenetic perspective. *Evolution* 50:2461–2470. [69]

Richman, A. D., and T. Price. 1992. Evolution of ecological differences in the Old World leaf warblers. *Nature* 355:817–821. [13, 87, 88]

Ricklefs, R. E. 1969. An analysis of nestling mortality in birds. *Smithsonian Contributions in Zoology* 9:1–48. [127, 137]

Ricklefs, R. E. 1987. Community diversity: Relative roles of local and regional processes. *Science* 235:167–171. [100]

Ricklefs, R. E. 2002. Splendid isolation: Historical ecology of the South American passerine fauna. *Journal of Avian Biology* 33:207–211. [115, 116]

Ricklefs, R. E. 2003. Global diversification rates of passerine birds. *Proceedings of the Royal Society of London Series B—Biological Sciences* 270:2285–2291. [104, 122]

Ricklefs, R. E. 2004. A comprehensive framework for global patterns in biodiversity. *Ecology Letters* 7:1–15. [99, 104]

Ricklefs, R. E. 2005a. Phylogenetic perspectives on patterns of regional and local species richness. Pp. 16–40 *in* E. Bermingham, C. Dick and C. Moritz, eds. *Tropical Rainforests: Past, Present, and Future.* University of Chicago Press, Chicago. [100, 116, 117]

Ricklefs, R. E. 2005b. Small clades at the periphery of passerine morphological space. *American Naturalist* 165:643–659. [122]

Ricklefs, R. E., and E. Bermingham. 1999. Taxon cycles in the Lesser Antillean avifauna. *Ostrich* 70:49–59. [148, 150–152]

Ricklefs, R. E., and E. Bermingham. 2001. Nonequilibrium diversity dynamics of the Lesser Antillean avifauna. *Science* 294:1522–1524. [148]

Ricklefs, R. E., and E. Bermingham. 2002. The concept of the taxon cycle in biogeography. *Global Ecology and Biogeography* 11:353–361. [147, 151, 153, 409]

Ricklefs, R. E., and G. W. Cox. 1972. Taxon cycles in the West Indian Avifauna. *American Naturalist* 106: 195–219. [147]

Riddle, O. 1919. Inheritance, fertility, and the dominance of sex and color in hybrids of wild species of pigeons. *Posthumous Works of Charles Otis Whitman.* The Carnegie Institution of Washington, Washington, DC. [378]

Rieseberg, L. H. 2001. Chromosomal rearrangements and speciation. *Trends in Ecology and Evolution* 16: 351–358. [345, 371, 380, 391, 392]

Ripley, S. D., and B. M. Beehler. 1990. Patterns of speciation in Indian birds. *Journal of Biogeography* 17: 639–648. [67]

Rising, J. D. 1983. The progress of oriole hybridization in Kansas. *Auk* 100:885–897. [365, 366]

Rising, J. D. 1996. The stability of the oriole hybrid zone in western Kansas. *Condor* 98:658–663. [366]

Rising, J. D., and G. F. Shields. 1980. Chromosomal and morphological correlates in two New World sparrows (Emberizidae). *Evolution* 34:654–662. [388]

Robbins, M. B., M. J. Braun, and E. A. Tobey. 1986. Morphological and vocal variation across a contact zone between the chickadees *Parus atricapillus* and *Parus carolinensis. Auk* 103:655–666. [308, 358]

Roberts, T. J. 1992. *The Birds of Pakistan.* Vol. 2. *Passeriformes.* Oxford University Press, Oxford. [359]

Roberts, V. 1997. *British Poultry Standards: Complete Specifications and Judging Points of All Standardized Breeds and Varieties of Poultry as Compiled by the Specialist Breed Clubs and Recognized by the Poultry Club of Great Britain.* Blackwell Science, Oxford and Cambridge, MA. [222]

Robinson, S. K., and J. Terborgh. 1995. Interspecific aggression and habitat selection by Amazonian birds. *Journal of Animal Ecology* 64:1–11. [342]

Rohwer, S. A. 1973. Significance of sympatry to behavior and evolution of Great Plains meadowlarks. *Evolution* 27:44–57. [114, 286–289, 306]

Rohwer, S. 1976. Specific distinctness and adaptive differences in southwestern meadowlarks. *Occasional Papers of the Museum of Natural History. University of Kansas*, Lawrence, Kansas 44:1–14. [29, 114]

Rohwer, S. 1982. The evolution of reliable and unreliable badges of fighting ability. *American Zoologist* 22: 531–546. [181, 182]

Rohwer, S., and M. S. Johnson. 1992. Scheduling differences of molt and migration for Baltimore and Bullock's orioles persist in a common environment. *Condor* 94:992–994. [95]

Rohwer, S., and J. Manning. 1990. Differences in timing and number of molts for Baltimore and Bullock's orioles: Implications to hybrid fitness and theories of delayed plumage maturation. *Condor* 92:125–140. [95]

Rohwer, S., and C. Wood. 1998. Three hybrid zones between hermit and Townsend's warblers in Washington and Oregon. *Auk* 115:284–310. [331, 344, 364]

Rohwer, S., C. Wood, and E. Bermingham. 2000. A new hybrid warbler *(Dendroica nigrescens × D. occidentalis)* and diagnosis of similar *D. townsendi × D. occidentalis* recombinants. *Condor* 102:713–718. [300]

Rohwer, S., E. Bermingham, and C. Wood. 2001. Plumage and mitochondrial DNA haplotype variation across a moving hybrid zone. *Evolution* 55:405–422. [28, 325, 331, 334, 335, 336, 340, 342, 343, 347, 364]

Rosenzweig, M. 1975. On continental steady states of species diversity. Pp. 121–140 *in* M. L. Cody and J. Diamond, eds. *Ecology and Evolution of Communities.* Belknap Press of Harvard University Press, Cambridge, MA. [120]

Rosenzweig, M. 1978. Competitive speciation. *Biological Journal of the Linnean Society* 10:275–289. [19]

Rosenzweig, M. L. 1992. Species diversity gradients—we know more and less than we thought. *Journal of Mammalogy* 73:715–730. [16, 99]

Rosenzweig, M. L. 1995. *Species Diversity in Space and Time.* Cambridge University Press, Cambridge and New York. [33, 34, 94, 397]

Rosenzweig, M. L., and Z. Abramsky. 1993. How are diversity and productivity related? Pp. 52–65 *in* R. E. Ricklefs and D. Schluter, eds. *Species Diversity in Ecological Communities.* University of Chicago Press, Chicago. [98]

Rothstein, S. I., and R. C. Fleischer. 1987. Brown-headed cowbirds learn flight whistles after the juvenile period. *Auk* 104:512–516. [280]

Rowe, C. 1999. Receiver psychology and the evolution of multicomponent signals. *Animal Behaviour* 58: 921–931. [172, 205, 282]

Rowe, C., and J. Skelhorn. 2004. Avian psychology and communication. *Proceedings of the Royal Society of London Series B—Biological Sciences* 271:1435–1442. [172]

Rowley, D. B., R. T. Pierrehumbert, and B. S. Currie. 2001. A new approach to stable isotope-based paleoaltimetry: Implications for paleoaltimetry and paleohypsometry of the High Himalaya since the Late Miocene. *Earth and Planetary Science Letters* 188:253–268. [119]

Roy, M. S. 1997. Recent diversification in African greenbuls (Pycnonotidae: *Andropadus*) supports a montane speciation model. *Proceedings of the Royal Society of London Series B—Biological Sciences* 264: 1337–1344. [102, 119]

Rundle, H. D., and P. Nosil. 2005. Ecological speciation. *Ecology Letters* 8:336–352. [14, 75, 83, 243]

Russello, M. A., and G. Amato. 2004. A molecular phylogeny of *Amazona:* Implications for Neotropical parrot biogeography, taxonomy, and conservation. *Molecular Phylogenetics and Evolution* 30:421–437. [102]

Rutgers, A., and K. A. Norris. 1972. *Encyclopedia of Aviculture.* Blandford Press, London. [222]

Rutgers, A., K. A. Norris, and C. H. Rogers. 1977. *Encyclopedia of Aviculture.* Blandford Press, London. [224]

Ryan, M. J., and E. A. Brenowitz. 1985. The role of body size, phylogeny, and ambient noise in the evolution of bird song. *American Naturalist* 126:87–100. [253, 254]

Ryan, M. J., and A. Keddy-Hector. 1992. Directional patterns of female mate choice and the role of sensory biases. *American Naturalist* 139:S4–S35. [173, 244]

Ryan, P. G., C. L. Moloney, and J. Hudon. 1994. Color variation and hybridization among *Nesospiza* buntings on Inaccessible Island, Tristan da Cunha. *Auk* 111:314–327. [86, 259]

Ryan, P. G., P. Bloomer, C. L. Moloney, T. J. Grant, and W. Delport. 2007. Ecological speciation in south Atlantic island finches. *Science* 315:1420–1423. [85, 86]

Sæther, S. A., G. P. Sætre, M. R. Servedio, N. Svedin, T. Borge, C. Wiley, J. Haavie, T. Veen, M. Hjernquist, S. Bures, M. Král, L. Gustafsson, J. Träff, and A. Qvarnström. 2007. Z-linked species recognition in fly-catcher hybrid zones facilitates reinforcement without sexual imprinting. Unpublished manuscript. [279]

Sætre, G. P., M. Král, and V. Bicik. 1993. Experimental evidence for interspecific female mimicry in sympatric *Ficedula* flycatchers. *Evolution* 47:939–945. [309, 310]

Sætre, G. P., T. Moum, S. Bures, M. Král, M. Adamjan, and J. Moreno. 1997. A sexually selected character displacement in flycatchers reinforces premating isolation. *Nature* 387:589–592. [310, 316, 356]

Sætre, G. P., M. Král, S. Bures, and R. A. Ims. 1999. Dynamics of a clinal hybrid zone and a comparison with island hybrid zones of flycatchers (*Ficedula hypoleuca* and *F. albicollis*). *Journal of Zoology* 247:53–64. [309, 312, 339, 356]

Sætre, G. P., T. Borge, J. Lindell, T. Moum, C. R. Primmer, B. C. Sheldon, J. Haavie, A. Johnsen, and H. Ellegren. 2001. Speciation, introgressive hybridization and nonlinear rate of molecular evolution in flycatchers. *Molecular Ecology* 10:737–749. [390]

Sætre , G. P., T. Borge, K. Lindroos, J. Haavie, B. C. Sheldon, C. Primmer, and A. C. Syvanen. 2003. Sex chromosome evolution and speciation in *Ficedula* flycatchers. *Proceedings of the Royal Society of London Series B—Biological Sciences* 270:53–59. [314, 339, 344, 345, 390]

Safran, R. J., C. R. Neuman, K. J. McGraw, and I. J. Lovette. 2005. Dynamic paternity allocation as a function of male plumage color in barn swallows. *Science* 309:2210–2212. [160]

Saino, N. 1992. Selection of foraging habitat and flocking by crow *Corvus corone* phenotypes in a hybrid zone. *Ornis Scandinavica* 23:111–120. [84, 355]

Saino, N., and A. M. Bolzern. 1992. Egg volume, chick growth and survival across a carrion/hooded crow hybrid zone. *Bollettino di Zoologia* 59:407–415. [339]

Saino, N., and L. Scatizzi. 1991. Selective aggressiveness and dominance among carrion crows, hooded crows and hybrids. *Bollettino di Zoologia* 58:255–260. [339, 342]

Saino, N., and S. Villa. 1992. Pair composition and reproductive success across a hybrid zone of carrion crows and hooded crows. *Auk* 109:543–555. [339, 355]

Saino, N., R. Lorenzini, G. Fusco, and E. Randi. 1992. Genetic variability in a hybrid zone between carrion and hooded crows (*Corvus corone corone* and *C. c. cornix,* Passeriformes, Aves) in north-western Italy. *Biochemical Systematics and Ecology* 7:605–613. [355

Saino, N., C. R. Primmer, H. Ellegren, and A. P. Møller. 1997. An experimental study of paternity and tail ornamentation in the barn swallow (*Hirundo rustica*). *Evolution* 51:562–570. [160, 161]

Saino, N., V. Bertacche, R. P. Ferrari, R. Martinelli, A. P. Møller, and R. Stradi. 2002. Carotenoid concentration in barn swallow eggs is influenced by laying order, maternal infection and paternal ornamentation. *Proceedings of the Royal Society of London Series B—Biological Sciences* 269:1729–1733. [162]

Salomon, M. 1987. Analyse d'une zone de contact entre deux formes parapatriques: Le cas des pouillots véloces *Phylloscopus c. collybita* et *P. c. brehmii. Revue d'Ecologie* 42:377–420. [361]

Salomon, M., J. Bried, A. J. Helbig, and J. Riofrio. 1997. Morphometric differentiation between male common chiffchaffs, *Phylloscopus [c.] collybita* Vieillot, 1817, and Iberian chiffchaffs, *P. [c.] brehmii* Homeyer, 1871, in a secondary contact zone (Aves; Sylviidae). *Zoologischer Anzeiger* 236:25–36. [361]

Salzburger, W., and A. Meyer. 2004. The species flocks of East African cichlid fishes: Recent advances in molecular phylogenetics and population genetics. *Naturwissenschaften* 91:277–290. [120]

Salzburger, W., J. Martens, and C. Sturmbauer. 2002. Paraphyly of the blue tit (*Parus caeruleus*) suggested from cytochrome *b* sequences. *Molecular Phylogenetics and Evolution* 24:19–25. [267]

Sammalisto, L. 1968. Variations in the selective advantage of hybrids in the Finnish population of *Motacilla flava* L. *Annales Zoolgica Fennica* 5:196–206. [84]

Sanderson, M. J. 2003. r8s: Inferring absolute rates of molecular evolution and divergence times in the absence of a molecular clock. *Bioinformatics* 19:301–302. [40]

Sasvari, L. 1985a. Keypeck conditioning with reinforcements in two different locations in thrush, tit and sparrow species. *Behavioural Processes* 11:245–252. [130, 131, 132]

Sasvari, L. 1985b. Different observational learning capacity in juvenile and adult individuals of congeneric bird species. *Zeitschrift für Tierpsychologie* 69:293–304. [128, 129]

Sato, A., C. O'Huigin, F. Figueroa, P. R. Grant, B. R. Grant, H. Tichy, and J. Klein. 1999. Phylogeny of Darwin's finches as revealed by mtDNA sequences. *Proceedings of the National Academy of Sciences of the United States of America* 96:5101–5106. [22, 23]

Sattler, G. D., and M. J. Braun. 2000. Morphometric variation as an indicator of genetic interactions between black-capped and Carolina chickadees at a contact zone in the Appalachian Mountains. *Auk* 117: 427–444. [345, 358]

Scherer, S., and T. Hilsberg. 1982. Hybridisierung und Verwandtschaftsgrade innerhalb der Anatidae—eine systematische und evolutionstheoretische Betrachtung. *Journal für Ornithologie* 123:357–380. [229, 231]

Schluter, D. 1986. Character displacement between distantly related taxa? Finches and bees in the Galápagos. *American Naturalist* 127:95–1002. [54]

Schluter, D. 1998. Ecological causes of speciation. Pp. 114–129 *in* D. Howard and S. Berlocher, eds. *Endless Forms: Species and Speciation*. Oxford University Press, Oxford. [120]

Schluter, D. 2000. *The Ecology of Adaptive Radiation*. Oxford University Press, Oxford. [21, 22, 42, 84, 243]

Schluter, D. 2001. Ecology and the origin of species. *Trends in Ecology and Evolution* 16:372–380. [14, 21, 22, 75, 83, 243]

Schluter, D., and P. R. Grant. 1984a. Determinants of morphological patterns in communities of Darwin's finches. *American Naturalist* 123:175–196. [52, 54, 76–78]

Schluter, D., and P. R. Grant. 1984b. Ecological correlates of morphological evolution in a Darwin's finch, *Geospiza difficilis. Evolution* 38:856–869. [127]

Schluter, D., and T. Price. 1993. Honesty, perception and population divergence in sexually selected traits. *Proceedings of the Royal Society of London Series B—Biological Sciences* 253:117–122. [246–248, 404]

Schluter, D., and R. E. Ricklefs. 1993. Convergence and the regional component of species diversity. Pp. 230–242 *in* R. E. Ricklefs and D. Schluter, eds. *Species Diversity in Ecological Communities*. University of Chicago Press, Chicago. [100, 101, 104]

Schluter, D., T. D. Price, and P. R. Grant. 1985. Ecological character displacement in Darwin's finches. *Science* 227:1056–1059. [52, 54, 77, 78]

Schmidt, H. A., K. Strimmer, K. M. Vingron, and A. von Haeseler. 2002. TREE-PUZZLE: Maximum likelihood phylogenetic analysis using quartets and parallel computing. *Bioinformatics* 18:502–504. [23, 39]

Schodde, R., and R. de Naurois. 1982. Patterns of variation and dispersal in the buff-banded rail *(Gallirallus philippensis)* in the south-west Pacific, with a description of a new subspecies. *Notornis* 29:131–142. [71]

Schoener, T. W. 1965. The evolution of bill size differences among sympatric congeneric species of birds. *Evolution* 19:189–213. [50, 86]

Schoener, T. W. 1971. Large-billed insectivorous birds: A precipitous diversity gradient. *Condor* 73:154–161. [98]

Schottler, B. 1995. Songs of blue tits *Parus caeruleus* from La Palma (Canary Islands)–a test of hypotheses. *Bioacoustics* 6:135–152. [190]

Schuetz, J. G. 2005. Reduced growth but not survival of chicks with altered gape patterns: Implications for the evolution of nestling similarity in a parasitic finch. *Animal Behaviour* 70:839–848. [295]

Schutz, F. 1965. Sexuelle Prägung bei Anatiden. *Zeitschrift für Tierpsychologie* 22:50–103. [278]

Scott, D. K., and T. H. Clutton-Brock. 1989. Mating systems, parasites, and plumage dimorphism in waterfowl. *Behavioral Ecology and Sociobiology* 26:261–273. [259, 260, 264]

Scott, D. M., C. D. Ankney, and C. H. Jarosch. 1976. Sapsucker hybridization in British Columbia: Changes in 25 years. *Condor* 78:253–257. [351]

Searcy, W. A. 1992. Song repertoire and mate choice in birds. *American Zoologist* 32:71–80. [163, 205, 212, 237]

Searcy, W. A., and K. Yasukawa. 1996. Song and female choice. Pp. 454–473 *in* D. Kroodsma and E. Miller, eds. *Ecology and Evolution of Acoustic Communication in Birds*. Cornell University Press, Ithaca, NY and London. [237]

Searcy, W. A., S. Nowicki, and M. Hughes. 1997. The response of male and female song sparrows to geographic variation in song. *Condor* 99:651–657. [190, 191]

Searcy, W. A., S. Nowicki, M. Hughes, and S. Peters. 2002. Geographic song discrimination in relation to dispersal distances in song sparrows. *American Naturalist* 159:221–230. [191]

Secondi, J., V. Bretagnolle, C. Compagnon, and B. Faivre. 2003. Species-specific song convergence in a moving hybrid zone between two passerines. *Biological Journal of the Linnean Society* 80:507–517. [360]

Seddon, N. 2005. Ecological adaptation and species recognition drives vocal evolution in neotropical suboscine birds. *Evolution* 59:200–215. [254, 314]

Seehausen, O. 2004. Hybridization and adaptive radiation. *Trends in Ecology and Evolution* 19:198–207. [323]

Seehausen, O. 2006. Conservation: Losing biodiversity by reverse speciation. *Current Biology* 16:R334–R337. [397]

Sefc, K. M., R. B. Payne, and M. D. Sorenson. 2005. Genetic continuity of brood-parasitic indigobird species. *Molecular Ecology* 14:1407–1419. [294]

Seger, J. 1985. Intraspecific resource competition as a cause of sympatric speciation. Pp. 43–53 *in* P. J. Greenwood, P. H. Harvey and M. Slatkin, eds. *Evolution: Essays in Honour of John Maynard Smith.* Cambridge University Press, Cambridge. [19]

Seiger, M. B. 1967. A computer simulation study of the influence of imprinting on population structure. *American Naturalist* 101:47–57. [273]

Selander, R. K. 1964. Speciation in wrens of the genus *Campylorhynchus. University of California Publications in Zoology* 74:1–259. [357, 358]

Selander, R. K. 1965. Hybridization of rufous-naped wrens in Chiapas, Mexico. *Auk* 82:206–214. [357, 358]

Selander, R. K., and R. F. Johnston. 1967. Evolution in the house sparrow. I. Intrapopulation variation in North America. *Condor* 69:217–258. [46]

Sellier, N., J.-M. Brun, M.-M. Richard, F. Batellier, V. Dupuy, and J.-P. Brillard. 2005. Comparison of fertility and embryo mortality following artificial insemination of common duck females *(Anas platyrhynchos)* with semen from common or Muscovy *(Cairina moschata)* drakes. *Theriogenology* 64:429–439. [372, 373]

Senar, J. C., J. Domenech, and M. Camerino. 2005. Female siskins choose mates by the size of the yellow wing stripe. *Behavioral Ecology and Sociobiology* 57:465–469. [160]

Servedio, M. R. 2000. Reinforcement and the genetics of nonrandom mating. *Evolution* 54:21–29. [302, 303]

Servedio, M. R. 2001. Beyond reinforcement: The evolution of premating isolation by direct selection on preferences and postmating, prezygotic incompatibilities. *Evolution* 55:1909-1920. [314]

Servedio, M. R., and R. Lande. 2006. Population genetic models of male and mutual mate choice. *Evolution* 60:674-685. [275, 312]

Servedio, M. R., and M. A. F. Noor. 2003. The role of reinforcement in speciation: Theory and data. *Annual Review of Ecology Evolution and Systematics* 34:339–364. [300]

Servedio, M. R., and G. P. Sætre. 2003. Speciation as a positive feedback loop between postzygotic and prezygotic barriers to gene flow. *Proceedings of the Royal Society of London Series B—Biological Sciences* 270:1473–1479. [391]

Shapiro, L. H., R. A. Canterbury, D. M. Stover, and R. C. Fleischer. 2004. Reciprocal introgression between golden-winged Warblers *(Vermivora chrysoptera)* and blue-winged Warblers *(V. pinus)* in eastern North America. *Auk* 121:1019–1030. [343]

Shapiro, B. J., D. Garant, T. A. Wilkin, and B. C. Sheldon. 2006. An experimental test of the causes of small-scale phenotypic differentiation in a population of great tits. *Journal of Evolutionary Biology* 19:176–183. [42, 44]

Sharp, S. P., A. McGowan, M. J. Wood, and B. J. Hatchwell. 2005. Learned kin recognition cues in a social bird. *Nature* 434:1127–1130. [283]

Sharpe, R. S., and P. A. Johnsgard. 1966. Inheritance of behavioral characters in F2 mallard × pintail *(Anas platyrhynchos* L. × *Anas acuta* L.) hybrids. *Behaviour* 27:259–272. [379, 390]

Sheldon, B. C., J. Merilä, A. Qvarnström, L. Gustafsson, and H. Ellegren. 1997. Paternal genetic contribution to offspring condition predicted by size of male secondary sexual character. *Proceedings of the Royal Society of London Series B—Biological Sciences* 264:297–302. [160]

Shephard, M. 1989. *Aviculture in Australia: Keeping and Breeding Aviary Birds.* Black Cockatoo Press, Prahran, Victoria. [222, 224, 226]

Sherry, T. W. 1990. When are birds dietarily specialized? Distinguishing ecological from evolutionary approaches. *Studies in Avian Biology* 13:337–352. [98, 131]

Shibusawa, M., S. Minai, C. Nishida-Umehara, T. Suzuki, T. Mano, K. Yamada, T. Namikawa, and Y. Matsuda. 2001. A comparative cytogenetic study of chromosome homology between chicken and Japanese quail. *Cytogenetics and Cell Genetics* 95:103–109. [385, 386]

Shibusawa, M., C. Nishida-Umehara, J. Masabanda, D. K. Griffin, T. Isobe, and Y. Matsuda. 2002. Chromosome rearrangements between chicken and guineafowl defined by comparative chromosome painting and FISH mapping of DNA clones. *Cytogenetic and Genome Research* 98:225–230. [385, 388]

Shibusawa, M., C. Nishida-Umehara, M. Tsudzuki, J. Masabanda, D. K. Griffin, and Y. Matsuda. 2004a. A comparative karyological study of the blue-breasted quail *(Coturnix chinensis,* Phasianidae) and California quail *(Callipepla californica,* Odontophoridae). *Cytogenetics and Cell Genetics* 106:82–90. [385, 386]

Shibusawa, M., C. Nishibori, C. Nishida-Umehara, M. Tsudzuki, J. Masabanda, D. K. Griffin, and Y. Matsuda. 2004b. Karyotypic evolution in the Galliformes: An examination of the process of karyotypic evolution by comparison of the molecular cytogenetic findings with the molecular phylogeny. *Cytogenetic and Genome Research* 106:111–119. [383]

Shields, G. F. 1973. Chromosomal polymorphism common to several species of *Junco* (Aves). *Canadian Journal of Genetics and Cytology* 15:461–471. [389]

Shields, G. F. 1976. Meiotic evidence for pericentric inversion polymorphism in *Junco* (Aves). *Canadian Journal of Genetics and Cytology* 18:747–751. [389]

Shiovitz, K. A. 1975. The process of species-specific song recognition by the indigo bunting, *Passerina cyanea*, and its relationship to the organization of avian acoustic behavior. *Behaviour* 55:128–179. [201]

Shiovitz, K. A., and W. L. Thompson. 1970. Geographic variation in song composition of the indigo bunting, *Passerina cyanea*. *Animal Behaviour* 18:151–158. [215]

Short, L. L. 1965. Hybridization in the flickers *(Colaptes)* of North America. *Bulletin of the American Museum of Natural History* 129:307–428. [352]

Short, L. L. 1969. Taxonomic aspects of avian hybridization. *Auk* 86:84–105. [300, 310, 311, 333]

Short, L. L., R. Schodde, and J. F. M. Horne. 1983a. 5-way hybridization of varied sittellas *Daphoenositta chrysoptera* (Aves, Neosittidae) in Central Queensland. *Australian Journal of Zoology* 31:499–516. [354]

Short, L. L., R. Schodde, R. A. Noske, and J. F. M. Horne. 1983b. Hybridization of white-headed and orange-winged varied sittellas, *Daphoenositta chrysoptera leucocephala* and *D. c. chrysoptera* (Aves, Neosittidae), in Eastern Australia. *Australian Journal of Zoology* 31:517–531. [354]

Shrigley, E. W. 1940. Qualitative and quantitative differences in the morphology of spermatozoa from ring doves, pearlnecks, and their F1 and backcross hybrids. *Journal of Experimental Zoölogy* 83:457–479. [381]

Shutler, D., and P. J. Weatherhead. 1990. Targets of sexual selection—song and plumage of wood warblers. *Evolution* 44:1967–1977. [259, 261]

Sibley, C. G. 1950. Species formation in the red-eye towhees of Mexico. *University of California Publications in Zoology* 50:109–194. [362]

Sibley, C. G., and J. E. Ahlquist. 1983. Phylogeny and classification of birds based on the data of DNA-DNA hybridization. Pp. 245–292 *in* R. F. Johnston, ed. *Current Ornithology*. Plenum Press, New York. [38]

Sibley, C. G., and J. E. Ahlquist. 1990. *Phylogeny and Classification of Birds: A Study in Molecular Evolution*. Yale University Press, New Haven, CT. [36, 38, 116, 117, 130]

Sibley, C. G., and B. L. Monroe. 1990. *Distribution and Taxonomy of Birds of the World*. Yale University Press, New Haven, CT. [2, 26, 33, 103, 130, 138, 187, 213, 233, 362, 381, 395]

Sibley, C. G., and D. A. West. 1959. Hybridization in the rufous-sided towhees of the Great Plains. *Auk* 76:326–338. [362]

Silverin, B. 1991. Behavioral, hormonal, and morphological responses of free-living male pied flycatchers to estradiol treatment of their mates. *Hormones and Behavior* 25:38–56. [164]

Simpson, G. G. 1953. *The Major Features of Evolution*. Columbia University Press, New York. [21]

Simpson, G. G. 1964. Species density of North American Recent mammals. *Systematic Zoology* 13:57–73. [16]

Slabbekoorn, H., and A. den Boer-Visser. 2006. Cities change the songs of birds. *Current Biology* 16:2326–2331. [256]

Slabbekoorn, H., and M. Peet. 2003. Ecology: Birds sing at a higher pitch in urban noise—great tits hit the high notes to ensure that their mating calls are heard above the city's din. *Nature* 424:267. [256]

Slabbekoorn, H., and T. B. Smith. 2000. Does bill size polymorphism affect courtship song characteristics in the African finch *Pyrenestes ostrinus*? *Biological Journal of the Linnean Society* 71:737–753. [85]

Slabbekoorn, H., and T. B. Smith. 2002. Bird song, ecology and speciation. *Philosophical Transactions of the Royal Society of London Series B—Biological Sciences* 357:493–503. [14, 212, 253, 269]

Slack, K. E., C. M. Jones, T. Ando, G. L. Harrison, R. E. Fordyce, U. Arnason, and D. Penny. 2006. Early penguin fossils, plus mitochondrial genomes, calibrate avian evolution. *Molecular Biology and Evolution* 23:1144–1155. [16, 384, 386]

Slagsvold, T. 2004. Cross-fostering of pied flycatchers *(Ficedula hypoleuca)* to heterospecific hosts in the wild: A study of sexual imprinting. *Behaviour* 141:1079–1102. [279]

Slagsvold, T., B. T. Hansen, L. E. Johannessen, and J. T. Lifjeld. 2002. Mate choice and imprinting in birds studied by cross-fostering in the wild. *Proceedings of the Royal Socity of London, B* 269:1449–1455. [276, 302]

Slater, P. J. B. 1981. Chaffinch song repertoires: Observations, experiments and a discussion of their significance. *Zeitschrift für Tierpsychologie* 56:1–24. [216]

Slater, P. J. B., and S. A. Ince. 1979. Cultural evolution in chaffinch song. *Behaviour* 71:146–166. [216–217]

Slater, P. J. B., and S. A. Ince. 1982. Song development in chaffinches: What is learnt and when? *Ibis* 124:21–26. [216]

Slater, P. J. B., S. A. Ince, and P. W. Colgan. 1980. Chaffinch song types: Their frequencies in the population and distribution between repertoires of different individuals. *Behaviour* 75:207–218. [216]

Slikas, B. 1997. Phylogeny of the avian family Ciconiidae (storks) based on cytochrome *b* sequences and DNA–DNA hybridization distances. *Molecular Phylogenetics and Evolution* 8:275–300. [102]

Slikas, B., I. B. Jones, S. R. Derrickson, and R. C. Fleischer. 2000. Phylogenetic relationships of Micronesian white-eyes based on mitochondrial sequence data. *Auk* 117:355–365. [148, 149, 151]

Slikas, B., S. L. Olson, and R. C. Fleischer. 2002. Rapid, independent evolution of flightlessness in four species of Pacific Island rails (Rallidae): An analysis based on mitochondrial sequence data. *Journal of Avian Biology* 33:5–14. [71]

Smith, G. 1912. Studies in the experimental analysis of sex. Part 9. On spermatogenesis and the formation of giant spermatozoa in hybrid pigeons. *Quarterly Journal of Microscopical Science* 58:159–170. [380, 381]

Smith, D. G. 1972. The role of epaulets in the red-winged blackbird, *Agelaius phoeniceus*, social system. *Behaviour* 41:251–267. [251]

Smith, T. B. 1987. Bill size polymorphism and intraspecific niche utilization in an African finch. *Nature* 329:717–719. [85]

Smith, T. B. 1993. Disruptive selection and the genetic basis of bill size polymorphism in the African finch *Pyrenestes*. *Nature* 363:618–620. [84, 85]

Smith, J. W., and C. W. Benkman. 2007. A coevolutionary arms race causes ecological speciation in crossbills. *American Naturalist* 169:455–465. [69, 78]

Smith, A. L., and V. L. Friesen. 2007. Differentiation of sympatric populations of the band-rumped storm petrel in the Galápagos Islands: An examination of genetics, morphology and vocalizations. *Molecular Ecology*. In press. [21]

Smith, G., and R. H. Thomas. 1913. On sterile and hybrid pheasants. *Journal of Genetics* 3:39–52. [375, 376, 380, 381]

Smith, T. B., L. A. Freed, J. K. Lepson, and J. H. Carothers. 1995. Evolutionary consequences of extinctions in populations of a Hawaiian honeycreeper. *Conservation Biology* 9:107–113. [47]

Smith, T. B., R. K. Wayne, D. J. Girman, and M. W. Bruford. 1997. A role for ecotones in generating rainforest biodiversity. *Science* 276:1855–1857. [16, 41, 63, 65]

Smith, T. B., C. J. Schneider, and K. Holder. 2001. Refugial isolation versus ecological gradients. *Genetica* 112–113:383–398. [1]

Smith, T. B., R. Calsbeek, R. K. Wayne, K. H. Holder, D. Pires, and C. Bardeleben. 2005. Testing alternative mechanisms of evolutionary divergence in an African rain forest passerine bird. *Journal of Evolutionary Biology* 18:257–268. [63, 65]

Smith, A. L., L. Monteiro, O. Hasegawa, and V. L. Friesen. 2007. Cryptic species in Atlantic populations of the band-rumped storm petrel. *Molecular Phylogenetics and Evolution*. In press. [21]

Smyth, R. J., Jr. 1990. Genetics of plumage, skin, and eye pigmentation in chickens. Pp. 109–167 *in* R. D. Crawford, ed. *Poultry Breeding and Genetics*. Elsevier, Oxford, New York and Tokyo. [226]

Snell, H. M., P. A. Stone, and H. L. Snell. 1995. Geographical characteristics of the Galápagos Islands. *Noticias de Galápagos* 55:18–24. [54]

Snow, D. 1976. *The Web of Adaptation: Bird Studies in the American Tropics*. Quadrangle, New York. [158, 270]

Snow, D. W. 1982. *The Cotingas: Bellbirds, Umbrellabirds, and Other Species*. Comstock Press, Ithaca, NY. [159]

Soha, J. A., and P. Marler. 2000. A species-specific acoustic cue for selective song learning in the white-crowned sparrow. *Animal Behaviour* 60:297–306. [201, 211, 281]

Sol, D. 2003. Behavioural innovation: A neglected issue in the ecological and evolutionary literature? Pp. 63–82 *in* K. Laland and S. M. Reader, eds. *Animal Innovation*. Oxford University Press, Oxford. [131, 133]

Sol, D., and L. Lefebvre. 2000. Behavioural flexibility predicts invasion success in birds introduced to New Zealand. *Oikos* 90:599–605. [131, 132]

Sol, D., S. Timmermans, and L. Lefebvre. 2002a. Behavioural flexibility and invasion sucess in birds. *Animal Behaviour* 63:495–502. [131. 132. 138]

Sol, D., S. Timmermans, and L. Lefebvre. 2002b. Behavioural flexibility and invasion success in birds (vol 63, pg 495, 2002). *Animal Behaviour* 64:516–516. [132]

Sol, D., R. P. Duncan, T. M. Blackburn, P. Cassey, and L. Lefebvre. 2005a. Big brains, enhanced cognition, and response of birds to novel environments. *Proceedings of the National Academy of Sciences of the United States of America* 102:5460–5465. [131]

Sol, D., D. G. Stirling, and L. Lefebvre. 2005b. Behavioral drive or behavioral inhibition in evolution: Subspecific diversification in holarctic passerines. *Evolution* 59:2669–2677. [131, 133]

Soler, M., and J. J. Soler. 1999. Innate versus learned recognition of conspecifics in great spotted cuckoos *Clamator glandarius*. *Animal Cognition* 2:97–102. [280]

Soligo, C. 2005. Anatomy of the hand and arm in *Daubentonia madagascariensis*: a functional and phylogenetic outlook. *Folia Primatologica* 76:262–300. [112]

Somes, R. G. 1990. Mutations and major variants in pheasants. Pp. 389–394 *in* R. D. Crawford, ed. *Poultry Breeding and Genetics*. Elsevier, Oxford, New York and Tokyo. [226]

Sorenson, M. D., and R. B. Payne. 2001. A single ancient origin of brood parasitism in African finches: Implications for host–parasite coevolution. *Evolution* 55:2550–2567. [36, 386]

Sorenson, M. D., A. Cooper, E. E. Paxinos, T. W. Quinn, H. F. James, S. L. Olson, and R. C. Fleischer. 1999. Relationships of the extinct moa-nalos, flightless Hawaiian waterfowl, based on ancient DNA. *Proceedings of the Royal Society of London Series B—Biological Sciences* 266:2187–2193. [113]

Sorenson, M. D., K. M. Sefc, and R. B. Payne. 2003. Speciation by host switch in brood parasitic indigobirds. *Nature* 424:928–931. [20, 294, 295, 386, 387, 396]

Sorenson, M. D., C. N. Balakrishnan, and R. B. Payne. 2004. Clade-limited colonization in brood parasitic finches *(Vidua* spp.). *Systematic Biology* 53:140–153. [294, 387]

Sorjonen, J. 1986. Mixed singing and interspecific territoriality—consequences of secondary contact of two ecologically and morphologically similar nightingale species in Europe. *Ornis Scandinavica* 17:53–67. [286, 357]

Sossinka, R. 1982. Domestication in birds. Pp. 373–403 *in* D. S. Farner, J. R. King and K. C. Parkes, eds. *Avian Biology.* Academic Press, New York. [225, 232]

Spalding, D. 1873. Instinct, with original observations on young animals. *Macmillan's Magazine* 27:282–293. [Reprinted (1954) in the *British Journal of Animal Behaviour* 2:2–11]. [133]

Sparling, D. W. 1980. Hybridization and taxonomic status of greater prairie chickens and sharp-tailed grouse (hybridization in grouse). *Prairie Naturalist* 12:92–102. [319]

Spear, L. B., P. Pyle, and N. Nur. 1998. Natal dispersal in the western gull: Proximal factors and fitness consequences. *Journal of Animal Ecology* 67:165–179. [68]

Starck, J. M., S. König, and E. Gwinner. 1995. Growth of stonechats *Saxicola torquata* from Africa and Europe: an analysis of genetic and environmental components. *Ibis* 137:519–531. [44]

Steadman, D. W. 1995. Prehistoric extinctions of Pacific island birds—biodiversity meets zooarchaeology. *Science* 267:1123–1131. [71, 142]

Steadman, D. W. 2006. *Extinction and Biogeography of Tropical Pacific Birds.* University of Chicago Press, Chicago. [71, 142, 148]

Steele, M. G., and G. J. Wishart. 1992. Evidence for a species-specific barrier to sperm transport within the vagina of the chicken hen. *Theriogenology* 38:1107–1114. [371, 372]

Stein, A. C., and J. A. C. Uy. 2006. Unidirectional introgression of a sexually selected trait across an avian hybrid zone: A role for female choice? *Evolution* 60:1476-1485. [337, 342, 346]

Stock, M., and H. H. Bergmann. 1988. Der Gesang des Rotkehlens *(Erithacus rubecula superbus)* von Teneriffa (Kanarische Inseln)—Struktur and Erkennen eines Inseldialekts. *Zoologische Jahrbücher: Abteilung für Allgemeine Zoologie und Physiologie der Tiere* 92:197–212. [267]

Stoddard, P. K. 1996. Vocal recognition of neighbors by territorial passerines. Pp. 356–374 *in* D. E. Kroodsma and E. H. Miller, eds. *Ecology and Evolution of Acoustic Communication in Birds.* Academic Press, New York. [284]

Storch, D., R. G. Davies, S. Zajˈcek, C. D. L. Orme, V. Olson, G. H. Thomas, T.-S. Ding, P. C. Rasmussen, R. S. Ridgely, P. M. Bennett, T. M. Blackburn, I. P. F. Owens, and K. J. Gaston. 2006. Energy, range dynamics and global species richness patterns: Reconciling mid-domain effects and environmental determinants of avian diversity. *Ecology Letters* 9:1308-1320. [108]

Studd, M. V., and R. J. Robertson. 1985. Sexual selection and variation in reproductive strategy in male yellow warblers *(Dendroica petechia). Behavioral Ecology and Sociobiology* 17:101–109. [181]

Studd, M. V., and R. J. Robertson. 1988. Differential allocation of reproductive effort to territorial establishment and maintenance by male yellow warblers *(Dendroica petechia). Behavioral Ecology and Sociobiology* 23:199–210. [181]

Summers-Smith, J. D. 1988. *The Sparrows: A Study of the Genus* Passer. Poyser, Calton, UK. [361]

Sutherland, W. J. 1998. Evidence for flexibility and constraint in migration systems. *Journal of Avian Biology* 29:441–446. [69, 93]

Suthers, R. A. 1990. Contributions to birdsong from the left and right sides of the intact syrinx. *Nature* 347: 473–477. [192]

Suthers, R. A., and S. A. Zollinger. 2004. Producing song—the vocal apparatus. *Annals of the New York Academy of Sciences* 1016:109–129. [212, 253]

Swan, M. A. 1985. Transmission electron microscopy of impaired spermatogenesis in an avian hybrid. *Gamete Research* 12:357–371. [381]

Swan, M. A., and L. Christidis. 1987. Impaired spermatogenesis in the finch hybrid *L. castaneothorax* by *L. punctulata:* Transmission electron microscopy and genetic analysis. *Gamete Research* 17:157–171. [381]

Swofford, D. L. 1998. *PAUP*: Phylogenetic Analysis Using Parsimony (*and Other Methods), Version 4.0.* Sinauer, Sunderland, MA. [23, 40]

Szymura, J. M., and N. H. Barton. 1991. The genetic structure of the hybrid zone between the fire-bellied toads *Bombina bombina* and *B. variegata:* Comparisons between transects and between loci. *Evolution* 45:237–261. [348]

Tanvez, A., N. Béguin, O. Chastel, A. Lacroix, and G. Leboucher. 2004. Sexually attractive phrases increase yolk androgens deposition in Canaries *(Serinus canaria). General and Comparative Endocrinology* 138: 113–120. [163]

Tao, Y., and D. L. Hartl. 2003. Genetic dissection of hybrid incompatibilities between *Drosophila simulans* and *D. mauritiana.* III. Heterogeneous accumulation of hybrid incompatibilities, degree of dominance, and implications for Haldane's rule. *Evolution* 57:2580–2598. [370, 379, 383]

Taper, M. L., and T. J. Case. 1985. Quantitative genetic models for the coevolution of character displacement. *Ecology* 66:355–371. [80, 81]

Tarr, C. L., and R. C. Fleischer. 1995. Evolutionary relationships of the Hawaiian honeycreepers (Aves: Drepanidinae). Pp. 147–159 *in* W. Wagner and V. Funk, eds. *Hawaiian Biogeography: Evolution in a Hot-spot Archipelago.* Smithsonian Institution Press, Washington, DC. [22]

Taylor, P. B. 1998. *A Guide to the Rails, Crakes, Gallinules, and Coots of the World.* Yale University Press, New Haven, CT. [71]

Taylor, E. B., J. W. Boughman, M. Groenenboom, M. Sniatynski, D. Schluter, and J. L. Gow. 2006. Speciation in reverse: Morphological and genetic evidence of the collapse of a three-spined stickleback *(Gasterosteus aculeatus)* species pair. *Molecular Ecology* 15:343–355. [398]

Tchernichovski, O., H. Schwabl, and F. Nottebohm. 1998. Context determines the sex appeal of male zebra finch song. *Animal Behaviour* 55:1003–1010. [163]

Tchernichovski, O., P. P. Mitra, T. Lints, and F. Nottebohm. 2001. Dynamics of the vocal imitation process: How a zebra finch learns its song. *Science* 291:2564–2569. [199]

Tebbich, S., M. Taborsky, B. Fessl, and D. Blomqvist. 2001. Do woodpecker finches acquire tool use through social learning? *Proceedings of the Royal Society of London Series B—Biological Sciences* 268:2189–2193. [134]

Temeles, E. 1994. The role of neighbours in territorial systems: when are they 'dear enemies'? *Animal Behaviour* 47:339–350. [284]

Templeton, A. R. 1980. The theory of speciation *via* the founder principle. *Genetics* 94:1011–1038. [49]

ten Cate, C., and P. Bateson. 1989. Sexual imprinting and a preference for supernormal partners in Japanese quail. *Animal Behaviour* 38:356–357. [277, 292]

ten Cate, C., and D. R. Vos. 1999. Sexual imprinting and evolutionary processes in birds: A reassessment. *Advances in the Study of Behavior* 28:1–31. [276]

ten Cate, C., D. R. Vos, and N. Mann. 1993. Sexual imprinting and song learning—2 of one kind. *Netherlands Journal of Zoology* 43:34-45. [281]

ten Cate, C., M. N. Verzijden, and E. Etman. 2006. Sexual imprinting can induce sexual preferences for exaggerated parental traits. *Current Biology* 16:1128–1132. [174, 175, 275, 276]

Terborgh, J. 1977. Bird species diversity on an Andean elevational gradient. *Ecology* 58:1007–1019. [99]

Terborgh, J. 1980. Causes of tropical species diversity. *Proceedings of the XVII International Ornithological Congress:* 955–961. [98]

Terborgh, J., and J. S. Weske. 1975. Role of competition in the distribution of Andean birds. *Ecology* 56: 562–576. [50]

Terrill, S. B., and P. Berthold. 1989. Experimental evidence for endogenously programmed differential migration in the blackcap *(Sylvia atricapilla). Experientia* 45:207–209. [94]

Thielcke, G. 1969. Die Reaktion von Tannen- und Kohlmeise *(Parus ater, P. major)* auf den Gesang nahverwandter Formen. *Journal für Ornithologie* 110:148–157. [195]

Thielcke, G. 1970. Lernen von Gesang als möglicher Schrittmacher der Evolution. *Zeitschrift für Zoologische Systematik und Evolutionsforschung* 8:309–320. [201, 207]

Thielcke, G. 1973a. Uniformierung des Gesangs der Tannenmeise *(Parus ater)* durch Lernen. *Journal für Ornithologie* 114:443–454. [201]

Thielcke, G. 1973b. On the origin of divergence of learned signals (songs) in isolated populations. *Ibis* 115: 511–516. [207]

Thielcke, G. 1983. Enstanden Dialekte des Zilpzalps *Phylloscopus collybita* durch Lernetzug? *Journal für Ornithologie* 124:333–368. [200, 207]

Thielcke, G., and K. Wüstenberg. 1985. Experiments on the origin of dialects in the short-toed tree-creeper *(Certhia brachydactyla). Behavioral Ecology and Sociobiology* 16:195–201. [190, 200, 201, 207]

Thielcke, G., and U. Zimmer. 1986. Early experience determines the song of the chiff-chaff. *Ethology* 73: 191–196. [207]

Thielcke, G., K. Wüstenberg, and P. H. Becker. 1978. Reaktionen von Zilpzalp und Fitis *(Phylloscopus collybita, Ph. trochilus)* auf verschiedene Gesangsformen des Zilpzalps. *Journal für Ornithologie* 119:213–226. [200]

Thomas, C. D., and J. J. Lennon. 1999. Birds extend their ranges northwards. *Nature* 399:213–213. [342]

Thompson, W. L. 1970. Song variation in a population of indigo buntings. *Auk* 87:58–71. [215]

Thompson, W. L. 1976. Vocalizations of the lazuli bunting. *Condor* 78:195–207. [215–216]

Thorneycroft, H. B. 1975. Cytogenetic study of the white-throated sparrow, *Zonotrichia albicollis* (Gmelin). *Evolution* 29:611–621. [388, 389]

Thorpe, W. H. 1958a. The learning of song patterns by birds, with especial reference to the song of the chaffinch *Fringilla coelebs*. *Ibis* 100:535–570. [189, 207, 216–217]

Thorpe, W. H. 1958b. Further studies on the process of song learning in the chaffinch *(Fringilla coelebs gengleri)*. *Nature* 182:554–557. [216–217]

Timmermans, S., L. Lefebvre, D. Boire, and P. Basu. 2000. Relative size of the hyperstriatum ventrale is the best predictor of feeding innovation rate in birds. *Brain, Behavior and Evolution* 56:196–203. [129]

Tinbergen, N. 1951. *The Study of Instinct*. Clarendon Press, Oxford. [167, 175, 179]

Tinbergen, N. 1953. *The Herring Gull's World*. Collins, London. [175, 284]

Tinbergen, N. 1959. Comparative Study of the Behaviour of Gulls (Laridae): A Progress Report. *Behaviour* 15:1–70. [175]

Titus, R. C. 1998. Short-range and long-range songs: Use of two acoustically distinct song classes by dark-eyed juncos. *Auk* 115:386–393. [200]

Todt, D., and M. Naguib. 2000. Vocal interactions in birds: The use of song as a model in communication. *Advances in the Study of Behaviour* 29:257–296. [205]

Tomback, D. F., D. B. Thompson, and M. C. Baker. 1983. Dialect discrimination by white crowned sparrows—reactions to near and distant dialects. *Auk* 100:452–460. [190]

Tomialojic, L. 1992. Colonization of dry habitats by the song thrush *Turdus philomelos:* Is the type of nest material an important constraint? *Bulletin of the British Ornithologists' Club* 112:27–34. [138]

Tonnis, B., P. R. Grant, B. R. Grant, and K. Petren. 2005. Habitat selection and ecological speciation in Galápagos warbler finches *(Certhidea olivacea* and *Certhidea fusca)*. *Proceedings of the Royal Society B—Biological Sciences* 272:819–826. [28, 132]

Torres, R., and A. Velando. 2003. A dynamic trait affects continuous pair assessment in the blue-footed booby, *Sula nebouxii*. *Behavioral Ecology and Sociobiology* 55:65–72. [162]

Trainer, J. M. 1983. Changes in song dialect distributions and microgeographic variation in song of white-crowned sparrows *(Zonotrichia leucophrys nuttalli)*. *Auk* 100:568–582. [307]

Trainer, J. M. 1989. Cultural evolution in song dialects of yellow-rumped caciques in Panama. *Ethology* 80:190–204. [193]

Tramontin, A. D., J. C. Wingfield, and E. A. Brenowitz. 1999. Contributions of social cues and photoperiod to seasonal plasticity in the adult avian song control system. *Journal of Neuroscience* 19:476–483. [164]

Trewick, S. A. 1997a. Flightlessness and phylogeny amongst endemic rails (Aves: Rallidae) of the New Zealand region. Philosophical Transactions of the *Royal Society of London, Series B Biological Sciences* 352:429–446. [71]

Trewick, S. A. 1997b. Sympatric flightless rails *Gallirallus dieffenbachii* and *G. modestus* on the Chatham Islands, New Zealand; morphometrics and alternative evolutionary scenarios. *Journal of the Royal Society of New Zealand*. 27:451–464. [68]

Tubaro, P. L., and D. A. Lijtmaer. 2002. Hybridization patterns and the evolution of reproductive isolation in ducks. *Biological Journal of the Linnean Society* 77:193–200. [376]

Tubaro, P. L., and B. Mahler. 1998. Acoustic frequencies and body mass in New World doves. *Condor* 100:54–61. [254]

Tubaro, P. L., and E. T. Segura. 1994. Dialect differences in the song of *Zonotrichia capensis* in the southern pampas: A test of the acoustic adaptation hypothesis. *Condor* 96:1084–1088. [258, 307]

Turelli, M., and H. A. Orr. 2000. Dominance, epistasis and the genetics of postzygotic isolation. *Genetics* 154:1663–1679. [367, 376, 382, 383, 390]

Turelli, M., N. H. Barton, and J. A. Coyne. 2001. Theory and speciation. *Trends in Ecology and Evolution* 16:330–343. [67]

Tuttle, E. M. 2003. Alternative reproductive strategies in the white-throated sparrow: Behavioral and genetic evidence. *Behavioral Ecology* 14:425–432. [388]

Utescher, T., V. Mosbrugger, and A. R. Ashraf. 2000. Terrestrial climate evolution in northwest Germany over the last 25 million years. *Palaios* 15:430–449. [100]

Uy, J. A. C., and G. Borgia. 2000. Sexual selection drives rapid divergence in bowerbird display traits. *Evolution* 54:273–278. [184, 185]

Vallet, E., I. Beme, and M. Kreutzer. 1998. Two-note syllables in canary songs elicit high levels of sexual display. *Animal Behaviour* 55:291–297. [163, 205]

van Rhijn, J. G., and R. Vodegel. 1980. Being honest about one's intentions: An evolutionary stable strategy for animal conflicts. *Journal of Theoretical Biology* 85:623–641. [283, 284]

van Riper III, C., S. G. van Riper, M. L. Goff, and M. Laird. 1986. The epizootiology and ecological significance of malaria in Hawaiian land birds. *Ecological Monographs* 56:327–344. [147]

Vaurie, C. 1957. Systematic notes on Palaearctic birds, no. 26, Paridae. *American Museum Novitates* 1833:1–15. [311]

Veen, T., T. Borge, S. C. Griffith, G. P. Sætre, S. Bures, L. Gustafsson, and B. C. Sheldon. 2001. Hybridization and adaptive mate choice in flycatchers. *Nature* 411:45–50. [309, 312, 339, 345, 356, 390]

Veen, T., N. Svedin, J. T. Forsman, M. B. Hjernqvist, A. Qvarnström, K. A. T. Hjernqvist, J. Träff, and M. Klaasen. 2007. Does migration of hybrids contribute to post-zygotic isolation in flycatchers? *Proceedings of the Royal Society, Series B—Biological Sciences* 274:707–712. [339]

Vehrencamp, S. L. 2001. Is song-type matching a conventional signal of aggressive intentions? *Proceedings of the Royal Society of London Series B—Biological Sciences* 268:1637–1642. [202]

Verner, J. 1964. Evolution of polygamy in the long-billed marsh wren. *Evolution* 18:252–261. [244, 248, 265]

Verzijden, M. N., E. Etman, C. van Heijningen, M. van der Linden, and C. ten Cate. 2007. Song discrimination learning in zebra finches induces highly divergent responses to novel songs. *Proceedings of the Royal Society B—Biological Sciences* 274:295–301. [205]

Voelker, G. 1999. Molecular evolutionary relationships in the avian genus *Anthus* (Pipits: Motacillidae). *Molecular Phylogenetics and Evolution* 11:84–94. [102]

Vos, D. R., J. Prijs, and C. ten Cate. 1993. Sexual imprinting in zebra finch males: A differential effect of successive and simultaneous experience with two colour morphs. *Behaviour* 128:1–14. [275]

Vriends, M. M. 1992. *The New Canary Handbook*. Barron's Educational Series, New York. [135, 222]

Vuilleumier, F. 1985. Forest birds of Patagonia: Ecological geography, speciation, endemism, and faunal history. *Ornithological Monographs* 36:255–304. [26]

Waas, J. R. 1990. Intraspecific variation in social repertoires: Evidence from cave- and burrow-dwelling little blue penguins. *Behaviour* 115:63–99. [168]

Wachtmeister, C. A. 2001. Display in monogamous pairs: A review of empirical data and evolutionary explanations. *Animal Behaviour* 61:861–868. [158, 161, 165]

Waddington, C. H. 1961. Genetic assimilation. *Advances in Genetics* 10:257–293. [133, 405]

Wagner, H. O. 1944. Notes on the life history of the emerald toucanet. *Wilson Bulletin* 56:65–76. [187]

Wake, D. B. 1991. Homoplasy: The result of natural selection, or evidence of design limitations. *American Naturalist* 138:543–567. [232]

Wallace, A. R. 1858. On the tendency of varieties to depart indefinitely from the original type. *Journal of the Proceedings of the Linnean Society: Zoology* 3:53–62. [232]

Wallace, A. R. 1891. *Natural Selection and Tropical Nature*. Macmillan, London and New York. [99, 261]

Wallschläger, D. 1980. Correlation of song frequency and body weight in passerine birds. *Experientia* 36:412. [252, 254]

Walther, B. A., D. H. Clayton, and R. D. Gregory. 1999. Showiness of Neotropical birds in relation to ectoparasite abundance and foraging stratum. *Oikos* 87:157–165. [260]

Warren, B. H., E. Bermingham, R. P. Prys-Jones, and C. Thebaud. 2005. Tracking island colonization history and phenotypic shifts in Indian Ocean bulbuls *(Hypsipetes:* Pycnonotidae). *Biological Journal of the Linnean Society* 85:271–287. [151, 233]

Warriner, C. C., W. B. Lemmon, and T. S. Ray. 1963. Early experience as a variable in mate selection. *Animal Behaviour* 11:221–224. [277]

Weir, J. T. 2006. Divergent patterns of species accumulation in lowland and highland neotropical birds. *Evolution* 60:842–855. [18, 38, 39, 40, 106, 107, 118, 119, 122]

Weir, J. T., and D. Schluter. 2004. Ice sheets promote speciation in boreal birds. *Proceedings of the Royal Society of London Series B—Biological Sciences* 271:1881–1887. [28, 39, 325, 335]

Weir, J. T., and D. Schluter. 2007. The latitudinal gradient in recent speciation and extinction rates. *Science* 315:1574–1576. [18, 25, 31, 32]

Wells, S., R. A. Bradley, and L. F. Baptista. 1978. Hybridization in *Calypte* hummingbirds. *Auk* 95:537–549. [317]

Werner, T. K., and T. W. Sherry. 1987. Behavioral feeding specialization in *Pinaroloxias inornata*, the Darwin's finch of Cocos Island, Costa Rica. *Proceedings of the National Academy of Sciences of the United States of America* 84:5506–5510. [19]

West, D. A. 1962. Hybridization in grosbeaks *(Pheucticus)* of the Great Plains. *Auk* 79:399–424. [364]

West-Eberhard, M. J. 1983. Sexual selection, social competition and speciation. *Quarterly Review of Biology* 58:155–183. [157, 264, 270]

West-Eberhard, M. J. 2003. *Developmental Plasticity and Evolution*. Oxford University Press, New York. [127, 135, 232, 409]

Whitfield, D. P. 1986. Plumage variability and territoriality in breeding turnstone *Arenaria interpres:* Status signalling or individual recognition? *Animal Behaviour* 34:1471–1482. [284]

Whitfield, D. P. 1987. Plumage variability, status signalling and individual recognition in avian flocks. *Trends in Ecology and Evolution* 2:13–18. [284]

Whitfield, D. P. 1988. The social significance of plumage variability in wintering turnstone *Arenaria interpres*. *Animal Behaviour* 36:408–415. [283, 284]

Whitlock, M. C. 1997. Founder effects and peak shifts without genetic drift: Adaptive peak shifts occur easily when environments fluctuate slightly. *Evolution* 51:1044–1048. [49]

Wiebe, K. L. 2000. Assortative mating by color in a population of hybrid northern flickers. *Auk* 117:525–529. [352]

Wienberg, J., and R. Stanyon. 1995. Chromosome painting in mammals as an approach to comparative genomics. *Current Opinion in Genetics and Development* 5:792–797. [384]

Wiens, J. J. 2004a. What is speciation and how should we study it? *American Naturalist* 163:914–923. [6, 73]

Wiens, J. J. 2004b. Speciation and ecology revisited: Phylogenetic niche conservatism and the origin of species. *Evolution* 58:193–197. [13, 15]

Wiles, G. J., J. Bart, R. E. Beck, and C. F. Aguon. 2003. Impacts of the brown tree snake: Patterns of decline and species persistence in Guam's avifauna. *Conservation Biology* 17:1350–1360. [137]

Wiley, R. H. 1973. The strut display of male sage grouse: A "fixed" action pattern. *Behaviour* 47:129–152. [160]

Wiley, R. H. 1991. Associations of song properties with habitats for territorial oscine birds of eastern North America. *American Naturalist* 138:973–993. [254, 256]

Wiley, R. H., and D. G. Richards. 1982. Adaptation for acoustic communication in birds: Sound transmission and signal detection. Pp. 131–181 *in* D. E. Kroodsma and E. H. Miller, eds. *Acoustic Communication in Birds*. Academic Press, New York. [256]

Wiley, C., N. Fogelberg, S. A. Sæther, T. Veen, N. Svedin, J. Vogel-Kehlenbeck, and A. Qvarnström. 2007. Direct benefits for hybridizing *Ficedula* flycatchers. *Journal of Evolutionary Biology*. In press. [312]

Williams, D. 1983. Mate choice in the mallard. Pp. 297–309 *in* P. Bateson, ed. *Mate Choice*. Cambridge University Press, Cambridge. [292]

Willis, E. O. 1974. Populations and local extinctions of birds on either side of Barro Colorado Island, Panama. *Ecological Monographs* 44:153–169. [70]

Wilson, E. O. 1961. The nature of the taxon cycle in the Melanesian ant fauna. *American Naturalist* 95:169–193. [147]

Wilson, J. D. 1992. Correlates of agonistic display by great tits, *Parus major*. *Behaviour* 121:168–214. [167, 177, 283]

Wilson, E. O., and W. L. Brown. 1953. The subspecies concept and its taxonomic application. *Systematic Zoology* 2:97–111. [3]

Wilson, D. S., and A. Hedrick. 1982. Speciation and the economics of mate choice. *Evolutionary Theory* 6:15–24. [304, 305, 408]

Witherby, H. F., F. C. R. Jourdain, N. F. Ticehurst, and B. W. Tucker. 1938. *The Handbook of British Birds,* Vol. 1. Witherby, London. [216]

Witte, K., U. Hirschler, and E. Curio. 2000. Sexual imprinting on a novel adornment influences mate preferences in the Javanese mannikin *Lonchura leucogastoides*. *Ethology* 106:349–363. [291, 292]

Woodruff, D. S. 2003. Neogene marine transgressions, palaeogeography and biogeographic transitions on the Thai–Malay Peninsula. *Journal of Biogeography* 30:551–567. [346]

Woolfit, M., and L. Bromham. 2005. Population size and molecular evolution on islands. *Proceedings of the Royal Society of London Series B—Biological Sciences* 272:2277–2282. [36]

Wright, T. F., and M. Dorin. 2001. Pair duets in the yellow-naped amazon (Psittaciformes: *Amazona auropalliata):* Responses to playbacks of different dialects. *Ethology* 107:111–124. [190, 193]

Wu, C. I. 1992. A note on Haldane's rule—hybrid inviability versus hybrid sterility. *Evolution* 46:1584–1587. [375]

Wu, C. I. 2001. The genic view of the process of speciation. *Journal of Evolutionary Biology* 14:851–865. [346, 391]

Wu, C. I., and A. W. Davis. 1993. Evolution of postmating reproductive isolation—the composite nature of Haldane's rule and its genetic basis. *American Naturalist* 142:187–212. [376, 379]

Wyles, J. S., J. G. Kunkel, and A. C. Wilson. 1983. Birds, behavior, and anatomical evolution. *Proceedings of the National Academy of Sciences (USA)* 80:4394–4397. [133]

Yamamoto, J. T., K. M. Shields, J. R. Millam, T. E. Roudybush, and C. R. Grau. 1989. Reproductive activity of force-paired cockatiels. *Auk* 106:86–93. [284]

Yang, S. Y., and R. K. Selander. 1968. Hybridization in the grackle *Quiscalus quiscula* in Louisiana. *Systematic Zoology* 17:107–143. [366]

Yang, Z. H. 1996. Among-site rate variation and its impact on phylogenetic analyses. *Trends in Ecology and Evolution* 11:367–372. [36, 39]

Zachos, J., M. Pagani, L. Sloan, E. Thomas, and K. Billups. 2001. Trends, rhythms, and aberrations in global climate 65 Ma to present. *Science* 292:686–693. [100]

Zahavi, A. 1975. Mate selection—a selection for a handicap. *Journal of Theoretical Biology* 53:205–214. [180]

Zann, R. A. 1996. *The zebra finch: A Synthesis of Field and Laboratory Studies*. Oxford University Press, Oxford and New York. [173, 222, 226, 227, 273]

Zeh, D. W., and J. A. Zeh. 1994. When morphology misleads—interpopulation uniformity in sexual selection masks genetic divergence in harlequin beetle-riding pseudoscorpion populations. *Evolution* 48: 1168–1182. [397]

Zink, R. M. 2002. A new perspective on the evolutionary history of Darwin's finches. *Auk* 119:864–871. [22]

Zink, R. M. 2004. The role of subspecies in obscuring avian biological diversity and misleading conservation policy. *Proceedings of the Royal Society of London Series B—Biological Sciences* 271:561–564. [3, 43]

Zink, R. M. 2006. Rigor and species concepts. *Auk* 123:887-891. [6]

Zink, R. M., and R. C. Blackwell. 1998. Molecular systematics of the scaled quail complex (genus *Callipepla*). *Auk* 115:394–403. [351]

Zink, R. M., and D. L. Dittmann. 1993. Gene flow, refugia, and evolution of geographic variation in the song sparrow *(Melospiza melodia)*. *Evolution* 47:717–729. [41, 43]

Zink, R. M., and J. V. Remsen. 1986. Evolutionary processes and patterns of geographic variation in birds. Pp. 1–69 *in* R. F. Johnston, ed. *Current Ornithology*. Plenum Press, New York. [41]

Zink, R. M., D. L. Dittmann, and W. L. Rootes. 1991. Mitochondrial DNA variation and the phylogeny of *Zonotrichia. Auk* 108:578–584. [258]

Zink, R. M., J. Klicka, and B. R. Barber. 2004. The tempo of avian diversification during the Quaternary. *Philosophical Transactions of the Royal Society of London Series B—Biological Sciences* 359:215–219. [107, 122]

Zuk, M., R. Thornhill, J. D. Ligon, and K. Johnson. 1990. Parasites and mate choice in red jungle fowl. *American Zoologist* 30:235–244. [180, 248]

Index of Common Names

Boldface indicates an illustration.

Akiapōlā'lau *Hemignathus munroi,* 112, **113**

Adjutant, Lesser *Leptoptilos javanicus,* 106

Antshrike, Variable *Thamnophilus caerulescens,* 326, 354

Avadavat, Red *Amandava amandava,* 387

Bellbird, Bare-throated *Procnias nudicollis,* **159**
    Bearded *Procnias averano,* **159**
    Three-wattled *Procnias tricarunculata,* **159**
    White *Procnias alba,* **159**

Blackbird, Eurasian *Turdus merula,* 42, 44, 130–132, 138, 189
    Red-winged *Agelaius phoeniceus,* 45, 177, 199, 251
    Yellow-headed *Xanthocephalus xanthocephalus,* 177

Blackcap *Sylvia atricapilla,* 44–46, 60, 92–94, 286–287

Bleeding-heart, Luzon *Gallicolumba luzonica,* 230

Bluebird, Western *Sialia mexicana,* 344

Bobwhite, Crested *Colinus cristatus,* 231
    Northern *Colinus virginianus,* 231, **318**

Booby, Blue-footed *Sula nebouxii,* 161
    Masked *Sula dactylatra,* 126

Bowerbird, Fawn-breasted *Chlamydera cerviniventris,* 185
    Great *Chlamydera nuchalis,* 185
    Satin *Ptilonorhynchus violaceus,* 168, 254
    Vogelkop *Amblyornis inornatus,* 184, 185

Budgerigar *Melopsittacus undulatus,* 161, 222–227

Bulbul, Garden *Pycnonotus barbatus,* 327, 359
    Himalayan *Pycnonotus leucogenys,* 359
    Red-eyed *Pycnonotus nigricans,* 327, 359
    Red-vented *Pycnonotus cafer,* 327, 359
    White-eared *Pycnonotus leucotis,* 327, 359

Bullfinch, Eurasian *Pyrrhula pyrrhula,* 372
    Lesser Antillean *Loxigilla noctis* **150**, 153

Bunting, Black-headed *Emberiza melanocephala,* 328
    Corn *Emberiza calandra,* 190
    Indigo *Passerina cyanea,* 188, 189, 192, 202, 204, 215–216, 286, 288, 307, 319, 328, 365
    Lazuli *Passerina amoena,* 215–216, 286, 288, 307, 308, 319, 328, 365
    Pine *Emberiza leucocephalos,* 328, 347, 362
    Red-headed *Emberiza bruniceps,* 328

Cacique, Yellow-rumped *Cacicus cela,* 193

Canary *Serinus canaria,* 162, 221–224, 227, 232, 280, 372, 381
    Island (see Canary)

Cardinal, Northern *Cardinalis cardinalis,* 191

Catbird, Gray *Dumetella carolinensis,* 201

Chaffinch *Fringilla coelebs* 177, 189, 203–204, 216–217, 265, 266, 268
    Blue *Fringilla teydea,* 189, 207, 217

Chickadee, Black-capped *Poecile atricapillus,* 195, 266, 327, 335, 340, 345, 357
    Carolina *Poecile carolinensis,* 43, 327, 335, 340, 345, 357

Chicken (*see* Junglefowl, Red)
    Greater Prairie- *Tympanuchus cupido,* 319

Chiffchaff *Phylloscopus collybita* **197**, 200, 207,208, 307, 327, 361, 391
    Common (*see* Chiffchaff)
    Iberian *Phylloscopus ibericus,* 197, 307, 327, 361, 391

Chingolo (see Sparrow, Rufous-collared)

Condor, California *Gymnogyps californianus,* 277

Cordonbleu, Red-cheeked *Uraeginthus bengalus,* 387

Cowbird, Brown-headed *Molothrus ater,* 191, 202, 207, 279–280

Creeper, Brown *Certhia americana,* 266
    Short-toed Tree- *Certhia brachydactyla,* 190, 200, 201, 207

Crossbill, Red *Loxia curvirostra,* 47, 69, 77, 78, **79**, 80, 99, 177

Crow, Carrion *Corvus corone,* 235, 327, 338, 339, 342, 354, 390
    Hooded *Corvus corone,* 235, 327, 338, 339, 342, 354, 390

Cuckoo, Great Spotted *Clamator glandarius,* 280

Curlew, Stone *Burhinus oedicnemus,* 384

Cut-throat *Amadina fasciata,* 387

Dodo *Raphus cucullatus,* 71, 299

Dove, African Collared- *Streptopelia roseogrisea,* 161, 163, 165, 167, 284, 381
    Bare-faced Ground- *Metriopelia ceciliae,* 230
    Carunculated Fruit- *Ptilinopus granulifrons,* 230
    Cloven-feathered *Drepanoptila holosericea,* 230
    Crested Quail- *Geotrygon versicolor,* 230
    Diamond *Geopelia cuneata,* 230
    European Turtle- *Streptopelia turtur,* 230
    Grey-chested *Leptotila cassini,* 230
    Jambu Fruit- *Ptilinopus jambu,* 230
    Orange *Ptilinopus victor,* 230
    Ring (see Dove, African Collared-), 161, 163, 165, 167, 284, 381
    Ring-necked *Streptopelia capicola,* 326, 353
    Rock *Columba livia,* 128, 161, 165, **220**–223, 226–230, 232, 277, 381
    Spotted *Streptopelia chinensis,* 381
    Stephan's *Chalcophaps stephani,* 230

Tambourine *Turtur tympanistria,* 230
Vinaceous *Streptopelia vinacea,* 326, 353
Duck, Falcated *Anas falcata,* 376
 Mandarin *Aix galericulata,* 176, 178, **179**
 Muscovy *Cairina moschata,* 222, 223, 226,
  372–373, 381
 Spot-billed *Anas poecilorhyncha,* 231
 Tufted *Aythya fuligula,* 231
 Wood *Aix sponsa,* 176, 178
Emu *Dromaius novaehollandiae,* 129, 130, 384
Fairywren, Variegated *Malurus lamberti,* 326, 354
Figbird, Timor *Sphecotheres viridis,* 327, 356
Finch, Bengalese *Lonchura striata,* 161, 205, 224,
  277–278, 293
 Black-throated *Poephila cincta,* 328, 361, 387
 Cassin's *Carpodacus cassinii,* 177
 Cocos *Pinaroloxias inornata,* 19, **23**
 Common Cactus- *Geospiza scandens,* **23**, 127,
  306, 324, 397
 Crimson *Neochmia phaeton,* 387
 Cuckoo *Anomalospiza imberbis,* 295
 Double-barred *Taeniopygia bichenovii,* 173, 174,
  387
 Green Warbler- *Certhidea olivacea,* **23**
 Grey Warbler- *Certhidea fusca,* **23**
 House *Carpodacus mexicanus,* 47, **136**, 137, 177,
  188, 193, 206, 266, 268
 Large Cactus- *Geospiza conirostris,* **23**, 262
 Large Ground- *Geospiza magnirostris,* 22, **23**, 24,
  53, 76, 90
 Large Tree- *Camarhynchus psittacula,* **23**
 Laysan *Telespiza cantans,* 47
 Long-tailed *Poephila acuticauda,* 173, 174
 Mangrove *Camarhynchus heliobates,* **23**
 Medium Ground- *Geospiza fortis,* 22, **23**, 24, 47,
  53, 72, 78, 82–84, 90, **91**, 253, 306, 316, 324,
  397
 Medium Tree- *Camarhynchus pauper,* **23**
 Painted *Emblema pictum,* 387
 Plum-headed *Neochmia modesta,* 387
 Red-headed *Amadina erythrocephala,* 387
 Sharp-beaked Ground- *Geospiza difficilis,* **23**,
  126, 191
 Small Ground- *Geospiza fuliginosa,* 22, **23**, 24,
  53, 54, 72, 76, 78, 79, 82–84, 90, 91
 Small Tree- *Camarhynchus parvulus,* **23**
 Star *Neochmia ruficauda,* 387
 Vegetarian *Camarhynchus crassirostris,* **23**
 Woodpecker *Camarhynchus pallidus,* 22, **23**, 24,
  112–**113**, 134, 135, 137, 140
 Zebra *Taeniopygia guttata,* 45, 162, 164,
  173–175, 180, 205, 222–224, 226, 227,
  273–278, 281, 284, 292, 318, 319, 386–388,
  392
Firecrest *Regulus ignicapillus,* 190, 266
Firefinch, Jameson's *Lagonosticta rhodopareia,* **296**
 Red-billed *Lagonosticta senegala,* 293–294, 387
Firetail, Diamond *Stagonopleura guttata,* 231, 387
 Painted (*see* Finch, Painted), 387

Flicker, Northern *Colaptes auratus,* 319, 326, 337,
  347, 352
 Red-shafted *Colaptes auratus,* 337
 Yellow-shafted *Colaptes auratus,* 337
Flycatcher, African Paradise- *Terpsiphone viridis,*
  310, 347
 Collared *Ficedula albicollis,* 292, 307, 308,
  309–315, 327, 339, 344–345, 356, 390
 Dusky-capped *Myiarchus tuberculifer,* 253
 European Pied (see Flycatcher, Pied)
 Pied *Ficedula hypoleuca,* 163, 279, 307, 308,
  309–315, 327, 336, 339, 344–345, 356, 390
Gadwall *Anas strepera,* 231, 376
Garganey *Anas querquedula,* 176, **179**
Gerygone, Western *Gerygone fusca,* 268
Goldcrest *Regulus regulus,* 190, 266
Goldfinch, American *Carduelis tristis,* 177
 European *Carduelis carduelis,* 381
Goose, Canada *Branta canadensis,* 42, 44
 Greylag *Anser anser,* 224
 Snow *Anser caerulescens,* 299
Grackle, Common *Quiscalus quiscula,* 328, 366
Grebe, Clark's *Aechmophorus clarkii,* 310, 319
 Great Crested *Podiceps cristatus,* **166**
 Junin *Podiceps taczanowskii,* 19, 48, **52**
 Silvery *Podiceps occipitalis,* 19, 46, 48, **52**
 Western *Aechmophorus occidentalis,* 310, 319
Greenbul, Little *Andropadus virens,* 63
Greenfinch *Carduelis chloris,* 47, 222–224
 European (see Greenfinch)
Grenadier, Purple *Uraeginthus ianthinogaster,* **296**
Grosbeak, Black-headed *Pheucticus melanocephalus,*
  177, 328, 364
 Pine *Pinicola enucleator,* 190
 Rose-breasted *Pheucticus ludovicianus,* 328,
  364
Grouse, Greater Sage *Centrocercus urophasianus,*
  160, **161**, 175, 248
 Sharp-tailed *Tympanuchus phasianellus,* 319
Guineafowl, Helmeted *Numida meleagris,* 221,
  377–378, 388
Gull, Glaucous-winged *Larus glaucescens,* 326, 338,
  353
 Herring *Larus argentatus,* 66, 94, 284
 Lesser Black-backed *Larus fuscus,* 94
 Western *Larus occidentalis,* 326, 338, 353
Heron, Fasciated Tiger- *Tigrisoma fasciatum,* 277
 Grey *Ardea cinerea,* 60
Honeyeater, Singing *Lichenostomus virescens,* 266
Hornbill, Red-billed *Tockus erthythrorhynchus,* 326,
  352
Huia *Heteralocha acutirostris,* 112, **113**
Hummingbird, Anna's *Calypte anna,* 317
 Costa's *Calypte costae,* 317
I'iwi *Vestiaria coccinea,* 47
Indigobird, Purple *Vidua purpurascens,* **296**
 Village *Vidua chalybeata,* 293–295
Jabiru *Jabiru mycteria,* 106
Jay, Eurasian *Garrulus glandarius,* 49

Junco, Dark-eyed *Junco hyemalis,* 33, 45, 47, 162,
    177, 200, 207, 266, 347, 389
    Guadalupe *Junco insularis,* 200, 267
    Yellow-eyed *Junco phaeonotus,* 33
Junglefowl, Grey *Gallus sonneratii,* 379
    Red *Gallus gallus,* 161, 180, 221–223, 225, 226
    377, 379, 382, 383–386, 388
Kingfisher, Common *Alcedo atthis,* 384
    Common Paradise- *Tanysiptera galatea,* 49
Kite, Black *Milvus migrans,* 127
    Black-winged *Elanus caeruleus,* 384
Lory, Cardinal *Chalcopsitta cardinalis,* 63
Lovebird, Fischer's *Agapornis fischeri,* 390–391
    Rosy-faced *Agapornis roseicollis,* 222, 223, 226,
    390–391
Lyrebird, Superb *Menura novaehollandiae,* 259
Magpie, Australian *Gymnorhina tibicen,* 327, 355
    Black-billed *Pica pica,* 280
Mallard *Anas platyrhynchos,* 44, 161, 176, 178, **179**,
    222, 223, 226, 231, 277, 278, 372–373, 379,
    381, 390, 392
Manakin, Golden-collared *Manacus vitellinus,* 326,
    338, 340–342, 353
    White-collared *Manacus candei,* 326, **338**,
    340–342, 346, 353
Meadowlark, Eastern *Sturnella magna,* 25, **27**, 114,
    287–289, 306, 390
    Western *Sturnella neglecta,* 25, **27**, 114, 287–289,
    306, 390
Metaltail, Coppery *Metallura theresiae,* 118
    Fire-throated *Metallura eopogon,* 118
    Neblina *Metallura odomae,* **118**
    Scaled *Metallura aeneocauda,* 118
    Tyrian *mtallura tyrianthina,* **118**
    Violet-throated *Metallura baroni,* 118
    Viridian *Metallura williami,* **118**
Mockingbird, Northern *Mimus polyglottos,* 161
Monarch, Black-winged *Monarcha frater,* 7
    Bougainville *Monarcha erythrostictus,* **5**
    Chestnut-bellied *Monarcha castaneiventris,* **5**, 6,
    7, 12, 409
    Island *Monarcha cinerascens,* 409
    White-capped *Monarcha richardsii,* **5**, 6, **7**
Munia, Chestnut-breasted *Lonchura castaneothorax,*
    227, 231, 381, 387
    Javan *Lonchura leucogastroides,* **290**–292, 297
    Scaly-breasted *Lonchura punctulata,* 231, 381
    White-rumped (*see* Finch, Bengalese), 161, 205,
    224, 277–278, 293
Myna, Common *Acridotheres tristis,* 46
    Hill *Gracula religiosa,* 190
Myzomela, Cardinal *Myzomela cardinalis,* 311
    Ebony *Myzomela pammelaena,* 46
    Scarlet-bibbed *Myzomela sclateri,* 46
    Sooty *Myzomela tristrami,* 311
Nightingale, Common *Luscinia megarhynchos,* 205,
    306, 327, 357
    Thrush *Luscinia luscinia,* 306, 327, 357
Oriole, Altamira *Icterus gularis,* **234**

Baltimore *Icterus galbula,* 95, **234**, 328, 365
    Bullock's *Icterus bullockii,* 95, 328, 365
    Hooded *Icterus cucullatus,* **234**
    Orchard *Icterus spurius,* **234**
Ostrich *Struthio camelus,* 277, 384, 388
Owl, Barn *Tyto alba,* 59
Parrot, Yellow-naped *Amazona auropalliata,* 190
Parrotfinch, Blue-faced *Erythrura trichroa,* 387
Partridge, Red-legged *Alectoris rufa,* 325, 350
    Rock *Alectoris graeca,* 325, 350
Peafowl, Indian *Pavo cristatus,* 161
Petrel, Band-Rumped Storm- *Oceanodroma castro,*
    21, 396
Pheasant, Common *Phasianus colchicus,* 221, 226,
    375, 381
    Japanese (*see* Pheasant, Common)
    Reeves's *Syrmaticus reevesii,* 375, 381
Pigeon, Bare-eyed *Columba corensis,* 230
    Common (*see* Dove, Rock), 128, 161, 165, 277
    Green Imperial- *Ducula aenea,* 230
    Nicobar *Caloenas nicobarica,* 230
    Passenger *Ectopistes migratorius,* 230
    Pheasant *Otidiphaps nobilis,* 230
    Seychelles Blue- *Alectroenas pulcherrima,* 230
    Snow *Columba leuconota,* 230
    Speckled Wood- *Columba hodgsonii,* 230
    Topknot *Lopholaimus antarcticus,* 230
    White-crowned *Columba leucocephala,* 230
    Wonga *Leucosarcia melanoleuca,* 381
Pintail, Northern *Anas acuta,* 379, 390, 392
Pochard, Common *Aythya ferina,* 231
    Ferruginous *Aythya nyroca,* 231
Pytilia, Green-winged *Pytilia melba,* 386, 387
    Red-winged *Pytilia phoenicoptera,* 386, 387
Quail, Blue-breasted *Coturnix chinensis,* 386
    California *Callipepla californica,* **318**, 325, 351
    Common *Coturnix coturnix,* 379
    Elegant *Callipepla douglasii,* 231
    Gambel's *Callipepla gambelii,* 231, **318**, 325, 351
    Japanese *Coturnix japonica,* 221, 289–290, 292,
    379, 385–386
    Mountain *Oreortyx pictus,* **318**
    Scaled *Callipepla squamata,* 231, **318**
Rail, Buff-banded *Gallirallus philippensis* 71
    Laysan *Porzana palmeri* 71
Redpoll, Common *Carduelis flammea,* 177
Robin, European *Erithacus rubecula,* 158, 266
    Red-capped *Petroica goodenovii,* 266
Rosella, Adelaide *Platycercus elegans,* 346, 347
    Crimson *Platycercus elegans,* 347
    Yellow *Platycercus elegans,* 347
Sapsucker, Red-breasted *Sphyrapicus ruber,* 28, 325,
    351
    Red-naped *Sphyrapicus nuchalis,* 28, 325, 351
    Yellow-bellied *Sphyrapicus varius,* 28
Scaup, Lesser *Aythya affinis,* 231
Seedcracker, Black-bellied *Pyrenestes ostrinus,* 84,
    85, 315
Serin, European *Serinus serinus,* 190

Shoveler, Australian *Anas rhynchotis,* 231
    Northern *Anas clypeata,* 231
    Red *Anas platalea,* 231
Silvereye *Zosterops lateralis,* 46, 47, 125, 311
Siskin, Eurasian *Carduelis spinus,* 138
Sittella, Varied *Daphoenositta chrysoptera,* 327, 354
Sparrow, Eurasian Tree *Passer montanus,* 130, 204, 268
    Field *Spizella pusilla,* 287
    Fox *Passerella iliaca,* 266
    House *Passer domesticus,* 45, 46, 127, 130, 328, 361
    Rufous-collared *Zonotrichia capensis,* **256**–258, 270
    Song *Melospiza melodia,* 41, 43, 190, 191, 199, 202–203, 288
    Swamp *Melospiza georgiana,* 190–192, 199, 205, 211
    White-crowned *Zonotrichia leucophrys,* 163, 190–194, 196, 198, 201, 204, 206, 211, 256
    White-throated *Zonotrichia albicollis,* 388, 389
    Willow *Passer hispaniolensis,* 328, 347, 361
Starling, Common *Sturnus vulgaris,* 46, 177
Stonechat, Common *Saxicola torquata,* 44
Stork, Abdim's *Ciconia abdimii,* 106
    Black *Ciconia nigra,* 106
    Black-necked *Ephippiorhynchus asiaticus,* 106
    Maguari *Ciconia maguari,* 106
    Marabou *Leptoptilos crumeniferus,* 106
    Milky *Mycteria cinerea,* 106
    Oriental *Ciconia boyciana,* 106
    Painted *Mycteria leucocephala,* 106
    Saddle-billed *Ephippiorhynchus senegalensis,* 106
    Storm's *Ciconia stormi,* 106
    White *Ciconia ciconia,* 106
    Wood *Mycteria americana,* 106
    Woolly-necked *Ciconia episcopus,* 106
    Yellow-billed *Mycteria ibis,* 106
Sunbird, Orange-tufted *Nectarinia bouvieri,* 190
Swallow, Barn *Hirundo rustica,* 69, 115, 160, 161, 180
Swiftlet, Mariana *Aerodramus vanikorensis,* 137
Tailorbird, Common *Orthotomus sutorius,* 139
Teal, Baikal *Anas formosa,* 231
    Cinnamon *Anas cyanoptera,* 231
    Common *Anas crecca,* 231
    Speckled *Anas flavirostris,* 178, 231
Thick-knee, Eurasian (*see* Curlew, Stone), 384
Thrasher, Brown *Toxostoma rufum,* 192
Thrush, Song *Turdus philomelos,* 128, 132, 138
Tit, Azure *Cyanistes cyanus,* 311
    Coal *Periparus ater,* 44, 128, 137, 194, **195**, 201, 207, 235
    Crested *Lophophanes cristatus,* 128
    European Blue *Cyanistes caeruleus,* 44, 68, 125, 128, 130, 190, 266, 311
    Great *Parus major,* 44, 58, 59, 66, 68, 130, 132, **167**, 168, 176, 177, 181, 182, 203, 253, 256, 279

Long-tailed *Aegithalos caudatus,* 283
Marsh *Poecile palustris,* 130, 132, 201
Willow *Poecile montanus,* 128, 207
Titmouse, Black-crested *Baeolophus atricristatus,* 327, 359
    Tufted *Baeolophus bicolor,* 327, 359
Towhee, Collared *Pipilo ocai,* 328, 362
    Eastern *Pipilo erythrophthalmus,* 328, 362
    Spotted *Pipilo maculatus,* 328, 362
Turkey *Meleagris gallopavo,* 222, 223
    Austratian Brush- *Alectura lathami,* 280
    Wild (*see* Turkey)
Turnstone, Ruddy *Arenaria interpres,* 284
Wagtail, Pied *Motacilla alba,* 285
    Yellow *Motacilla flava,* 235
Warbler, Adelaide's *Dendroica adelaidae,* **150**
    Black-throated Green *Dendroica virens,* 28
    Blue-winged *Vermivora pinus,* 288, 319, 328, 335, 343, 363
    Buff-barred *Phylloscopus pulcher,* 88, 178
    Dusky *Phylloscopus fuscatus,* 137
    Garden *Sylvia borin,* 286–287
    Golden-spectacled *Seicercus burkii,* **316**
    Golden-winged *Vermivora chrysoptera,* 288, 319, 328, 335, 343, 363
    Great Reed *Acrocephalus arundinaceus,* 68, 251, 252
    Greenish *Phylloscopus trochiloides,* 65, 66, 87, 88, 92, 93, 178, 236–**237**
    Hermit *Dendroica occidentalis,* 28, 320, 328, **330**, 340–341, 343–344, 364
    Hooded *Wilsonia citrina,* 284
    Hume's Leaf- *Phylloscopus humei,* 88, 178
    Icterine *Hippolais icterina,* 327, 359
    Japanese Bush- *Cettia diphone,* 264, 266
    Large-billed Leaf- *Phylloscopus magnirostris,* 88, 178
    Lemon-rumped *Phylloscopus chloronotus,* 88, 178
    Marsh *Acrocephalus palustris,* 251
    Melodious *Hippolais polyglotta,* 327, 359
    Sedge *Acrocephalus schoenobaenus,* 251
    Seychelles *Acrocephalus sechellensis,* 264
    Tickell's Leaf- *Phylloscopus affinis,* 88, 178
    Townsend's *Dendroica townsendi,* 28, 320, 328, **330**, 340–341, 343–344, 364
    Tytler's Leaf- *Phylloscopus tytleri,* 88, 178
    Western Crowned- *Phylloscopus occipitalis,* 87, 88, 178
    Whistler's *Seicercus whistleri,* **316**
    Willow *Phylloscopus trochilus,* 327, 336, 360
    Yellow-rumped *Dendroica coronata,* 328, 363
Waxbill, Common *Estrilda astrild,* 295, 387
    Fawn-breasted *Estrilda paludicola,* 387
Weaver, Parasitic (*see* Finch, Cuckoo)
Wheatear, Black-eared *Oenanthe hispanica,* 327, 357
    Pied *Oenanthe pleschanka,* 327, 357
    Variable *Oenanthe picata,* 347

Whistler, Golden *Pachycephala pectoralis,* **60**, 62, 271, 314, 347
White-eye, Bridled *Zosterops conspicillatus,* 149
    Golden-green *Zosterops nigrorum,* 149
    Slender-billed *Zosterops tenuirostris,* 311
    Yap Olive *Zosterops oleagineus,* 149
Whydah, Straw-tailed *Vidua fischeri,* **296**
Wigeon, Eurasian *Anas penelope,* 176
Woodcreeper, Wedge-billed *Glyphorynchus spirurus,* 31
Woodpecker, Great Spotted *Dendrocopos major,* 311
    Green-backed *Campethera cailliautii,* 326, 336, 352
    Syrian *Dendrocopos syriacus,* 311

Woodswallow, Black-faced *Artamus cinereus,* 327, 356
Wren, Banded *Thryothorus pleurostictus,* 202–203
    Bewick's *Thryomanes bewickii,* 266
    Carolina *Thryothorus ludovicianus,* 286
    House *Troglodytes aedon,* 266
    Marsh *Cistothorus palustris,* 211
    Rufous-naped *Campylorhynchus rufinucha,* 327, 357
    Winter *Troglodytes troglodytes,* 59, 190, 265
Yellowhammer *Emberiza citrinella,* 177, 191, 212, 213, 328, 362

Index of Scientific Names

Boldface indicates an illustration.

Acridotheres tristis Common Myna, 46
Acrocephalus arundinaceus Great Reed Warbler, 68,
    251, 252
    palustris Marsh Warbler, 251
    schoenobaenus Sedge Warbler, 251
    sechellensis Seychelles Warbler, 264
Aechmophorus clarkii Clark's Grebe, 310, 319
    occidentalis Western Grebe, 310, 319
Aegithalos caudatus Long-tailed Tit, 283
Aerodramus vanikorensis Mariana Swiftlet, 137
Agapornis fischeri Fischer's Lovebird, 390–391
    roseicollis Rosy-faced Lovebird, 222, 223, 226,
    390–391
Agelaius phoeniceus Red-winged Blackbird, 45, 177,
    199, 251
Aix galericulata Mandarin Duck, 176, 178, **179**
    sponsa Wood Duck, 176, 178
Alcedo atthis Common Kingfisher, 384
Alectoris graeca Rock Partridge, 325, 350
    rufa Red-legged Partridge, 325, 350
Alectroenas pulcherrima Seychelles Blue-Pigeon, 230
Alectura lathami Australian Brush-turkey, 280
Amadina erythrocephala Red-headed Finch, 387
    fasciata Cut-throat, 387
Amandava amandava Red Avadavat, 387
Amazona auropalliata Yellow-naped Parrot, 190
Amblyornis inornatus Vogelkop Bowerbird, 184, 185
Anas acuta Northern Pintail, 379, 390, 392
    clypeata Northern Shoveler, 231
    crecca Common Teal, 178, 251
    cyanoptera Cinnamon Teal, 231
    falcata Falcated Duck, 376
    flavirostris Speckled Teal, 231
    formosa Baikal Teal, 231
    penelope Eurasian Wigeon, 176
    platalea Red Shoveler, 231
    platyrhynchos Mallard, 44, 161, 176, 178, **179**,
    222, 223, 226, 231, 277, 278, 372, 373, 379,
    381, 390, 392
    poecilorhyncha Spot-billed Duck, 231
    querquedula Garganey, 176, **179**
    rhynchotis Australian Shoveler, 231
    strepera Gadwall, 231, 376
Andropadus virens Little Greenbul, 63
Anomalospiza imberbis Cuckoo Finch, or Parasitic
    Weaver, 295
Anser anser Greylag Goose, 224
    caerulescens Snow Goose, 299
Ardea cinerea Grey Heron, 60
Arenaria interpres Ruddy Turnstone, 284

Artamus cinereus Black-faced Woodswallow, 327,
    356
Aythya affinis Lesser Scaup, 231
    ferina Common Pochard, 231
    fuligula Tufted Duck, 231
    nyroca Ferruginous Pochard, 231
Baeolophus atricristatus Black-crested Titmouse,
    327, 359
    bicolor Tufted Titmouse, 327, 359
Branta canadensis Canada Goose, 42, 44
Burhinus oedicnemus Stone Curlew, or Eurasian
    Thick-knee, 384
Cacicus cela Yellow-rumped Cacique, 193
Cairina moschata Muscovy Duck, 222, 223, 226,
    372–373, 381
Callipepla californica California Quail, **318**, 325, 351
    douglasii Elegant Quail, 231
    gambelii Gambel's Quail, 231, **318**, 325, 351
    squamata Scaled Quail, 231, **318**
Caloenas nicobarica Nicobar Pigeon, 230
Calypte anna Anna's Hummingbird, 317
    costae Costa's Hummingbird, 317
Camarhynchus crassirostris Vegetarian Finch, **23**
    heliobates Mangrove Finch, **23**
    pallidus Woodpecker Finch, 22, **23**, 24, 112, **113**,
    134, 135, 137, 140
    parvulus Small Tree-Finch, **23**
    pauper Medium Tree-Finch, **23**
    psittacula Large Tree-Finch, **23**
Campethera cailliautii Green-backed Woodpecker,
    326, 336, 352
Campylorhynchus rufinucha Rufous-naped Wren,
    327, 357
Cardinalis cardinalis Northern Cardinal, 191
Carduelis carduelis European Goldfinch, 381
    chloris European Greenfinch, or Greenfinch 47,
    222–224
    flammea Common Redpoll, 177
    spinus Eurasian Siskin, 138
    tristis American Goldfinch, 177
Carpodacus cassinii Cassin's Finch, 177
    mexicanus House Finch, 47, **136**, 137, 177, 188,
    193, 206, 266, 268
Centrocercus urophasianus Greater Sage Grouse, 160,
    161, 175, 248
Certhia americana Brown Creeper, 266
    brachydactyla Short-toed Tree-Creeper, 190, 200,
    201, 207
Certhidea fusca Grey Warbler-Finch, **23**
    olivacea Green Warbler-Finch, **23**
Cettia diphone Japanese Bush-Warbler, 264, 266
Chalcophaps stephani Stephan's Dove, 230

Chalcopsitta cardinalis Cardinal Lory, 63
Chlamydera cerviniventris Fawn-breasted
    Bowerbird, 185
    nuchalis Great Bowerbird, 185
Ciconia abdimii Abdim's Stork 106
    boyciana Oriental Stork 106
    ciconia White Stork 106
    episcopus Woolly-necked Stork 106
    maguari Maguari Stork 106
    nigra Black Stork 106
    stormi Storm's Stork 106
Cistothorus palustris Marsh Wren 211
Clamator glandarius Great Spotted Cuckoo, 280
Colaptes auratus Northern Flicker, Red-shafted
    Flicker and Yellow-shafted Flicker, 319, 326,
    337, 347, 352
Colinus cristatus Crested Bobwhite, 231
    virginianus Northern Bobwhite, 231, **318**
Columba corensis Bare-eyed Pigeon, 230
    hodgsonii Speckled Wood-Pigeon, 230
    leucocephala White-crowned Pigeon, 230
    leuconota Snow Pigeon, 230
Columba livia Rock Dove, or Common Pigeon, 128,
    161, 165, **220**–223, 226–230, 232, 277, 381
Corvus corone Carrion Crow and Hooded Crow,
    235, 327, 338, 339, 342, 354, 390
Coturnix chinensis Blue-breasted Quail, 386
    coturnix Common Quail, 379
    japonica Japanese Quail, 221, 289–290, 292, 379,
    385–386
Cyanistes caeruleus European Blue Tit, 44, 68, 125,
    128, 130, 190, 266, 311
    cyanus Azure Tit, 311
Daphoenositta chrysoptera Varied Sittella, 327, 354
Dendrocopos major Great Spotted Woodpecker, 311
    syriacus Syrian Woodpecker, 311
Dendroica adelaidae Adelaide's Warbler, **150**
    coronata Yellow-rumped Warbler, 328, 363
    occidentalis Hermit Warbler, 28, 320, 328, **330**,
    340–341, 343–344, 364
    townsendi Townsend's Warbler, 28, 320, 328,
    330, 340–341, 343–344, 364
    virens Black-throated Green Warbler, 28
Drepanoptila holosericea Cloven-feathered Dove,
    230
Dromaius novaehollandiae Emu, 129, 130, 384
Ducula aenea Green Imperial-Pigeon, 230
Dumetella carolinensis Gray Catbird, 201
Ectopistes migratorius Passenger Pigeon, 230
Elanus caeruleus Black-winged Kite, 384
Emberiza bruniceps Red-headed Bunting, 328
    calandra Corn Bunting, 180
    citrinella Yellowhammer, 177, 191, 212, 213, 328,
    362
    leucocephalos Pine Bunting, 328, 347, 362
    melanocephala Black-headed Bunting, 328
Emblema pictum Painted Finch, or Painted Firetail,
    387
Ephippiorhynchus asiaticus Black-necked Stork, 106
    senegalensis Saddle-billed Stork, 106

Erithacus rubecula European Robin, 158, 266
Erythrura trichroa Blue-faced Parrotfinch, 387
Estrilda astrild Common Waxbill, 295, 387
    paludicola Fawn-breasted Waxbill, 387
Ficedula albicollis Collared Flycatcher, 292, 307, 308,
    309–315, 327, 339, 344–345, 356, 390
    hypoleuca Pied Flycatcher, or European Pied
    Flycatcher, 163, 279, 307, 308, **309**–315, 327,
    336, 339, 344–345, 356, 390
Fringilla coelebs Chaffinch, 177, 189, 203–204,
    216–217, 265, 266,268
    teydea Blue Chaffinch, 189, 207, 217
Gallicolumba luzonica Luzon Bleeding-heart, 230
Gallirallus philippensis Buff-banded Rail, 71
Gallus gallus Chicken, or Red Junglefowl, 161, 180,
    221–223, 225, 226, 377, 379, 382, 383–385,
    388
    sonneratii Grey Junglefowl, 379
Garrulus glandarius Eurasian Jay, 49
Geopelia cuneata Diamond Dove, 230
Geospiza conirostris Large Cactus-Finch, **23**, 262
    difficilis Sharp-beaked Ground-Finch, **23**, **126**,
    191
    fortis Medium Ground-Finch, 22, **23**, 24, 47, 53,
    72, 78, 82–84, 90, **91**, 253, 306, 316, 324, 397
    fuliginosa Small Ground-Finch, 22, **23**, 24, 53,
    54, 72, 76, 78, 79, 82–84, 90, 91
    magnirostris Large Ground-Finch, 22, **23**, 24, 53,
    76, 90
    scandens Common Cactus-Finch, **23**, 127, 306,
    324, 397
Geotrygon versicolor Crested Quail-Dove, 230
Gerygone fusca Western Gerygone, 268
Glyphorynchus spirurus Wedge-billed Woodcreeper,
    31
Gracula religiosa Hill Myna, 190
Gymnogyps californianus California Condor, 277
Gymnorhina tibicen Australian Magpie, 327, 355
Hemignathus munroi Akiapōlā'lau, 112, **113**
Heteralocha acutirostris Huia, 112, **113**
Hippolais icterina Icterine Warbler, 327, 359
    polyglotta Melodious Warbler, 327, 359
Hirundo rustica Barn Swallow, 69, 115, 160, 161, 180
Icterus bullockii Bullock's Oriole, 95, 328, 365
    cucullatus Hooded Oriole, **234**
    galbula Baltimore Oriole, 95, **234**, 328, 365
    gularis Altamira Oriole, **234**
    spurius Orchard Oriole, **234**
Jabiru mycteria Jabiru, 106
Junco hyemalis Dark-eyed Junco, 33, 45, 47, 162,
    177,200, 207, 266, 347, 389
    insularis Guadalupe Junco, 200
    phaeonotus Yellow-eyed Junco 33
Lagonosticta rhodopareia Jameson's Firefinch, **296**
    senegala Red-billed Firefinch, 293–294, 387
Larus argentatus Herring Gull, 66, 94, 284
    fuscus Lesser Black-backed Gull, 94
    glaucescens Glaucous-winged Gull, 326, 338,
    353
    occidentalis Western Gull, 326, 338, 353

Leptoptilos crumeniferus Marabou Stork, 106
    javanicus Lesser Adjutant, 106
Leptotila cassini Grey-chested Dove, 230
Leucosarcia melanoleuca Wonga Pigeon, 381
Lichenostomus virescens Singing Honeyeater, 266
Lonchura castaneothorax Chestnut-breasted Munia,
    227, 231, 381, 387
    leucogastroides Javan Munia, **290**–292, 297
    punctulata Scaly-breasted Munia, 231, 381
    striata Bengalese Finch, or White-rumped
        Munia 161, 205, 224, 277–278, 293
Lopholaimus antarcticus Topknot Pigeon, 230
Lophophanes cristatus Crested Tit, 128
Loxia curvirostra Red Crossbill, 47, 69, 77–78, **79**, 80,
    99, 177
Loxigilla noctis Lesser Antillean Bullfinch, **150**, 153
Luscinia luscinia Thrush Nightingale, 306, 327, 357
    megarhynchos Common Nightingale, 205, 306,
        327, 357
Malurus lamberti Variegated Fairywren, 326, 354
Manacus candei White-collared Manakin, 326, **338**,
    340–342, 346, 353
    vitellinus Golden-collared Manakin, 326, **338**,
        340–342, 353
Meleagris gallopavo Wild Turkey, or Turkey, 222, 223
Melopsittacus undulatus Budgerigar, 161, 222–227
Melospiza georgiana Swamp Sparrow, 190–192, 199,
    205, 211
    melodia Song Sparrow, 41, 43, 190, 191, 199
Menura novaehollandiae Superb Lyrebird, 259
Metallura aeneocauda Scaled Metaltail, 118
    baroni Violet-throated Metaltail, 118
    eupogon Fire-throated Metaltail, 118
    odomae Neblina Metaltail, **118**
    theresiae Coppery Metaltail, 118
    tyrianthina Tyrian Metaltail, **118**
    williami Viridian Metaltail, **118**
Metriopelia ceciliae Bare-faced Ground-Dove, 230
Milvus migrans Black Kite, 127
Mimus polyglottos Northern Mockingbird, 161
Molothrus ater Brown-headed Cowbird, 191, 202,
    207, 279–280
Monarcha castaneiventris Chestnut-bellied
    Monarch, **5**, 6, **7**, 12, 409
    cinerascens Island Monarch, 409
    erythrostictus Bougainville Monarch, **5**
    frater Black-winged Monarch, 7
    richardsii White-capped Monarch, **5**, 6, **7**
Motacilla alba Pied Wagtail, 285
    flava Yellow Wagtail, 235
Myiarchus tuberculifer Dusky-capped Flycatcher,
    253
Mycteria americana Wood Stork, 106
    cinerea Milky Stork, 106
    ibis Yellow-billed Stork, 106
    leucocephala Painted Stork, 106
Myzomela cardinalis Cardinal Myzomela, 311
    pammelaena Ebony Myzomela, 46
    sclateri Scarlet-bibbed Myzomela, 46
    tristrami Sooty Myzomela, 311

Nectarinia bouvieri Orange-tufted Sunbird, 190
Neochmia modesta Plum-headed Finch, 387
    phaeton Crimson Finch, 387
    ruficauda Star Finch, 387
Numida meleagris Helmeted Guineafowl, 221,
    377–378, 388
Oceanodroma castro Band-rumped Storm-Petrel,
    21, 396
Oenanthe hispanica Black-eared Wheatear, 327, 357
    picata Variable Wheatear, 347
    pleschanka Pied Wheatear, 327, 357
Oreortyx pictus Mountain Quail, **318**
Orthotomus sutorius Common Tailorbird, 139
Otidiphaps nobilis Pheasant Pigeon, 230
Pachycephala pectoralis Golden Whistler, **60**, 62, 271,
    314, 347
Parus major Great Tit, 44, 58, 59, 66, 68, 130, 132,
    167, 168, 176, 177, 181, 182, 203, 253, 256,
    279
Passer domesticus House Sparrow, 45, 46, 127, 130,
    328, 361
    hispaniolensis Willow Sparrow, 328, 347, 361
    montanus Eurasian Tree Sparrow, 130, 204, 268
Passerella iliaca Fox Sparrow, 266
Passerina amoena Lazuli Bunting, 215–216, 286,
    288, 307, 308, 319, 328, 365
    cyanea Indigo Bunting, 188, 189, 192, 202, 204,
        215–216, 286, 288, 307, 319, 328, 365
Pavo cristatus Indian Peafowl, 161
Periparus ater Coal Tit, 44, 128, 137, 194, **195**, 201,
    207, 235
Petroica goodenovii Red-capped Robin, 266
Phasianus colchicus Common Pheasant, or Japanese
    Pheasant, 221, 226, 375, 381
Pheucticus ludovicianus Rose-breasted Grosbeak,
    328, 364
    melanocephalus Black-headed Grosbeak, 177,
        328, 364
Phylloscopus affinis Tickell's Leaf-Warbler, 88, 178
    ibericus Iberian Chiffchaff, 197, 307, 327, 361,
        391
    chloronotus Lemon-rumped Warbler, 88, 178
    collybita Common Chiffchaff, or Chiffchaff, **197**,
        200, 207, 208, 307
    fuscatus Dusky Warbler, 137
    humei Hume's Leaf-Warbler, 88, 178
    magnirostris Large-billed Leaf-Warbler, 88, 178
    occipitalis Western Crowned-Warbler, 87, 88,
        178
    pulcher Buff-barred Warbler, 88, 178
    trochiloides Greenish Warbler, 65, 66, 87, 88, 92,
        93, 178, 236–**237**
    trochilus Willow Warbler, 327, 336, 354
    tytleri Tytler's Leaf-Warbler, 88, 178
Pica pica Black-billed Magpie, 280
Pinaroloxias inornata Cocos Finch, 19, **23**
Pinicola enucleator Pine Grosbeak, 190
Pipilo erythrophthalmus Eastern Towhee, 328, 362
    maculatus Spotted Towhee, 328, 362
    ocai Collared Towhee, 328, 362

Platycercus elegans Adelaide Rosella, Crimson
    Rosella, and Yellow Rosella, 346, 347
Podiceps cristatus Great Crested Grebe, **166**
    occipitalis Silvery Grebe, 19, 46, 48, **52**
    taczanowskii Junin Grebe, 19, 48, **52**
Poecile atricapillus Black-capped Chickadee, 195,
    266, 327, 335, 340, 345, 357
        carolinensis Carolina Chickadee, 43, 327, 335,
        340, 345, 357
        montanus Willow Tit, 128, 207
        palustris Marsh Tit, 130, 132, 201
Poephila acuticauda Long-tailed Finch, 173, 174
    cincta Black-throated Finch, 328, 261, 387
Porzana palmeri Laysan Rail, 71
Procnias alba White Bellbird, **159**
    averano Bearded Bellbird, **159**
    nudicollis Bare-throated Bellbird, **159**
    tricarunculata Three-wattled Bellbird, **159**
Ptilinopus granulifrons Carunculated Fruit-Dove,
    230
        jambu Jambu Fruit-Dove, 230
        victor Orange Dove, 230
Ptilonorhynchus violaceus Satin Bowerbird, 168, 254
Pycnonotus barbatus Garden Bulbul, 327, 359
    cafer Red-vented Bulbul, 327, 359
    leucogenys Himalayan Bulbul, 359
    leucotis White-eared Bulbul, 327, 359
    nigricans Red-eyed Bulbul, 327, 359
Pyrenestes ostrinus Black-bellied Seedcracker, 84,85,
    315
Pyrrhula pyrrhula Eurasian Bullfinch, 372
Pytilia melba Green-winged Pytilia, 386, 387
    phoenicoptera Red-winged Pytilia, 386, 387
Quiscalus quiscula Common Grackle, 328, 366
Raphus cucullatus Dodo, 71, 299
Regulus ignicapillus Firecrest, 190, 266
    regulus Goldcrest, 190, 266
Saxicola torquata Common Stonechat, 44
Seicercus burkii Golden-spectacled Warbler, **316**
    whistleri Whistler's Warbler, **316**
Serinus canaria Canary, or Island Canary, 162,
    221–224, 227, 232, 280, 372, 381
        serinus European Serin, 190
Sialia mexicana Western Bluebird, 344
Sphecotheres viridis Timor Figbird, 327, 356
Sphyrapicus nuchalis Red-naped Sapsucker, 28, 325,
    351
        ruber Red-breasted Sapsucker, 28, 325, 351
        varius Yellow-bellied Sapsucker, 28
Spizella pusilla Field Sparrow, 287
Stagonopleura guttata Diamond Firetail, 231, 387
    capicola Ring-necked Dove, 326, 353
    chinensis Spotted Dove, 381
    roseogrisea African Collared-Dove, or Ring
        Dove, 161, 163, 165, 167, 284, 381
Streptopelia turtur European Turtle-Dove, 230
    vinacea Vinaceous Dove, 230
Struthio camelus Ostrich, 277, 384, 388
Sturnella magna Eastern Meadowlark, 25, **27**, 114,
    287–289, 306, 390

    neglecta Western Meadowlark, 25, **27**, 114,
        287–289, 306, 390
Sturnus vulgaris Common Starling, 46, 177
Sula dactylatra Masked Booby, 126
    nebouxii Blue-footed Booby, 161
Sylvia atricapilla Blackcap, 44–46, 60, 92–94,
    286–287
        borin Garden Warbler, 286–287
Syrmaticus reevesii Reeves's Pheasant, 375, 381
Taeniopygia bichenovii Double-barred Finch, 173,
    174, 387
        guttata Zebra Finch, 45, 161, 163, **173**, 174, 175,
        180, 205, 222–224, 226, 227, 273–278, 281,
        284, 292, 318, 319, 386–388, 392
Tanysiptera galatea Common Paradise Kingfisher, 49
Telespiza cantans Laysan Finch, 47
Terpsiphone viridis African Paradise-Flycatcher, 310,
    347
Thamnophilus caerulescens Variable Antshrike, 326,
    354
Thryomanes bewickii Bewick's Wren, 266
    ludovicianus Carolina Wren, 286
    pleurostictus Banded Wren, 202–203
Tigrisoma fasciatum Fasciated Tiger-Heron, 277
Tockus erthythrorhynchus Red-billed Hornbill, 326,
    352
Toxostoma rufum Brown Thrasher, 192
Troglodytes aedon House Wren, 266
    troglodytes Winter Wren, 59, 190, 265
Turdus merula Eurasian Blackbird, 42, 44, 130–132,
    138, 189
        philomelos Song Thrush, 128, 132, 138
Turtur tympanistria Tambourine Dove, 230
Tympanuchus cupido Greater Prairie-Chicken, 319
    phasianellus Sharp-tailed Grouse, 319
Tyto alba Barn Owl, 59
Uraeginthus bengalus Red-cheeked Cordonbleu, 387
    ianthinogaster Purple Grenadier, **296**
Vermivora chrysoptera Golden-winged Warbler, 288,
    319, 328, 335, 343, 363
        pinus Blue-winged Warbler, 288, 319, 328, 335,
        343, 363
Vestiaria coccinea I'iwi, 47
Vidua chalybeata Village Indigobird, 293–295
    fischeri Straw-tailed Whydah, **296**
    purpurascens Purple Indigobird, **296**
Wilsonia citrina Hooded Warbler, 284
Xanthocephalus xanthocephalus Yellow-headed
    Blackbird, 177
Zonotrichia albicollis White-throated Sparrow, 388,
    389
        capensis Chingolo, or Rufous-collared Sparrow,
        256–258, 270
        leucophrys White-crowned Sparrow, 163,
        190–194, 196, 198, 201, 204, 206, 211, 256
Zosterops conspicillatus Bridled White-eye, 149
    lateralis Silvereye, 46, 47, 125, 311
    nigrorum Golden-green White-eye, 149
    oleagineus Yap Olive White-eye, 149
    tenuirostris Slender-billed White-eye, 311

Subject Index

A

acoustic adaptation, 252–258

adaptive radiation, 21–25, 34, 35, 252–261
    in species other than birds, 120, 400–401
    models of, 120

adaptive surface, 76–87, 93–94, 134, 136

Africa, 26, 107, 115, 116, 119

allospecies, 3–5, 10, 25–29, 31–34, 61–62, 114,
    141–155, 196–197
    age of, 31–33, 153

Andes, 50, 97, 101, 118, 119

assortative mating, 77, 94, 209, 273–281, 290–292,
    300–306, 308–310, 313–317
    in hybrid zones, 333, 336

Australia, 26, 324

B

behavioral flexibility, 125–139
    correlate with speciation, 129–131

behavioral innovations, 127, 129

Bergmann's rule, 43

biological species, 12, 22, 24, 398

biological species concept, 2–3, 397

body size, divergence in, 87–89

boreal zone, 28–29

bounded hybrid superiority, 332, 337–338

bowerbirds, 138, 184–185

brain size
    correlate with foraging behavior, 129
    correlate with introduction success, 131–132
    correlate with nesting behavior, 138
    correlate with speciation, 130–131

brood parasitism, 20, 138, 279–280, 293–296

C

carotenoids, 135–137

character displacement
    ecological, 80–82, 110–112, 122, 396
    reproductive, 302, 313–314

chromosome, 371, 383–389
    paint, 383–384
    inversion, 383–389
    polymorphisms, 388–389

climate change in the past, 28–29, 100

Cocos Island, 19

coevolution, 80

color polymorphism, 315, 388

community drift, 57

comparative map, 384–385

comparative methods, 58–59

competitive exclusion, 111, 114, 122

condition dependence, 180–181, 246

condition indicators, 181

contingency, *see* historical contingency

convergence
    in morphology and ecology, 82
    in plumage, 306
    in song, 307–308

convergent evolution, 112–114

co-operative breeding, 125–126

courtship displays, 160–166

Crusoe experiment, 133

cultural drift in song, 205–210, 264–265, 269–270

cultural mutation, 198–204

D

Darwin's Finches, 22–24, 53–54, 76–80, 84, 90–91,
    126–127, 134–135, 306, 308, 323–324, 373,
    397

dialect, song, 193–194, 256–258

dispersal distance
    estimates of, 59, 207, 336
    correlate with speciation, 125–126

DNA–DNA hybridization, 38, 115–117

Dobzhansky–Muller incompatibilities, 368–370

domestic breeds, 220–228
    genetic basis to differences among, 226

Drosophila, 1, 367, 370–371, 379, 382

E

ecological competition, 48–49, 53–57, 67–69, 80–82,
    109–112, 122, 331–333

ecological controls model, 104, 108–112

ecological opportunity, 33–34, 121

ecological speciation, 14, 75–76, 82–95, 152–154,
    243, 397
    comparison with ecological controls, 120–121

ecotones, 16, 63, 332

endemics on islands, 141–144

evolution, rapid, 46–47

extinction, 106–107, 144–152
    rate, 148–152
    on Hawaiian Islands, 147
    in Pacific Ocean, 142

F

filial imprinting, 281–283

fitness surface, 76–80, 84, 136

flightlessness, evolution of, 68–69, 70–71

fossils, 16, 18, 38
    use in calibrations, 15–16, 18, 38

founder effect speciation, 49

frequency-dependent selection, 202, 205–206

G

Galápagos Islands, 1, 19, 21–25, 34, 48, 53–56,
76–78, 80, 84–85, 90–91, 112, 125–126,
134–135, 147, 261–262, 400
gene flow, effect on differentiation, 30, 57–63,
329–331
generalist
diet, 131
habitat, 131
generalization, 198, 282, 285–287
genetic assimilation, 133–135, 211, 289
genetic drift, 49, 395
geographical variation
factors favoring, 42–57
factors opposing, 57–63
in color patterns, 58–63, 240, 330–340
in morphology, 42–57
in songs, 192–196
great speciator, 60
GTR model, 40

H

habitat, divergence in, 82, 87–89, 326–328, 337–340
Haldane's rule, 375, 382–383
Hawaiian honeycreepers, 18, 22, 85, 147
Hawaiian Islands, 18, 22, 25, 29, 71, 85, 112–113,
147, 400
Himalayas, 87–89
historical contingency
in artificial breeding, 224–225
in song, 210, 213, 224
in plumage, 176, 178, 224–225
HKY model, 39
honesty, 180–182
hybrid
intrinsic loss of fitness in, 83, 342, 373–383,
390–391
natural selection against, 83–85, 320, 342
sexual selection against, 304, 317–320, 342
hybrid index, 325, 330
hybrid speciation, 347
hybrid swarm, 302, 329
hybrid zone, 323–366
historical movement of, 340–342
width of, 329–340
hybridization
creative role for, 345–347
frequency of, 300, 311, 336
in adaptive radiation, 323–324

I

immigrants, lowered fitness of, 68–69
immigration, 68–69
imprinting
filial, 281–283
sexual, 276–278, 291–292
India, 4, 67
Indian Ocean, 142–143
individual recognition, 283–284

innovations, role in speciation of, 129–137
intrinsic incompatibility, 373–383
introduced species, colonization success of, 131–132
introgression, 342–345, 391–392
island biogeography, 144–148
island patterns
in color, 60–63, 261–264
in morphology, 53–54
in songs, 264–268

J

Jamaica, 33

K

kin recognition, 283

L

Lake Junin, 19, 48, 52
language
in birds, 167
in humans, 208–210
parallels with birdsong, 208–210
latitudinal gradient, 97–100
leapfrog pattern, 235
learning
associative, 282, 285–287
observational, 128
social, 128, 273, 280–281
trial-and-error, 129
leks, 160, 248, 250
Lesser Antilles, 150–152
lineage through time plots, 101–108, 116
lizard tails, eating of, 127

M

Madagascar, 33
magic trait, 20, 21, 94
mass extinction, 15
mate choice, direct benefits of, 180–182, 246–248
mate preferences, refinement during speciation of,
301–304
mate stimulation, 161–166
maternal effects, 44–45, 369–370, 383
meiotic drive, 30, 369, 382–383
melanism, 232
milk bottles, 128
misimprinting, 276–277, 306
molecular clock, 18, 36–38
monogamous species, sexual selection in, 160–161,
245–246, 249–250

N

nest-building, correlates of speciation with, 137–139
neural networks, 169–172, 182–185
New Guinea, 33, 109–112
niche differences, order of evolution of, 87–89
North America, 26–29, 117, 324
northern Melanesia, 5, 7, 26, 60–63, 143–144, 153

P

Pacific Ocean, 142–143
pair bonds, long term, 264
paleoaltitudes, 119
paleoclimates, 28, 100, 118
parallel evolution, 232–240
    in colors, 232–235
    in songs, 235
parasitic finches, 20, 36, 293–296, 396, 399–400
parasitism, 180, 259
    brood, 20, 138, 279–280, 293–296
Passerines, diversification of, 138–139
phenotypic plasticity, tests of, 42, 44–45
phylogenetic species, 6–8, 11–12, 22, 24
phylogenetic species concept, 6–8, 11–12, 398
phylogeny, construction methods, 38–40
phylogroup, 31–32
Pleistocene epoch, 27–28, 118–119, 152, 236, 334, 346
polygynous species, sexual selection in, 160, 245–252
postmating isolation
    ecological causes of, 83–85, 320, 342
    social causes of, 304, 317–320, 342
    intrinsic, 83, 342, 373–383, 390–391
postmating prezygotic isolation, 371–373
predation, 137–139
premating isolation
    and ecological character displacement, 82–83, 314, 396
    and reinforcement, 301–304, 313–317
    ecological causes of, 82–83
    social causes of, 273–281, 284–290
pterosaurs, 16
pure birth model, 104, 106

R

rails, 68–69, 71
range expansions, importance of, 14, 112–114, 131
range limits, 342
recombination, 20, 345, 371, 383, 389, 391
refugia model, 15, 27–28
regiolect, 196
reinforcement, 301–304, 313–317, 399
    sexual selection model and, 302–303
reproductive character displacement, 302, 313–314
ring species, 66, 93, 236–238
runaway sexual selection, 239

S

search costs, 248, 304–306, 310–312
seasonal migration, 69, 92–95
sex chromosomes, 225, 227, 382–383
    reduced introgression of, 345, 391
sex linkage, 225, 227, 345, 382–383, 391
sexual conflict, 180, 369
sexual imprinting, 276–278, 291–292
sexual selection
    and ecology, 243–268

and parasites, 180, 259
    by female choice, 160–161
    by male competition, 166–168, 308–310
    role in speciation of, 270–272
sister species, ecological differences between, 28–29, 87–89
small populations, 49, 54–57, 261–265
social selection, 157–186
song dialects, 193–194, 256–258
song learning
    prevalence of, 187
    role in speciation of, 213
South America, 26, 115–120
speciation
    allochronic, 21
    allo-parapatric, 72
    allopatric, 8–9, 13–17, 21–31
    by distance, 65–66
    clinal, 65–67
    cultural, 293–296
    ecological, 14, 75–76, 82–95, 152–54, 243, 397
    hybrid, 347
    in mountainous regions, 16, 33, 117–119
    island model, 67–73
    nonecological models of, 29–31, 265, 268–270, 395–396
    parapatric, 65–73
    sympatric, 19–21, 34, 396
    time to, 31–33
speciation clock, 376
speciation conflict, 303
speciation machine, 71
speciation rate
    apparent decline in, 102–108
    estimate of, 102–103
species
    biological, 12, 22, 24, 398
    definition as used in book, 4–5, 395
    genetic basis of differences, 228, 232, 344, 371
    local numbers, 97–101
    phylogenetic, 6–8, 11–12, 22, 24
    regional numbers, 97–101
suboscines
    diversification of, 116
subspecies, 3, 41, 43, 58–60
    correlates with dispersal, 58–60
substitution models, 36, 39–40
superspecies, 3–5, 25–29, 34, 61–62, 118, 159

T

taxon cycle, 147–148
tension zone, 329, 331, 335
territoriality, interspecific, 114, 286–287, 306–310
threat displays, 166–168, 176–178
transference hypothesis, 185
Tres Marías Islands, 48
trial-and-error learning, 129
Tristan da Cunha, 85, 86, 89

U
urban habitats, 100–101, 131, 138

W
W chromosome, 382–383
West Indies, 24, 55–56, 150–153
white-eyes, 24, 46, 89, 125, 149, 311
withdrawal of learning, 206–207

Z
Z chromosome, 225, 227, 382–383
zoogeographic species, 61–62

"This is an insightful and original work, comprehensive and up to date, covers many interesting ideas, and is particularly good on inclusion of recent genetic information on the process of speciation in birds. It will be the best work available on its topic, the behavioral and genetic causes and consequences of speciation in birds."—**Robert Payne, Curator of Birds and Professor of Zoology, University of Michigan Museum of Zoology**

"An insightful and thought-provoking treatise on speciation and its consequences in birds, the taxon that brought you the biological species concept and the doctrine of allopatric speciation. This book delivers a grand update that expands our understanding of the role of ecology and behavior." —**Dolph Schluter, Professor and Canada Research Chair, University of British Columbia, and the author of** *The Ecology of Adaptive Radiation*

"As the literature in any field explodes there is simultaneously an increasing need for synthesis yet an increasing difficulty in achieving it. This is certainly true for the ever-popular subject of ornithology. Trevor Price takes up the challenge to explain how birds speciate, and succeeds magnificently. It is a comprehensive review of all the major ideas, beautifully illustrated with pictures of birds. More than 1300 works are cited, but more impressive is the range of subjects, from genetics to biogeography, from the reconstruction of phylogeny to ecology and the causes of reproductive isolation, all discussed with admirable clarity. If they were alive today Ernst Mayr would bestow patrician approval on this work of scholarship, and Theodosius Dobzhansky would applaud from the sidelines." —**Peter R. Grant, Class of 1877 Professor of Zoology, Princeton University, and the author of** *Ecology and Evolution of Darwin's Finches*

"I enjoyed reading this book. It is packed with new ideas presented in a rich context of natural history. The engaging text will stimulate and guide graduate students and researchers alike in a fundamentally important area of research."—**Robert E. Ricklefs, Curators' Professor of Biology, University of Missouri at St. Louis, and the author of** *The Economy of Nature*

"*Speciation in Birds* is a significant advance in our understanding of the genesis and maintenance of biological diversity. It will be a "must read" not only for ornithologists, but for evolutionary biologists of all stripes—this is the most synthetic and innovative treatment of speciation ever published for any group of organisms. The cliché is certainly overused, but Trevor Price is truly someone who thinks "outside the box." He continually looks at questions in new ways and comes up with novel perspectives on important issues in the field." —**Jonathan B. Losos, Curator in Herpetology in the MCZ and Monique and Philip Lehner Professor for the Study of Latin America, Harvard University**

On the cover:
Male White-collared Manakin, *Manacus candei (bottom left)*, male Golden-collared Manakin, *Manacus vitellinus (top right)*, and a male hybrid between the two species. All three males can be found displaying at some locations in Panama (see p. 338).

Illustration by Emiko-Rose Paul from photographs by Robb Brumfield and Adam Stein.

ROBERTS AND COMPANY PUBLISHERS
www.roberts-publishers.com

ISBN 0-9747077-8-3

9 780974 707785

Printed in China

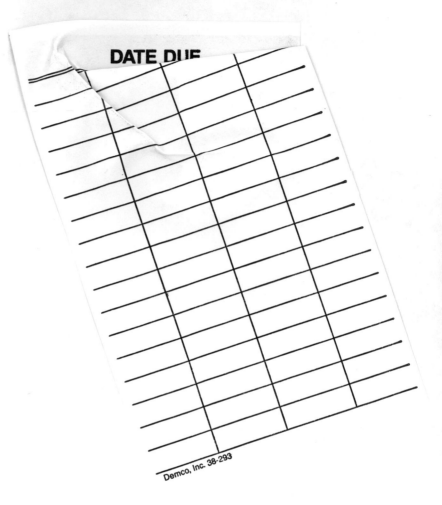

DATE DUE

Demco, Inc. 38-293